Advances in Intelligent Systems and Computing

Volume 1126

The series “Advances in Intelligent Systems and Computing” contains publications on theory, applications, and design methods of Intelligent Systems and Intelligent Computing. Virtually all disciplines such as engineering, natural sciences, computer and information science, ICT, economics, business, e-commerce, environment, healthcare, life science are covered. The list of topics spans all the areas of modern intelligent systems and computing such as: computational intelligence, soft computing including neural networks, fuzzy systems, evolutionary computing and the fusion of these paradigms, social intelligence, ambient intelligence, computational neuroscience, artificial life, virtual worlds and society, cognitive science and systems, Perception and Vision, DNA and immune based systems, self-organizing and adaptive systems, e-Learning and teaching, human-centered and human-centric computing, recommender systems, intelligent control, robotics and mechatronics including human-machine teaming, knowledge-based paradigms, learning paradigms, machine ethics, intelligent data analysis, knowledge management, intelligent agents, intelligent decision making and support, intelligent network security, trust management, interactive entertainment, Web intelligence and multimedia.

The publications within “Advances in Intelligent Systems and Computing” are primarily proceedings of important conferences, symposia and congresses. They cover significant recent developments in the field, both of a foundational and applicable character. An important characteristic feature of the series is the short publication time and world-wide distribution. This permits a rapid and broad dissemination of research results.

**** Indexing: The books of this series are submitted to ISI Proceedings, EI-Compendex, DBLP, SCOPUS, Google Scholar and Springerlink ****

More information about this series at http://www.springer.com/series/11156

Zhengbing Hu · Sergey Petoukhov · Matthew He
Editors

Advances in Artificial Systems for Medicine and Education III

Springer

Editors
Zhengbing Hu
School of Educational
Information Technology
Central China Normal University
Wuhan, Hubei, China

Sergey Petoukhov
Mechanical Engineering Research Institute
Russian Academy of Sciences
Moscow, Russia

Matthew He
Halmos College of Natural Sciences
and Oceanography
Nova Southeastern University
Davie, FL, USA

ISSN 2194-5357 ISSN 2194-5365 (electronic)
Advances in Intelligent Systems and Computing
ISBN 978-3-030-39161-4 ISBN 978-3-030-39162-1 (eBook)
https://doi.org/10.1007/978-3-030-39162-1

This Springer imprint is published by the registered company Springer Nature Switzerland AG
The registered company address is: Gewerbestrasse 11, 6330 Cham, Switzerland

Contents

Advances in Mathematics and Bio-mathematics

Development of an Intelligent System for Predicting the Forest Fire Development Based on Convolutional Neural Networks

Tatiana S. Stankevich(✉)

Kaliningrad State Technical University,
1, Sovietsky prospect, 236022 Kaliningrad, Russian Federation
tatiana.stankevich@klgtu.ru

Abstract. Forests are a natural renewable resource and can meet the needs of the society, provided that they are used for a multiple, rational, continuous and sustainable use. Forest fires are a natural component of forest ecosystems and cannot be completely eliminated. However, in recent decades, there has been a tendency to transform forest fires from a natural regulatory factor into a catastrophic phenomenon causing significant economic, environmental and social damage. It is critical to understand the relationships between the underlying environmental factors and spatial behaviour of a forest fire in order to develop effective and scientifically sound forest fire management plans. The key objective of this study is to enhance the efficiency of the formation of a real-time forest fire forecast under the unsteady and uncertain conditions. In the article, the author proposes to develop an intelligent system for predicting the forest fire development based on artificial intelligence and deep computer-aided learning. A key element of the system is forest fire propagation models that recognise data from successive images, predict the forest fire dynamics and generate an image with a fire propagation forecast. It is proposed to build forest fire propagation models by using a real-time forest fire forecasting method. In the article, the author presented a structural diagram of an intelligent system to forecast the dynamics of a forest fire and described the functional structure of the system by constructing its functional models in the form of IDEF0 diagrams.

Keywords: Artificial intelligence · Deep machine learning · Convolutional neural network (CNN) · Wildfire · Forest fire · Real-time forecast · Big data

1 Introduction

Forests are ecological systems and are a natural renewable resource that allows satisfying the needs of the society, provided that the forests are used in a multiple, rational, continuous and sustainable way. One of the key principles of forest management is to ensure that forests are conserved and protected, also from forest fires.

Forest fires are uncontrolled movements of fire across the forest and are one of the most destructive natural disasters and forces [1]. Moreover, forest fires are a natural component of forest ecosystems, which cannot be fully eliminated [2].

Z. Hu et al. (Eds.): AIMEE 2019, AISC 1126, pp. 3–12, 2020.
https://doi.org/10.1007/978-3-030-39162-1_1

However, in recent decades, there has been a tendency for forest fires to turn from a natural regulatory factor into a catastrophic phenomenon causing significant economic, environmental and social damage [3]. For example, the statistics of the Federal Forestry Agency clearly confirm the significant negative effect of forest fires for the Russian Federation [4]: for ten years, from 2009 to 2018 there was recorded an increase in forest land areas affected by forest fires by approximately three times; from 1992 to 2018 a decrease in the number of forest fires in Russia by 53% was recorded. In the Russian Federation, from 2016 to 2018, there were no civilians who died and suffered from forest fires; among forest firefighters, during the same period, 14 people died, 31 people were injured in fighting forest fires. In the five southern EU member states (Portugal, Spain, France, Italy and Greece), the number of people killed in putting out fires in 2016 was 2, and 16 people were injured; in 2017, 86 people and 68 people, respectively [5]. Thus, against the backdrop of the observed reduction in the number of forest fires, the social and material damage from forest fires is growing. As can be seen from Figs. 1 and 2, the Russian trend is similar to the situation in the USA [6] and Europe [5].

In other words, although the occurrence and development of forest fires are due to regional meteorological and climatic characteristics as well as the regional forest vegetation type, at present, the global forest fire statistics show a trend towards a decrease in the number of fires with a parallel increase in damage.

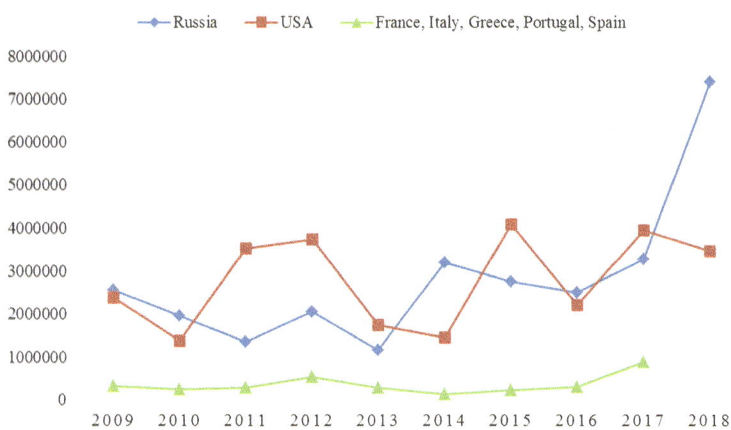

Fig. 1. Dynamics of forest land destruction by fires

Thus, it is very relevant for the national economy to ensure that the necessary forest fire safety level, the adequate level of live support for the population and the environmental situation in Russia and globally are created. Although efforts to prevent fires are essential [7], it is also important to have tools that allow you to make effective decisions when a fire has already occurred and must be suppressed [8]. It is critical to understand the relationships between the underlying environmental factors and the spatial behaviour of a forest fire in order to develop effective and scientifically sound forest fire management plans.

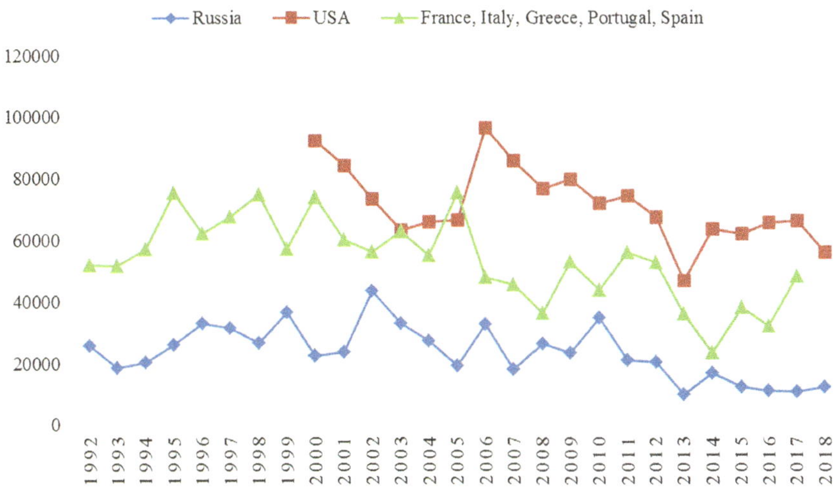

Fig. 2. Dynamics of forest fires

The existing traditional forest fire forecasting models have a number of significant drawbacks. The author proposes to conduct a research aimed at increasing the efficiency of a real-time forest fire forecast in the context of unsteadiness and uncertainty. To this end, the author proposes to develop an intelligent system to forecast in real time the forest fire dynamics in order to take into account the effects of environmental factors, the nature of forest plantations and the type of fire on the forecast. The author suggests that visual data, including satellite orbital images, are used as input as this is the most effective and less expensive way to solve the problem in countries with a large territory, such as Russia.

2 Theoretical Overview of the Problem: Field of Study and Related Works

Forest fires result from the interaction of various elements of a socio-economic, political and cultural nature, which is influenced by climatic factors determining the scale and intensity of the fire behaviour [9].

These are key challenges for developing fire management strategies to predict the occurrence and spread of forest fires. The scientists' key research focuses on the construction of models for predicting the occurrence of fires. Forest fires are the key focus in the works [10–15]. In the last decade, scientists have been actively using up-to-date information technology to predict the occurrence of fires. Many researchers propose the use of geographic information system and remote sensing technologies to predict forest fires [16]. Researchers have proposed the use of artificial neural networks [17], fuzzy logic [18], and ANFIS [19]. However, it is necessary to have tools to predict the behaviour of a forest fire depending on environmental factors, which will make it possible to make effective decisions in fighting fire. Understanding the spatial patterns of fire spread is the key to improving forest safety management, especially in the face of global climate change.

It is difficult to model the forest fire dynamics due to two key reasons [10]:

(1) the extreme complexity of the physical phenomenon due to the heterogeneity of fuel and many influential environmental factors;
(2) the significant complexity of conducting full-scale experiments to validate the developed models.

Currently, researchers have developed an extensive set of models based on various methods for predicting fire behaviour [10–15]:

(1) Empirical and quasi-empirical models based on the results of statistical analysis of experimental data for determining statistical dependencies between input and output parameters.
(2) Physical and quasi-physical models based on the fundamental chemistry and/or physics methods to describe the processes occurring in a forest fire.
(3) Mathematical models (including simulation and wave models) that use formulas to describe the fire dynamics, in some cases, involving statistical data.

Some of the models considered are integrated into computer systems and are widely used in practice. For example, in the forest fire dynamics forecasting systems Prometheus [20] and FlamMap [21], the wave models of fire are used. The fire development model is based on a concept similar to Prometheus, ad is used by the US National Park Service, the USDA Forest Service and other federal and state land administration agencies in the FlamMap forest fire simulation system. The application of the Van Wagner model and quasi-empirical Rothermel model is based on fire dynamics forecasting systems, such as FARSITE [22]. FARSITE widely used by the U.S. National Park Service, the U.S. Department of Agriculture Forest Service, and other federal and state land administration agencies allows you to simulate the spread of forest fires.

The existing traditional forest fire forecasting models have a number of significant drawbacks (limited functionality in the unsteady and uncertain conditions; low forecast accuracy; significant time and computational costs making them inapplicable in the real-time forecasting conditions; taking into account only a limited set of environmental factors) [23].

Thus, despite the wide variety of models for predicting the forest fire dynamics, some application limitations are identified.

In recent years, researchers have gained unprecedented opportunities to improve the fire safety of forests through the use of artificial intelligence, large data processing systems and deep machine learning. The author proposes to increase the efficiency in producing real-time forest fire dynamics forecasts under the unsteady and uncertain conditions through the use of a convolutional neural network (CNN). Although CNNs are used for solving recognition and classification problems (for image classification [24], automatic speech recognition etc.), they can also be used for forecasting.

The use of a convolutional neural network for real-time forest fire dynamics forecasting makes it possible to formulate a forecast under complex conditions (with uncertainty and unsteadiness) and minimise time costs due to the parallelisation of high-performance computing. Thus, a convolutional neural network is an effective tool for obtaining a real-time forecast of the spread of a forest fire if used in real life.

3 Features of the Development of an Intelligent System for Forecasting the Forest Fire Dynamics

The intelligent system is designed to predict the forest fire dynamics depending on the influence of environmental factors, the nature of forest plantations and the type of fire in complex conditions (under the unsteady and uncertain conditions and with a shortage of temporary resources).

In implementing an intelligent system, a structural diagram of the system was constructed (Fig. 3). The system consists of the following subsystems, an information subsystem, a intelligent subsystem and a user interface subsystem.

3.1 User Interface Subsystem

The user interface subsystem includes a user interface and allows the user to interact with the subsystems of the system (an intelligent subsystem and an information subsystem).

To implement the user friendliness principle, it is necessary to build a user interface by taking into account the ergonomic requirements of hardware and software in the field of human-system interaction in accordance with ISO/IEC 25010: 2011 [25]. The user interface designed by taking into account the above requirements will interact between the user and system elements through dialog boxes for solving management problems in acquiring knowledge and explaining the results.

3.2 Information Subsystem

The information subsystem includes a visual database on the forest fire dynamics. Visual data were obtained from heterogeneous sources, the fire propagation data through the NASA FIRMS resource management system [26]; the data on the nature of forest vegetation from the European Space Agency Climate Change Initiative's global annual Land Cover Map [27]; the data on environmental factors, namely, air temperature at a height of 2 m above the surface of the earth, relative humidity, wind speed at a height of 10 m above ground; the data on the nature of forest vegetation with Ventusky InMeteo [28]. At present, a set of more than 26,000 images has been formed which makes it possible to attribute the visual data to Big Data and a corresponding database has been developed.

The main tasks solved by the information subsystem are data collection and storage; retrieval in a convenient form of the required data for the user; data exchange between subsystems of the system.

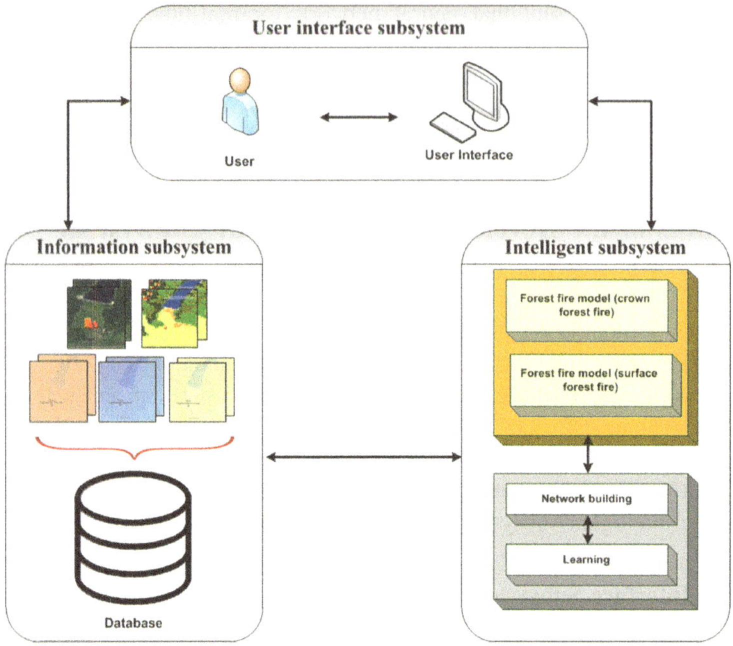

Fig. 3. Structural diagram of the intelligent forest fire forecasting system

3.3 Intelligent Subsystem

The intelligent subsystem shown in Fig. 3 is a working module of the system and contains models of the forest fire development dynamics developed with convolutional neural networks, and a unit for constructing networks based on a database from a visual database. Models designed to form an operational forecast under complex conditions (with uncertainty and unsteadiness, subject to a shortage of time) are based on the use of CNNs. To build and configure artificial neural networks, it is proposed to use the constructed visual database on the forest fire dynamics.

After determining the main elements, the system was examined with a system approach, the system elements (units) were selected for consideration and determination of their functional purpose, features of interaction with other system units as well as for identifying input and output information flows. To solve this problem, a functional model of the system in the form of IDEF0was constructed. Figure 4 shows diagram A-0 'Perform forecasting forest fire dynamics' and diagram A0 'Perform forecasting forest fire dynamics', where Ti – the air temperature at a height of 2 m above the ground; Wi – relative air humidity; SWi – wind speed at a height of 10 m above the ground; Si – the area of forest fire; FT – the type of forest vegetation; F – a real-time forecast of the forest fires dynamics.

The general logical model of the forest fire development dynamics developed with convolutional neural networks includes the following steps:

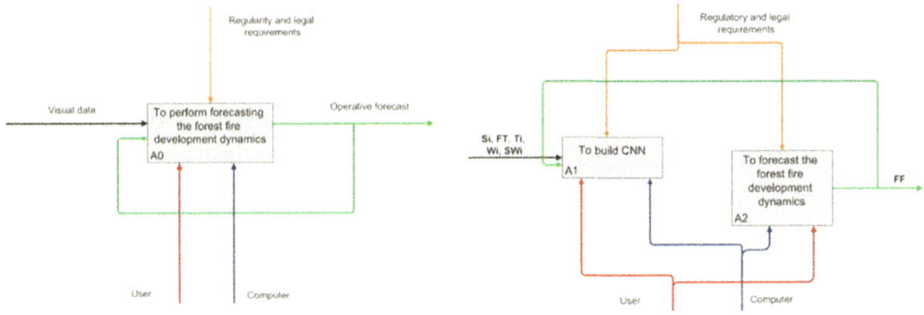

a)
Diagram A-0 'Perform forecasting the forest fire development dynamics'

b)
Diagram A0 'Perform forecasting the forest fire development dynamics'

Fig. 4. IDEF0 diagram

1. Input data in the form of images in JPEG format.
2. Input data pre-processing (including the format verification, input data size verification and noise removal; modified median filter from [29] was applied).
3. Recognition of objects with convolutional neural networks according to the formulas [30], where $C^i_{m,n}$ output on i-map C-layer in m, n; $\varphi = A \cdot \tanh(B \cdot p)$ is position, where $A = 1,7159$, $B = 2/3$, p is weighted sum; b is shifting; Q_i is set of map indices of the previous layer associated with the map C^i; K_C is the size of the square of the of the receptor field for the neuron $C^i_{m,n}$; $Z^q_{k,l}$ is part of custom features responsible for interacting with the q-map of the previous layer; D is a set of neurons on a subsequent map ($\lambda + 1$ layer) connected with a neuron in m, n; $w_i^{\lambda+1}$ is map index of the S-layer, where connected with the map C-layer; $\delta^{\lambda}_{m,n}$ is balance for a neuron with coordinates m, n in the layer map λ; q is the part of the kernel of custom functions for which gradient components are obtained; $SizeC$ is map size of the C-layer; $y_{m,n}$ is network output value; $X^q_{m+k,n+l}$ is input values for the neuron $C^i_{m,n}$:

$$y_{m,n} = C^i_{m,n} = \varphi(p) = \varphi(b + \sum_{q \in Q_i} \sum_{k=0}^{K_{C-1}} \sum_{l=0}^{K_{C-1}} X^q_{m+k,n+l} \cdot Z^q_{k,l}),$$

$$\delta^{\lambda}_{m,n} = \sum_{i \in D} \delta_i^{\lambda+1} \cdot w_i^{\lambda+1}[m,n] \cdot \varphi'(p^{\lambda}_{m,n}),$$

$$\frac{\partial E}{\partial (Z^{\lambda}_{k,l})^q} = \sum_{m=0}^{SizeC} \sum_{n=0}^{SizeC} \delta^{\lambda}_{m,n} \cdot y^{\lambda-1}_{m+k,n+l}.$$

3.1. Fire data recognition: a pre-processed colour image (a three-channel image) with a resolution of 400 × 400 pixels is input. A convolutional neural network for recognising objects in an image (the forest fire data) contains an input, convolutional layers, pooling layers, fully connected layers, and an output. In this case, the core size for each convolutional layer is 3 × 3, and the function ReLu (x) was used as the activation function [31]. In the pooling layers, a 2 × 2 filter with a step of 2 was used, and max-pooling was chosen as the pooling method. At the output of the convolutional neural network, the Object Recognition method was used.

3.2. Recognition of the data on environmental factors (air temperature at a height of 2 m above the ground, relative humidity, wind speed at a height of 10 m above the ground): the content of item 3.2 corresponds to item 3.1, however, the purpose of recognition is to arrange for determining the background colour (and not an object as for the convolutional neural network described above). To solve this problem, a convolutional neural network similar to the network from item 3.1 is developed, however, a distinctive feature is the use of Semantic segmentation at the network output instead of Object recognition. It is proposed to complete the construction of an ensemble of three convolutional neural networks. One network performs background recognition to evaluate air temperature 2 m above the ground. The second convolutional neural network performs background recognition to evaluate the wind speed at a height of 10 m above the ground. The third convolutional neural network performs background recognition to assess relative air humidity.

3.3. Recognition of the data on the nature of forest vegetation: the content of item 3.3 corresponds to item 3.2.

4. Forecasting the forest fire development dynamics: the creation of real-time forecast under the conditions of uncertainty and unsteadiness depending on the influence of environmental parameters with formulas [32]:

$$z = enc(X) = Pool(\mathrm{Re}\,Lu(Conv\ldots(Pool(\mathrm{Re}\,Lu(Conv(X,\ K_1))), K_2))),$$

$$X' = dec(z) = \sigma(Deconv(\mathrm{Re}\,Lu\ldots(Deconv(z,\ K_3))), K_4))).$$

To build a forecast, a network has been developed that is similar in structure to the auto-encoder network (an artificial neural network that reproduces input data at the output) and contains convolutional and developmental layers:

5. The retrieval of the generated image with a real-time forecast in the form of a map of the area with a selected area and the coordinates of the fire propagation over time.

Thus, the development of mathematical models for the forest fire propagation with unsteadiness and uncertainty through the use of elements of artificial intelligence, artificial neural networks. Currently, it is planned to additionally configure the artificial neural network models included in the composition and submit an application for state registration of the database.

4 Conclusions

Thus, to solve the managerial task of localising and eliminating a forest fire under complex conditions, it was proposed to develop and implement an intelligent system for predicting the forest fire dynamics based on the use of artificial intelligence and deep machine learning elements. The structural diagram of the intellectual system for predicting the forest fire dynamics has been completed. The paper describes the functional structure of the system by constructing its functional models in the form of

IDEF0 diagrams, i.e. diagrams A-0 'Perform forecasting the forest fire dynamics' and diagrams A0 'Perform forecasting the forest fire dynamics'.

As the final product, which will be performed with the scientific research output, we consider an intelligent system for predicting the forest fire dynamics depending on the influence of environmental factors, the nature of forest vegetation and the type of fire given that there is unsteadiness and uncertainty.

Acknowledgements. The reported study was funded by RFBR according to the research project № 18-37-00035 «mol_a».

References

1. Satir, O., Berberoglu, S., Donmez, C.: Mapping regional forest fire probability using artificial neural network model in a Mediterranean forest ecosystem. Geomat. Nat. Hazards Risk **7**, 1645–1658 (2016)
2. Dimopoulou, M., Giannikos, I.: Spatial optimization of resources deployment for forestfire management. Int. Trans. Oper. Res. **8**, 523–534 (2001)
3. Byram, G.M.: Combustion of forest fuels. In: Davis, K.P. (ed.) Forest Fire: Control and Use, pp. 61–89. McGraw-Hill, New York (1959)
4. UISIS. https://fedstat.ru. Accessed 29 July 2019
5. EFFIS. http://effis.jrc.ec.europa.eu. Accessed 29 July 2019
6. US Wildfires. https://www.ncdc.noaa.gov. Accessed 29 July 2019
7. Martínez, J., Vega-García, C., Chuvieco, E.: Human-caused wildfire risk rating for prevention planning in Spain. J. Environ. Manag. **90**, 1241–1252 (2009)
8. Martell, D.L.: A review of recent forest and wildland fire management decision support systems research. Curr. Forestry Rep. **1**, 128–137 (2015)
9. Sánchez, J.: Los incendios forestales y las prioridades de investigación en México. In: Congreso Forestal Mexicano, México, pp. 719–723 (1989)
10. Silva, F.R., Guijarro, M., Madrigal, J., Jimenez, E., Molina, J.R., Hernando, C., Velez, R., Vega, J.A.: Assessment of crown fire initiation and spread models in Mediterranean conifer forests by using data from field and laboratory experiments. Forest Syst. **26**(2), 14 (2017). https://doi.org/10.5424/fs/2017262-10652
11. Sullivan, A.L.: Wildland surface fire spread modelling, 1990–2007. 1: physical and quasi-physical models. Int. J. Wildland Fire **18**, 349–368 (2009)
12. Sullivan, A.L.: Wildland surface fire spread modelling, 1990–2007. 2: empirical and quasi-empirical models. Int. J. Wildland Fire **18**, 369–386 (2009)
13. Sullivan, A.L.: Wildland surface fire spread modelling, 1990–2007. 3: simulation and mathematical analogue models. Int. J. Wildland Fire **18**, 387–403 (2009)
14. Perminov, V., Goudov, A.: Mathematical modeling of forest fires initiation, spread and impact on environment. Int. J. Geomate **13**(35), 93–99 (2017). http://www.geomatejournal.com/sites/default/files/articles/93-99-6704-Valeriy-July-2017-35-a1.pdf
15. Shi, Y.: A probability model for occurrences of large forest fires. Int. J. Eng. Manuf. (IJEM) **1**, 1–7 (2012). https://doi.org/10.5815/ijem.2012.01.01
16. Adab, H., Kanniah, K.D., Solaimani, K.: Modeling forest fire risk in the northeast of Iran using remote sensing and GIS techniques. Nat. Hazards **65**, 1723–1743 (2013)
17. Safi, Y., Bouroumi, A.: Prediction of forest fires using artificial neural networks. Appl. Math. Sci. **7**, 271–286 (2013)

18. Agarwal, P.K., Patil, P.K., Mehal, R.: A methodology for ranking road safety hazardous locations using analytical hierarchy process. Proc. - Soc. Behav. Sci. **104**, 1030–1037 (2013)
19. Angayarkkani, K., Radhakrishnan, N.: An effective technique to detect forest fire region through ANFIS with spatial data. In: 3rd International Conference on Electronics Computer Technology (ICECT), Kanyakumari, India, p. 2430 (2011). https://doi.org/10.1109/ICECTECH.2011.5941794
20. Prometheus. http://www.firegrowthmodel.ca/prometheus/overview_e.php. Accessed 29 July 2019
21. FlamMap. https://www.firelab.org/project/flammap. Accessed 29 July 2019
22. FARSITE. https://www.firelab.org/project/farsite. Accessed 29 July 2019
23. Stankevich, T.S.: Operational prediction of the forest fire dynamics. In: VI International Baltic Maritime Forum 2018: XVI International Scientific Conference "Innovation in Science, Education and Entrepreneurship-2018", pp. 1079–1087. Izdatelstvo BGARF, Kaliningrad, Russia (2018). (in Russian)
24. Hussain, M., Dey, E.K.: Remote sensing image scene classification. J. Manuf. Sci. Eng. **4**, 13–20 (2018)
25. SO/IEC 25010:2011: Systems and software engineering. Systems and Software Quality Requirements and Evaluation (SQuaRE). System and Software Quality Models. Standartinform, Mocsow (2009)
26. FIRMS. https://firms.modaps.eosdis.nasa.gov/map/#z:3.0;c:44.286,17.596. Accessed 29 July 2019
27. Land Cover Map ESA/CCI. http://maps.elie.ucl.ac.be/CCI/viewer/. Accessed 29 July 2019
28. Ventusky InMeteo. https://www.ventusky.com. Accessed 29 July 2019
29. Matlab Answers. https://www.mathworks.com/matlabcentral/answers/33104-need-code-for-median-filtering-on-color-images. Accessed 29 July 2019
30. Jay Kuo, C.-C.: Understanding convolutional neural networks with a mathematical model. J. Vis. Commun. Image Represent. **41**, 406–413 (2016)
31. Nemkov, R.M., et al.: Using of a convolutional neural network with changing receptive fields in the tasks of image recognition. In: Proceedings of the First International Scientific Conference, IITI 2016, pp. 15–25. Springer, Switzerland (2016)
32. Unsupervised methods. Diving deep into autoencoders. https://www.cl.cam.ac.uk/~pv273/slides/UCLSlides.pdf. Accessed 29 July 2019

Application of Intelligent Algorithms for the Development of a Virtual Automated Planning Assistant for the Optimal Tourist Travel Route

Natalia Yanishevskaya[1], Larisa Kuznetsova[1], Ksenia Lokhacheva[1], Lubov Zabrodina[1], Denis Parfenov[1,2](✉), and Irina Bolodurina[1,2]

[1] Orenburg State University, Orenburg 460018, Russia
parfenovdi@gmail.com
[2] Federal State Scientific Institution «Federal Research Centre of Biological Systems and Agro-Technologies of the Russian Academy of Sciences», Orenburg 460000, Russia

Abstract. The article considers an approach based on the use of the production model of knowledge representation, as well as the algorithm of the ant colony simulation method for finding the optimal route in a loaded graph taking into account the time of stops and sightseeing. At the first stage of the system, the intelligent module, based on a small survey of users, selects the most interesting objects for the user, taking into account his preferences regarding recreation, mode of travel, as well as time and budget constraints. In the second stage, the route planning module builds the optimal route between the places proposed by the system in the first stage. The results of the study show that the proposed software-algorithmic solution is relevant and allows the user to build the optimal route for a tourist trip between objects.

Keywords: Tourism · Travel · Intellectual recommendation systems · Optimization · Route planning · Coordinate descent method · TSP · The production model

1 Introduction

Currently, due to the decrease of tourist trips abroad, domestic and incoming tourism has begun to develop. It is known, that the Russian Federation is a unique platform for the development of national tourism potential due to the length of the territory and the historically established ethnic diversity. However, as of 2019, the share of tourism in Russia's GDP is only 3.5%, while in the leading tourist countries it is about 10%. According to the Deputy Minister of Economic Development of the Russian Federation, S. Galkin, the main medium-term strategic goal of the development of the tourism industry in Russia is to increase its rate to 6%. The growth of tourist traffic should occur both in existing points of attraction and in new directions. Therefore, the task of developing regional tourism in the Russian Federation becomes important.

Z. Hu et al. (Eds.): AIMEE 2019, AISC 1126, pp. 13–22, 2020.
https://doi.org/10.1007/978-3-030-39162-1_2

Lots of Russian regions have a high tourist potential, represented by a variety of natural, cultural and spiritual treasures, developed infrastructure, a variety of tourist routes, which attract more and more tourists every year. However, provision of the relevant information on the recreational opportunities of the regions to the population is significantly behind the pace of development of cultural sites. As a result, the development of the competitive tourist industry slows down, as there is no reliable source of complete information about all types of recreation available to the tourist.

In these circumstances, the development of an intelligent virtual assistant that can become a unique software product that combines an innovative approach to the aggregation of information from various sources and features of the proposed algorithmic and technical solutions becomes relevant.

Also, the development of the proposed intellectual virtual assistant will improve the competitiveness of the domestic tourism, and will attract public and private investments, which are the basis for the emergence of a diverse tourism product.

As a part of the study, an approach based on the use of the production model of knowledge representation and the algorithm of the ant colony imitation for finding the optimal travel route in a loaded graph, taking into account the time for stops and sightseeing, was proposed.

This research is a part of a bigger research, which is developing a creation of models and methods of planning travel routes based on user preferences. The main goal of this research is developing a method of finding optimal route between several places based on using algorithm of the ant colony simulation method for finding the optimal route in a loaded graph taking into account the time of stops and sightseeing.

This paper is organized as follows. Section 2 presents the results of related works review devoted to the consideration of various approaches to automating the process of creating a tourism product. Section 3 describes the main idea of the combined approach and the mathematical formulation of subtasks for constructing an automated travel planning system. Section 4 contains a comparative analysis of classical and heuristic methods for finding the optimal route using the example of the ant colony algorithm and the Dijkstra algorithm. Section 5 describes the research results.

2 Related Works

Basically, the biggest part of the budget is spent on travel and transportation expenses. Therefore, it is important to determine the number of places to visit, as well as to make the best route between them to allocate the budget effectively during the trip. Nowadays intelligent travel planning systems are used for these purposes [1].

Thus, article [2] proposes a practical application for the Autonomous Region of Andalusia. This system takes into account wishes and needs of the particular tourist, including interesting activities and characteristics of the area. The system provides recommendations based on multi-criteria optimization methods. The user has the opportunity to select dates of the beginning and the ending of the journey, clarify the places of interest, and events for visiting by marking them on an interactive map.

Authors of the research [3] have designed an automatic travel itinerary planning system for the Taiwan domestic area (ATIPS), which uses an algorithmic framework for automatic travel route planning. The system uses an approach that effectively combines the five most important factors considered by the traveler (user preferences, popularity, time, distance and cost). The greedy algorithm is used to identify the best tourist spot at the current stage requiring the tourist spot selection to be within a certain radius of the current position. In this way, the travel route is generated automatically according to the user's preferences.

The study of Xie [4] describes a system that uses leverages rating information from underlying recommender systems, allows flexible package configuration and incorporates users' cost budgets on both time and money. Also, the described CompRec-Trip system has a graphical user interface that allows users to customize the returned composite recommendations and take into account external local information.

The main disadvantage of the systems discussed above is the impossibility of building the most profitable route, which is quite an important criterion while planning a tourist trip. Since the problem of finding the optimal path (the traveling salesman problem) belongs to the set of NP-complete decision problems, classical methods of approximation are used for it. That is why Dijkstra's algorithm [5, 6], a breadth-first search algorithm [7], a Floyd-Warshall algorithm [8], a sweep algorithm or an A* algorithm [9] are among the most popular. Dijkstra's algorithm is the most frequently used since it is considered one of the best classical algorithms for finding the shortest path. However, some studies claim that particular combinations of the algorithms described above produce a result that is closer to an exact solution [10–14].

To solve the traveling salesman problem new optimization approaches are being developed, for example, Local algorithms, Genetic algorithms (Goldberg 1989) [14, 16], Tabu Search (Fiechter 1994) [17, 18], Ant Colony (Angus and Hendtlass 2005) [17, 19].

Nowadays there is a big amount of relevant services for travel planning. An overview of existing solutions with the indication of advantages and disadvantages is given in Table 1.

Table 1. Overview of existing solutions.

Service name	User preferences	Building the best route
Tripomatic	–	+
Waytips	–	+
TripAdvisor	–	–
Youroute	–	–
Trivago	+	–
Aviasales	+	–
OnlineTours	+	–
Level Travel	+	–

As a result of the analysis of the existing services functionality, the following disadvantages were found:

- providing incomplete information about available opportunities (in particular, the lack of information about the sights of some regions of the country);
- lack of the functionality of the planning route visualization.

A survey of the researches has shown that most of the existing solutions allow us either to optimize the time characteristics of the journey, or to choose the travel destination that best suits the user's wishes. Also, none of the systems offers a route for traveling through the territory of Russia as a whole and the Orenburg region in particular.

In this regard, this study aims to develop a software and algorithmic solution for the problem of automatically travel route planning based on combining a production model of knowledge representation to determine the most preferred travel places and a modified ant colony method to find the best route between them.

3 Approach to Solution

It is necessary to plan every day of the trip for effective travel organization. Every trip is limited by the total number of days, as well as by finances and the wishes of the traveler to visit various places and sights of Russian regions. At the same time, it is important to provide the tourist with information about the directions that most satisfy his needs, taking into account the peculiarities of existing routes, and the cost of travel and entertainment options with the respect to the fact that the route must be optimal [20].

For the formation of the transport and logistics structure, the following tasks are solved within the project:

- identification of possible tourist destinations based on user preferences;
- planning time resources.

Consider them in more detail.

3.1 Consideration of User's Preferences for Components of the Trip and Formation of the Travel Route

One of the main components of the system is an intelligent module of the travel places selection [6, 21], which is based on the production model of knowledge representation.

In the general case, the production model can be represented as follows:

$$u = <S; L; A \rightarrow B; Q> \tag{1}$$

where S is the description of the situations class; L is the condition for product activation; $A \rightarrow B$ is the product core; Q is the post condition of the production rule.

At this stage, the user selects the preferred types of activities (beaches, ski resorts, historical sites, etc.), the way of travelling (car, bus, train), including time and budget restrictions. As a result of analyzing the provided information, the intelligent virtual assistant will create a list of cities/towns, types of activity available in them and popular places for recreation, most appropriate for the given parameters. After that, the user will be asked to choose the most favorite options, the optimal route between which will be built as it given in Sect. 3.2.

3.2 Construction of the Optimal Route

At this stage, the intelligent assistant prepares a trip plan in the form of a tourist route. The tourist route includes a complete description of all the places planned for visiting, marking them on the interactive map of the uploaded GIS-service.

To solve the problem we need information about the starting and the ending points of the route (which were defined at the previous step), as well as the time for moving between places of interest. Yandex Maps web mapping service allows getting all necessary information. For this purpose, a set of freely distributed GMap.NET cross-platform open source libraries was used in the program.

Nowadays both exact and approximate solution algorithms are used to find the optimal route (to solve the traveling salesman problem). The exact algorithms include the brute force algorithm and selection of the optimal route among all the found. Methods that reduce full enumeration of paths are related to approximate methods of finding the optimal route, which are divided into classical (greedy algorithm, modification of Dijkstra's algorithm, branch and bound method) and heuristic (genetic algorithm, simulation annealing, ant algorithms). Let us compare the lengths of routes found using heuristic optimization methods. The results of the comparison are shown in Table 2.

Table 2. The comparison of heuristic route optimization methods.

	The task of a traveling salesman for 50 points	The task of a traveling salesman for 75 points
Imitation annealing	443	580
Genetic algorithms	428	545
Ant algorithms	425	535

Note that ant algorithms achieve greater accuracy in finding the optimal route. Therefore, to solve the problem, we will use a modified algorithm of the ant colony.

Since the problem we are solving takes into account not only the distance between the sights but also the priority of their visit, both of these parameters should influence equally on the probability of the k-th ant moving from point i to point j on the t-th iteration. This result can be achieved with the normalization of the priorities of the visited places:

$$pr_i = \frac{pr_1}{pr_i \cdot S_{pr}}, i = \overline{1, n} \quad (2)$$

where n is the number of vertices of the graph, pr_i is the priority to visit the i-th vertex, and S_{pr} is the normalization factor calculated by the formula:

$$S_{pr} = \sum_{pr_j \neq pr_h}^{n} \frac{pr_1}{pr_j}, \forall h < j, h \in N. \tag{3}$$

Let $M_k(t)$ be the set of numbers of vertices visited by the k-th ant at the t-th iteration. We suppose that the route with the biggest value of $F = \sum_{i \in M_k(t)} pr_i$ is the best route.

In addition, the probability-proportional rule, which determines the probability of the k-th ant moving from point i to point j at the t-th iteration, plays an important role in the algorithm:

$$\begin{cases} P_{ik,k}(t) = \dfrac{[\tau_{ij}(t)]^{\alpha} \cdot [\eta_{ij}]^{\beta}}{\sum_{l \in J_{i,k}} [\tau_{ij}(t)]^{\alpha} \cdot [\eta_{ij}]^{\beta}}, & if \quad j \in J_{i,k}, \\ P_{ik,k}(t) = 0, & if \quad j \notin J_{i,k}, \end{cases} \tag{4}$$

where α and β are adjustable parameters that specify the weight of the pheromone trace and visibility while choosing a route. If $\alpha = 0$, then the nearest place will be chosen, which corresponds to the greedy algorithm of the classical optimization theory. If $\beta = 0$, only pheromone amplification works, which causes the routes to degenerate to a single suboptimal solution.

Since hours of the sights visiting are limited, there could be selected only those paths, the length of which would not exceed this value. If a longer path was found, the search should be stopped in advance. Additionally, the time for the return trip is taken into account automatically. Also, it is necessary to make a loop at all the vertices (considering time for sightseeing).

After each iteration, the amount of pheromone deposited on the edge is determined by the formula:

$$\Delta\tau_{ij,k}(t) = \begin{cases} \frac{Q \cdot \sqrt[p]{f_{k,t}}}{L_k(t) \cdot F}, & if (i,j) \in T_k(t), \\ 0, & if (i,j) \notin T_k(t), \end{cases} \tag{5}$$

where $T_k(t)$ is the route traveled by ant k at iteration t; $L_k(t)$ is the length of this route; $F = \max_{k,t} f_{k,t}$; Q is the adjustable parameter ($S_{\max}$ is the maximum duration of a trip per day).

4 Experimental Studies

Let us compare the results obtained using the heuristic algorithm of the ant colony imitation and the classical Dijkstra's algorithm. Let it be necessary to build an optimal route between settlements and sights of the Orenburg region: Orenburg city, Buzuluksky bor, Sol-Iletsk city, Zmeinaya Gora (p. Mikhailovka), p. Aksakovo, p. Saraktash where Orenburg city is both the starting and the ending point of the route. Figure 1 shows the initial loaded graph of selected places, taking into account their transport accessibility and distance between them.

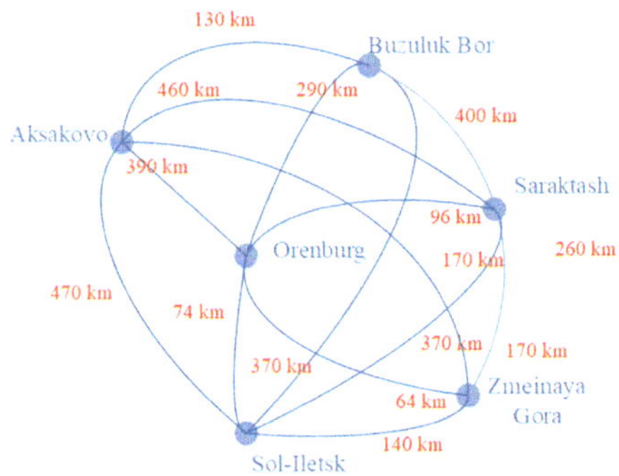

Fig. 1. Loaded graph

The weights of the edges are the distances between places. The time spent on moving from one place of visit to another, is calculated depending on the chosen method of movement. So, traveling by bus or train, the user knows the exact time spent on the road. When traveling by car, travel time is calculated as the distance divided by the speed, we will consider it as a constant. We suppose that the user always moves between the two selected objects on the same type of transport. Thereby, we reduce the time costs description to the calculation of the total length of the travel route.

Let us find the optimal route between these places using the Dijkstra algorithm. We obtain the following result: Orenburg - Aksakovo - Buzuluksky bor - Zmeinaya Gora - Saraktash - Sol-Iletsk - Orenburg. The loaded graph with the selection of the route is shown in Fig. 2.

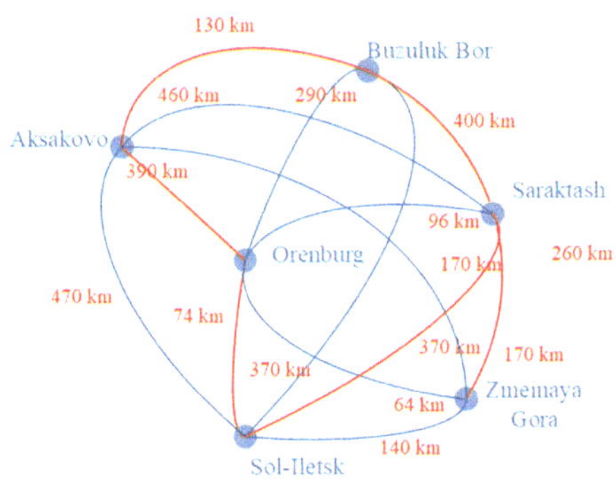

Fig. 2. Loaded graph with route highlighting

The optimal route obtained using the ant colony algorithm is as follows: Orenburg - Aksakovo - Buzuluksky bor - Zmeinaya Gora - Sol-Iletsk - Saraktash - Orenburg. The loaded graph with the selection of the route is shown in Fig. 3.

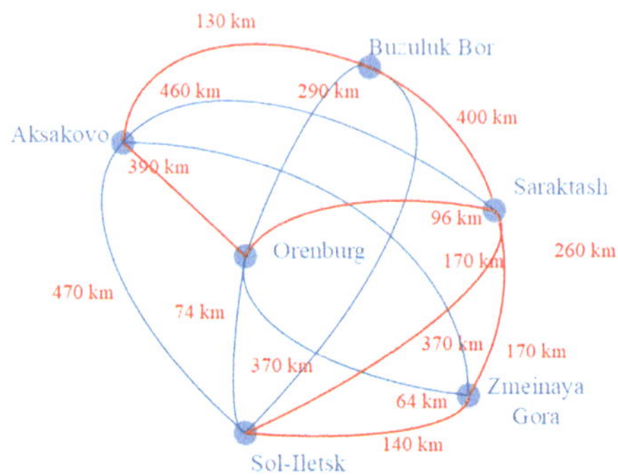

Fig. 3. Loaded graph with route highlighting

A comparative analysis of methods for finding the shortest route showed that the route length obtained using the modified ant colony algorithm is shorter than the route length obtained using the Dijkstra algorithm, which are 1,186 km and 1,194 km, respectively.

Thus, the chosen method for solving the subtask of constructing the optimal route (the traveling salesman problem)—the ant colony method—allows one to obtain an approximate solution with the least error, i.e. find the best route to follow.

5 Summary and Conclusion

In this study, a comparative analysis of heuristic optimization methods for searching of an optimal route is held. It is revealed that the results gained using the ant colony method reach the closest to the exact value. That is why this method was chosen to solve the problem of finding the optimal route between the most suitable for a particular user places of travel. While developing the algorithm for solving the problem, such opportunities as specifying the priority of visiting different places of interest, as well as the time for viewing attractions and stops were taken into account.

Thus, the problem of creating an optimal route with the above assumptions is the Traveling Salesman Problem with Time Windows. To solve this problem, a modified algorithm for ant colony method has been developed in this study.

The results of the study show that the proposed software and algorithmic solution is relevant and allows to build the best route for a tourist trip between the objects most interesting to the user, taking into account preferences about the types of activities, the way of traveling, as well as the time and budget constraints.

Acknowledgment. The study was conducted with the support of the Ministry of Education of the Orenburg region in the framework of the research "Intellectual virtual assistant for planning trips to the sights of the Orenburg region" (project no. 3 on 14 August 2019). The studies were performed in accordance with the R & D plan for 2019–2020 at the Federal State Scientific Institution «Federal Research Centre of Biological Systems and Agro-technologies of the Russian Academy of Sciences» (# 0761-2019-0004).

References

1. Shakhovska, N., Shakhovska, K., Fedushko, S.: Some aspects of the method for tourist route creation. In: Proceedings of the International Conference of Artificial Intelligence, Medical Engineering, Education, pp. 527–537. Springer, Cham (2018)
2. Rodríguez, B., Molina, J., Pérez, F., et al.: Interactive design of personalized tourism routes. J. Tour. Manag. **33**(4), 926–940 (2002)
3. Chang, H.-T., Chang, Y.-M., et al.: ATIPS: automatic travel itinerary planning system for domestic areas. J. Comput. Intell. Neurosci. **2016**, 13 (2016)
4. Xie, M., Lakshmanan, L.V.S., Wood P.T.: CompRec-Trip: a composite recommendation system for travel planning. In: Proceedings of the IEEE 27th International Conference on Data Engineering, ICDE 2011, pp. 1352–1355. IEEE (2011)
5. Wang, H., Zhang, F., Cui, P.: A parking lot induction method based on Dijkstra algorithm. In: Proceedings of the 2017 Chinese Automation Congress (CAC), pp. 5247–5251. IEEE (2017)
6. Miah, Md.S.U., Masuduzzaman, Md., Sarkar, W., Islam, H.M.M., Porag, F., Hossain, S.: Intelligent tour planning system using crowd sourced data. Int. J. Educ. Manag. Eng. (IJEME) **8**(1), 22–29 (2018)
7. Hsu, C.-M., Lian, F.-L., Ting, J.-A., et al: A road detection based on bread-first search in urban traffic scenes. In: Proceedings of the 2011 8th Asian Control Conference (ASCC), pp. 1393–1397. IEEE (2011)
8. Hougardy, S.: The Floyd-Warshall algorithm on graphs with negative cycles. J. Inf. Process. Lett. **110**(8–9), 279–281 (2010)
9. Cui, S.-G., Wang, H., Yang, L.: A simulation study of A-star algorithm for robot path planning. In: Proceedings of the 16th International Conference on Mechatronics Technology, pp. 506–509. IEEE (2012)
10. Djojo M. A., Karyono K.: Computational load analysis of Dijkstra, A*, and Floyd-Warshall algorithms in mesh network. In: Proceedings of the 2013 International Conference on Robotics, Biomimetics, Intelligent Computational Systems, pp. 104–108. IEEE (2013)
11. Furculita, A.G., Ulinic, M.V., Rus, A.B., et al: Implementation issues for modified Dijkstra's and Floyd-Warshall algorithms in OpenFlow. In: Proceedings of the 2013 RoEduNet International Conference 12th Edition: Networking in Education and Research, pp. 1–6. IEEE (2013)
12. Dela Cruz, J.C., Magwili, G.V., Mundo, J.P.E., et al: Items-mapping and route optimization in a grocery store using Dijkstra's, Bellman-Ford and FloydWarshall algorithms. In: Proceedings of the IEEE Region 10 Annual International Conference, pp. 243–246. IEEE (2017)
13. Risald, R., Mirino, A., Suyoto: Best route selection using Dijkstra and Floyd-Warshall algorithm. In: Proceedings of the 2017 11th International Conference on Information & Communication Technology and System, pp. 155–158. IEEE (2017)

14. Zulfiqar, L.O.M., Isnanto, R.R., Nurhayati, O.D.: Optimal distribution route planning based on collaboration of Dijkstra and sweep algorithm. In: Proceedings of the 2018 10th International Conference on Information Technology and Electrical Engineering, pp. 371–375. IEEE (2018)
15. Liu, J., Li, W.: Greedy permuting method for genetic algorithm on traveling salesman problem. In: Proceedings of the 2018 8th International Conference on Electronics Information and Emergency Communication, pp. 47–51. IEEE (2018)
16. Gupta, I.K., Choubey, A., Choubey, S.: Randomized bias genetic algorithm to solve traveling salesman problem. In: Proceedings of the 2017 8th International Conference on Computing, Communication and Networking Technologies (ICCCNT), pp. 1–6. IEEE (2017)
17. Chen, H., et al: Ant colony optimization with tabu table to solve TSP problem. In: Proceedings of the 2018 37th Chinese Control Conference (CCC), pp. 2523–2527. IEEE (2018)
18. Yang, N., Ma, X., Li, P.: An improved angle-based crossover tabu search for the larger-scale traveling salesman problem. In: Proceedings of the 2009 WRI Global Congress on Intelligent Systems, pp. 584–587. IEEE (2009)
19. Liu, Y., Shen, X., Chen, H.: An adaptive ant colony algorithm based on common information for solving the traveling salesman problem. In: Proceedings of the 2012 International Conference on Systems and Informatics, ICSAI 2012, pp. 763–766. IEEE (2012)
20. Bolodurina, I., Parfenov, D.: The optimization of traffic management for cloud application and services in the virtual data center. In: Proceedings of the International Conference on Parallel Computing Technologies, pp. 418–426. Springer, Cham (2017)
21. Dennouni, N., Yvan, P., Lancieri, L., Slama, Z.: Towards an incremental recommendation of POIs for mobile tourists without profiles. Int. J. Intell. Syst. Appl. (IJISA) **10**(10), 42–52 (2018)

The Use of Convolutional Polycategories in Problems of Artificial Intelligence

Georgy K. Tolokonnikov(✉)

Federal Scientific Agro-Engineering Center VIM, Russian Academy of Sciences, 1st Institute Passage, 5, Moscow, Russia
admcit@mail.ru

Abstract. Convolutional polycategories introduced by the author find application in the general theory of systems, in the theory of artificial neural networks, in other areas of artificial intelligence. The report gives further applications of convolutional polycategories in algebraic biology and logical calculi, covering the classical and intuitionistic predicate calculus. Based on the formalism of convolutional polycategories, a new categorical definition of information is given, reflecting its semantic component. This definition finds application in algebraic biology with a basic code example of DNA and RNA molecules in a cell. A polycategorical model is given for the method of typical quantifiers used in AI and stronger than the Robinson method and the inverse Maslov method. The model reveals the categorical basis of calculus and is used to study the properties of the calculus of typical quantifiers. The main results are of a fundamental theoretical nature for algebraic biology and AI.

Keywords: Genetic code · DNA · Tensor product · System theory · Categories · Topoi · Logic · Quantifiers · Predicate calculus · Artificial intelligence

1 Introduction

Convolution polycategories introduced by the author [1] find applications in the categorical theory of systems, in the theory of artificial neural networks [2], in other areas of artificial intelligence. The report gives further applications of convolutional polycategories in algebraic biology and logical calculus, covering the classical and intuitionistic predicate calculus in an approach based on the theory of typical quantifiers, which has numerous applications in artificial intelligence and mathematical systems theory [3]. For the logical calculations of typical quantifiers, a model is proposed in the form of nonassociative convolutional polycategories, which allows progress in studying the properties of these calculi. A convolutional polycategory is a set of poly-arrows with generalizing the usual composition of arrows, multi-arrows, and poly-arrows in the traditional theory of categories by convolutions. In the categorical theory of systems, the system is a poly-arrow in a suitable convolutional polycategory.

The formation of systems from subsystems is formalized as a functor with convolution, which formalizes the system-forming factor of the informal theory of functional systems according to P.K. Anokhin.

Z. Hu et al. (Eds.): AIMEE 2019, AISC 1126, pp. 23–32, 2020.
https://doi.org/10.1007/978-3-030-39162-1_3

The principle of isomorphism is modeled by the notion of similarity, the principle of hierarchy is modeled by the procedure of forming a system from subsystems. The categorical theory of systems formalized the basic principles of the theory of functional systems and made it possible to solve a number of long-standing problems of the theory of functional systems itself. One of the basic principles of the theory of functional systems, consisting of the presence of a system-forming factor in formalization, pushes out from the usual set-theoretic mathematics since the principle postulates the construction of the system as a process going from whole to parts. Set-theoretic mathematics cannot do without the elements from which the sets are built, so we have to go into the categorical paradigm, for which the constructions begin with arrows without the need for elements in objects.

Based on convolutional polycategories, it is possible to construct a categorical model for artificial neural networks, which, as was proved in [4], are associative convolutional polycategories with convolutions of the corona type.

Thus, it was possible to clarify the nature of the branch points of the axons of neurons of an artificial neural network, which made it possible to correct a number of inaccuracies in the theory of neural networks, in particular, the backpropagation method in the form proposed by S. Osovsky. We hope that this model will be useful for the rapidly developing direction of neural networks, including applications in genetics [5–9]. The report proposes further applications of convolutional polycategories in algebraic biology [10–12] and typical quantifiers theory [3] which provides, in particular, the logical J-calculus, basic after the Robinson resolution method and the inverse Maslov method and numerous applications in artificial intelligence [3].

2 Algebraic Biology and the Concept of Information in Categorical Systems

The author joined the work on algebraic biology, which arose initially in the form of matrix genetics, and developed by prof. Petukhov and his school [10–12].

Each accomplished science has its own subject, tasks, and methods for their solution. The main subject of study in algebraic biology is the deoxyribonucleic acid (DNA) molecule adopted in this science as the basis for all living things, which provides storage, transmission from generation to generation, and implementation of a genetic program for the development and functioning of living organisms. However, DNA in algebraic biology comes together with the environment (cell, etc.) that can read in-formation encoded in the sequence of nucleotides that make up DNA and provide other service needs for its functioning. Algebraic biology studies DNA in living systems, the chemical and physical properties of DNA are as interesting as they reflect the properties of DNA as subsystems in a living system of a unicellular or multicellular organism. Already in the description of the subject of research in algebraic biology, we involve the systems approach, talking about subsystems and the interaction of subsystems with the superior system and among themselves. Algebraic biology is a deep system science. The main task in algebraic biology is to elucidate the mechanisms of unfolding the properties of organisms and their populations based on a given DNA molecule with its genetic code. This also includes the task of identifying the properties themselves due to the genetic code embedded in the DNA. However, the deployment of the genome in the functions and properties of the organism is impossible without the interaction of the organism as a system with other systems, including individuals of the same population, which are referred to the concept of external conditions and the external environment. Thus, the task of algebraic biology is considered in the context of the development and interaction of systems, and systems of animate and inanimate nature. In other words, algebraic biology is considered in the context of the general theory of systems.

We note in particular that the methods of solving problems used in algebraic biology and which are strictly mathematical (algebra, etc.), turning this science into a strictly mathematical discipline. As already mentioned, the formalization of the theory of functional systems, in view of the principle of the presence of the system-forming factor and the process of building a system that flows from whole to parts, determined by it, leads to categorical mathematics, which took shape with the completion of the basic sections of the topos theory.

Adding to the already used algebra in algebraic biology also categorical methods of the categorical theory of systems greatly enriches its methods and serves as an important application of the theory of convolution polycategories. We will dwell on this application in another place, and here we will consider the definition of the concept of information, initiated by algebraic biology, reflecting its substantive aspect.

Given the application of the categorical theory of systems, algebraic biology is presented as one of the fundamental exact sciences, strictly mathematically unfolding from the genetic code the properties and functions of individuals corresponding to the genome of the population. The significant successful steps are already taken in this direction, and the very formulation of the problem in the author's opinion is unprecedented.

For the following, we consider the definition and properties of convolutional polycategories known (see [1]). Informally, a convolutional polycategory can be thought of as a set of poly-arrows with convolutions, generalizing the usual categorical composition of arrows, multi-arrows, and poly-arrows. The quantitative aspects of the concept of information are studied in detail, starting with the works of C. Shannon and the works of Holevo [13] (regarding quantum information) and other scientists. However, the content aspect of the transmitted signals remains unformalized. Below, in the framework of the categorical system theory, a definition of the concept of information is proposed, which takes into account its objective content. The main example that initiated the above definition of the concept of information is taken from algebraic biology and is associated with the DNA code of living organisms.

The first thing we will rely on is that information is always associated with certain objects and their properties. In the paradigm of categorical systems, we understand these objects as categorical systems. So, if there is a system or systems, then information about them is information about their properties. Thus, the information in our approach is inextricably linked with the systems whose properties are described in this information. The properties of a system, if a person describes them, are given by single predicates in the language of classical logic.

However, we will be more interested in, so to speak, objective information that is not related to the consciousness of a human researcher, although both cases will be included in the general scheme. The objective and most important example mentioned above is the information encoded in the DNA about the object (mammals, populations, bacteria, etc.). We will immediately speak about the system (DNA system), referring to the categorical system description of these bio objects.

The building material for any code is the alphabet - a set of some molds, with which you can stamp the letters of the alphabet themselves in the quantity necessary for the code. One example is the alphabets of various calculi depicted on paper (or on a computer monitor), another example of the alphabet is nitrogen bases (adenine - A, cytosine - C, guanine - G, thymine - T) from which the DNA code is assembled.

The sequence of letters written out (or somewhere located) one after another, is called by a word. There is a first deductive system that builds a set of words corresponding to the system, the so-called correctly constructed formulas, for not every word of the possible words in the alphabet is needed for a given specific system.

We postulate that for systems with an informational description there is a convolutional functor that translates languages as sets of systems into a set of source systems. For source systems, there is some procedure that manipulates material objects, which recreates the original set of systems using information systems. An example of such a functor for the DNA systems of mammals and humans is the process of reconstructing from the zygotes of the whole living organism in various environmental conditions.

Now we are able to give a precise definition of information in the categorical theory of systems.

Definition
A system is called a system with a complete informational description if there are a corresponding language and a functor F that realizes an isomorphism from a language as a system into the original system. Suppose there is a set of systems $S = \{S_\alpha\}, \alpha \in I$, I is an index set, with a full informational description of the languages $L = \{L_\alpha\}$, then the pair $Inf_S = \{L, F\}, F : L \to S$ is called the information about these systems.

What is the important information? It can be transmitted in time and space, without transferring the material systems themselves. According to the information obtained, if possible, you can completely restore the original set of systems, for example, as real material objects. Information transfer in categorical systems theory relies on a commutative diagram (Fig. 1).

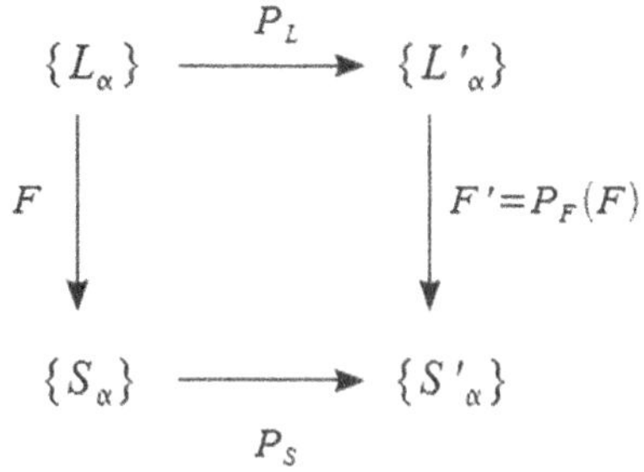

Fig. 1. Commutative information transfer diagram

Information transfer Inf_S is modeled by the mapping $P = \{P_L, P_F, P_S\}$, which is often functorial, and the set of functors, in this case, can be different categories of convolutional functors. At the same time, various constructions are possible: from the category of convolution functors with arrows in the form of natural transformations to cubic categories of convolution functors. We clearly see the presence of a unit in the form of the identity map R.

It is of interest to consider the case when the functor F is not an isomorphism when the set of formal language systems is not enough to fully recover. This corresponds to incomplete information on the set of systems. Let us turn to the main example of algebraic biology, which initiated the given definition of the concept of information.

Consider the zygote of a mammal M, for example, a mouse, as a categorical system. Suppose we fully know the set of codons in the zygote DNA. In this case, we can build the language L_M, based on the system forming factor, consisting of the fact that it should be a mouse zygote, not an elephant. It is easy to get the language, both deductive systems for proteins purely theoretically.

But looking at the zygote DNA, we see objectively the built language L_M already objectively without a person. Thus, the DNA molecule itself is a system with the L_M language. If we place a zygote in a vessel with a low temperature, preventing it from division, we will receive almost all the information Inf_M in pure form, namely, the code (which is usually considered information and, which is transmitted via communication channels) and the zygote itself part of the functor $F : L_M \to S_M$, capable, as it is, obviously, of reproducing a mouse as the S_M system under the presence of suitable environmental conditions. If a person does not intervene, then the zygote develops successfully, referring to the DNA code, in which case we have the full functor $F : L_M \to S_M$ and full information about the S_M system, that is, about the mouse living in certain conditions.

As we can see, the categorical language of the theory of systems provides a formal basis for departing from the intuitive and mystical speculations about what "information" or even some "valuable information" is of the living [14]. "From an informational point of view, living organisms are informational entities" [10], this intuitive idea acquires a completely accurate implementation in the categorical theory of systems.

3 Logical Calculus in the Context of Convolutional Polycategories

The main task of artificial intelligence (AI) is to search for a conclusion ("path") from the initial state of the system to its target state (inference rules, states, etc. are formalized in a certain language of "description of knowledge"). The well-known method for solving this problem is the Robinson resolutions method (which is the basis of the Prolog programming language) and the inverse method of S. Maslov. Academician S. N. Vasilyev proposed another method based on the theory of typical quantifiers developed by him. This method turned out to be in many cases stronger than the two, found numerous applications for intelligent AI systems [3]. The theory of typical quantifiers itself has not yet been studied sufficiently. Therefore, its categorical model given below is useful for studying it. For type quantifiers, the L_0 language of first-order logic and axiomatic theories (the usual language of first-order theories) is used, which we consider known. Let us proceed to the construction of the language of generalized TK-formulas, denoted here by L_1.

Definition

Let the above set of formulas L_0 (the presence of algebraic and other structures on this set in this definition is irrelevant) given in the alphabet $\{L_0, \exists, \forall, x_i, :\}$, to which are also added parentheses and a comma. The L_1 language consists of the following words in the alphabet, called TK-formulas

(1) $\exists : A, \forall : A, \exists x_i : A, \forall x_j : A, A \in L_0$- TK-formulas;
(2) if F is a TK-formula and $w \in \{\exists : A, \forall : A, \exists x_i : A, \forall x_j : A, A \in L_0\}$, then w (F) is the TK-formula;
(3) if $(F_1, \ldots, F_k), k \geq 2$ is a set of TK-formulas, then $w(F_1, \ldots, F_k)$ is the TK-formula;
(4) there are no other TK-formulas.

The designation $w_1 w_2 \ldots w_k \overset{\text{def}}{=} w_1(w_2(\ldots w_k)\ldots)$ (k-1 closing bracket) is used, for example, we write $w_\beta w_\alpha F$ instead of $w_\beta(w_\alpha(F))$ or $w_\gamma w_\beta w_\alpha$ instead of $w_\gamma(w_\beta(w_\alpha))$.

In the theory of positively formed formulas, the following operation is used: from $w(F_1, \ldots, F_i, F_{i+1}, \ldots, F_n)$ go to $w(F_1, \ldots, F_i, G, F_{i+1}, \ldots, F_n)$. Connection of vectors, poly arrows, etc. usually uses the composition, however, in it, the factors should have the same "input-output", which, moreover, disappear after the composition in general. The above operation is radically different from the composition operation. It is found in the theory of ω-hypergraph constructions in the form of an algebra of inscribed disks [1].

Now we give the definition of the mapping of the set of TK-formulas to the set of formulas L_0.

Definition

We define the ip mapping of TK-formulas to formulas from L_0.

(1) $(\exists x : A)^{ip} = \exists x A, A \in L_0$;
$(\forall x : A)^{ip} = \forall x(\neg A)$;
$(\exists : A)^{ip} = A$;
$(\forall : A)^{ip} = \neg A$;
(2) let $A \in L_0, B \in L_1$,
$(\exists x : A(B))^{ip} = \exists x(A \wedge B^{ip})$;
$(\forall x : A(B))^{ip} = \forall x(A \to B^{ip})$;
$(\exists : A(B))^{ip} = A \wedge B^{ip}$;
$(\forall : A(B))^{ip} = A \to B^{ip}$;

(3) let $F_1, \ldots, F_n \in L_1$,

$$(\exists x : A(F_1, \ldots, F_n))^{ip} = \exists x\left(A \wedge \left(\wedge_{i=1}^{n} F_i^{ip}\right)\right), \wedge_{i=1}^{n} F_i^{ip} \overset{\text{def}}{=} (F_1^{ip} \wedge (F_2^{ip} \wedge \ldots) \ldots);$$
$$(\forall x : A(F_1, \ldots, F_n))^{ip} = \forall x\left(A \rightarrow \left(\vee_{i=1}^{n} F_i^{ip}\right)\right), \vee_{i=1}^{n} F_i^{ip} \overset{\text{def}}{=} (F_1^{ip} \vee (F_2^{ip} \vee \ldots) \ldots);$$
$$(\exists : A(F_1, \ldots, F_n))^{ip} = A \wedge \left(\wedge_{i=1}^{n} F_i^{ip}\right);$$
$$(\forall : A(F_1, \ldots, F_n))^{ip} = A \rightarrow \left(\vee_{i=1}^{n} F_i^{ip}\right).$$

Two TK-formulas are called equivalent if they go into the same formula from L_0, for example, $(\exists : A\exists : B)^{ip} = A \wedge B = (\exists : (A \wedge B))^{ip}$, that is, $\exists : A\exists : B, \exists : (A \wedge B)$ are equivalent.

We introduce the following model of a TK-tree with nodes named by formulas from L_0: the Q typical quantifier: $Q : A\ Q : A, A \in L_0, Q \in \{\exists, \forall, \exists x, \forall x\}$ is interpreted as a multi-arrow with one output and a countable number of inputs. All inputs and outputs are the same, the very same element of multi-arrows, as is customary in the theory of operads, is not displayed on the diagrams. We enumerate the inputs of multi-arrows, after which we can explicitly indicate componentwise operations of the composition $\delta^{(i)}$ (Fig. 2), which are implemented by the correspondence maps of TK-formulas and PI-formulas from the second part of the definition of TK-formulas, for example (Fig. 3),

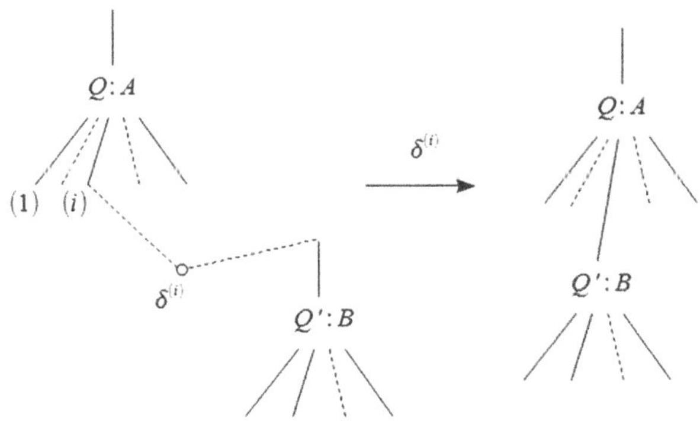

Fig. 2. Operations of composition $\delta^{(i)}$

Theorem

The operations $\left(\delta_A^{(i)}(Q' : B)\right)(Q : A) = (Q : A)\delta^{(i)}(Q' : B)$ are associative and commutative "horizontally", namely,

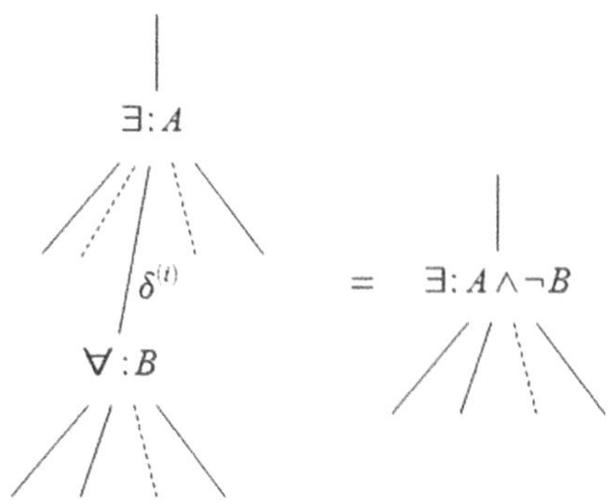

Fig. 3. An example of applying the operation of composition $\delta^{(i)}$

$$\delta_A^{(i)}(Q' : B)\delta_A^{(j)}(Q'' : B') = \delta_A^{(j)}(Q'' : B')\delta_A^{(i)}(Q' : B),$$
$$\left(\delta_A^{(i)}(Q' : B)\delta_A^{(j)}(Q'' : B')\right)\delta_A^{(k)}(Q''' : B'') =$$
$$= \delta_A^{(j)}(Q'' : B')\left(\delta_A^{(i)}(Q' : B)\delta_A^{(k)}(Q''' : B'')\right), i, j, k \in \mathbb{N},$$

but when trying to propagate to a different arrangement of brackets than in the definition, they turn out to be non-associative "vertically".

Proof
We restrict ourselves to the second part of the statement. Vertically, the order of the brackets is fixed in the definition, for example, $\exists : xA\forall : B\exists : C \overset{\text{def}}{=} \exists : xA(\forall : B(\exists : C))$.

A different order of parentheses gives a different result, to demonstrate which it suffices to give the example Fig. 4.

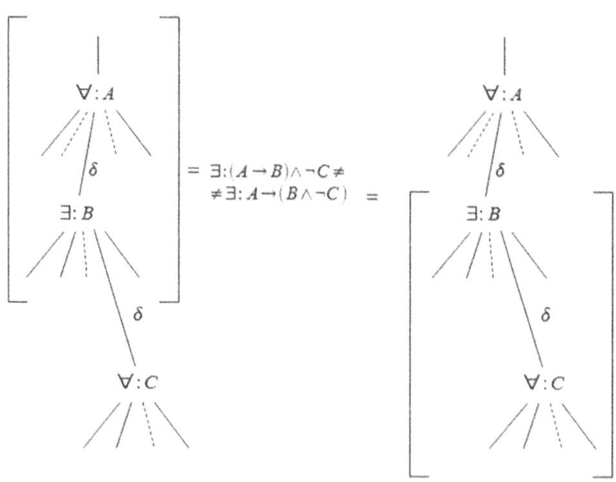

Fig. 4. The example of non-associative operation of composition

The theorem is proved.

Theorem
An inductively defined mapping of TK-formulas into an IP-formula is defined everywhere on the set of TK-formulas. In other words, every TK-formula is translated according to the steps specified in the definition of the mapping to an element of a free algebra with generators in the form of an IP-formula and signature operations "{"¬,∧,∨, → ,∃x_i,∀x_i "}", the second, third and fourth of which are binary, the rest are unary.

Theorem
Let a set of TK-formulas with its mapping of TK-formulas into the formulas of the predicate calculus language from the previous theorem be given. Then the previous theorems and definitions define a convolutional multicategory with a single object and with a countable number of arrows inputs. Explicitly defined mappings of TK-formulas to IP-formulas are defined by multiplication of multi-arrows.

The proof is carried out by explicit construction of the multicategory given in the theorem according to the relations given in the previous definitions and theorems, which models the typical quantifier language from the definitions given. This allows by categorical methods to answer questions about the structure and properties of typical quantifier language.

Theorem
A set of TK-multicategories forms a category of TK-multicategories if we take the functors from one TK-multicategory to another as arrows, preserving the operations δ with their properties.

4 Conclusion

In the work, on the basis of the formalism of convolutional polycategories developed by the author, a new categorical definition of information is given, which reflects its semantic component, which is a known difficulty for existing definitions. This definition finds application in algebraic biology with a basic code example of DNA and RNA molecules in a cell. The polycategorical model of typical quantifiers used in AI for the method of typical quantifiers which more powerful than the Robinson method and the inverse Maslov method is given. The model reveals the polycategorical basis of calculus and is used to study the properties of the calculus of typical quantifiers. The main results are of a fundamental theoretical nature for algebraic biology and AI. It is planned to use these results for problems of the transfer of genetic information, as well as to strengthen the computational capabilities of the method of calculating typical quantifiers.

References

1. Tolokonnikov, G.K.: Mathematical foundations of the theory of biomachsystems. In: Biomachsystems. Theory and Applications, pp. 31–213. Rosinformagrotekh, Moscow (2016)
2. Tolokonnikov, G.K.: Convolution polycategories and categorical splices for modeling neural networks. In: Advances in Intelligent Systems and Computing, Volume 938, Advances in Computer Science for Engineering and Education II, Conference Proceedings ICCSEEA, pp. 259–267 (2019). ISSN 2194-5357. ISSN 2194-5365
3. Vasiliev, S.N., Zherlov, A.K., Fedosov, E.A., Fedunov, B.E.: Intelligent control of dynamic systems, 352 p. Fizmatlit, Moscow (2000)
4. Tolokonnikov, G.K.: Manifest: neurographs, neurocategories and categorical splices. Biomachsystems **1**(1), 59–146 (2017)
5. Karande, A.M., Kalbande, D.R.: Weight assignment algorithms for designing fully connected neural network. Int. J. Intell. Syst. Appl. (IJISA) **10**(6), 68–76 (2018)
6. Dharmajee Rao, D.T.V., Ramana, K.V.: Winograd's inequality: effectiveness for efficient training of deep neural networks. Int. J. Intell. Syst. Appl. (IJISA) **6**, 49–58 (2018)
7. Hu, Z., Tereykovskiy, I.A., Tereykovska, L.O., Pogorelov, V.V.: Determination of structural parameters of multilayer perceptron designed to estimate parameters of technical systems. Int. J. Intell. Syst. Appl. (IJISA) **10**, 57–62 (2017)
8. Awadalla, M.H.A.: Spiking neural network and bull genetic algorithm for active vibration control. Int. J. Intell. Syst. Appl. (IJISA) **10**(2), 17–26 (2018)
9. Abuljadayel, A., Wedyan, F.: An approach for the generation of higher order mutants using genetic algorithms. Int. J. Intell. Syst. Appl. (IJISA) **10**(1), 34–35 (2018)
10. Petoukhov, S.: Matrix genetics, algebra of genetic code, noise immunity. RHD, Moscow (2008)
11. Petoukhov, S., Petukhova, E., Hazina, L., Stepanyan, I., Svirin, V., Silova, T.: The genetic coding, united-hypercomplex numbers and artificial intelligence. In: Hu, Z., Petoukhov, S., He, M. (eds.) Advances in Artificial Systems for Medicine and Education, AIMEE. Advances in Intelligent Systems and Computing, vol. 658. Springer, Cham (2017)
12. He, M., Hu, Z.B., Petoukhov, S.V.: Triply stochastic cubes associated with genetic code numerical mappings. In: Advances in Intelligent Systems and Computing, vol. 754, pp. 606–616 (2018)
13. Holevo, A.S.: Introduction to quantum information theory. MTSNMO, Moscow (2002)
14. Chernavsky, D.S.: The problem of the origin of life and thinking from the point of view of modern physics. UFN **170**(2), 157–183 (2000)

Development of Matrix Methods for Genetic Analysis and Noise-Immune Coding

Nikolay A. Balonin[1(✉)], Mikhail B. Sergeev[1], and Sergey V. Petoukhov[2]

[1] Saint Petersburg State University of Aerospace Instrumentation, 67, B. Morskaia Street, 190000 St. Petersburg, Russian Federation
korbendfs@mail.ru, mbse@mail.ru

[2] Mechanical Engineering Research Institute of the Russian Academy of Sciences, 4, Malyi Kharitonievsky pereulok, 101990 Moscow, Russian Federation
spetoukhov@gmail.com

Abstract. The article discusses new methods of studying and using Hadamard matrices. This class of matrices is commonly used in developing AI systems, noise-immune coding of data and simulation of bioinformational entities. The relations between Hadamard matrices and hyper-complex numbers are discussed. A special attention is given to the new method of pre-orthogonal sequences for building multi-block Hadamard matrices of large orders. This method, due to its mathematical characteristics, should also be useful in modeling the noise-immunity features of genetic coding.

Keywords: Hadamard matrices · Cyclic shifts · Noise immunity · Bioinformatics

1 Introduction

Mathematical matrices play an important role in digital technologies of artificial intelligence and noise-immune coding of data; they also serve as the basis for matrix genetics, studying structural features of genetic systems [1–4]. They generalize the concept of number, being widely used in university education. In computers, data are stored in the form of matrices. Some sets of matrices serve as matrix representations of multidimensional (complex or hypercomplex) numbers, which play a prominent role in modern science. Our article is devoted to new achievements in building and using special types of matrices and vector orthogonal bases related to the famous Hadamard matrices.

According to its definition, a Hadamard matrix of order n is a square $(n \times n)$-matrix $\mathbf{H}(n)$ with elements +1 and −1, satisfying the following condition:

$$\mathbf{H}(n) \times \mathbf{H}(n)^T = n\mathbf{I}_n, \quad (1)$$

where $\mathbf{H}(n)^{\mathrm{T}}$ is the transposed matrix and $\mathbf{I}_n$ is the identity matrix of order n. Tensor (Kronecker) powers of any Hadamard matrix generate new Hadamard matrices of enlarged orders.

Z. Hu et al. (Eds.): AIMEE 2019, AISC 1126, pp. 33–42, 2020.
https://doi.org/10.1007/978-3-030-39162-1_4

Hadamard matrices play a significant part in theories of discrete signals, orthogonal transformations, orthogonal and bi-orthogonal codes, and other areas with applications in digital electronics, communications, coding, control, image recognition, diagnostics, filtering, quantum computer development, etc. [1, 2, 5]. These same matrices allow you to model the features of structured genetic alphabets DNA and RNA, producing new approaches to understanding the phenomenon of noise-immune genetic coding in living organisms [3, 4].

The great interest to Hadamard matrices is caused by their mathematical features associated with orthogonal transformations and orthogonal systems of functions, and therefore demanded in modern digital communications and control. Tens of thousands of publications are devoted to the features and applications of Hadamard matrices and Walsh functions represented by Hadamard matrix rows.

In our article, we discuss the relations between certain families of Hadamard matrices and hypercomplex numbers, as well as new approaches to building multi-block Hadamard matrices of large orders, promising for their application in noise-immune data coding or genetic system structure modelling.

2 Hadamard Matrices and Hypercomplex Numbers

Multi-dimensional number systems (complex numbers, hyperbolic numbers, dual numbers, hypercomplex numbers) are important tools in modern mathematical natural science. They also attract the attention of biological problem researchers [6, 7]. For example, the article with a peculiar title "Is the Brain a 'Clifford Algebra Quantum Computer'?" [7] demonstrated the use of hypercomplex algebras in visual perception modelling in order to effectively solve the problem of biological pattern recognition in a multi-spectral environment.

In this connection, it is interesting that during matrix modelling of genetic coding structures, some types of Hadamard matrices were discovered, which have a connection with multi-dimensional number systems [8]. Figure 1 gives you an example of such a (4 × 4) Hadamard matrix H_4, which can be represented as a sum of four sparse matrices $H_4 = H_{40} + H_{41} + H_{42} + H_{43}$. The set of these sparse matrices is closed under multiplication, corresponding to the multiplication table of Hamiltonian quaternions. The first sparse matrix H_{40} is an identity matrix, represented in the multiplication table by the number 1.

This means that a Hadamard matrix is a matrix representation of a sum of basic elements in Hamiltonian quaternion algebra. In other words, it is a Hamiltonian quaternion with unit coordinates.

Raising H_4 matrix to tensor powers produces new Hadamard matrices of a larger order, which are an algebraic extension of Hamiltonian quaternions, being a sum of basic elements of the respective algebras. Hamiltonian quaternions are closely related to Pauli matrices, electromagnetic field theory (James Maxwell used Hamiltonian quaternions as the language for his equations), special relativity, spin theory, quantum mechanical theory of chemical valence, etc. Only in the 20th century thousands of papers were devoted to quaternions [9], and now quaternions have come out in the system of genetic alphabets.

For some reason, our physical space is arranged in accordance with Hamiltonian quaternions. It turns out now that creating a system of molecular genetic alphabets also depends on them. Matrix approach in genetics has built an important bridge between physics, biology and computer science for their cross-fertilization.

$$\begin{bmatrix} 1 & 1 & -1 & 1 \\ -1 & 1 & 1 & 1 \\ 1 & -1 & 1 & 1 \\ -1 & -1 & -1 & 1 \end{bmatrix} = \begin{vmatrix} 1&0&0&0 \\ 0&1&0&0 \\ 0&0&1&0 \\ 0&0&0&1 \end{vmatrix} + \begin{vmatrix} 0&1&0&0 \\ -1&0&0&0 \\ 0&0&0&1 \\ 0&0&-1&0 \end{vmatrix} + \begin{vmatrix} 0&0&-1&0 \\ 0&0&0&1 \\ 1&0&0&0 \\ 0&-1&0&0 \end{vmatrix} + \begin{vmatrix} 0&0&0&1 \\ 0&0&1&0 \\ 0&-1&0&0 \\ -1&0&0&0 \end{vmatrix}$$

	1	H_{41}	H_{42}	H_{43}
1	1	H_{41}	H_{42}	H_{43}
H_{41}	H_{41}	- 1	H_{43}	$-H_{42}$
H_{42}	H_{42}	$-H_{43}$	- 1	H_{41}
H_{43}	H_{43}	H_{42}	$-H_{41}$	- 1

Fig. 1. Decomposition of a Hadamard (4 × 4) matrix $H_4 = H_{40} + H_{41} + H_{42} + H_{43}$ to a sum of four sparse matrices, and a multiplication table of these sparse matrices.

3 Pre-orthogonal Sequence Method and Multi-block Hadamard Matrices

Let us consider the problem of building multi-block Hadamard matrices of large order, which bring new opportunities for noise-immune data coding and genetic structure modelling. Finding and building such multi-block orthogonal matrices is a complicated and laborious task. Here we will briefly discuss the new method of "pre-orthogonal" sequences, which considerably simplifies the search for such matrices and has already led to finding multi-block orthogonal matrices of record high order.

The condition of vector orthogonality is important for noise-immune data coding. It has been known that if a system of orthogonal vectors is transferred via a communication channel, it is highly possible to restore distorted signals coming to the receiver. This is why Hadamard matrices with their mutually orthogonal rows represented as Walsh functions are used in noise-immune data coding.

But how can we build orthogonal Hadamard matrices of various large orders (for example, order 236) and find orthogonal bases in vector spaces of large order? The difficulties of this challenge can be illustrated by a matrix of order 92, which was found in the middle of the 20th century by joint efforts of several prominent mathematicians [10]. They relied on the idea proposed in 1940s [11] about the decomposition of a matrix to square blocks **A**, **B**, **C** and **D**, supported by Lagrange's four-square theorem. The Eq. (1) is a quadratic equation. Taking into account the nature of the block matrix elements, the search for it can be reduced to an illustrative decomposition exercise. In this particular case, it will be the decomposition of the matrix order to sums of squares produced by the blocks [12, 13]. The blocks in the matrix are arranged in a fixed way called an array.

To simplify the search for matrices, symmetric blocks were used in [11], placed half symmetrically and half asymmetrically (having opposite signs with respect to the diagonal). This was supposed to ensure that the matrix would be orthogonal. We should point out that the idea stopped working as soon as on the order 140 [14], but for those years it was a breakthrough.

In this article, we will focus on cyclic blocks whose rows are transformations of the first row of a block, obtained by cyclic shifts of its components. An example of such a block is a block based on the first row [a_1, a_2, a_3; a_3, a_1, a_2; a_2, a_3, a_1]. Below, we consider cyclic blocks **A**, **B**, **C** and **D** based on vector sequences *a*, *b*, *c*, *d* as their first rows.

In signal processing, self-multiplication of a vector shifted in a cycle (without moving the displaced elements to a new vacant place [a_1, a_2, a_3] □ [0, a_1, a_2] □ [0, 0, a_1]) is called autocorrelation function (AF). Self-multiplication of a vector shifted in a cycle when the displaced elements move to a new vacant place freed by the sequence is called periodic autocorrelation function (PAF). The method of building multi-block orthogonal Hadamard matrices on the base of pre-orthogonal sequences relies on operations with PAF.

This method uses the opportunity to replace the orthogonality condition (1) by a simplified condition $\mathbf{A}a + \mathbf{B}b + \mathbf{C}c + \mathbf{D}d = n\mathbf{I}$, which implicitly is a sum of periodic autocorrelation functions:

$$\mathrm{PAF}(a) + \mathrm{PAF}(b) + \mathrm{PAF}(c) + \mathrm{PAF}(d) = n\mathbf{e}, \tag{2}$$

where $\mathbf{e} = (1, 0, \ldots 0)$.

Example. Let us consider a sequence $a = [a_1, a_2, a_3]$ along with its respective cyclic matrix and periodic autocorrelation function PAF(a):

$$\mathbf{A} = \begin{pmatrix} a_1 & a_2 & a_3 \\ a_3 & a_1 & a_2 \\ a_2 & a_3 & a_1 \end{pmatrix}, \ \mathrm{PAF}(a) = \mathbf{A}a = \begin{pmatrix} a_1 & a_2 & a_3 \\ a_3 & a_1 & a_2 \\ a_2 & a_3 & a_1 \end{pmatrix} \begin{pmatrix} a_1 \\ a_2 \\ a_3 \end{pmatrix} = \begin{pmatrix} a_1a_1 + a_2a_2 + a_3a_3 \\ a_3a_1 + a_1a_2 + a_2a_3 \\ a_2a_1 + a_3a_2 + a_2a_3 \end{pmatrix}$$

For comparison, a common autocorrelation function based on a triangular matrix looks like

$$A_0 = \begin{pmatrix} a_1 & a_2 & a_3 \\ 0 & a_1 & a_2 \\ 0 & 0 & a_1 \end{pmatrix}, \ \mathrm{AF}(a) = A_0a = \begin{pmatrix} a_1 & a_2 & a_3 \\ 0 & a_1 & a_2 \\ 0 & 0 & a_1 \end{pmatrix} \begin{pmatrix} a_1 \\ a_2 \\ a_3 \end{pmatrix} = \begin{pmatrix} a_1a_1 + a_2a_2 + a_3a_3 \\ a_1a_2 + a_2a_3 \\ a_1a_3 \end{pmatrix}.$$

Matrices with elements 1 and –1 were introduced by the founder of matrix mathematics J. Sylvester who specified the family of Hadamard matrices by a recursive sequence, most well-known in Hadamard matrix theory: $\mathbf{H} = \begin{pmatrix} \mathbf{A} & \mathbf{A} \\ \mathbf{A} & -\mathbf{A} \end{pmatrix}$, where **A** is a Hadamard matrix at the previous step of the recursion.

You can start from $\mathbf{A} = 1$, as Sylvester did. Hadamard himself found two more initial non-symmetric matrices of orders 12 and 20 [15].

Later, constructions on two (not necessarily orthogonal) blocks **A** and **B** came into use, for example, $\mathbf{H} = \begin{pmatrix} \mathbf{A} & \mathbf{A} \\ \mathbf{B}^{\mathrm{T}} & -\mathbf{A}^{\mathrm{T}} \end{pmatrix}$, for symmetric matrices $\mathbf{A} = \mathbf{A}^{\mathrm{T}}$. For such a pair of blocks, the orthogonality condition as a whole is rewritten as PAF(a) + PAF (b) = $n\mathbf{e}$. Since the first element of **e** vector is standard, the formula can be simplified by brackets, which mean dropping the first element of the PAF. Then we have:

$$[\mathrm{PAF}(a)] = -[\mathrm{PAF}(b)]. \tag{3}$$

It should be noted here that all PAFs look like a sequence {v, some numbers}; after the first element v is dropped, we stay with [PAF(a)] = {some numbers}. For example, PAF(a) = {4, 1, −2, 1}, then [PAF(a)] = {1, −2, 1}.

The equality (3) brings us closer to understanding what pre-orthogonality is. In orthogonal two-circulant matrices (made of two cyclic blocks), periodical autocorrelation functions change in antiphase, being equal in amplitude. This rule allows us to organize a sufficiently fast search for orthogonal matrices.

Since not all matrix orders can be decomposed to sums of two squares, the above-mentioned order 92 is unattainable with such structuring. Two-circulant matrices, as it was discovered, are not flexible enough to be symmetric on orders larger than 32, either [16]. You will recall that for the case of one-block monocycles, Ryser [17] specified the largest order of such a matrix as 4.

This makes it especially interesting to discuss a symmetric three-block Hadamard matrix relying on Gauss's theorem about the decomposition of any integer to a sum of three triangular numbers:

$$\mathbf{H} - \begin{pmatrix} \mathbf{A} & -\mathbf{BR} & -\mathbf{CR} & -\mathbf{DR} \\ -\mathbf{CR} & \mathbf{RD} & -\mathbf{A} & -\mathbf{RB} \\ -\mathbf{BR} & -\mathbf{A} & -\mathbf{RD} & \mathbf{RC} \\ -\mathbf{DR} & -\mathbf{RC} & \mathbf{RB} & -\mathbf{A} \end{pmatrix}, \tag{4}$$

where $\mathbf{A} = \mathbf{A}^{\mathrm{T}}$ is a symmetric block. The symmetry of this array is ensured by the condition **B** = **C**. Iterative extension of Hadamard matrices by Sylvester's rule ensures that symmetric Hadamard matrices have orders with even value v. Therefore, arrays with an odd size of blocks v are more primary, when $r = (v–1)/2$ is the array number (4).

For example, when the blocks have a unit size $v = 1$, this is an initial array $r = 0$, producing a Hadamard matrix of order $n = 4v = 4$. The next odd size of a block $v = 3$ brings an array with number $r = 1$, producing a Hadamard matrix of order $n = 4v = 12$. Between these two symmetric matrices of orders 4 и 12 there is a symmetric Hadamard matrix of order 8 with even value $v = 2$, which can be easily obtained by J. Sylvester's recursive formula mentioned above from a symmetric Hadamard matrix of order 4. The next odd block size $v = 5$ produces an array with number $r = 2$, producing a Hadamard matrix of order $n = 4v = 20$; between it and the matrix of order 12 there is a symmetric Hadamard matrix of order 16 with even value $v = 4$. We can obtain it from a symmetric matrix of order 8. Whatever size an even block has, you can always build a respective matrix from the matrices built earlier.

In [12, 13], the authors explain the relation between the triangular numbers of decomposition $r = T_x + 2T_y + T_z$ (which is always possible, according to Gauss's theorem supplemented by Liouville for weighted sums; the sources are listed in the appendix) and the amounts of −1 in rows *a*, *b*, *d* of the three blocks **A**, **B** and **D**. As a matter of fact, this connection reduces the discussion of Hadamard's conjecture about the existence of all possible Hadamard matrices to theorems proved in number theory.

In research literature, the array (4) is called Balonin–Seberry array. It is also known as Propus, which is the traditional name of a triple star Eta Geminorum in the constellation of Gemini.

The orthogonality condition (2), due to its symmetry, leads to equality PAF(a) + 2PAF(b) + PAF(d) = $n\mathbf{e}$ and then to [PAF(a) + PAF(d)] = −2[PAF(b)]. It facilitates the search for *b* sequences and does not seem to provide the search for *a* and *d* sequences. It is evident that the sum of their periodic autocorrelation functions over amplitude should be an even number (up to sign), and this limits the values of absolute value sequences denoted as $|a|$, $|d|$, by magnitudes 1, 3, 5, 7 with respect to base 8.

Example. Let us represent a typical pre-orthogonal combination

$$[\mathrm{PAF}(|\mathrm{a}|) \bmod 8] = [1, 1, 5, 1, 5, 7, 3, 3, 7, 1, 3, 3, 3, 3, 1, 7, 3, 3, 7, 5, 1, 5, 1, 1],$$
$$[\mathrm{PAF}(|\mathrm{d}|) \bmod 8] = [3, 3, 1, 5, 1, 3, 1, 7, 3, 5, 1, 7, 7, 1, 5, 3, 7, 1, 3, 1, 5, 1, 3, 3].$$

Note that in these two numerical sequences number 1 always confronts numbers 3 and 5 but never 7. In other words, we unexpectedly have a certain correspondence or complementarity in the set of four numbers 1, 3, 5 and 7.

Since the times of G.W. Leibniz and Bernard Bolzano it has been known that cognition is a search for analogies. Now, what known phenomenological facts could possibly be similar to this mathematical complementarity in a set of four numbers? One that springs to mind is the complementarity in the set of four nitrogenous bases in DNA molecules. In the DNA double helix, these four bases are naturally split to two complementary pairs: adenine is always paired with thymine, and cytosine with guanine. RNA molecules, instead of thymine, use uracil, which is similar, forming a complementary pair with adenine. DNA molecules carry genetic information transmitted from generation to generation, which is highly noise-immune. The relation mentioned in the beginning of our paper between structured DNA alphabets and Hadamard matrices and Walsh functions attests to the fact that noise-immune data coding supported by orthogonal bases can be used to simulate and study the wonderful features of genetic coding. The pre-orthogonal sequence method, which has unexpectedly led us to complementarity in a four-number set, provides new opportunities for studying (on the base of analogies) the structural features of a genetic system, from the perspective of search for orthogonal bases adequate for modelling noise-immune features of genetic information processing.

Note also that the pre-orthogonal sequence method can be considered as a certain "code", which detects orthogonal sequences within millions of other sequences. PAF (a) does not fix a sequence rigidly. It allows you, for example, to shift a in a cycle without changing this function. It can be called a "shadow" of the sequence. PAF (a) with respect to modulo 8 is still less rigid in terms of sequence fixation. Therefore, the pre-orthogonal sequence method leads to a set of non-trivial versions of orthogonal bases in vector spaces of high dimensions. Instead of dealing with all possible blocks orthogonalizing them in the same time, the method allows you to follow a much easier way, rejecting a huge amount of unnecessary combinations: you make block pairs pre-orthogonal and then use these pairs to build up a multi-block symmetric Hadamard matrix.

The problem of the number of symmetric Hadamard matrix variants for $n < 2 \times 10^5$ has a computing solution shown in Fig. 2 in the form of two qualitative diagrams. The one on the right copies the characteristic shown on the left, but along the horizontal axis it uses the square root of n. The double borders on the top and bottom specify the areas of low probability values.

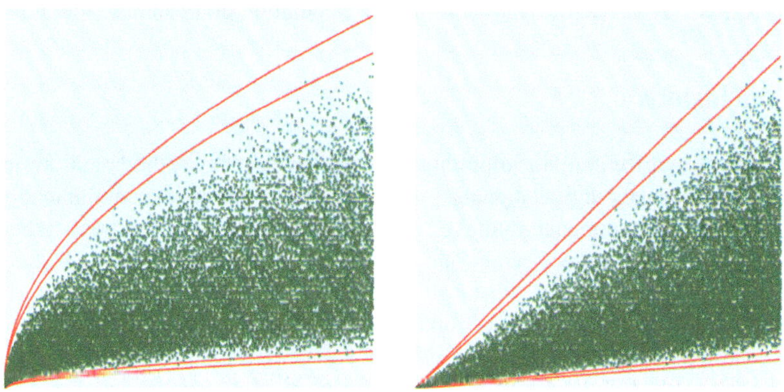

Fig. 2. Qualitive diagram of symmetric Hadamard matrix variants

As you can see from Fig. 2, unlike two-circulant matrices, the number of symmetric matrices only grows. The researchers from Distributed Computing System Lab at Saint Petersburg State University of Aerospace Instrumentation, along with the researchers from University of Waterloo, Department of Pure Mathematics and Institute for Quantum Computing, Waterloo, Ontario, N2L 3G1, Canada (the facilities of the Shared Hierarchical Academic Research Computing Network (SHARCNET) and Compute/Calcul Canada), have set a peculiar record by finding a symmetric Hadamard matrix of order 236 [18] shown in Fig. 3. This order had not been solved so far, being especially interesting, as the research program [19] could not find Williamson matrices on it.

Fig. 3. Symmetric Hadamard matrix of order 236. Red cells correspond to elements with value −1; white cells correspond to elements with value 1.

Note that orders 92, 116, … , 236, etc. are among the orders, which are unattainable by Kronecker products or similar combinatorial methods, therefore Hadamard matrices were searched for on computers. Symmetric Propuses of orders 92, 116, 172 were used as test matrices by Olivia de Mateo during her calculation on quantum computers [20].

4 Conclusions

Mathematical matrices are an important tool for developing methods and systems of artificial intelligence, digital signals processing, noise-immune communication and bioinformational phenomena studies. The world of matrices in not yet studied well enough, still having many secrets and hidden useful abilities. The wide family of Hadamard matrices is one of the most important classes of matrices, having various scientific and technological applications. In our article, we have discussed some new results and approaches in studying and applications of Hadamard matrices, including the promising method of pre-orthogonal sequences. We believe that the discussed results and approaches will help the further development of matrix analysis and its applications in numerous directions presented, for example, in works [21–28]. We use the obtained results to create new digital devices for artificial intelligence and new approaches to studying genetic systems.

The conclusion section looks good in general, apart from this, the author may emphasize the highlights of this paper and talk about what these highlights may bring to us (academic and application values). Bare in mind that future work should be connected to the current limitations.

Acknowledgement. The authors express their gratitude for long-standing collaboration and support to professors Jennifer Seberry and Dragomir Đoković. In technical work with the manuscript and references, T.V. Balonina was very helpful (find a more complete list of works on http://mathscinet.ru/tamara).

The work has been carried out with the support of Ministry of Education and Science of the Russian Federation for research within the development part of the scientific governmental task #2.2200.2017/4.6.

References

1. Ahmed, N.U., Rao, K.R.: Orthogonal Transforms for Digital Signal Processing. Springer, New York (1975)
2. Sklar, B.: Digital Communication. Fundamentals and Applications. Prentice Hall, Upper Saddle River (2001)
3. Petoukhov, S.V.: Matrix genetics, algebras of the genetic code, noise-immunity, 316 p. Regular and Chaotic Dynamics, Moscow (2008). (in Russian)
4. Petoukhov, S.V., He, M.: Symmetrical Analysis Techniques for Genetic Systems and Bioinformatics: Advanced Patterns and Applications. IGI Global, Hershey (2010)
5. Seberry, J., Wysocki, B.J., Wysocki, T.A.: On some applications of Hadamard matrices. Metrica **62**, 221–239 (2005)
6. Petoukhov, S.V., Petukhova, E.S.: On genetic unitary matrices and quantum-algorithmic genetics. In: Hu, Z., Petoukhov, S., He, M. (eds.) Advances in Artificial Systems for Medicine and Education II, AIMEE 2018. Advances in Intelligent Systems and Computing, vol. 902, pp. 103–115. Springer, Cham (2020)
7. Labunets, V., Rundblad, E., Astola, J.: Is the brain a 'Clifford Algebra Quantum Computer'? In: Dorst, L., Doran, C., Lasenby, J. (eds.) Applications of Geometric Algebra in Computer Science and Engineering. Birkhäuser, Boston (2002)
8. Petoukhov, S.V.: Symmetries of the genetic code, hypercomplex numbers and genetic matrices with internal complementarities. Symmetry: Cult. Sci. **23**(3–4), 275–301 (2012)
9. Gsponer, A., Jean-Pierre Hurni, J.-P.: Quaternions in mathematical physics (2): Analytical bibliography. Submitted 6 July 2008. http://arxiv.org/abs/math-ph/0511092
10. Baumert, L., Golomb, S.W., Marshall Jr., M.: Discovery of an Hadamard matrix of order 92. Bull. Amer. Math. Soc. **68**, 237–238 (1962). Communicated by F. Bohnenblust, California Institute of Technology
11. Williamson, J.: Hadamard's determinant theorem and the sum of four squares. Duke Math. J. **11**, 65–81 (1944)
12. Balonin, N.A., Sergeev, M.B.: Helping Hadamard conjecture to become a theorem. Part 1. Informatsionnoupravliaiushchie sistemy [Inf. Control Syst.] (6), 2–13 (2018). https://doi.org/10.31799/1684-8853-2018-6-2-13. (in Russian)
13. Balonin, N.A., Sergeev, M.B.: Helping Hadamard conjecture to become a theorem. Part 2. Informatsionnoupravliaiushchie sistemy [Inf. Control Syst.] (1), 2–10 (2019). https://doi.org/10.31799/1684-8853-2019-1-2-10. (in Russian)
14. Đoković, D.Ž.: Williamson matrices of order 4n for n = 33; 35; 39. Discrete Math. **115**, 267–271 (1993)
15. Hadamard, J.: Résolution d'une question relative aux déterminants. Bulletin des sciences mathématiques **17**, 240–246 (1893)
16. Balonin, N.A., Đoković, D.Ž.: Symmetric Hadamard matrices of orders 268, 412, 436 and 604. Informatsionnoupravliaiushchie sistemy [Inf. Control Syst.] (4), pp. 2–8 (2018). https://doi.org/10.31799/1684-8853-2018-4-2-8. Accessed 23 Mar 2018. arXiv:1803.08787
17. Ryser, H.J.: Combinatorial Mathematics. The Carus Mathematical Monographs, no. 14, 162 p. The Mathematical Association of America/Wiley, New York/Hoboken (1963)
18. Abuzin, L.V., Balonin, N.A., Đoković, D.Ž., Kotsireas, I.S.: Hadamard matrices from Goethals-Seidel difference families with a repeated block. Informatsionnoupravliaiushchie sistemy [Inf. Control Syst.] (2019, in print)
19. Holzmann, W.H., Kharaghani, H., Tayfeh-Rezaie, B.: Williamson matrices up to order 59. Des. Codes Cryptogr. **46**, 343–352 (2008)

20. Di Matteo, O.: Methods for parallel quantum circuit synthesis, fault-tolerant quantum RAM, and quantum state tomography. A thesis for the degree of PhD in Physics-Quantum Information, Waterloo, Ontario, Canada, 38 p (2019). https://uwspace.uwaterloo.ca/bitstream/handle/10012/14371/DiMatteo_Olivia.pdf?sequence=3&isAllowed=y
21. Angadi, S.A., Hatture, S.M.: Biometric person identification system: a multimodal approach employing spectral graph characteristics of hand geometry and palmprint. Int. J. Intell. Syst. Appl. (IJISA) (3), 48–58 (2016). http://www.mecs-press.org/ijisa/ijisa-v8-n3/IJISA-V8-N3-6.pdf
22. Sahana, S.K., AL-Fayoumi, M., Mahanti, P.K.: Application of modified ant colony optimization (MACO) for multicast routing problem. Int. J. Intell. Syst. Appl. (IJISA) (4), 43–48 (2016). http://www.mecs-press.org/ijisa/ijisa-v8-n4/IJISA-V8-N4-5.pdf
23. Algur, S.P., Bhat, P.: Web video object mining: a novel approach. Int. J. Intell. Syst. Appl. (IJISA) (4), 67–75 (2016). http://www.mecs-press.org/ijisa/ijisa-v8-n4/IJISA-V8-N4-8.pdf
24. Hata, R., Akhand, M.A.H., Islam, M.Md., Murase, K.: Simplified real-, complex-, and quaternion-valued neuro-fuzzy learning algorithms. IJISA **10**(5), 1–13 (2018). https://doi.org/10.5815/ijisa.2018.05.01
25. Awadalla, M.H.A.: Spiking neural network and bull genetic algorithm for active vibration control. Int. J. Intell. Syst. Appl. (IJISA) **10**(2), 17–26 (2018). https://doi.org/10.5815/ijisa.2018.02.02
26. Abuljadayel, A., Wedyan, F.: An approach for the generation of higher order mutants using genetic algorithms. Int. J. Intell. Syst. Appl. (IJISA) **10**(1), 34–35 (2018)
27. Petoukhov, S.V.: The system-resonance approach in modeling genetic structures. Biosystems **139**, 1–11 (2016)
28. Biamonte, J., Wittek, P., Pancotti, N., Rebentrost, P., Wiebe, N., Lloud, S.: Quantum machine learning. Nature **549**, 195–202 (2017)

Analysis of Oscillator Behavior Under Multi-frequency Excitation for Oscillatory Neural Networks

M. M. Gourary(✉) and S. G. Rusakov

Institute for Design Problems in Microelectronics of Russian Academy of Sciences (IPPM RAS), Moscow, Russia
gourary@yandex.ru

Abstract. Problems of the oscillator behavior under quasiperiodic multi-frequency excitation are considered in the paper. The new concept of the multi-frequency synchronization is proposed as the natural extension of the traditional synchronization concept for the periodically excited oscillator. Derived analytical expressions and performed simulation examples demonstrate the consistency of the proposed concept. The investigations are based on the modified Kuramoto model of a single oscillator under a narrowband quasiperiodic excitation. Analytical results are obtained by multiple timescales methods.

Keywords: Oscillators · Kuramoto model · Phase macromodels · Synchronization · Dynamical system · Transfer factor

1 Introduction

Artificial Neural Networks have been widely used in many engineering [1], medical [2] and financial [3] applications. One of the promising directions of artificial neural networks design is the development of oscillatory neural networks [4] representing the system of coupled oscillators with a rich set of dynamic behavior modes [5–7]. Especially synchronization conditions in oscillator ensembles are of research interest [8]. The problem of simulating synchronization phenomena in the system of weakly coupled oscillator elements is the leading problem in the development of ONN. The problem to estimate the synchronization properties of a large set of coupled oscillators is related to the simulation problem with high complexity [9].

But the synchronization cannot fully characterize the behavior of the oscillator ensemble because partially synchronized (clustered) states can exist in the ensemble [10]. Such states are defined by quasi-periodic oscillations and can exist even in ensembles of identical oscillators (chimeras states) [11].

Each oscillator in the ensemble is exposed to the impact of other devices, so in general case, it is under quasi-periodic excitation. Thus, such elementary configuration can be the basis for the analysis of more complicated systems of interacting oscillators.

However, there is an insufficient number of publications on this issue. Some old papers [12, 13] analyze specific types of oscillators (e.g. Van der Pol oscillator). Papers devoted to laser devices [14] suffer from a lack of generality. Many papers consider a

Z. Hu et al. (Eds.): AIMEE 2019, AISC 1126, pp. 43–54, 2020.
https://doi.org/10.1007/978-3-030-39162-1_5

quasiperiodic behavior of periodically excited multi-frequency oscillator [15], which does not apply to the problem under consideration.

In [16] structural transformations of a frequency-modulated informational signal by a single oscillatory neuron are investigated. The neuron is presented by a threshold device based on van der Pol oscillator. Besides the excitation is defined as the sum of simple non-stationary signals modulated by Gaussian curve. Thus, the method cannot analyze an arbitrary oscillator under stationary quasi-periodic excitation.

This paper is aimed at the analysis of specific properties of the arbitrary periodic oscillator under multi-frequency excitation. We introduce a new concept of the oscillator quasi-periodic synchronization, which occurs when the number of the oscillator fundamental frequencies becomes equal to the number of the excitation fundamentals. The validity of the concept is confirmed by numerical experiments that used the modified Kuramoto model proposed in the article. Experiments also confirmed the obtained asymptotic expressions for some modes of single- and two-frequency excitation.

The rest of the paper is organized as follows. Section 2 presents the synchronization concept, generalized to the case of multi-frequency excitation. Section 3 describes the modified Kuramoto model. Sections 4, 5 contain the derivation of some analytical expressions for single- and two-frequency excitation correspondingly. Numerical results presented in Sect. 6 for the modulated excitation demonstrate the switching of the generator into synchronization mode.

2 Principles of the Oscillator Behavior Under External Excitation

Periodic waveforms appear in two types of the oscillatory systems – in the systems under the periodic excitation and in the free-running oscillators. The phase and the frequency of forced oscillations are fully defined by the excitation (Fig. 1a). In free-running oscillators the frequency is defined by the system properties, and ambiguous solutions with an arbitrary phase exist in the system (Fig. 1b). If a sufficiently large periodic excitation is applied to the oscillator, then synchronization occurs, in which the behavior of the oscillator is similar to the behavior of a system with forced oscillations - the frequency coincides with the excitation frequency, and fixed oscillation phase is defined by the excitation (Fig. 1c).

This can be explained as follows. Autonomous oscillations arise because of the system internal forces, but large excitation overpowers these forces, and the oscillator behaves like a non-oscillatory system. This effect is especially noticeable when the excitation frequency is close to the oscillator eigenfrequency (natural frequency of the free-running oscillator).

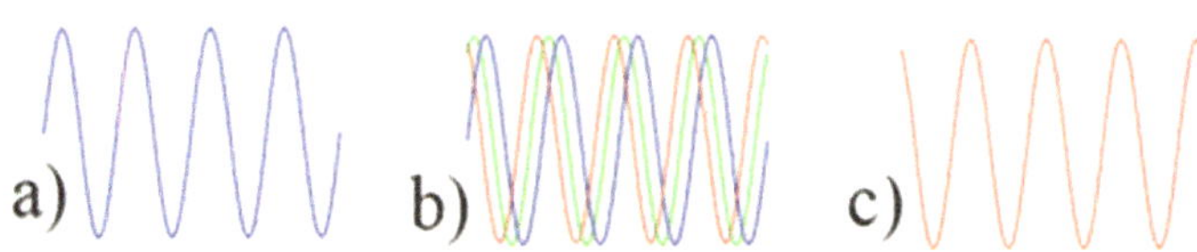

Fig. 1. Periodic waveforms in the oscillatory systems: (a) forced systems, (b) free running oscillators, (c) synchronized oscillators.

If the excitation is less than the synchronization level, the oscillator exhibits quasi-periodic oscillations with two fundamentals. One of them is always the excitation frequency and the other one is an intrinsic fundamental depending on both the excitation magnitude and the oscillator properties (Fig. 2a). At the bifurcation point (synchronization magnitude) the intrinsic fundamental merges with the external frequency providing single-frequency oscillations of the synchronization mode. The set of bifurcation points for different excitation frequencies form the border of the synchronization region (Arnold's tongue) shown in Fig. 2b.

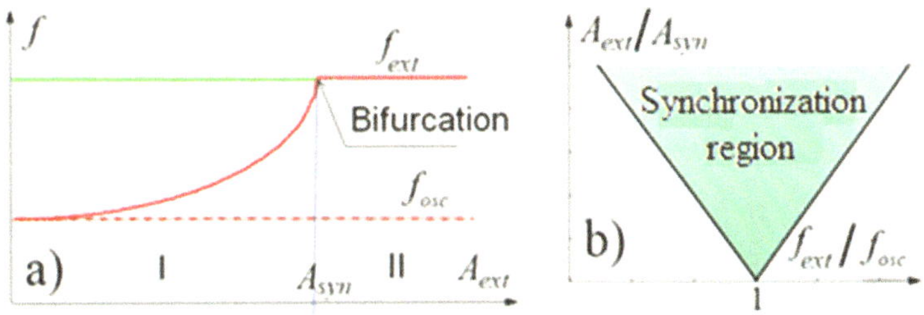

Fig. 2. The oscillator with eigenfrequency f_{osc} under the excitation frequency f_{ext}: (a) dependence of the oscillator fundamentals on the excitation magnitude A_{ext}, (b) Arnold's tongue for small excitations.

We suppose that the behavior of the oscillator under multi-frequency quasi-periodic excitation can be considered as the following extension of the presented properties of periodically (single-frequency) excited oscillator:

1. For the excitation with K fundamentals the quasi-periodic oscillator solution can contain either K or $K+1$ fundamentals.
2. The solution with $K+1$ fundamentals contains K excitation fundamentals and additional intrinsic fundamental which depends on the oscillator properties and magnitudes of excitation harmonics. Phases of solution harmonics contain arbitrary phase shift of the harmonic of the intrinsic fundamental.
3. The oscillator behavior under the solution with K fundamentals is similar to the behavior of forced oscillations in the nonoscillatory system: the solution fundamentals coincide with the excitation fundamentals and all phases of solution harmonics are unambiguously defined by the excitation phases.
4. The disappearance of intrinsic fundamental from the oscillator spectrum can be considered as the effect of *quasi-periodic synchronization*. Proposed definition is the natural extension of the synchronization concept for periodically excited oscillator to the oscillator under quasi-periodic excitation.
5. The transfer to synchronized state occurs when the excitation magnitude achieves the value of the bifurcation point. If the excitation is characterized by a number of parameters then the surface of bifurcation points can be defined in the parameters space.

Below we present some analytical and simulation examples which demonstrate the consistency of the proposed concept.

3 Modified Kuramoto Model for a Single Oscillator Under a Narrowband Quasiperiodic Excitation

Kuramoto model (KM) [17, 18] is the initial mathematical model for behavior of a set of coupled oscillators. KM for M coupled sinusoidal oscillators and N external excitation sources is presented in general form by the following system of M ordinary differential equations [17]:

$$\frac{d\theta_m}{dt} = \omega_m + \sum\nolimits_{n=1}^{M+N} K_{mn}\sin(\theta_m - \theta_n). \tag{1}$$

Here $m = 1, 2, \ldots, M, \theta_m, \omega_m$ is the eigenfrequency of m-th oscillator, θ_m is the instantaneous phase either of m-th oscillator ($m \leq M$) or of (m-M)-th external excitation ($m >$ M), K_{mn} determine coupling factor from n-th to m-th oscillator. For one oscillator with parameters ω, θ (1) is presented as

$$\frac{d\theta}{dt} = \omega + \sum\nolimits_{n=1}^{N} K_n \sin(\theta - \theta_n). \tag{2}$$

A similar equation was obtained in known Adler paper [19] for electrical LC oscillator with the eigenfrequency close to the excitation frequency. Thus (2) correctly represents an oscillator behavior in the narrow frequency band near its eigenfrequency. So it is expedient to transform (2) in the form with respect to the deviation of θn and θ from the phase of eigenfrequency oscillations (ωt):

$$\varphi(t) = \theta(t) - \omega t \text{ or } \theta(t) = \omega t + \varphi(t). \tag{3}$$

Substituting (3) into (2) leads to

$$\frac{d\varphi}{dt} = \sum\nolimits_{n=1}^{N} K_n \sin(\varphi - \varphi_n). \tag{4}$$

Sinusoidal excitations in (4) are presented as

$$V_n = A_n \sin(\omega t + \Delta\omega_n)t. \tag{5}$$

In Adler model, the coupling factor with the excitation is proportional to the excitation magnitude. So we can assume $K_n = bA_n$, where b is the oscillator sensitivity to the excitation. Besides, adding up a narrowband set of sinusoids (5) results in amplitude-phase modulated waveform [20, 21]

$$V(t) = \sum\nolimits_{n=1}^{N} A_n \sin(\omega t + \Delta\omega_n)t = A(t)\sin(\omega t + \varphi_{ex}(t)). \tag{6}$$

Here $A(t)$ and $\varphi_{ex}(t)$ is the amplitude and phase modulation waveforms. Substituting the time-varying magnitude and phase from (6) into (4) leads to the final expression

$$\frac{d\varphi}{dt} = bA(t)\sin(\varphi(t) - \varphi_{ex}(t)). \tag{7}$$

4 Basic Expressions for Single-Frequency Excitation

Before analyzing an oscillator under quasiperiodic excitation we consider some expressions for the oscillator excited by the periodic perturbation with magnitude A and circular frequency $\omega + \Delta\omega$, where $\Delta\omega$ is the deviation of the excitation frequency from the oscillator eigenfrequency In this case (7) with $A(t) = A$, $\varphi_{ex}(t) = \Delta\omega t$ results in

$$\frac{d\varphi}{dt} = bA\sin(\varphi - \Delta\omega t). \tag{8}$$

Synchronized oscillator behavior corresponds with the solution of (8) in the form

$$\varphi(t) = \varphi_0 + \Delta\omega t, \tag{9}$$

where constant φ_0 is the oscillator phase shift. After substituting (9) into (8) we obtain the equation with respect to φ_0

$$\Delta\omega = bA \cdot \sin\varphi_0. \tag{10}$$

The solution of (9) exists if the synchronization condition $\Delta\omega \leq bA$ is true. So the synchronization amplitude (A_{syn}) in the bifurcation point is determined as

$$A_{syn} = \Delta\omega/b. \tag{11}$$

If the excitation magnitude is less than A_{syn} (11) then two-frequency waveforms define a beat mode of the oscillator. The detailed analysis of beat mode for Adler oscillator in [22] gives implicit expressions for phase waveforms. Here we derive the simple explicit formula for small excitation magnitude $A \ll A_{syn}$ or $bA \ll 1$.

To analyze the beat mode one can apply (9) replacing constant φ_0 by time-varying phase $\psi(t) : \varphi(t) = \psi(t) + \Delta\omega t$. After its substituting into (8) we have

$$\frac{d\psi}{dt} + \Delta\omega = bA \cdot \sin\psi. \tag{12}$$

Due to the small value of A one can neglect the right hand side of (12) to obtain the first iteration ($\overline{\psi}$) of solving (8): $d\overline{\psi}/dt + \Delta\omega = 0$, that leads to $\overline{\psi}(t) \approx -\Delta\omega t$. The deviation of the oscillator instantaneous frequency is evaluated as the time derivative of the phase: $\Delta\omega_{inst} = d\varphi/dt = d\psi/dt + \Delta\omega$. After substituting $\Delta\omega_{inst}$ and $\overline{\psi}$ into left- and right-hand sides of (12) correspondingly, one obtains:

$$\Delta\omega_{inst} = -bA \sin \Delta\omega t. \tag{13}$$

Thus small sinusoidal excitation with frequency $\omega_{ex} = \omega + \Delta\omega$ results in the sine wave of the instantaneous frequency. The sine wave has a frequency of $\Delta\omega$ and an amplitude proportional to the amplitude of the excitation with the transfer factor equal to the oscillator sensitivity to the excitation.

The results of the numerical simulation of (8) are presented in Fig. 1. Simulation parameters are the following: $b = 100$, $\omega = 100$ Hz, $\Delta\omega = 5$ Hz. From (11) one can obtain $A_{syn} = 0.05$. The simulation was performed for different values of the excitation magnitude defined by relative parameter $a = A/A_{syn}$. For small excitation magnitudes ($a = 0.1, 0.2. 0.4$) the waveforms (Fig. 3a) are close to sine waves and their magnitudes satisfy linear dependence (13). Waveforms in Fig. 3b corresponding to $a = 0.7, 0.9$ notably differ from sine waves and their periods increase with the growth of the relative magnitude a. Spectra of the waveforms (Fig. 3c) confirm the drift of the frequency from almost $\Delta\omega = 5$ Hz for a = 0.1 to 2.15 Hz for a = 0.9 that corresponds with Fig. 1a.

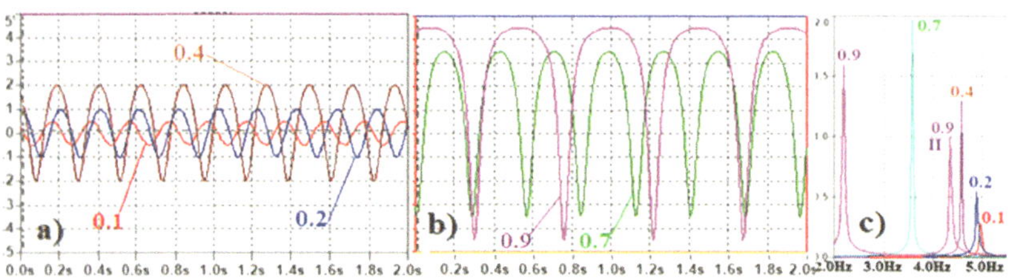

Fig. 3. Single-frequency excitations. Waveforms of the instantaneous frequency deviation: (a) small magnitudes $a = 0.1, 0.2. 0.4$, (b) moderate magnitudes $a = 0.7, 0.9$. (c) Spectra of all waveforms in the neighborhood of their first harmonics.

Note that if the spectrum of the instantaneous frequency of the quasiperiodic waveform contains n fundamentals then the waveform spectrum contains additional (n +1)th fundamental equal to the average value of the instantaneous frequency.

5 Asymptotic Expressions for Two-Frequency Excitation

Any two-frequencies oscillation can be defined through the torus function $u(\tau_1, \tau_2)$ that is 2π-periodic in each of the variables τ_1, τ_2 [23]. Then quasiperiodic waveform is expressed as $x(t) = u(\omega_1 t, \omega_2 t)$. We consider here modulated oscillation with fundamentals $\omega_1 =$

$_{car}$, (carrier frequency), $\omega_2 = \omega_{mod}$, (modulation frequency). If the sinusoidal modulation is considered then (6) corresponds to the torus function $\varphi_{ex}(\tau_{mod}) = \tau_{mod} + \mu \sin(\tau_{mod})$ for phase modulation (PM) with modulation index μ and

$$A(\tau_{mod}) = A(1 + m \sin(\tau_{mod})) \tag{14}$$

for amplitude modulation (AM) with modulation index m.

The solution of (7) for modulation (14) cannot be obtained in the closed form, so only some asymptotic cases will be considered. Our asymptotic approaches are based on the property of narrowband modulated waveform

$$\omega_{mod} \ll \omega_{car}, \tag{15}$$

which gives the way to apply multiple timescales methods [23].

For the small excitation magnitude, one can apply (13) to each harmonic of the excitation and due to constant transfer factor in (13) it is equivalent to the replacement the fixed excitation magnitude in (13) by its time-varying expression

$$\Delta\omega_{inst} = -bA(t) \sin \Delta\omega t. \tag{16}$$

Expression (16) can be presented through the solution of (12) $\Delta\tilde{\omega}_{inst} = \frac{d\psi}{dt}$, which is dependent on the magnitude $A_0 - \Delta\tilde{\omega}_{inst}(A_0)$. Thus

$$\Delta\omega_{inst} = A(t)\Delta\tilde{\omega}_{inst}(A_0)/A_0. \tag{17}$$

Testing results for both (16), (17) are combined on the same plots in Fig. 4. Two types of AM were applied: the standard one with $m = 0.5$ and AM with a suppressed carrier. The plots of excitations are shown in the top row for $a = 0.3$ and $a = 0.8$. Each other plot represents waveforms of (16) and (17). For $a = 0.3$ both curves practically coincide, and for $a = 0.8$ forms of the curves essentially differ but their magnitudes have close values.

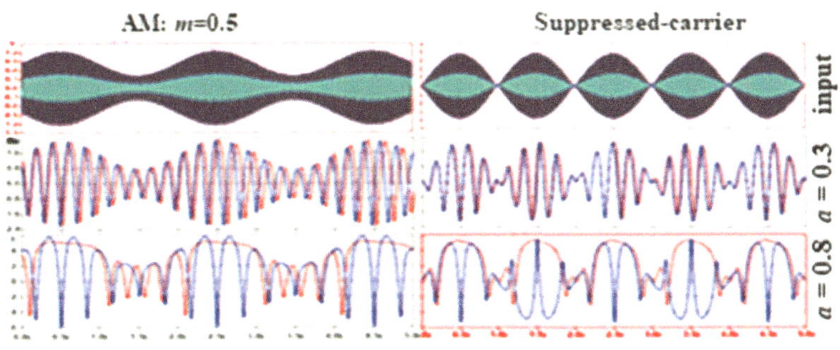

Fig. 4. AM excitations. Left column: $m = 0.5$, right column: suppressed carrier. Comparison with the modulated single-frequency waveforms. Top row: excitation waveforms. Middle row: results for small $a = 0.3$. Bottom row: results for moderate $a = 0.8$.

Another asymptotic solution of (7) can be obtained under the large oscillator excitation $A(t) \gg A_{syn}$ for all t. To obtain the result we can consider the oscillator behavior as slow-varying synchronization mode and replace fixed φ_0 in (10) by slow function $\psi(t)$

$$\Delta\omega = bA(t) \cdot \sin\psi(t). \tag{18}$$

Differentiating (18) by time, we get

$$b\frac{dA}{dt} \cdot \sin\psi(t) + bA(t) \cdot \cos\psi(t) \cdot \frac{d\psi}{dt} = 0, \tag{19}$$

From (19) the instantaneous frequency $\Delta\omega_{inst} = d\psi/dt + \Delta\omega$ can be obtained as

$$\Delta\omega_{inst} - \Delta\omega = -\frac{\Delta\omega \cdot dA/dt}{A(t) \cdot \sqrt{b^2A(t)^2 - \Delta\omega^2}} \approx -\frac{\Delta\omega \cdot dA/dt}{b \cdot A(t)^2}. \tag{20}$$

For sinusoidal AM (14) we have from (20)

$$\Delta\omega_{inst} - \Delta\omega = -\frac{\Delta\omega \cdot \cos\omega_{mod}t}{b \cdot A(1 + m\sin\omega_{mod}t)}. \tag{21}$$

Simulation results (Fig. 5) confirm inverse proportionality to A in (21).

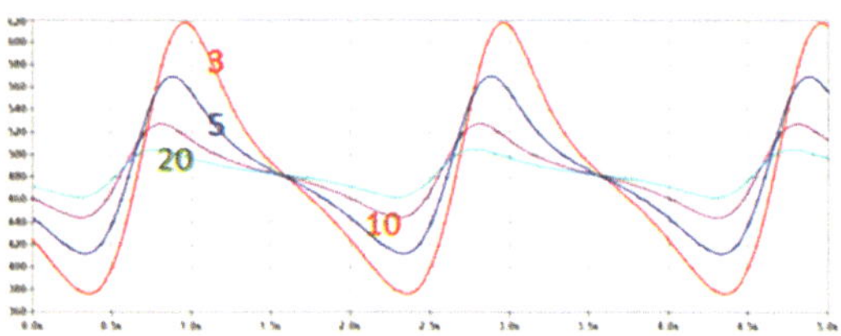

Fig. 5. Simulation results for large AM excitation with a = 3, 5, 10, 20.

6 Detection of Bifurcation Points by Numerical Experiments

We derived expressions for small (15) and large (18) excitation magnitudes based on the slow-varying character of the modulation waveform. Such a derivation fundamentally can not be applied to obtain a solution if the modulation waveform may approach the synchronization level at some points in time. Then the intrinsic fundamental is close to the excitation fundamental that leads to the very small difference frequency less than the modulation frequency. In such a case, the modulation can not be considered as the slow-varying waveform and numerical experiments are required to perform the analysis of the oscillator behavior.

Simulation results for AM excitations with $m = 0.5$ and $a = 0.5, 0.55, 0.6, 0.7, 0.8$ are shown in Fig. 6. One can see from the spectral plots that for $a = 0.7, 0.8$ harmonics are evenly distributed with the step equal to the carrier frequency (0.5 Hz). Thus the waveforms represent periodic oscillations. Irregular harmonic distribution for $a = 0.5$, 0.55, 0.6 do not correspond to periodic oscillations and demonstrate two-frequency quasiperiodic behavior. Therefore one can conclude that the bifurcation point is located in the interval of magnitudes $a = [0.6, 0.7]$.

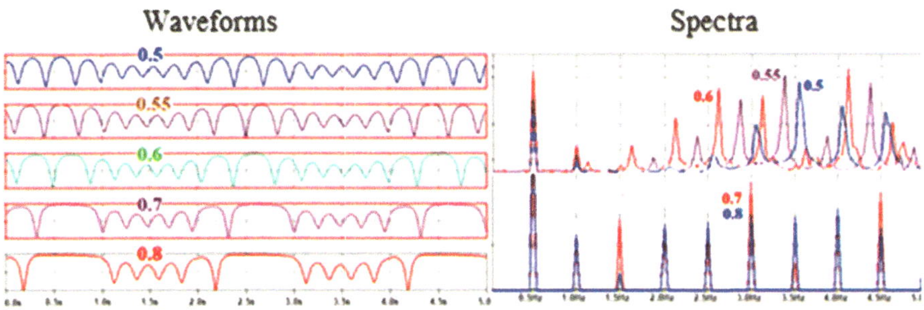

Fig. 6. Waveforms and spectra for AM large excitation with $a = 0.5, 0.55, 0.6, 0.7, 0.8$.

More accurate detection of the bifurcation point is shown in Fig. 7 which defines the narrow interval around the point: $a \in [0.65, 0.67]$. In Fig. 6c the spectrum of the instantaneous frequency for $a = 0.67$ contains multiples of the carrier frequency (0.5 Hz) only. For $a = 0.65$ also harmonics of mixed (with the intrinsic fundamental) frequencies are seen that evidence two-frequency quasiperiodic waveform.

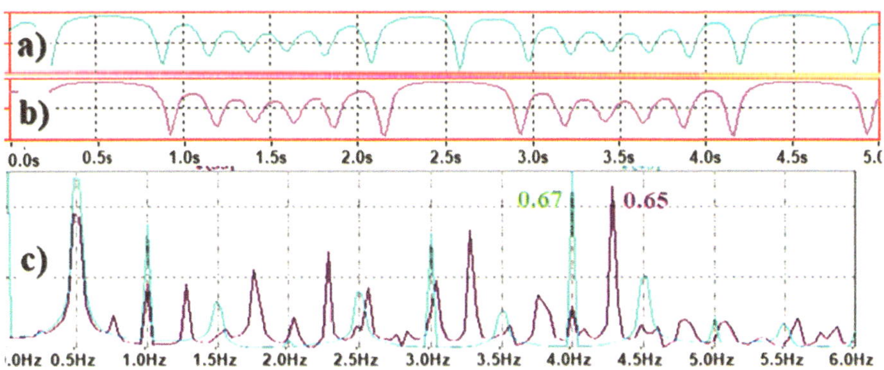

Fig. 7. Detection of the bifurcation point for AM excitation with the modulation index $m = 0.5$. Waveforms for (a) $a = 0.65$ and (b) $a = 0.67$; (c) spectra for both magnitudes.

A similar analysis was performed for the oscillator under AM excitation with a suppressed carrier (see Fig. 4, top right plot). Waveforms for $a = 0.7, 0.898, 0.899, 1.1$ are shown in Fig. 8. The difference between spectra corresponding to the magnitudes below (Fig. 7a, b) and above (Fig. 7c, d) the bifurcation point is clearly seen. Thus the interval around the bifurcation point is defined as a ∈ [0.898, 0.899].

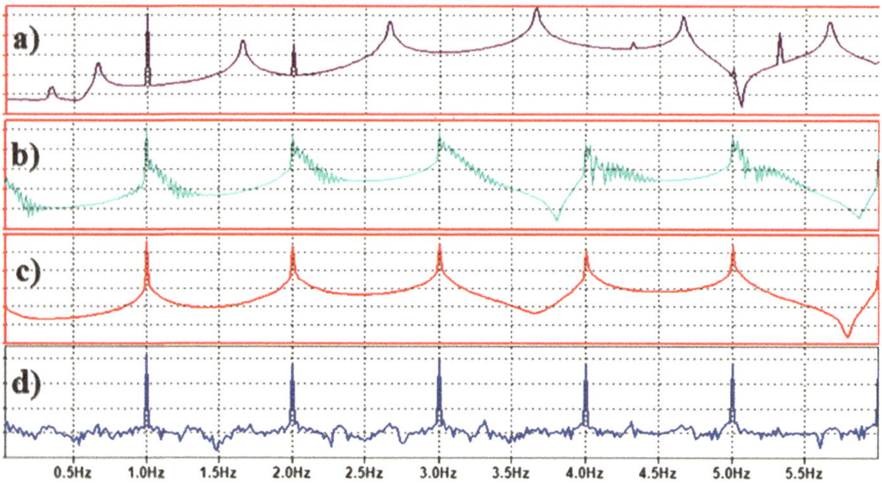

Fig. 8. Detection of the bifurcation point for AM excitation with a suppressed carrier. Spectra for (a) $a = 0.7$ (b) $a = 0.898$ (c) $a = 0.899$, (d) $a = 1.1$.

7 Conclusions

The paper proposes a natural extension of the concept of oscillator synchronization under periodic excitation to the case of quasiperiodic excitation. The multi-frequency oscillator synchronization is formulated as the coincidence of the fundamentals in the oscillator and excitation spectra. The behavior of quasiperiodically synchronized oscillator is similar to the behavior of a nonoscillatory system.

For the numerical investigation of the oscillator under quasiperiodic excitation, the modification of the Kuramoto model was performed that takes into account the excitation magnitude.

Numerical experiments with the oscillator under modulated excitation confirmed the existence of bifurcation points corresponding to the switch of the oscillator into the synchronization mode.

Some other numerical experiments provided the validation of the asymptotic expressions derived in the paper to evaluate oscillator under periodic and modulated excitations.

Acknowledgements. The reported study was funded by RFBR, project number 19-29-03012.

References

1. Kharola, A.: Artificial neural networks based approach for predicting LVDT output characteristic. Int. J. Eng. Manuf. (IJEM) **8**(4), 21–28 (2018). https://doi.org/10.5815/ijem.2018.04.03
2. Mohsen, A.A., Alsurori, M., Aldobai, B., Mohsen, G.A.: New approach to medical diagnosis using artificial neural network and decision tree algorithm: application to dental diseases. Int. J. Inf. Eng. Electron. Bus. (IJIEEB) **11**(4), 52–60 (2019). https://doi.org/10.5815/ijieeb.2019.04.06
3. Gupta, D.K., Goyal, S.: Credit risk prediction using artificial neural network algorithm. Int. J. Mod. Educ. Comput. Sci. (IJMECS) **10**(5), 9–16 (2018). https://doi.org/10.5815/ijmecs.2018.05.02
4. Kuzmina, M., Manykin, E., Grichuk, E.: Oscillatory neural networks. In: Problems of Parallel Information Processing, p. 160. Walter de Gruyter GmbH, Berlin/Boston (2014)
5. Hoppensteadt, F.C., Izhikevich, E.M.: Pattern recognition via synchronization in phase-locked loop neural networks. IEEE Trans. Neural Netw. **11**(3), 734–738 (2000)
6. Maffezzoni, P., Bahr, B., Zhang, Z., Daniel, L.: Oscillator array models for associative memory and pattern recognition. IEEE Trans. Circuits Syst. I Regul. Pap. **62**(6), 1591–1598 (2015)
7. Bonnin, M., Corinto, F., Gilli, M.: Periodic oscillations in weakly connected cellular nonlinear networks. IEEE Trans. Circuits Syst. I Regul. Pap. **55**(6), 1671–1684 (2008). https://doi.org/10.1109/TCSI.2008.916460
8. Ashwin, P., Coombes, S., Nicks, R.: Mathematical frameworks for oscillatory network dynamics in neuroscience. J. Math Neurosci. **6**(1), 2 (2016). https://doi.org/10.1186/s13408-015-0033-6
9. Bhansali, P., Roychowdhury, J.: Injection locking analysis and simulation of weakly coupled oscillator networks. In: Li, P., Silveira, L.M., Feldmann, P. (eds.) Simulation and Verification of Electronic and Biological Systems, pp. 71–93. Springer, Dordrecht (2011). https://doi.org/10.1007/978-94-007-0149-6_4
10. Kumar, P., Verma, D., Parmananda, P.: Partially synchronized states in an ensemble of chemo-mechanical oscillators. Phys. Lett. A. **381**(29), 2337–2343 (2017). https://doi.org/10.1016/j.physleta.2017.05.032
11. Frolov, N.S., Goremyko, M.V., Makarov, V.V., Maksimenk, V.A., Hramov, A.E.: Numerical and analytical investigation of the chimera state excitation conditions in the Kuramoto-Sakaguchi oscillator network. In: Proceedings of SPIE 10063, Dynamics and Fluctuations in Biomedical Photonics XIV, 100631H (2017). https://doi.org/10.1117/12.2251702
12. Asfar, K.R., Nayfeh, A.H., Mook, D.T.: Response of self-excite oscillation to multifrequency excitations. J. Sound Vib. **79**(4), 589–604 (1981)
13. El-Bassiouny, A.F.: Parametrically excited nonlinear systems: a comparison of two methods. Int. J. Math. Math. Sci. **32**(12), 739–761 (2002). https://doi.org/10.1155/S0161171202007019
14. Malinowski, M., et al.: Towards on-chip self-referenced frequency-comb sources based on semiconductor mode-locked lasers. Micromachines **10**(6), 391 (2019). https://doi.org/10.3390/mi10060391
15. Kuznetsov, A.P., Sataev, I.R., Tyuryukina, L.V.: Synchronization of quasi-periodic oscillations in coupled phase oscillators. Tech. Phys. Lett. **36**(5), 478–481 (2010). https://doi.org/10.1134/S1063785010050263

16. Peleshchak, R., Lytvyn, V., Bihun, O., Peleshchak, I.: Structural transformations of incoming signal by a single nonlinear oscillatory neuron or by an artificial nonlinear neural network. Int. J. Intell. Syst. Appl. (IJISA) **11**(8), 1–10 (2019). https://doi.org/10.5815/ijisa.2019.08.01
17. Acebrón, J.A., Bonilla, L.L., Vicente, C.J.P., Ritort, F., Spigler, R.: The Kuramoto model: a simple paradigm for synchronization phenomena. Rev. Mod. Phys. **77**(1), 137–185 (2005). https://doi.org/10.1103/RevModPhys.77.137
18. Gourary, M.M., Rusakov, S.G.: Analysis of oscillator ensemble with dynamic couplings. In: Hu, Z., Petoukhov, S., He, M. (eds.) Advances in Artificial Systems for Medicine and Education II, AIMEE 2018. Advances in Intelligent Systems and Computing, vol. 902. Springer, Cham (2019). https://doi.org/10.1007/978-3-030-12082-5_15
19. Adler, R.: A study of locking phenomena in oscillators. Proc. IEEE **61**(10), 1380–1385 (1973)
20. Schilder, F., Vogt, W., Schreiber, S., Osinga, H.M.: Fourier methods for quasi-periodic oscillations. Int. J. Numer. Methods Eng. **67**(5), 629–671 (2006)
21. Langella, R., Testa, A.: Amplitude and phase modulation effects of waveform distortion in power systems. Electr. Power Qual. Util. J. **13**(1), 25–32 (2007)
22. Razavi, B.: A study of injection locking and pulling in oscillators. IEEE J. Solid-State Circuits **39**(9), 1415–1424 (2004)
23. Desroches, M., et al.: Mixed-mode oscillations with multiple time scales. SIAM Rev. **54**(2), 211–288 (2012). https://doi.org/10.1137/100791233

New Mathematical Approaches to the Problems of Algebraic Biology

Georgy K. Tolokonnikov[1(✉)] and Sergey V. Petoukhov[2]

[1] Federal Scientific Agro-Engineering Center VIM, Russian Academy of Sciences, 1st Institute Passage, 5, Moscow, Russia
admcit@mail.ru

[2] Mechanical Engineering Research Institute, Russian Academy of Sciences, M.Kharitonievsky pereulok, 4, Moscow, Russia
spetoukhov@gmail.com

Abstract. The analysis of algebraic biology in its current state, as a separate independent science with its subject of study, tasks and methods for their solution, is carried out. The necessity of applying the systems approach in algebraic biology in its modern version using the categorical theory of systems and the categorical language for algebraic methods for studying DNA and its properties is shown. On this way, authors hope, in particular, to find new approaches to the creation of artificial intelligence systems and effective biotechnologies of regenerative medicine. Examples of everyday structures that are not reducible to sets are considered.

Keywords: Hypercomplex numbers · Genetic code · DNA · Tensor product · System theory · Categories · Topos

1 Introduction

Each completed science has its own subject, tasks and methods for their solution. We will try to outline the subject, tasks and methods of algebraic biology, systematically developed over the years in the works of one of the co-authors of the article and his school ([1–5], many works are available at http://petoukhov.com). At present, attempts are being made to use the systems approach in the further development of algebraic biology in the form of the categorical theory of systems that has arisen in recent years, developed by the second co-author [6]. The introduction briefly describes the subject, tasks and methods of algebraic biology, in the following text they are dealt with in more details, the presentation focuses mainly on research methods.

The initial subject of study in algebraic biology is the structure of genetic molecules of DNA and RNA, which are the custodians of genetic information transmitted from generation to generation and which determine much in the inherited properties of living organisms. At the same time, living organisms are considered, first of all, as informational entities, whose main task is to transfer hereditary information along a chain of generations. Modeling and studying the properties of the genetic system is conducted

Z. Hu et al. (Eds.): AIMEE 2019, AISC 1126, pp. 55–64, 2020.
https://doi.org/10.1007/978-3-030-39162-1_6

by well-known algebraic methods of academic science in order to include biology in the future in the field of advanced mathematical natural science.

Algebraic biology considers a variety of genetic and other inherited biological subsystems in their relationship, and therefore is a deeply system science. The main task in algebraic biology is the modeling and elucidation of the mechanisms of noise-resistant genetic coding of hereditary information, as well as the realization of the inherited properties of organisms and their populations based on information in DNA and RNA molecules, taking into account the interaction of developing organisms with the environment. On this way, we hope, in particular, to find new approaches to the creation of artificial intelligence systems and effective biotechnologies of regenerative medicine. To analyze such problems on the basis of algebraic modeling with their wide statement, it is advisable to involve the general theory of systems. The very general theory of systems originally originated in biology and neurobiology (Anokhin [7], Bertalanffy [8] and others).

Algebraic biology occupies a special place in theoretical and mathematical biology through the use of rigorous algebraic methods - in the broad sense of the word - of academic science for modeling and analyzing biological phenomena including molecular-genetic systems. The effectiveness of algebraic methods has been long proven its effectiveness in science and earlier led to many unexpected and surprising natural scientific discoveries.

The main task of the report is to substantiate the possibility and expediency of applying in the field of algebraic biology new algebraic methods related to the categorical theory of systems.

2 The Effectiveness of Algebra in Studying the Properties of the DNA of Organisms

The key observation that led to numerous consequences and discoveries of the properties of hidden symmetries of DNA, first within the framework of so called "matrix genetics" [1] and further in algebraic biology, was the revealing tensor family of genetic matrices on the basis of an alphabet of DNA nitrogenous bases: adenine A, cytosine C, guanine G, tymine T. Based on the study of algebraic properties, the connection of the mentioned genomatrices with the Pythagorean musical system, the golden section, the matrix of dyadic shifts from the theory of discrete signals, metric tensors of Riemannian geometry. It was possible to create a genetic musical scales of the "golden wurf" (or Fibonacci-stages scales), akin to the Pythagorean musical scales [1, 2]; to reveal algebraic connections between matrix genetics and the Punnet squares on inherited traits in Mendelian genetic [3]; to develop the concept of multi-resonance genetics [4] and also to reveal connections between the genomatrices and Hadamard matrices, which play a prominent role in the theory of discrete signals and in quantum computers.

[C, A; T, G] =

C	A
T	G

$[C, A; T, G]^{(2)}$ =

CC	CA	AC	AA
CT	CG	AT	AG
TC	TA	GC	GA
TT	TG	GT	GG

$[C, A; T, G]^{(3)}$ =

CCC Pro	CCA Pro	CAC His	CAA Gln	ACC Thr	ACA Thr	AAC Asn	AAA Lys
CCT Pro	CCG Pro	CAT His	CAG Gln	ACT Thr	ACG Thr	AAT Asn	AAG Lys
CTC Leu	CTA Leu	CGC Arg	CGA Arg	ATC Ile	ATA Met	AGC Ser	AGA Stop
CTT Leu	CTG Leu	CGT Arg	CGG Arg	ATT Ile	ATG Met	AGT Ser	AGG Stop
TCC Ser	TCA Ser	TAC Tyr	TAA Stop	GCC Ala	GCA Ala	GAC Asp	GAA Glu
TCT Ser	TCG Ser	TAT Tyr	TAG Stop	GCT Ala	GCG Ala	GAT Asp	GAG Glu
TTC Phe	TTA Leu	TGC Cys	TGA Trp	GTC Val	GTA Val	GGC Gly	GGA Gly
TTT Phe	TTG Leu	TGT Cys	TGG Trp	GTT Val	GTG Val	GGT Gly	GGG Gly

Fig. 1. Tensor family of genetic matrices $[C, A; T, G]^{(n)}$ (n = 1, 2, 3) [3]. The dark cells of the matrix contain 32 triplets with strong roots and 8 doublets that play the role of such strong roots. White cells of the matrix will contain 32 triplets with weak roots, as well as 8 doublets that play the role of such weak roots. Encoded amino acids and stop signals of protein synthesis are shown for the case of the mitochondrial vertebrate genetic code, which is the most symmetrical among the known variants of the genetic code.

Taking into account the numeric specificity of hydrogen bonds in complementary pairs of DNA nucleobases, these symbolic genomatrices (Fig. 1) are transformed into numeric matrices showing hidden algebraic relations of DNA parameters with the famous golden section f = $(1 + 5^{0.5})/2$ = 1, 618…, which is well known in aesthetics of proportions and in many inherited physiological systems [1, 2]. Simultaneously hidden algebraic relations of molecular-genetic systems with Fibonacci numbers and biological laws of phyllotaxis are discovered. The tensor family of these numeric genomatrices are connected with formalisms of some hypercomplex numbers [1, 2].

Extensive applications of algebraic constructions, using, in particular, Kronecker's product of matrices, created the basis for algebraic biology that arose on this way. Note that the Kronecker product of matrices is an operation of the monoidal category of vector spaces, many properties of which are revealed on a categorical basis (Morita's structural theorem for algebras on the equivalence of the corresponding categories of modules, etc.).

3 Systems Movement and Categorical Systems Theory

The theory of systems arose in the works of the neurophysiologist Anokhin [7] in the 1930s, which included the basics of cybernetics long before the appearance of the works of Wiener [9], who publicly recognized the priority of Anokhin. The boom in systemic movement began in the 1960s as a result of the development of the work of the biologist Bertalanffy [8] and other scientists who began to develop approaches to systems. The most common in the 1970s was the concept of a system as a relationship on the Cartesian product of a set of outputs and a set of inputs (black box). The concept of a system was generalized to systems of processes and systems described by their properties, expressed in the language of the predicate calculus of classical or intuitionistic logic [10].

In the work [10], the logical approach led to the theory of type quantifiers, which made it possible to construct new calculi that are being developed nowadays, which have found numerous practical applications. In recent years, biomachsystems [6] and the categorical theory of systems have appeared, which made it possible to formalize the basic principles of Anokhin-Sudakov of the theory of functional systems, which we briefly discuss [6].

The system is built from some formations, which then become subsystems. The construction itself is determined by the system-forming factor, as a rule, it is the result that the emerging system will strive for. The energy of the system-forming factor is expended on two main actions on these entities: they change and then connect in an optimal way for an integrated system, becoming subsystems with their own system-forming factor each, reflecting the assistance of the subsystems to achieving the result of the main system. After the result is achieved, the system is inhibited by other systems, or it disintegrates. The organism in the process of life moves from one functional system to another; the life cycle of organisms consists from such system-quantum steps. In the last years of his life, Anokhin began writing about the general theory of systems and actually built it on an informal level [7], using the successes of the theory of functional systems (all human physiology was rebuilt on the basis of functional systems, etc.). He put in its basis - in addition to the presence of a system-forming factor - also the principles of isomorphism and the hierarchy of systems. An important aspect of Anokhin's theory was the awareness and justification of the principle of building systems in the process from the whole to the parts, from the defining of the system-forming factor to the very construction of the system, arising from holistic ideas about it.

The last principle of the general theory of systems pushes it from the paradigm of set-theoretic mathematics and, in fact, under formalization, it unambiguously leads to the use of category theory as an adequate one for modern general theory of systems.

Categorical science uses the concept of arrows in categories, multi-arrows in multicategories, and poly-arrows in Szabo polycategories [11] according to convolutional polycategories [12, 13]. The types of arrows have a composition operation, generalized in the theory of convolutional polycategories to convolution operations, which allowed formalizing the most general concept of a system in the categorical language.

Categorical morphisms are given by a set of functions preserving composition and convolution of functions, called a functor. Categories, multicategories and polycategories by Szabo [11] are special cases of convolutional polycategories that find numerous applications, in particular, in the theory of artificial neural networks [13] and systems theory (Figs. 1 and 2). We note in particular that these categorical formations are algebraic objects, have an algebraic nature, and their proposed use in algebraic biology is consistent with the name of this science (Fig. 3).

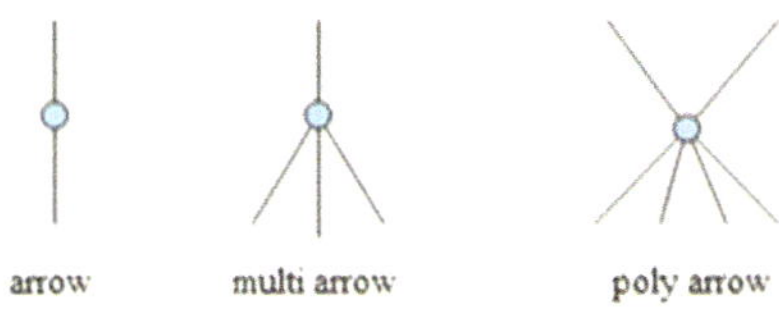

Fig. 2. Examples of categorical arrows (directions not indicated)

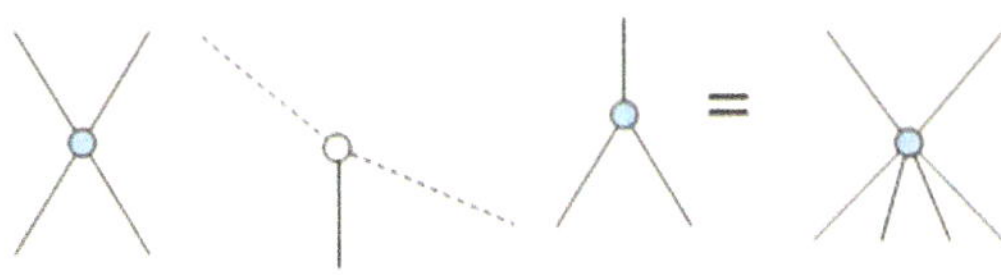

Fig. 3. Convolution example (dotted line)

In the categorical theory of systems, a system is a poly-arrow in a given convolutional polycategory. The principle of isomorphism corresponds to the concept of similarity; the principle of hierarchy is implemented by building systems from subsystems using convolutions. The role of the system-forming factor is played by a functor paired with a convolution: the functor itself changes subsystems, and the convolution connects the modified subsystems into a new integral system, just as it is formulated in Anokhin's theory on an informal level.

The categorical theory of systems not only formalized, in particular, the theory of functional systems, but also made it possible to solve a number of problems that could not be solved for decades in the theory of functional systems [6].

4 Set-Theoretic, Categorical and Systemic Paradigms

The most important moment, which takes into account the use of the categorical theory of systems in algebraic biology, is the fixation of the system paradigm ade-quate to the systems approach. Before the categorical theory of systems, all constructions were carried out by approaches on the basis of set theory. The set-theoretic paradigm was clearly constructed by N. Bourbaki in the well-known multivolume presentation of mathematics based on the theory of sets: all concepts (matrices, integrals, …) were built from elements; from such standpoint, the set does not exist without its elements. But in the theory of systems, we go from the whole to the parts, when there is no even talk about elements. Moreover, having no idea about the elements, we, nevertheless, for the future system write out its connections with other systems. Is it possible, on the basis of links (arrows!), somehow to complete the object to its elements? Such a program, which was surprising at the time, has been implemented in the theory of topos, categories closest to sets. As the classics write, "*the position of specialists in category theory, consisting in the fact that the function, and not the relation of belonging of an element to a set, should serve as the basic mathematical concept, is fully justified*" [14]. With the advent of the book [15], which summed up the development of the theory of topoi, mathematics began to develop on a new categorical basis. This meant the formation of a new categorical paradigm, when all concepts of mathematics were reduced not to sets, but to categories.

It seems that for developing algebraic biology one should use this new systemic paradigm step by step in parallel with traditional set-theoretical standpoints, which are well known for biologists. Toposes (objects of topoi which are categories) are a generalization of sets: ordinary sets that make up the category Set are a very, very special case of objects in topos. Rethinking the results previously obtained in mathematical and theoretical biology from the standpoint of the new systemic paradigm is not an easy task, but the development of the foundations of mathematics always entails similar tasks of rethinking what was already known in the natural sciences and leads to new achievements. Remaining in the set-theoretic paradigm, we lose a huge variety of topos universums, not to mention other types of categories and their further generalizations, and, consequently, categorical systems.

One should emphasize that the categorical theory of systems offers the following systemic paradigm; now all the formations in models are systems, and the categorical systems described by convolutional polycategories. In the categorical paradigm, informal constructions of the general theory of systems built by Anokhin are realized.

A categorical analysis of logic showed [14, 15] that the theory of sets itself gives rise to classical logic, within which reasoning is carried out, and a generalization to topoi led to intuitionistic logic: the logic defined by topos is generally intuitionistic. So, the new mathematics by necessity begins to use non-classical logic and these directions have been developing rapidly since the 1970s. Unfortunately, in higher education institutions neither topos, nor the theory of categories, and, therefore, no new mathematics is taught. However, among programmers of a sufficiently high level there are quite a few specialists in category theory, functional programming languages do not fit into classical mathematics, constructive tasks also require non-classical logics (intuitionistic, first of all, see [10], etc.), quantum computing gave rise to categorical quantum mechanics [16]. Moreover, in the work [17], the new mathematics is considered as the basis of modern physics.

5 New Methods in Algebraic Biology

Comprehension of the systems approach in algebraic biology leads to the use of modern algebra and category theory, which was discussed in some detail in previous sections. Here we will build a topos, which is not classical but describes a fragment of the model of each of the living objects. However, the universum, which turns out naturally for aggregates of living objects, turns out to be non-classical. In other words, we will show that a new mathematics, not based on the theory of sets, appears in algebraic biology even without a discussion of "high matters" of the general theory of systems.

Every living system in its life cycle passes from one result, that is, the corresponding functional system, to another result. Models of results or functional systems usually contain a number of parameters (blood pH, pressure, etc.) characterizing the functional system, we denote them $a_1, a_2, \ldots, a_k, \ldots$. The parameters run through some ranges of values, both in the process of achieving the result $a_1 \in \Delta_1, a_2 \in \Delta_2, \ldots, a_k \in \Delta_k, \ldots$. by the system, and when the result reached $a_1 \in \Delta_1^0 \subseteq \Delta_1, a_2 \in \Delta_2^0 \subseteq \Delta_2, \ldots, a_k \in \Delta_k^0 \subseteq \Delta_k, \ldots$ In this case, it is assumed that we can answer the question of whether the parameter satisfies the result or not ($a_i \in \Delta_i^0 -$ "yes", $a_i \notin \Delta_i^0 -$ "no"), and that there are mechanisms for transferring parameters to the result area (the system achieves the result). The parameter itself may be from several components. In other words, we have the sets Δ_i, which run through the parameters a_i and the action of the monoid; on X_i of the set (Boolean $X_i = P(\Delta_i)$) of the subsets of the set Δ_i; 1 goes into the function $\lambda_1(\xi) = \xi \in X_i$ and 0 goes into the function $\lambda_0(\xi) = \Delta_i^0 \in X_i$.

In other words, the action of a monoid is the process of achieving a result by a functional system. It turns out that the set of such actions is a non-classical topos, as we will show now.

So, we have a series of sets X_i (since the choice of systems is not limited, we can assume that these sets are arbitrary) with pairs of functions $\lambda = (\lambda_1, \lambda_2)$. Of the functions between such objects (X_i, λ), we will consider only those that leave the diagram commutative, that is, the compositions $\lambda^{(j)\circ} f = f^{\circ} \lambda^{(i)}$ are equal (Fig. 4).

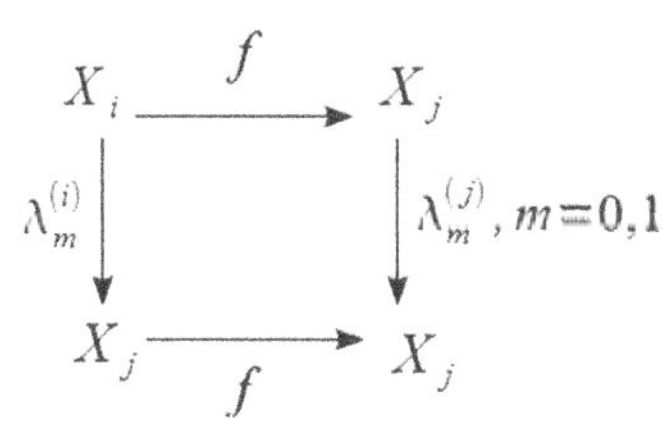

Fig. 4. Diagram defining arrow f.

It turns out that the series of such objects (X_i, λ) - with the indicated f arrows - forms a topos, a category (we will call it topos M_2), all the requirements to which are fulfilled on the category of sets, but very different from the category of sets.

It can be shown that every non-zero object of topos M_2 is not empty, that topos M_2 is two-valued, the elements of the Ω subobject classifier are $\{\emptyset, \{\emptyset\}\}$ and $\emptyset$, and that

$$\text{И} = (\text{И}_1, \text{И}_2), \text{Л} = (\text{Л}_1, \text{Л}_2) : (\{\emptyset\}, \lambda_0) \rightarrow \Omega, \text{И}_1(\emptyset) = \{\{\emptyset\}, \emptyset\}, \text{Л}_1(\emptyset) = \emptyset.$$

However, the topos M_2 is not classic.

To make sure of this, consider the following arrow

$$f_\Omega = (f_{\Omega,1}, f_{\Omega,2}) : \Omega \to \Omega,$$

$$f_{\Omega,1} : L_{M_2} \to L_{M_2}, f_{\Omega,1} \neq 1_\Omega, f_{\Omega,1}(\{\emptyset\},\emptyset) = (\{\emptyset\},\emptyset),$$

$$f_{\Omega,1}\{\emptyset\} = (\{\emptyset\},\emptyset), f_{\Omega,1}\emptyset = \emptyset, f_{\Omega,2}\omega_0 = \omega_0, f_{\Omega,2}\omega_1 = \omega_1,$$

arrow (see diagram Fig. 5, for clarity, arrow [*И*, *Л*] is indicated by dotted) [*И*, *Л*] coproduct $[И,Л]_1(\{\emptyset\}') = (\{\emptyset\},\emptyset), [И,Л]_1(\{\emptyset\}'') = \emptyset$ and their composition.

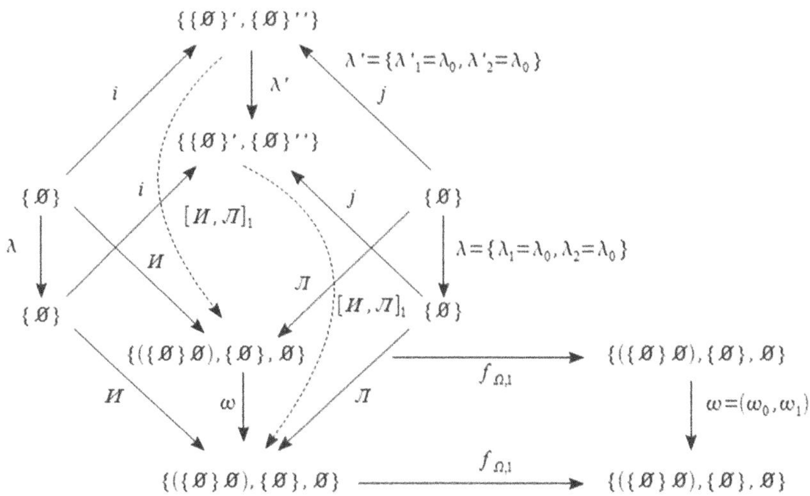

Fig. 5. Diagram showing that arrow [*И*, *Л*] (indicated by dotted) is not an isomorphism

So, we have $f_{\Omega,1} \neq 1_{\Omega,1}$ and $f_{\Omega,1}\circ[И,Л]_1 = 1_{\Omega,1}\circ[И,Л]_1$ then there is no shortening by $[И,Л]_1$ and, therefore, the arrow [*И*, *Л*] is not epimorphic and, therefore, an isomorphic arrow: the topos M_2 is not classical.

The principle of extensionality for the arrows also turns out to be unfulfilled:

$$f_\Omega\circ И = 1_\Omega\circ И, f_\Omega\circ Л = 1_\Omega\circ Л,$$

that is, no element in Ω distinguishes between different arrows $f_\Omega \neq 1_\Omega$.

At this, the differences between the topos M_2 and the category of sets do not end far; the key ones take place in the description of the logic of the theory (the algebra of subobjects, the analogue of the algebra of subsets from the Boolean, is non-zero, etc.).

Let us summarize the result that is important for our analysis: the topos M_2 most strongly violates the properties of sets and the universum based on it cannot be reduced to description in the category of sets, while this topos simulates, as we have seen, the totality of the processes of achieving the result planned by the system.

We believe that methods of categorical system theory and the theory of convolution polycategories can be used in future for developing many biological and biotechnological approaches described for example in [18–23].

6 Conclusions

The results achieved in algebraic biology, including some discoveries made, allow us to speak of it as a separate independent science with its own subject, tasks, methods, and a significant perspective. It is substantiated that algebraic biology is deeply system science, in which, as was shown (using the example of topos and others), one can use new mathematics that emerged after the formation of the platform of topos in category theory in the 1980s. For development of algebraic biology, methods of categorical system theory and the theory of convolution polycategories are proposed, which can be used in parallel with traditional mathematical methods known in biological sciences. On the proposed way, one should expect mutual enrichment of mathematics and theoretical biology in the interests of development of biotechnology, medical engineering, and fundamentally new artificial intelligence systems. The rapid development of the mathematical sciences, bioinformatics and computer technology provides great new opportunities for the implementation of such mutual enrichment.

References

1. Petoukhov, S.V.: Matrix genetics, algebra of genetic code, noise immunity. RHD, Moscow (2008)
2. Petoukhov, S.V., He, M.: Symmetrical Analysis Techniques for Genetic Systems and Bioinformatics: Advanced Patterns and Applications. IGI Global, Hershey (2009)
3. Petoukhov, S.V.: Matrix genetics and algebraic properties of the multi-level system of genetic alphabets. Neuroquantology **9**(4), 60–81 (2011)
4. Petoukhov, S.V.: The system-resonance approach in modeling genetic structures. Biosystems **139**, 1–11 (2016)
5. Petoukhov, S., Petukhova, E., Hazina, L., Stepanyan, I., Svirin, V., Silova, T.: The genetic coding, united-hypercomplex numbers and artificial intelligence. In: Hu, Z.B., Petoukhov, S., He, M. (eds.) Advances in Artificial Systems for Medicine and Education. Advances in Intelligent Systems and Computing, vol. 658. Springer, Cham (2017)
6. Chernoivanov, V.I., Sudakov, S.K., Tolokonnikov, G.K.: Biomachsystems, functional systems, categorical theory of systems. Institute of Normal Physiology, Moscow (2018)
7. Anokhin, P.K.: The fundamental questions of the general theory of functional systems. In: Principles of System Organization of Functions, Moscow, Nauka, pp. 5–61 (1973)
8. von Bertalanffy, L.: General System Theory. A Critical Review, General Systems, vol. VII, pp. 1–20 (1962)
9. Wiener, N.: Cybernetics or Control and Communication in the Animal and the Machine. The Technology Press and Wiley, New York (1948)
10. Vasiliev S.N., Zherlov A.K., Fedosov E.A., Fedunov B.E. Intelligent control of dynamic systems. Moscow, Fizmatlit (2000)
11. Szabo, M.: Polycategories. Comm. Algebra **3**(8), 663–689 (1975)

12. Tolokonnikov, G.K.: Manifest: neurographs, neurocategories and categorical gluing. Biomashsystems **1**(1), 59–146 (2017)
13. Tolokonnikov, G.K.: Convolution polycategories and categorical splices for modeling neural networks. In: Advances in Intelligent Systems and Computing, Volume 938, Advances in Computer Science for Engineering and Education II, Proceedings ICCSEEA 2019, pp. 259–267 (2019). ISSN 2194-5357. ISSN 2194-5365
14. Goldblat, R.: Topoi: Categorical Analysis of Logic. Dover Publications, Mineola (2006). ISBN-10 0486450260
15. Johnston, P.T.: Theory of Topos. Dover Publications, Mineola (2014). ISBN-13 978-0486493367
16. Abramsky, S., Coecke, B.: Categorical quantum mechanics. In: Handbook of Quantum Logic and Quantum Structures: Quantum Logic, pp. 261–324. Elsevier, North-Holland (2009)
17. Doring, A., Isham, C.J.: A topos foundation for theories of physics. J. Math. Phys. **49**(5), 053515 (2008)
18. Karande, A.M., Kalbande, D.R.: Weight assignment algorithms for designing fully connected neural network. Int. J. Intell. Syst. Appl. (IJISA) **10**(6), 68–76 (2018)
19. Dharmajee Rao, D.T.V., Ramana, V.: Winograd's inequality: effectiveness for efficient training of deep neural networks. Int. J. Intell. Syst. Appl. (IJISA) **6**, 49–58 (2018)
20. Hu, Z., Tereykovskiy, I.A., Tereykovska, L.O., Pogorelov, V.V.: Determination of structural parameters of multilayer perceptron designed to estimate parameters of technical systems. Int. J. Intell. Syst. Appl. (IJISA) **10**, 57–62 (2017)
21. Awadalla, M.H.A.: Spiking neural network and bull genetic algorithm for active vibration control. Int. J. Intell. Syst. Appl. (IJISA) **10**(2), 17–26 (2018)
22. Abuljadayel, A., Wedyan, F.: An approach for the generation of higher order mutants using genetic algorithms. IJISA **10**(1), 34–35 (2018)
23. Kumar, A., Sharma, R.: A genetic algorithm based fractional fuzzy PID controller for integer and fractional order systems. Int. J. Intell. Syst. Appl. (IJISA) **10**(5), 23–32 (2018)

Analysis of Changes in Topological Relations Between Spatial Objects at Different Times

Sergey Eremeev(✉)

Vladimir State University, Vladimir, Russia
sv-eremeev@yandex.ru

Abstract. There are many problems with situations in which the relationships between elements change over time. The initial data can be images of some area for a different period of time or from different scales. The solution of these problems is necessary for a detailed analysis of the map. In the article the problem of analysis of topological relations between spatial objects for different periods of time is considered. It is proposed to use the methods of temporal graph theory to present information about the relations between objects taking into account time. A mathematical model for storing information about topological relations is demonstrated. The relationship matrix contains information about the topology of the map for different periods of time. An algorithm for the analysis of unchanged objects for a given period of time is developed. An algorithm to determine the areas of the map that have changed the maximum number of times is also developed. The results of experiments on the division of the map into 4 and 16 sectors are shown. Screenshots of map fragments and matrix of changes of topological connections of temporal graph are given. These algorithms can be used in the modeling of environmental disasters, environmental planning, for the analysis of real estate in municipal GIS.

Keywords: Topological relations · Temporal graphs · Spatial objects · GIS

1 Introduction

There are many problems with situations in which the relationships between elements change over time [1, 2]. The initial data can be images of the area for a different period of time or from different scales. The solution of these problems is necessary for a detailed analysis of the topographic characteristics of the territory [3], assessment and monitoring of the environment [4], simulation of emergency situations, etc. The task of change analysis of topological relations between spatial objects in different period of time is important. Its solution will allow us to find the most stable and the most changeable areas of the map. For example, the analysis of changes in topological relations in municipal GIS will help to show the dynamism of the territory [5].

A useful and effective tool in GIS is the application of graph theory. In GIS, the features represented on the map are defined as vertices. The lines connecting them represent the edges. Such graphs can be analyzed to find the shortest path, the best route or the least cost. Relations in graph theory are constant and do not change over

Z. Hu et al. (Eds.): AIMEE 2019, AISC 1126, pp. 65–75, 2020.
https://doi.org/10.1007/978-3-030-39162-1_7

time. These relationships are called static. A static graph is a set of vertices and their connecting edges. Each edge stores a weight that corresponds to the type of relationship between the vertices. However, static graphs are not suitable for the analysis of topological relations in time. Therefore, it is advisable to use a new approach based on temporal graphs which was proposed in the article [6]. Temporal graphs are widely used in machine learning [7, 8], tasks in the field of taxonomy [9], etc. There are also own problems in temporal graphs, for example, for path optimization [10, 11].

A temporal graph is a triple $G = (X, \{Y_t\}, T)$, where X is a set of vertices of the graph with the number of vertices $|X| = n$, $T = \{1, 2, \ldots, N\}$ is a set of natural numbers that determine discrete time, $\{Y_t\}$ is a family maps of the set of vertices in itself at time $t \in T$ so that $\forall (t \in T)\ Y_t : X \to X$.

The aim of the work is to develop and study new algorithms for the analysis of topological relations between spatial objects over different periods of time.

2 Related Work

Currently, spatial-temporal GIS are widely developed. Problems of spatial-temporal GIS development and data model for access to information from different sources are shown in [12]. The main difficulty of spatio-temporal analysis is different representation of data of the same territory. An important direction is the analysis of satellite images of the territory. The problem of classification of the study area into different categories is solved in [13, 14]. It is necessary for land use planners to ensure sustainable development of the region. There are also methods that use 4D data modeling. For example, historical images are processed in [15] to obtain spatio-temporal information. This approach uses the time parameter in combination with the methods of reconstructing images in 3D. Three-dimensional simulation over time allows us to identify and analyze changes in individual spatial scenes and landscapes. Thus, many developments for the analysis of spatio-temporal information use remote sensing data. The search for changes in the sequences of maps is performed by digital image processing.

However, the analysis of changes in topological relations is quite promising and an open problem. Solving this problem requires the creation of new approaches and models to store map topology between spatio-temporal objects. Developments to create an algebra that allows us the processing of spatio-temporal data as a whole are explored in this direction. A new map algebra approach that uses topology is introduced in [16]. This topology based map algebra includes spatio-temporal topological operators for specifying spatio-temporal topological operations between the objects over time. In addition, GRASS GIS is developed. Another algebra for processing spatio-temporal GIS is shown in [17]. Unified spatio-temporal analysis is based on Clifford algebra. The software package CAUSA (Clifford Algebra based Unified Spatial-Temporal Analysis) is developed. It is a useful tool for studying and modeling the characteristics of the dynamic process of complex geographical phenomena within a single spatio-temporal structure.

3 Methodology

3.1 Constructing an Adjacency Matrix of a Temporal Graph

Initially, we need to construct an adjacency matrix of a temporal graph. In the adjacency matrix of graphs, topological relations are denoted as: "disjoint" is 0, "meet" is 1, "overlap" is 2, "inside" is 3, "contains" is 4. Create the adjacency matrix for the next set of objects (see Fig. 1). It is constructed at the first moment of time and shown in Table 1.

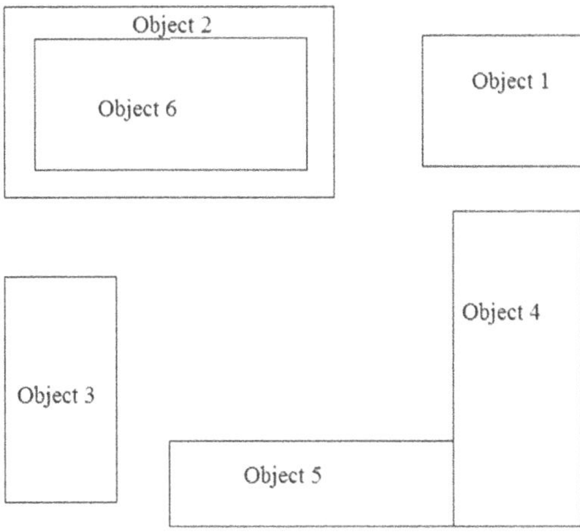

Fig. 1. Test object set.

Table 1. Adjacency matrix of the topological relationships between objects in Fig. 1

	Object 1	Object 2	Object 3	Object 4	Object 5	Object 6
Object 1	0	0	0	0	0	0
Object 2	0	0	0	0	0	4
Object 3	0	0	0	0	0	0
Object 4	0	0	0	0	1	0
Object 5	0	0	0	1	0	0
Object 6	0	3	0	0	0	0

After several experiments at the sixth moment of time, the objects on the map and the adjacency matrix contain changes (see Fig. 2 and Table 2).

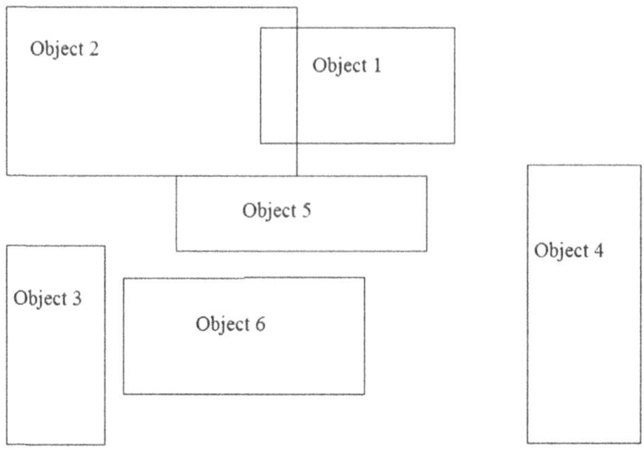

Fig. 2. Set of test objects at the sixth moment of time.

Table 2. Adjacency matrix of the objects in Fig. 2 at the sixth moment of time.

	Object 1	Object 2	Object 3	Object 4	Object 5	Object 6
Object 1	000000	002222	000000	000000	000000	000000
Object 2	002222	000000	000000	000000	004041	400000
Object 3	000000	000000	000000	000000	000000	000000
Object 4	000000	000000	000000	000000	120200	000000
Object 5	000000	003031	000000	120200	000000	000000
Object 6	000000	300000	000000	000000	000000	000000

At the sixth moment of time, the topological relationships of the objects contain changes. It is only object 3 without change. At all times, it has no topological relationships with other objects.

3.2 Construction of a Change Matrix in Topological Relationships Based on the Adjacency Matrix of a Temporal Graph

Len C be a change matrix in topological relationships:

$$C = \begin{bmatrix} c_{1,1}, c_{1,2}, \ldots, c_{1,n} \\ c_{2,1}, c_{2,2}, \ldots, c_{2,n} \\ c_{n,1}, c_{n,2}, \ldots, c_{n,n} \end{bmatrix} (i, j = 1, 2, \ldots, n) \tag{1}$$

where $c_{i,j}$ is the number of object changes.

The matrix (1) shows how many changes in topological relationships of objects occurred in a certain number of moments of time. Elements $c_{i,j}$ are calculated in the expression (2):

$$c_{i,j} = c_{i,j} + 1, \text{ if } b_t \neq b_{t+1}, (i,j = 1, 2, \ldots, n) \tag{2}$$

where b_t is a type of topological connection such that $b_t \in \{0, 1, 2, 3, 4\}$.

Set $\{0, 1, 2, 3, 4\}$ denotes the possible values of the types of topological connections, t is the moment of time.

When constructing a matrix (1), topological relations in the adjacency matrix of a temporal graph are iterated and compared. If in the current and next moments of time topological relations have different values, it means that in the matrix of changes in topological relations at the corresponding positions the value of the number of changes will increase by one, otherwise it will remain the same.

Based on the adjacency matrix of a temporal graph in Table 2, its matrix of changes in topological relations is constructed and shown in Table 3.

Table 3. Change matrix of temporal graph calculated on the basis of expression (2) and Table 2.

	Object 1	Object 2	Object 3	Object 4	Object 5	Object 6
Object 1	0	1	0	0	0	0
Object 2	1	0	0	0	4	1
Object 3	0	0	0	0	0	0
Object 4	0	0	0	0	4	0
Object 5	0	4	0	4	0	0
Object 6	0	1	0	0	0	0

According to the change matrix of the temporal graph in Table 3, we can say how many times the topological relations of objects have been changed. The line with the selected color shows that object 3 has no changes.

3.3 Algorithm for Determining Objects that Have a Minimum Number of Changes

Let $\min_c$ be a minimum element in the change matrix that is equal to:

$$\min_c = \min(c_{i,j}) \quad (i,j = 1, 2, \ldots, n) \tag{3}$$

If some object has changed shape or position in space relative to another with a change in the type of topological relationship, then the relationship between the two objects has changed. After analyzing the values in the rows and columns of the change matrix, select the objects that have the minimum values.

Let A be a set of all objects that have minimum number of changes:

$$A = \{a_k \in X : c_{i,j} = \min_c\} \; (i,j = 1,2,\ldots,n) \tag{4}$$

where $k \in \{1,2,\ldots,n\}$, a_k are objects with minimal changes.

A special case of this algorithm is the finding of unchanged objects whose topological connections remained constant at all times. This means that the number of object changes must be zero. Table 3 shows an example of a matrix of changes in topological relationships in which object 3 is not changed. To find such objects, it is proposed to use the following expression:

$$A' = \left\{ a_k \in X : \sum_{j=1}^{n} c_{k,j} = 0 \right\} \quad k \in \{1,2,\ldots,n\} \tag{5}$$

3.4 Algorithm for Determining the Areas of the Map that Have the Maximum Number of Changes

Divide the map into sections that are equal sectors of square shape and their number is arbitrary.

Let Q be a matrix of sectors:

$$Q = \begin{bmatrix} q_{1,1}, q_{1,2}, \ldots, q_{1,m} \\ q_{2,1}, q_{2,2}, \ldots, q_{2,m} \\ q_{m,1}, q_{m,2}, \ldots, q_{m,m} \end{bmatrix} \tag{6}$$

where $q_{k,l}$ $(k,l = 1,2,\ldots,m)$ is a sector that contains a subset of objects in a set X.

Changing the topological relationships of objects affects the change in the map area. Therefore, changing the coordinates of an object without changing the topological relationship with other objects means that the sector has not changed.

Let G be a matrix of sector changes:

$$G = \begin{bmatrix} g_{1,1}, g_{1,2}, \ldots, g_{1,m} \\ g_{2,1}, g_{2,2}, \ldots, g_{2,m} \\ g_{m,1}, g_{m,2}, \ldots, g_{m,m} \end{bmatrix} \tag{7}$$

where $g_{k,l}$ $(k,l = 1,2,\ldots,m)$ contains the number of changes.

The algorithm iterates over the objects in each sector and counts the number of changes in it by the change matrix. The result of the algorithm is a matrix corresponding to the number of map sectors each cell of which contains the sum of changes in the sector:

$$g_{k,l} = \sum_{i=1}^{n} \sum_{j=1}^{n} c_{i,j} : a_i, a_j \in q_{k,l} \; (i,j = 1,2,\ldots,n;\; k,l = 1,2,\ldots,m) \tag{8}$$

Let $\max_g$ be an element in the matrix of sector changes G with the maximum value. To determine the map areas that have changed the maximum number of times we must evaluate the expression:

$$\max_g = \max(g_{i,j}) \quad (i,j = 1,2,\ldots,n) \tag{9}$$

4 Experiments

The initial data for the study is a map that consists of polygon objects. Some of these objects are marked with the letters a, b, c, d. We use 3 time periods to model the situation. At each time period, spatial objects change their position. It leads to a change in topological relationships between objects. For a more detailed simulation, we divide the map first into 4 sectors (Fig. 3), and then on 16 (Fig. 4). After that, we analyze the sectors and calculate the matrix of changes in topological relations in each sector (Table 4).

Table 4. Number of change in 4 sectors.

12	8
2	0

At time t_1 the sector [1] contains 10 objects. Objects "a" and "b" are combined by the link "overlap". The number of changes is 0. At time t_2 connection between objects "a" and "b" disappeared. The matrix (7) takes into account 2 changes. The topological relations "meet" and "overlap" appeared:

1. between "b" and "c",
2. between "a" and "d".

In total, there were 6 changes in the sector [1]. At time 3, the 4 relations broke up and the relation between objects "a" and "b" appeared again adding 2 more links to the total number of changes at all times.

Thus, the sector [1] is the most changeable. Objects have changed in it 12 times in 3 points of time.

Figure 4 shows a map with objects divided into 16 sectors, and Table 5 shows the result.

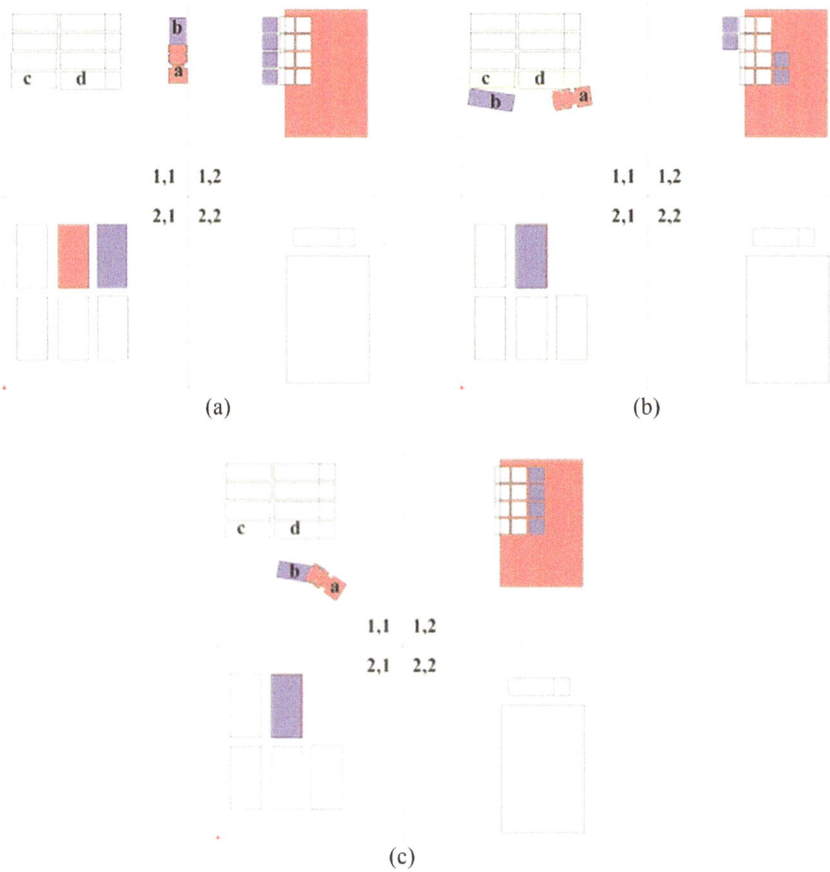

Fig. 3. Map with objects divided into 4 sectors: at time t_1 (a), at time t_2 (b), at time t_3 (c).

Figure 4 and Table 5 demonstrate much more detailed information on topological relationships in sectors. Table 5 shows that two sectors contain the maximum number of changes.

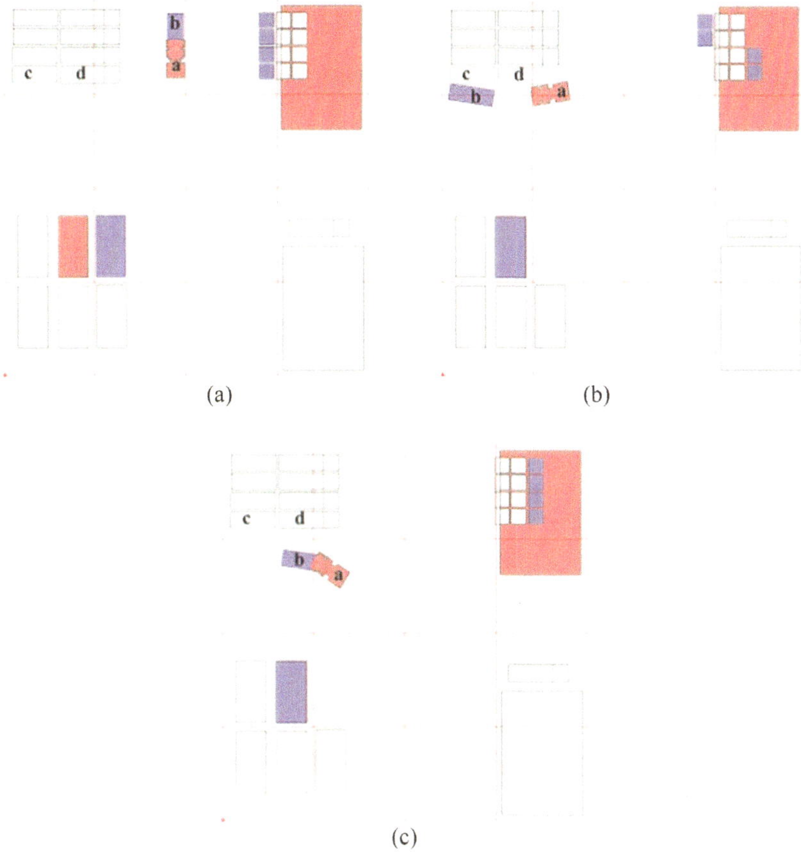

Fig. 4. Map with objects divided into 16 sectors: at time t_1 (a), at time t_2 (b), at time t_3 (c).

Table 5. Number of change in 16 sectors.

2	4	4	8
4	8	0	0
2	0	0	0
0	0	0	0

5 Conclusions

Two algorithms designed to analyze changes in topological relations are developed in this paper. The first algorithm allows us to determine the features on the map that have no changes over a period of time. The second algorithm finds the areas on the map that have changed the maximum number of times. The algorithms are based on the model of establishing topological relations between objects on the map and the adjacency matrix of a temporal graph. The result is the construction and analysis of the matrix of changes. The experiments on the division of the map into 4 and 16 sectors are studied. The developed algorithms can be used in the modeling of environmental disasters, environmental planning, for the analysis of real estate in municipal GIS.

Acknowledgment. The reported study was funded by RFBR and Vladimir region according to the research project No. 17-47-330387.

References

1. Sitanggang, I., Roseli, S., Syaufina, L.: Spatial co-location patterns on weather and forest fire data. Int. J. Inf. Technol. Comput. Sci. (IJITCS) **10**(9), 13–20 (2018). https://doi.org/10.5815/ijitcs.2018.09.02
2. Mamoria, P., Raj, D.: An analysis of fuzzy and spatial methods for edge detection. Int. J. Inf. Eng. Electron. Bus. (IJIEEB) **8**(6), 62–68 (2016)
3. Eremeev, S., Seltsova, E.: Algorithms for topological analysis of spatial data. In: Hu, Z., Petoukhov, S., He, M. (eds.) Advances in Artificial Systems for Medicine and Education II, AIMEE2018 2018. AISC, vol. 902, pp. 81–92. Springer, Cham (2020)
4. Sitanggang, I., Shofiana, D., Sihombing, B.: Hotspot sequence patterns with an improvement in spatial feature. Int. J. Eng. Manuf. (IJEM) **8**(6), 13–25 (2018)
5. Eremeev, S., Andrianov, D., Kovalev, Y. Kuptsov, K.: Algorithm for encoding nD spatial objects into GIS. In: Proceedings of the International Conference Information Technology and Nanotechnology. Session Image Processing and Earth Remote Sensing, ITNT-2018, Samara, Russia, pp. 149–155 (2018)
6. Kostakos, V.: Tempotal graphs. Phys. A **6**, 1007–1023 (2009)
7. Jain, A., Zamir, A.R., Savarese, S., Saxena, A.: Structural-RNN: deep learning on spatio-temporal graphs. In: CVPR (2016)
8. Chen, X., Liu, Y., Liu, H., Carbonell, J.: Learning spatial-temporal varying graphs with applications to climate data analysis. In: Proceedings of the Twenty-Fourth AAAI Conference on Artificial Intelligence, AAAI-10, pp. 425–430 (2010)
9. Sridhar, M., Cohn, A.G., Hogg, D.C.: Discovering an event taxonomy from video using qualitative spatio-temporal graphs. In: Coelho, H., Suder, R., Wooldridge, M. (eds.) 19th European Conference on Artificial Intelligence, ECAI 2010, pp. 1103–1104. IOS Press, Lisbon (2010)
10. Erlebach, T., Hoffmann, M., Kammer, F.: On temporal graph exploration. In: Halldórsson, M., Iwama, K., Kobayashi, N., Speckmann, B. (eds.) Automata, Languages, and Programming, ICALP 2015. LNCS, vol. 9134, pp. 444–455. Springer, Heidelberg (2015)
11. Mertzios, G., Michail, O., Spirakis, P.: Temporal network optimization subject to connectivity constraints. Algorithmica **81**(4), 1416–1449 (2019)
12. Ferreira, K., de Oliveira, A.G., Monteiro, A, de Almeida, D.B.F.C: Temporal GIS and spatiotemporal data sources. In: Proceedings XVI GEOINFO, pp. 1–13 (2015)
13. Choudhary, K., Boori, M., Kupriyanov, A.: Spatio-temporal analysis through remote sensing and GIS in Moscow region. In: Russia Proceedings of the International conference Information Technology and Nanotechnology. Session Image Processing, Geoinformation Technology and Information Security, pp. 42–46 (2017)
14. Dewan, A.M., Yamaguchi, Y.: Land use and land cover change in Greater Dhaka, Bangladesh: using remote sensing to promote sustainable urbanization. Appl. Geogr. **29**, 390–401 (2009)

15. Nocerino, E., Menna, F., Remondino, F.: Multi-temporal analysis of landscapes and urban areas. In: International Archives of the Photogrammetry, Remote Sensing and Spatial Information Sciences, XXXIX-B4, pp. 85–90 (2012)
16. Gebbert, S., Leppelt, T., Pebesma, E.: A topology based spatio-temporal map algebra for big data analysis. Data **4**, 86 (2019)
17. Yuan, L., Yu, Z., Chen, S., Luo, W., Wang, Y., Lü, G.: CAUSTA: clifford algebra based unified spatio temporal analysis. Trans. GIS **14**(s1), 59–83 (2010)

Many-Parameter Quaternion Fourier Transforms for Intelligent OFDM Telecommunication System

Valeriy G. Labunets[1(✉)] and Ekaterina Ostheimer[2]

[1] Ural State Forest Engineering University, 37, Sibirskiy Trakt, 620100 Ekaterinburg, Russian Federation
vlabunets05@yahoo.com
[2] Capricat LLC, Pompano Beach, FL, USA

Abstract. In this paper, we aim to investigate the superiority and practicability of many-parameter quaternion Fourier transforms (MPQFT) from the physical layer security (PHY-LS) perspective. We propose novel Intelligent OFDM-telecommunication system (Intelligent-OFDM-TCS), based on MPFT. New system uses inverse MPQFT for modulation at the transmitter and direct MPQFT for demodulation at the receiver. The purpose of employing the MPFTs is to improve the PHY-LS of wireless transmissions against to the wide-band anti-jamming communication. Each MPQFT depends on finite set of independent parameters (angles), which could be changed independently one from another. When parameters are changed, multi-parametric transform is also changed taking form of a set known (and unknown) orthogonal (or unitary) transforms. We implement the following performances as bit error rate (BER), symbol error rate (SER), the Shannon-Wyner secrecy capacity (SWSC) for novel Intelligent-MPWT-OFDM-TCS. Simulation results show that the proposed Intelligent OFDM-TCS have better performances than the conventional OFDM system based on DFT against eavesdropping

Keywords: Many-parameter transforms · Quaternion fourier transform · OFDM · Telecommunication system · Anti-eavesdropping communication

1 Introduction

With most of the data transmission systems nowadays use orthogonal frequency division multiplexing telecommunication system (OFDM-TCS) based on the discrete Fourier transform (DFT). Some versions of it is: digital audio broadcast (DAB), digital video broadcast (DVB), and wireless local area network (WLAN), standards such as IEEE802.11g and long term evolution (LTE and its extension LTE-Advanced, Wi-Fi (IEEE 802.11), worldwide interoperability for microwave ACCESS (WiMAX IEEE 802.16) or ADSL [1]. The concept of using parallel data broadcast by means of frequency division multiplexing (FDM) was printed in mid 60s [2]. The conventional OFDM is a multi-carrier modulation technique that is basic technology having high-speed transmission capability with bandwidth efficiency and robust performance in multipath fading environments. OFDM divides the available spectrum into a number of

Z. Hu et al. (Eds.): AIMEE 2019, AISC 1126, pp. 76–92, 2020.
https://doi.org/10.1007/978-3-030-39162-1_8

parallel orthogonal sub-carriers and each sub-carrier is then modulated by a low rate data stream at different carrier frequency. In OFDM system, the modulation and demodulation can be applied easily by means of inverse and direct discrete Fourier transforms (DFT). The conventional OFDM will be denoted by the symbol $\mathcal{F}_N$-**OFDM**. All sub-carriers $\{\mathbf{subc}_k(n)\}_{k=0}^{N-1} = \{e^{j2\pi kn/N}\}_{k=0}^{N-1}$ form matrix of discrete orthogonal Fourier transform $\mathcal{F}_N = [\mathbf{subc}_k(n)]_{k,n=0}^{N-1} \equiv [e^{j2\pi kn/N}]_{k,n=0}^{N-1}$. At the time, the idea of using the fast algorithm of different orthogonal transforms $\mathcal{U}_N = [\mathbf{subc}_k(n)]_{k,n=0}^{N-1}$ for a software-based implementation of the OFDM's modulator and demodulator, transformed this technique from an attractive. OFDM-TCS, based on arbitrary orthogonal (unitary) transform $\mathcal{U}_N$ will be denoted as $\mathcal{U}_N$-**OFDM**. The idea which links $\mathcal{F}_N$-**OFDM** and $\mathcal{U}_N$-**OFDM** is that, in the same manner that the complex exponentials $\{e^{j2\pi kn/N}\}_{k=0}^{N-1}$ are orthogonal to each-other, the members of a family of $\mathcal{U}_N$-sub-carriers $\{\mathbf{subc}_k(n)\}_{k=0}^{N-1}$ will satisfy the same property.

The $\mathcal{U}_N$−**OFDM** reshapes the multi-carrier transmission concept, by using carriers $\{\mathbf{subc}_k(n)\}_{k=0}^{N-1}$ instead of OFDM's complex exponentials $\{e^{j2\pi kn/N}\}_{k=0}^{N-1}$. There are a number of candidates for orthogonal function sets used in the OFDM-TCS: discrete wavelet sub-carriers [3], Golay complementary sequences [4], Walsh functions [5], pseudo random sequences [6].

In this work, we propose a simple and effective anti-eavesdropping and anti-jamming Intelligent OFDM system, based on quaternion many-parameter transform (QMPT) $\mathcal{QU}_N(\varphi_1, \varphi_2, \ldots, \varphi_q)$ instead of ordinary DFT $\mathcal{F}_N$.

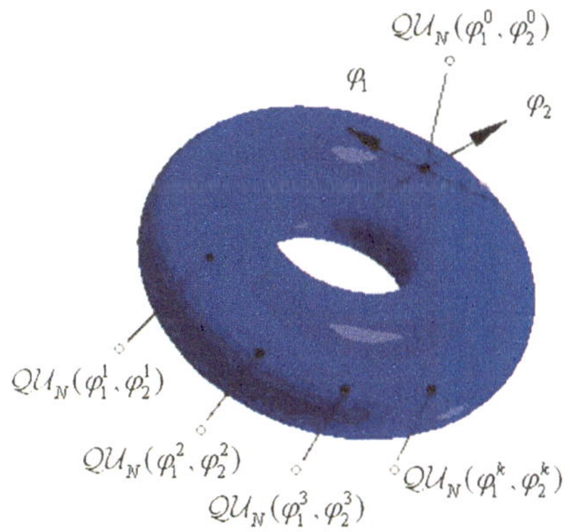

Fig. 1. q-D torus $\mathbf{Tor}_q[0, 2\pi] = [0, 2\pi]^q$

Each QMPT depends on finite set of free parameters $\boldsymbol{\theta} = (\varphi_1, \varphi_2, \ldots, \varphi_q)$, and each of them can take its value form 0 to 2π. When parameters are changed, QMPT is changed too taking form of known (and unknown) quaternion transforms. The vector of parameters $\boldsymbol{\theta} = (\varphi_1, \varphi_2, \ldots, \varphi_q) \in \mathbf{Tor}_q[0, 2\pi] = [0, 2\pi]^q$ belongs to the q-D torus (see Fig. 1). When the vector of parameters $(\varphi_1, \varphi_2, \ldots, \varphi_q)$ runs completely the q-D torus $\mathbf{Tor}_q[0, 2\pi]$, the ensemble of the orthogonal quaternion transforms is created. Intelligent OFDM system uses some concrete values of the parameters $\varphi_1 = \varphi_1^0, \varphi_2 = \varphi_2^0, \ldots, \varphi_q = \varphi_q^0$, *i.e.*, it uses a concrete realization of QMPT $\mathcal{QU}_N^0 \equiv \mathcal{QU}_N(\varphi_1^0, \varphi_2^0, \ldots, \varphi_q^0)$. The vector $(\varphi_1^0, \varphi_2^0, \ldots, \varphi_q^0)$ plays the role of some analog key (see Fig. 2), whose knowing is necessary for entering into the TCS with the aim of intercepting the confidential information.

Fig. 2. Key of parameters $(\varphi_1, \varphi_2, \ldots, \varphi_q)$

Quantity of parameters can achieve the values $p = 10\,000$. So, searching the vector key by scanning the 10000-dimensional torus $[0, 2\pi]^{10\,000}$ with the aim of finding the working parameters $(\varphi_1^0, \varphi_2^0, \ldots, \varphi_q^0)$ is very difficult problem for the enemy cyber-means. But if, nevertheless, this key were found by the enemy in the cyber attack, then the system could change values of the working parameters for rejecting the enemy attack. If the system is one of the TCP type, then in such a case, it will transmit the confidential information on the new sub-carriers (*i.e.*, in the new orthogonal basis). As a result, the system will counteract against the enemy radio-electronic attacks.

The enemy problem is also complicated by the fact that the QMPT is additionally arranged with digital key that is joined with the non-commutativity of the quaternion multiplication operation. In the QMPT matrix multiplication on the data vector, each element of the vector can be multiplied on the matrix element either from the left or from the right. Let the symbol "0" means multiplication of the data vector $\vec{\mathbf{v}} = (v_1, \ldots, v_k, \ldots, v_N)$ on the matrix element from the left (L) and the symbol "1" means the multiplication from the right (R). Then the N-D binary vector $(b_1, b_2, \ldots, b_N)$ is the digital key (see Fig. 3) showing onto the way, by which the multiplication of the QMPT matrix on the data vector has to be implemented:

$$\left[\mathrm{qmp}_k^{b_k}(n|\theta_1, \theta_2, \ldots, \theta_p)\right] \cdot \vec{\mathbf{v}} \Rightarrow \begin{cases} \mathrm{qmp}_k^{b_k}(n) \cdot v_k, & b_k = 0, \\ v_k \cdot \mathrm{qmp}_k^{b_k}(n), & b_k = 1, \end{cases}.$$

where $\left\{\mathrm{qmp}_k^{b_k}(n)\right\}_{k,n=1}^{N}$ are the set of matrix elements of quaternion transform, *i.e.*, $\mathcal{QU}^{(b_1,b_2,\ldots,b_N)}(\varphi_1, \varphi_2, \ldots, \varphi_q) = \left[\mathrm{qmp}_k^{b_k}(n|\theta_1, \theta_2, \ldots, \theta_p)\right]_{k,n=1}^{N}$. So, the number of such keys is equal to 2^N. They form the Boolean cube $\mathbf{B}_2^N$. Knowing this key is necessary to enter into the Intelligent OFDM TCS.

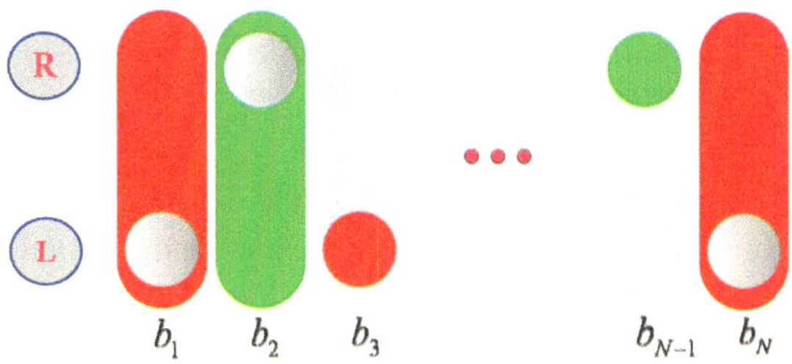

Fig. 3. N-D binary vector-key $(b_1, b_2, \ldots, b_N)$

Thus, the space of keys is the Descartes product $\mathbf{Tor}_q[0, 2\pi] \times \mathbf{B}_2^N$. Searching the pair of keys $(\varphi_1, \varphi_2, \ldots, \varphi_q) \in \mathbf{Tor}^q[0, 2\pi]$ and $(b_1, b_2, \ldots, b_N) \in \mathbf{B}_2^N$ in the space of keys is very difficult problem for the enemy, especially, when we have the OFDM TCS. Such a system can also defend itself by changing the values of the working parameters and crypto key accordingly to some law that is known for the transmitter and receiver in advance. The law can be deterministic or the pseudo-random (similarly to the law, by which the contemporary TCSs change their working frequency). The main advantage of using QMPT in OFDM TCS is that it is a very flexible anti-eavesdropping and anti-jamming Intelligent OFDM TCS. These TCS have additional advantages in comparison with the classic TCS: the multi-parametric transforms allow one to optimize (and as a result, to enhance) the technical characteristics of the system (by changing its parameters) such as the PARP (peak to average power ratio), BER (bit error rate), SER (symbol error rate), and the ISI (inter-symbol interference).

The paper is organized as follows. Section 2 of the paper presents a brief introduction to the quaternion algebra. Sections 3 and 4 present quaternion Fourier transforms.

2 Quaternions

The space of quaternions denoted by $\mathbb{H}(\mathbf{R})$ were first invented by W.R. Hamilton in 1843 as an extension of the complex numbers into four dimensions [7]. General information on quaternions may be obtained from [8].

Definition 1. Numbers of the form ${}^4\mathbf{q} = a\mathbf{1} + b\mathbf{i} + c\mathbf{j} + d\mathbf{k}$, where $a, b, c, d \in \mathbf{R}$ are called quaternions, where (1) $\mathbf{1}$ is the real unit; (2) $\mathbf{i}, \mathbf{j}, \mathbf{k}$ are three imaginary units. We speck that quaternions ${}^4\mathbf{q} = a + b\mathbf{i} + c\mathbf{j} + d\mathbf{k}$ are written in the standard format.

The addition and subtraction of two quaternions ${}^4\mathbf{q}_1 = a_1 + x_1\mathbf{i} + y_1\mathbf{j} + z_1\mathbf{k}$ and non–zero element ${}^4\mathbf{q}_2 = a_2 + x_2\mathbf{i} + y_2\mathbf{j} + z_2\mathbf{k}$ are given by

$$^4\mathbf{q}_1 \pm {}^4\mathbf{q}_2 = (a_1 \pm a_2) + (b_1 \pm b_2)\mathbf{i} + (b_1 \pm b_2)\mathbf{j} + (b_1 \pm b_2)k.$$

The product of quaternions for the standard format Hamilton defined according as:

$$^4\mathbf{q}_1 \circ {}^4\mathbf{q}_2 = (a_1 + b_1\mathbf{i} + c_1\mathbf{j} + d_1\mathbf{k}) \circ (a_2 + b_2\mathbf{i} + c_2\mathbf{j} + d_2\mathbf{k}) = (a_1a_2 - b_1b_2 - c_1c_2 - d_1d_2) +$$
$$+ (a_1b_2 + b_1a_2 + c_1d_2 - d_1c_2)\mathbf{i} + (a_1c_2 + c_1a_2 + d_1b_2 - b_1d_2)\mathbf{j} + (a_1d_2 + d_1a_2 + b_1c_2 - c_1b_2)\mathbf{k},$$

where $\mathbf{i}^2 = \mathbf{j}^2 = \mathbf{k}^2 = -1$; $\mathbf{i} \circ \mathbf{j} = -\mathbf{i} \circ \mathbf{j} = \mathbf{k}$, $\mathbf{i} \circ \mathbf{k} = -\mathbf{k} \circ \mathbf{i} = \mathbf{j}$, $\mathbf{j} \circ \mathbf{k} = -\mathbf{k} \circ \mathbf{j} = \mathbf{i}$. The set of quaternions with operations of multiplication and addition forms 4-D algebra $\mathbb{H}(\mathbf{R}) = \mathbb{H}(\mathbf{R}|1, \mathbf{i}, \mathbf{j}, \mathbf{k}) := \mathbf{R} + \mathbf{R}\mathbf{i} + \mathbf{R}\mathbf{j} + \mathbf{R}\mathbf{k}$ over the real field $\mathbf{R}$.

Number component a and direction component $^3\mathbf{r} = b\mathbf{i} + c\mathbf{j} + d\mathbf{k} \in \mathbf{R}^3$ were called the *scalar* and 3-D *vector* parts of quaternion, respectively. Now these components are denoted as $S(^4\mathbf{q}) = a \in \mathbf{R}$ and $V(\mathbf{q}) = {}^3\mathbf{r} = b\mathbf{i} + c\mathbf{j} + d\mathbf{k}$. A non–zero element $^3\mathbf{r} = bi + cj + dk$ is called pure vector quaternion. Hence, according to W. Hamilton every quaternion is the sum of a scalar number and a pure vector quaternion $^4\mathbf{q} = a + (b\mathbf{i} + c\mathbf{j} + d\mathbf{k}) = a + {}^3\mathbf{r} = S(\mathbf{q}) + V(\mathbf{q})$, where $a = S(^4\mathbf{q})$, $V(^4\mathbf{q}) = {}^3\mathbf{r}$. Since $\mathbf{i} \circ \mathbf{j} = \mathbf{k}$, then a quaternion $^4\mathbf{q} = a + b\mathbf{i} + c\mathbf{j} + d\mathbf{k} = (a1 + b\mathbf{i}) + (c\mathbf{j} + d\mathbf{i} \circ \mathbf{j}) = (a + b\mathbf{i}) + (c + d\mathbf{i}) \circ \mathbf{j} = \mathbf{z} + \mathbf{w} \circ \mathbf{j}$ is the sum of two complex numbers $\mathbf{z} = a + b\mathbf{i}$, $\mathbf{w} = c + d\mathbf{i}$ with a new imaginary unit $\mathbf{j}$. So, every quaternion can be represented in several ways:

(1) as a 4-D hypercomplex number $^4\mathbf{q} = a + b\mathbf{i} + c\mathbf{j} + d\mathbf{k} = (a, b, c, d)$, $a, b, c, d \in \mathbf{R}$ (standard 4-D format);
(2) as a sum of a scalar and vector parts $\mathbf{q} = a + {}^3\mathbf{r} = (a, {}^3\mathbf{r})$ (1, 3-D hypercomplex format);
(3) as a 2-D hypercomplex numbers $^{2,2}\mathbf{q} = \mathbf{z} + \mathbf{w} \circ \mathbf{j} = (\mathbf{z}, \mathbf{w})$, $\mathbf{z}, \mathbf{w} \in \mathbf{C}$ (2, 2-D complex format).

The product of quaternions for the last two forms Hamilton defined as:

$$^4\mathbf{q}_1 \circ {}^4\mathbf{q}_2 = (\mathbf{z}_1 + \mathbf{w}_1 \circ \mathbf{j}) \circ (\mathbf{z}_2 + \mathbf{w}_2 \circ \mathbf{j}) = (\mathbf{z}_1\mathbf{z}_2 - \mathbf{w}_1\bar{\mathbf{w}}_2) + (\mathbf{w}_1\bar{\mathbf{z}}_2 + \mathbf{z}_1\mathbf{w}_2) \circ \mathbf{j},$$
$$^4\mathbf{q}_1 \circ {}^4\mathbf{q}_2 = (a_1 + \vec{\mathbf{r}}_1) \circ (a_2 + \vec{\mathbf{r}}_2) = (a_1a_2 - (\vec{\mathbf{r}}_1, \vec{\mathbf{r}}_2)) + (a_1\vec{\mathbf{r}}_2 + a_2\vec{\mathbf{r}}_1 + [\vec{\mathbf{r}}_1 \times \vec{\mathbf{r}}_2]),$$

where $S(^4\mathbf{q}_1 \circ {}^4\mathbf{q}_2) = a_1a_2 - \langle {}^3\mathbf{r}_1 | {}^3\mathbf{r}_2 \rangle$, $V(^4\mathbf{q}_1 \circ {}^4\mathbf{q}_2) = a_1{}^3\mathbf{r}_2 + a_2{}^3\mathbf{r}_1 + {}^3\mathbf{r}_1 \times {}^3\mathbf{r}_2$. Here $\langle {}^3\mathbf{r}_1 | {}^3\mathbf{r}_2 \rangle = b_1b_2 + c_1c_2 + d_1d_2$ and $^3\mathbf{r}_1 \times {}^3\mathbf{r}_2 = \mathbf{i}(c_1d_2 - d_1c_2) - \mathbf{j}(b_1d_2 - d_1b_2) + \mathbf{k}(b_1c_2 - c_1b_2)$ are scalar and vector products, respectively.

Definition 2. Let $^4\mathbf{q} = a + b\mathbf{i} + c\mathbf{j} + d\mathbf{k} \in \mathbb{H}(\mathbf{R})$ be a quaternion $(a, b, c, d \in \mathbf{R})$. Then $^4\bar{\mathbf{q}} = \overline{a + b\mathbf{i} + c\mathbf{j} + d\mathbf{k}} = a - b\mathbf{i} - c\mathbf{j} - d\mathbf{k}$, $^4\bar{\mathbf{q}} = \overline{a + {}^3\mathbf{r}} = a - {}^3\mathbf{r}$ is the conjugate of $^4\mathbf{q}$, $N(^4\mathbf{q}) = ||^4\mathbf{q}|| = \sqrt{a^2 + b^2 + c^2 + d^2} = \sqrt{^4\bar{\mathbf{q}} \circ {}^4\mathbf{q}} = \sqrt{^4\mathbf{q} \circ {}^4\bar{\mathbf{q}}}$ is the norm of $^4\mathbf{q}$, and $tr(^4\mathbf{q}) = 2a = {}^4\mathbf{q} + {}^4\bar{\mathbf{q}}$ is the trace of $^4\mathbf{q}$. Therefore $^4\mathbf{q}^2 - tr(^4\mathbf{q})^4\mathbf{q} + N^2(^4\mathbf{q}) = 0$.

Proposition 1. We have $\overline{^4\mathbf{q}_1 \circ {}^4\mathbf{q}_2} = {}^4\bar{\mathbf{q}}_2 \circ {}^4\bar{\mathbf{q}}_1$ and $N(^4\mathbf{q}_1 \circ {}^4\mathbf{q}_2) = N(^4\mathbf{q}_1) \cdot N(^4\mathbf{q}_2)$ f or every $^4\mathbf{q}_1, {}^4\mathbf{q}_2 \in \mathbb{H}(\mathbf{R})$. Note that $||1|| = 1$, $||\mathbf{i}|| = ||\mathbf{j}|| = ||\mathbf{k}|| = 1$.

Definition 3. Quaternions $\{{}^4\boldsymbol{\rho}|N({}^4\boldsymbol{\rho}) = 1\}$ of unit norm are called unit quaternions.

The unit quaternions $\boldsymbol{\rho}$ form a 3D hypersphere $\mathbb{S}^3 \subset \mathbb{H}(\mathbf{R}) \sim \mathbf{R}^4$. For each quaternion ${}^4\mathbf{q}$ with nonzero norm the following quaternion

$$
\begin{aligned}
{}^4\boldsymbol{\rho} &= \frac{{}^4\mathbf{q}}{\|{}^4\mathbf{q}\|} = \frac{a + {}^3\mathbf{r}}{\|{}^4\mathbf{q}\|} = \frac{a}{\|{}^4\mathbf{q}\|} + \frac{{}^3\mathbf{r}}{\|{}^4\mathbf{q}\|} = \frac{a}{\|{}^4\mathbf{q}\|} + \frac{\|{}^3\mathbf{r}\|}{\|{}^4\mathbf{q}\|}\frac{{}^3\mathbf{r}}{\|{}^3\mathbf{r}\|} = \frac{a}{\|{}^4\mathbf{q}\|} + \frac{\|{}^3\mathbf{r}\|}{\|{}^4\mathbf{q}\|}{}^3\boldsymbol{\mu} \\
&= \cos\,\alpha + {}^3\boldsymbol{\mu}\sin\,\alpha = \cos\,\alpha + (\mu_1\mathbf{i} + \mu_2\mathbf{j} + \mu_3\mathbf{k})\sin\,\alpha
\end{aligned}
$$

is an unit quaternion, where $\|{}^3\mathbf{r}\| = \sqrt{b^2 + c^2 + d^2}$, ${}^3\boldsymbol{\mu} = {}^3\mathbf{r}/\|{}^3\mathbf{r}\|$, $\cos\,\alpha = a/\|{}^4\mathbf{q}\|$, $\sin\,\alpha = \|{}^3\mathbf{r}\|/\|{}^4\mathbf{q}\|$, $\mu_1 = b/\|{}^3\mathbf{r}\|$, $\mu_2 = c/\|{}^3\mathbf{r}\|$, $\mu_3 = d/\|{}^3\mathbf{r}\|$ and $\boldsymbol{\mu} = \mu_1\mathbf{i} + \mu_2\mathbf{j} + \mu_3\mathbf{k}$. Obviously,

$$
{}^4\mathbf{q} = \|{}^4\mathbf{q}\| \cdot \left[\cos\,\alpha + {}^3\boldsymbol{\mu}(\gamma, \theta)\sin\,\alpha\right] = \|{}^4\mathbf{q}\| \cdot [\cos\,\alpha + (\mu_1\mathbf{i} + \mu_2\mathbf{j} + \mu_3\mathbf{k})\sin\,\alpha].
$$

where ${}^3\boldsymbol{\mu}(\gamma, \theta = \mathbf{i}\cos\,\gamma + \mathbf{j}\sin\,\gamma\cos\,\theta + \mathbf{k}\sin\,\gamma\,\sin\,\theta \in \mathbf{S}^2$, $\theta, \varphi \in [0, \pi]$, $\alpha \in [0, 2\pi)$ are the polar coordinates on $\mathbb{S}^3$. Obviously,

$$
\begin{aligned}
&a = \|{}^4\mathbf{q}\|\cos\,\alpha,\; b = \left(\|{}^4\mathbf{q}\|\cos\,\gamma\right)\sin\,\alpha,\; c = \left(\|{}^4\mathbf{q}\|\sin\,\gamma\,\cos\,\theta\right)\sin\,\alpha, \\
&d = \left(\|{}^4\mathbf{q}\|\sin\,\gamma\,\sin\,\theta\right)\sin\,\alpha.
\end{aligned}
$$

In particular, for ${}^4\mathbf{q}_1 \equiv {}^3\mathbf{r}_1 = b_1\mathbf{i} + c_1\mathbf{j} + d_1\mathbf{k}$ and ${}^4\mathbf{q}_2 \equiv {}^3\mathbf{r}_2 = b_2\mathbf{i} + c_2\mathbf{j} + d_2\mathbf{k}_2$ we obtain

$$
{}^3\mathbf{r}_1 \circ {}^3\mathbf{r}_2 = -\langle{}^3\mathbf{r}_1)|{}^3\mathbf{r}_2\rangle + \left[{}^3\mathbf{r}_1 \times {}^3\mathbf{r}_2)\right],\; {}^3\mathbf{r}^2 = {}^3\mathbf{r} \circ {}^3\mathbf{r} = -\langle{}^3\mathbf{r}|{}^3\mathbf{r}\rangle = -\|{}^3\mathbf{r}\|^2,
$$

and for a pure quaternion ${}^3\boldsymbol{\mu} \in \mathbb{S}^2 \subset \mathbf{R}^3$ with unity norm $\|{}^3\boldsymbol{\mu}\| = 1$ we have ${}^3\boldsymbol{\mu}^2 = -\|{}^3\boldsymbol{\mu}\|^2 = -1$, where $\mathbb{S}^2$ denotes the unit 2-D sphere in 3-D space $\mathbf{R}^3$. This unit-vector product identity represents the generalization of the complex-variable identity $i^2 = -1$.

This means that, if in the ordinary theory of complex numbers there are only two different square roots of negative unity ($+i$ and $-i$) and they differ only in their signs, then in the quaternion theory there are infinite numbers of different square roots of negative unity

$$
{}^3\boldsymbol{\mu} = {}^3\boldsymbol{\mu}(\gamma, \theta) = \left(\mu_x\mathbf{i} + \mu_y\mathbf{j} + \mu_z\mathbf{k}\right) = (\cos\,\gamma \cdot \mathbf{i} + \sin\,\gamma\cos\,\theta \cdot \mathbf{j} + \sin\,\gamma\,\sin\,\theta \cdot \mathbf{k}) \in \mathbf{S}^2,
$$

which gives ${}^3\boldsymbol{\mu}^2 = {}^3\boldsymbol{\mu}^2(\gamma, \theta) = -1$. Here ${}^3\boldsymbol{\mu}(\gamma, \theta) = (\cos\,\gamma, \sin\,\gamma\,\cos\,\theta, \sin\,\gamma\,\sin\,\theta)$ being still that point on the spherical surface, which has for its rectangular coordinates $\cos\,\gamma, \sin\,\gamma\cos\,\theta, \sin\,\gamma\,\sin\,\theta$ (see Fig. 4). In the feature we will omit left index: $\boldsymbol{\mu}(\gamma, \theta) \equiv {}^3\boldsymbol{\mu}(\gamma, \theta)$.

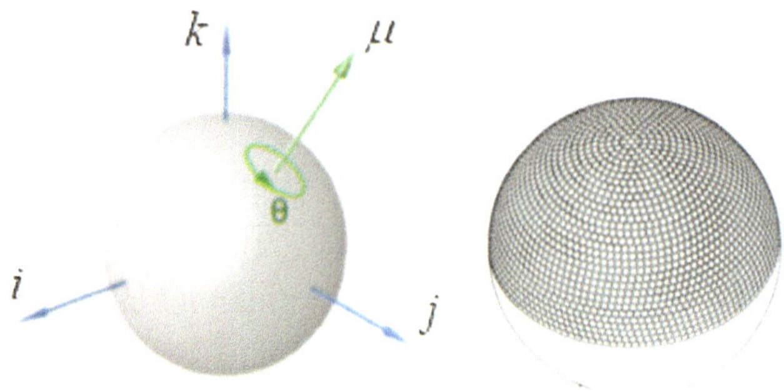

Fig. 4. Each 3D vector $\vec{\mu} \in \mathbb{S}^2$ of unit length can play a role of classical imaginary unit. For example, the special elements ${}^3\mathbf{i}, {}^3\mathbf{j}, {}^3\mathbf{k}$ are such elements.

Definition 4. A functions ${}^4\mathbf{f}(n) : [0, N-1] \to \mathbb{H}(\mathbf{R})$ are called quaternion-valued discrete functions. They have the following form:

$$ {}^4\mathbf{f}(n) = f_0(n) + f_1(n)\mathbf{i} + f_2(n)\mathbf{j} + f_3(n)\mathbf{k} = (f_0(n), f_1(n), f_2(n), f_3(n)). $$

The exponential function is $\exp({}^4\mathbf{q}) = 1 + {}^4\mathbf{q} + \frac{{}^4\mathbf{q}^2}{2!} + \ldots + \frac{{}^4\mathbf{q}^m}{m!} + \ldots = \sum_{m=0}^{\infty} \frac{{}^4\mathbf{q}^m}{m!}$.

Theorem 1 [14, 15]. For ${}^4\mathbf{q} = a + {}^3\mathbf{r} \in \mathbb{H}(\mathbf{R})$ we have

$$ \exp(a + {}^3\mathbf{r}) = e^a \exp({}^3\mathbf{r}) = e^a \left(\cos(\|{}^3\mathbf{r}\|) + \frac{{}^3\mathbf{r}}{\|{}^3\mathbf{r}\|} \sin(\|{}^3\mathbf{r}\|) \right). $$

Obviously, $\|\exp({}^3\mathbf{r})\| = 1$ and $\|\exp({}^3\mathbf{r})\| = \|\exp(a + {}^3\mathbf{r})\| = e^a$. In general case $\exp({}^4\mathbf{q}_1) \circ \exp({}^4\mathbf{q}_2) \neq \exp({}^4\mathbf{q}_2) \circ \exp({}^4\mathbf{q}_1)$ and $\exp({}^4\mathbf{q}_1 + {}^4\mathbf{q}_2) \neq \exp({}^4\mathbf{q}_1) \circ \exp({}^4\mathbf{q}_2) \neq \exp({}^4\mathbf{q}_2) \circ \exp({}^4\mathbf{q}_1)$.

3 Quaternion Fourier Transforms

3.1 Historical Remarks

Before defining the quaternion Fourier transform, we briefly outline its relationship with Clifford Fourier transformations. Quaternions and Clifford hypercomplex number were first simultaneously and independently applied to quaternion-valued Fourier and Clifford-valued Fourier transforms by Labunets [9] and Sommen [10, 11], respectively, at the 1981. The Labunets quaternion transforms were over quaternion with real and Galois coefficients (i.e., over $\mathbb{H}[\mathbf{R}]$ and $\mathbb{H}[\mathbf{GF}(p)]$). They generalize both classical and co-called number theoretical transforms (NNTs) and proposed for application to fast signal processing. Ernst [12] and Delsuc [13] in the late 1981s, seemingly without knowledge of the earlier works of Labunets and Sommen proposed bicomplex Fourier transforms over 4D commutative hypercomplex algebra of bicomplex numbers ($\mathbf{C} \oplus \mathbf{C}$). Note that the bicomplex algebra is quite different from the quaternion algebra; among general things, bicomplex multiplication is commutative, but quaternion one is noncommutative. For this reason, the Ernst and Delsuc transforms are direct sum of ordinary Fourier transforms (*i.e.*, duplex Fourier transform). They are a little bit similar in kind to quaternion Fourier transforms. Ernst and Delsuc's transforms were two-dimensional and proposed for application to nuclear magnetic resonance (NMR) imaging.

Two new ideas emerged in 1998–1999 in a paper by Labunets [14] and Sangwine [15]. These were, firstly, the choice of a general root ${}^3\boldsymbol{\mu}$ of -1 (a unit quaternion with zero scalar part) rather than a basis unit ($\mathbf{i}, \mathbf{j}$ or $\mathbf{k}$) of the quaternion algebra, and secondly, the choice of a general roots ${}^3\boldsymbol{\mu}_0 = \boldsymbol{\mu}_0(\gamma_0, \theta_0), {}^3\boldsymbol{\mu}_1 = \boldsymbol{\mu}_1(\gamma_1, \theta_1), \ldots, {}^3\boldsymbol{\mu}_{N-1} = \boldsymbol{\mu}_{N-1}(\gamma_{N-1}, \theta_{N-1})$ of -1 (see cloud of imaginary units on Fig. 1) in Clifford algebra to create multi-parameter and fractional Fourier-Clifford transforms (with eigenvalues $e^{-\boldsymbol{\mu}_0(\gamma_0,\theta_0)}, e^{-\boldsymbol{\mu}_1(\gamma_1,\theta_1)}, \ldots\ldots, e^{-{}^3\boldsymbol{\mu}_{N-1}(\varphi_{N-1},\theta_{N-1})}, \ldots$).

Labunets, Rundblad-Ostheimer and Astola [16–18] used the classical and number theoretical quaternion Fourier and Fourier-Clifford transforms for fast invariant recognition of 2D, 3D and nD color and hyperspectral images, defined on Euclidean and non-Euclidean spaces. These publications give useful interpretation of quaternion and Cliffordean Fourier coefficients: they are relative quaternion- or Clifford-valued invariants of hyperspectral images with respect to Euclidean and non-Euclidean rotations and motions of physical and hyperspectral spaces. It removes the veil of mysticism and mystery from quaternion- and Clifford-valued Fourier coefficients. In the works of scientists Brackx, De Schepper, Sommen, and De Bie [19–22] mathematical theory of Fourier-Clifford transforms accepted the final completeness, beauty and elegance.

3.2 Quaternion Fourier Transforms

According to Theorem 1, for non-zero $\alpha \in \mathbf{R}$ and a non-zero ${}^4\mathbf{q} = a + {}^3\boldsymbol{\mu}$

$$\exp({}^4\mathbf{q}\alpha) = \exp((a + {}^3\boldsymbol{\mu})\alpha) = e^{a\alpha}\left(\cos\left(||{}^3\boldsymbol{\mu}||\alpha\right) + \frac{{}^3\boldsymbol{\mu}}{||{}^3\boldsymbol{\mu}||}\sin\left(||{}^3\boldsymbol{\mu}||\alpha\right)\right).$$

In particular case, for ${}^4\mathbf{q} \equiv {}^3\boldsymbol{\mu} = \boldsymbol{\mu}(\gamma, \theta)$ we have $e^{\boldsymbol{\mu}(\gamma,\theta)\alpha} = \cos(\alpha) + \boldsymbol{\mu}(\gamma, \theta)\sin(\alpha)$. For $\alpha = \alpha_k = 2\pi k/N\ (k = 0, 1, \ldots, N-1)$ we obtain quaternion-valued discrete harmonics

$$e^{\boldsymbol{\mu}(\gamma_k,\theta_k)\frac{2\pi}{N}kn} = \boldsymbol{\varepsilon}_k^{kn} = \cos\left(\frac{2\pi}{N}kn\right) + \boldsymbol{\mu}(\gamma_k, \theta_k)\sin\left(\frac{2\pi}{N}kn\right),$$

where each quaternion harmonic $\boldsymbol{\varepsilon}_k^{-kn} = \exp(-2\pi\boldsymbol{\mu}(\gamma_k, \theta_k)kn/N)$ has its own imaginary unit $\boldsymbol{\mu}_k := \boldsymbol{\mu}(\gamma_k, \theta_k) = (\cos\gamma_k \cdot \mathbf{i} + \sin\gamma_k \cos\theta_k \cdot \mathbf{j} + \sin\gamma_k \sin\theta_k \cdot \mathbf{k}) \in \mathbf{S}^2$, $k = 0, 1, \ldots, N-1$. Due to the non-commutative property of quaternion multiplication, there are two different types of quaternion Fourier transforms (QFTs).

These QFTs are the left- and right-sided QFTs (LS-QFT and RS-QFT), respectively.

Definition 4. The direct discrete quaternion Fourier transforms of $\mathbf{f}(n) : [0, N-1] \to \mathbb{H}(\mathbf{R})$ are defined as

$$
\begin{aligned}
{}^4\mathbf{QF}(k|\gamma_k, \theta_k) = \mathcal{QF}^{(\gamma,\theta)}\{{}^4\mathbf{f}(n)\} &= \frac{1}{\sqrt{N}} \sum_{n=0}^{N-1} e^{-\boldsymbol{\mu}(\gamma_k,\theta_k)\frac{2\pi}{N}kn} \circ {}^4\mathbf{f}(n) \\
&= \frac{1}{\sqrt{N}} \sum_{n=0}^{N-1} \boldsymbol{\varepsilon}_k^{-kn} \circ {}^4\mathbf{f}(n),
\end{aligned}
\tag{1}
$$

$$
\begin{aligned}
{}^4\mathbf{FQ}(k|\varphi_k, \theta_k) = \mathcal{FQ}^{(\varphi,\theta)}\{{}^4\mathbf{f}(n)\} &= \frac{1}{\sqrt{N}} \sum_{n=0}^{N-1} {}^4\mathbf{f}(n) \circ e^{-\boldsymbol{\mu}(\gamma_k,\theta_k)\frac{2\pi}{N}kn} \\
&= \frac{1}{\sqrt{N}} \sum_{n=0}^{N-1} {}^4\mathbf{f}(n) \circ \boldsymbol{\varepsilon}_k^{-kn},
\end{aligned}
\tag{2}
$$

where $\mathcal{QF}, \mathcal{FQ}$ are LS-QFT and RS-QFT, $\boldsymbol{\gamma} = (\gamma_0, \gamma_1, \ldots, \gamma_{N-1})$, $\boldsymbol{\theta} = (\theta_0, \theta_1, \ldots, \theta_{N-1})$.

Definition 5. The inverse quaternion Fourier transforms are defined as

$$
{}^4\mathbf{f}(n) = 2\frac{1}{\sqrt{N}} \sum_{k=0}^{N-1} \frac{\boldsymbol{\mu}(\gamma_k, \theta_k)}{[\boldsymbol{\mu}(\gamma_{-k}, \theta_{-k}) + \boldsymbol{\mu}(\gamma_k, \theta_k)]_L} \circ \boldsymbol{\varepsilon}_k^{kn} \circ {}^4\mathbf{QF}(k|\gamma_k, \theta_k),
\tag{3}
$$

$$
{}^4\mathbf{f}(n) = 2\frac{1}{\sqrt{N}} \sum_{k=0}^{N-1} {}^4\mathbf{QF}(k|\gamma_k, \theta_k) \circ \boldsymbol{\varepsilon}_k^{kn} \circ \frac{\boldsymbol{\mu}(\gamma_k, \theta_k)}{[\boldsymbol{\mu}(\gamma_{-k}, \theta_{-k}) + \boldsymbol{\mu}(\gamma_k, \theta_k)]_R}.
\tag{4}
$$

We see, that $\mathcal{QF} = \mathcal{QF}^{(\gamma,\theta)}$ and $\mathcal{FQ} = \mathcal{FQ}^{(\gamma,\theta)}$ depend on $2N$ parameters (γ_k, θ_k), $k \in \{0, 1, \ldots, N-1\}$.

4 Classical Fractional and Many-Parameter Fourier Transforms

With most of the data transmission systems nowadays use orthogonal frequency division multiplexing telecommunication system (OFDM-TCS) based on the discrete Fourier transform. It is a unitary operator $\mathcal{F}_N = \left[e^{j2\pi kn/N}\right]_{k,n=1}^{N-1}$: $F(k) = (\mathcal{F}_N f)(k)$. Relevant properties are that the square $(\mathcal{F}^2 f)(n) = f(\underset{N}{\ominus} n)$ is the inversion operator modulo N, and that its fourth power $(\mathcal{F}^4 f)(n) = f(n)$, is the identity; hence $\mathcal{F}^3 = \mathcal{F}^{-1}$. The operator $\mathcal{F}$ thus generates the cyclic Fourier group of order 4: $\mathbf{Gr}_4(\mathcal{F}) = \{\mathcal{F}^a\}_{a\in\{0,1,2,3\}} = \{I, \mathcal{F}^1, \mathcal{F}^2, \mathcal{F}^3\}$. The idea of fractional powers of the

Fourier operator $\mathcal{F}$ appears in the mathematical literature [23–31]. This idea is to consider the eigenvalue decomposition of the Fourier $(N \times N)$–matrix

$$\mathcal{F}_N = \left[e^{j2\pi kn/N}\right] = \left[\sum_{s=0}^{N-1} \lambda_s |h_s(k)\rangle\langle h_s(n)|\right] = \left[\sum_{s=0}^{N-1} e^{2\pi js/4} |h_s(k)\rangle\langle h_s(n)|\right] \\ = \mathbf{U}\{diag(\lambda_0, \ldots, \lambda_{N-1})\}\mathbf{U}^{-1} = \mathbf{U}\Lambda\mathbf{U}^{-1}, \quad (5)$$

where $\lambda_s = e^{j2\pi s/4} = j^s$ and $h_l(n)$ are eigenvalues and eigenfunctions in the form of the discrete Hermite functions, $|h_s(k)\rangle$ and $\langle h_s(n)|$ are vector-column and vector-row, respectively, $\mathbf{U} = [|h_0(k)\rangle, |h_1(k)\rangle, \ldots, |h_{N-1}(k)\rangle]$ is the matrix of eigenvectors.

The continuously family of FrFT $\{\mathcal{F}^a\}_{a\in[0,4)}$ is constructed by replacing the s-th eigen-value $\lambda_s = e^{j2\pi s/4}$ by its a-th power $\lambda_s^a = e^{jsa2\pi/4}$, for a between 0 and 4, or $\lambda_s^\alpha = e^{jsa\pi/2} = e^{js\alpha}$ for α between 0 and 2π, where $\alpha = a\pi/2$.

The eigenvalues of the standard DFT matrix $\mathcal{F}_N$ are the fourth roots of unity, to be denoted by $\lambda_s \in \{e^{j2\pi s/4}\}_{s=0}^{3} \in \{\pm 1, \pm j\}$. This divides the space of N-point complex signals into four Fourier invariant subspaces whose dimensions N_s are the multiplicities of the eigenvalues λ_s, which have a modulo 4 recurrence in the dimension $N = 2^N = 4M$ given by $N_0 = M+1, N_1 = M-1,\ N_2 = M,\ N_3 = M$. Let a function $s(n)$: $\{0, 1, 2, \ldots, N-1\} \to \{0, 1, 2, 3\}$ describes of the distribution of eigenvalues along main diagonal $\mathbf{Diag}(e^{j\pi s(n)a/2}) = \mathbf{Diag}(e^{js(n)\alpha})$. This function takes $M+1$ times value 0, $M-1$ times value 1, and M times values 2 and 3.

Definition 6 [23–31]. The discrete classical fractional Fourier transform are defined as

$$\mathcal{F}^\alpha = \left[e_k^{(\alpha)}(n)\right] := \mathbf{U}\left\{\mathbf{Diag}\left(e^{js(m)\alpha}\right)\right\}\mathbf{U}^{-1} = \sum_{m=0}^{N-1} e^{js(m)\alpha} |h_m(k)\rangle\langle h_m(n)|, \quad (6)$$

and if $s(m) = m$ then we obtain the Bargmann fractional Fourier transform [26]

$$\mathcal{BF}^\alpha = \left[be_k^{(\alpha)}(n)\right] := \mathbf{U}\{\mathbf{Diag}(e^{jm\alpha})\}\mathbf{U}^{-1} = \sum_{m=0}^{N-1} e^{jm\alpha} |h_m(k)\rangle\langle h_m(n)|, \quad (7)$$

The identical and classical Fourier transformations are both the special cases of the FrFTs. They correspond to $\alpha = 0$ ($\mathcal{F}^0 = I$) and $\alpha = \pi/2$ ($\mathcal{F}^{\pi/2} = \mathcal{F}$), respectively.

Definition 7. The discrete classical-like and Bargmann-like $(s(m) = m)$ many-parameter DFT we define by the following way

$$\mathcal{F}^{(\boldsymbol{\alpha})} = \mathcal{F}^{(\alpha_0, \alpha_1, \ldots, \alpha_{N-1})} = \mathbf{U}\left\{diag\left(e^{js(m)\alpha_m}\right)\right\}\mathbf{U}^{-1} = \sum_{m=0}^{N-1} e^{js(m)\alpha_m} |h_m(k)\rangle\langle h_m(n)|, \quad (8)$$

where $\boldsymbol{\alpha} = (\alpha_0, \alpha_1, \ldots, \alpha_{N-1})$.

5 Fractional and Many-Parameter Quaternion Fourier Transforms

If quaternion harmonics have equivalent imaginary units ${}^3\boldsymbol{\mu}(\gamma_k,\theta_k)\equiv{}^3\boldsymbol{\mu}(\gamma,\theta)$, $\forall k=0,1,\ldots,N-1$), then quaternion Fourier matrices $\mathcal{QF},\mathcal{FQ}$ contains commutative entries $e^{-{}^3\boldsymbol{\mu}(\gamma_k,\theta_k)\frac{2\pi}{N}kn}$. For this reason $\mathcal{QF}$ and $\mathcal{FQ}$ have the same real-valued eigenfunction as ordinary DFT but with quaternion-valued eigenvalues $\left\{e^{2\pi\cdot{}^3\boldsymbol{\mu}(\gamma,\theta)s/4}\right\}_{s=0}^{3}=\left\{{}^3\boldsymbol{\mu}^s(\gamma,\theta)\right\}_{s=0}^{3}$. Therefore,

$$\left[e^{-\boldsymbol{\mu}(\gamma,\theta)\frac{2\pi}{N}kn}\right]=\left[\sum_{s=0}^{N-1}e^{2\pi\cdot{}^3\boldsymbol{\mu}(\gamma,\theta)s/4}|h_s(k)\rangle\langle h_s(n)|\right]=\left[\sum_{s=0}^{N-1}{}^3\boldsymbol{\mu}^s(\gamma,\theta)|h_s(k)\rangle\langle h_s(n)|\right]$$

where $\{h_s(n)\}_{s=0}^{N-1}$ is the set of discrete real-valued Hermite functions. Hence, we can define fractional and many parameter quaternion Fourier transforms.

Definition 8. For single parameter $\alpha\in\mathbf{Tor}_{2\pi}^1$ we introduce fractional quaternion Fourier transforms (FrQFT) of classical and Bargmann ($s(m)=m$) structures as

$$\mathcal{F}_Q^{\alpha}(\boldsymbol{\gamma},\boldsymbol{\theta})=\left[\sum_{m=0}^{N-1}e^{{}^3\boldsymbol{\mu}(\gamma_m,\theta_m)\cdot s(m)\cdot\alpha}|h_m(k)\rangle\langle h_m(n)|\right]=\mathbf{U}\cdot\mathbf{Diag}\left(e^{{}^3\boldsymbol{\mu}(\gamma_m,\theta_m)\cdot s(m)\cdot\alpha}\right)\cdot\mathbf{U}^{-1}. \tag{9}$$

Definition 9. For N-parameter $(\alpha_0,\ldots,\alpha_{N-1})\in\mathbf{Tor}_{2\pi}^N$ we introduce many-parameter quaternion Fourier transforms (MPQFT) of classical and Bargmann structures as

$$\mathcal{F}_Q^{\boldsymbol{\alpha}}(\boldsymbol{\gamma},\boldsymbol{\theta})=\left[\sum_{m=0}^{N-1}e^{{}^3\boldsymbol{\mu}(\gamma_m,\theta_m)\cdot s(m)\cdot\alpha_m}|h_m(k)\rangle\langle h_m(n)|\right]=\mathbf{U}\cdot\mathbf{Diag}\left(e^{{}^3\boldsymbol{\mu}(\gamma_m,\theta_m)\cdot s(m)\cdot\alpha_m}\right)\cdot\mathbf{U}^{-1}, \tag{10}$$

where $\boldsymbol{\gamma}=(\beta_0,\beta_1,\ldots,\beta_{N-1})$, $\boldsymbol{\theta}=(\theta_0,\theta_1,\ldots,\theta_{N-1})$, $\boldsymbol{\alpha}=(\alpha_0,\alpha_1,\ldots,\alpha_{N-1})$. Obviously, they are $3N$-parameter transforms.

Due to the non-commutative property of quaternion multiplication, there are left- and right-sided transforms (LS-FrQFTs, LS-MPQFTs and RS-FrQFTs, RS-MPQFTs).

Definition 10. The direct discrete LS-FrQFTs, LS-MPQFTs and RS-FrQFTs, RS-MPQFTs of $\mathbf{f}(n):[0,N-1]\to\mathbb{H}(\mathbf{R})$ are defined as

$$\begin{aligned}|\mathbf{QF}^{\alpha}(k|\boldsymbol{\gamma},\boldsymbol{\theta})\rangle&=\mathcal{QF}^{\alpha}(\boldsymbol{\gamma},\boldsymbol{\theta})|\mathbf{f}(n)\rangle=\sum_{m=0}^{N-1}\left(e^{{}^3\boldsymbol{\mu}_m(\gamma_m,\theta_m)\cdot s(m)\cdot\alpha}\circ|h_m(k)\rangle\right)\langle h_m(n)|\mathbf{f}(n)\rangle,\\|\mathbf{FQ}^{\alpha}(k|\boldsymbol{\gamma},\boldsymbol{\theta})\rangle&=\mathcal{FQ}^{\alpha}(\boldsymbol{\gamma},\boldsymbol{\theta})|\mathbf{f}(n)\rangle=\sum_{m=0}^{N-1}\left(\langle h_m(n)|\mathbf{f}(n)\rangle\circ e^{{}^3\boldsymbol{\mu}_m(\gamma_m,\theta_m)\cdot s(m)\cdot\alpha}\right)|h_m(k)\rangle,\end{aligned} \tag{11}$$

$$\begin{aligned}
\left|\mathbf{QF}^{\boldsymbol{\alpha}}(k|\boldsymbol{\gamma},\boldsymbol{\theta})\right\rangle &= \mathcal{QF}^{\boldsymbol{\alpha}}(\boldsymbol{\gamma},\boldsymbol{\theta})|\mathbf{f}(n)\rangle = \sum_{m=0}^{N-1} e^{3\boldsymbol{\mu}_m(\gamma_m,\theta_m)\cdot s(m)\cdot\alpha_m} \circ |h_m(k)\rangle\langle h_m(n)|\mathbf{f}(n)\rangle, \\
\left|\mathbf{FQ}^{\boldsymbol{\alpha}}(k|\boldsymbol{\gamma},\boldsymbol{\theta})\right\rangle &= \mathcal{FQ}^{\boldsymbol{\alpha}}(\boldsymbol{\gamma},\boldsymbol{\theta})|\mathbf{f}(n)\rangle = \sum_{m=0}^{N-1} \langle h_m(n)|\mathbf{f}(n)\rangle \circ e^{3\boldsymbol{\mu}_m(\gamma_m,\theta_m)\cdot s(m)\cdot\alpha_m}|h_m(k)\rangle,
\end{aligned} \tag{12}$$

6 Anti-eavesdropping: Bob and Alice vs. Eve

The system model that is going to be used in this work is known as the wiretap channel model, that was introduced by Schannon [32] and Wyner [33] (see Fig. 5). This model is composed of two legitimate users, named Alice and Bob, while the passive eavesdropper named Eve attempts to eavesdrop the information. A legitimate user (Alice) transmits her confidential messages to a legitimate receiver (Bob), while Eve is trying to eavesdrop Alice's information. We suppose that the eavesdropper knows the frame of OFDM signal of the legitimate Intel-OFDM-TCS (i.e. knows initial values of parameters $\boldsymbol{\theta}^0 = \left(\varphi_0^0, \varphi_1^0, \ldots, \varphi_q^0\right)$ at the time t_0) and has the capability to demodulate OFDM signals. Hence, the legitimate transmitter/receiver (Alice/Bob) and eavesdropper (Eva) use identical parameters of Intel-OFDM-TCS which remain constant over several time slots.

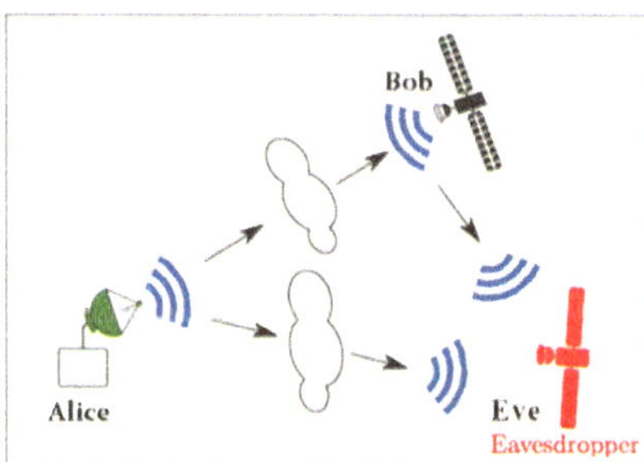

Fig. 5. Eavesdropping attack.

Alice transmits her data using OFDM with N quaternion-valued sub-carriers $\left\{\mathbf{qsubc}_k\left(n|\boldsymbol{\theta}^0\right)\right\}_{k=0}^{N-1}$, i.e., she uses the quaternion transform $\mathcal{QU}_N^0 \equiv \mathcal{QU}^{\mathbf{b}^0}\left(\boldsymbol{\theta}^0\right)$ with fixed keys $\boldsymbol{\theta}^0 = \left(\varphi_0^0, \varphi_1^0, \ldots, \varphi_q^0\right)$, $\mathbf{b}^0 = (b_1, b_2, \ldots, b_N)$. When sub-carriers (i.e. unitary transform $\mathcal{QU}^{\mathbf{b}^0}(\boldsymbol{\theta}^0)$) of Alice and Bob Intelligent-OFDM-TCS are identified by Eva, this TCS can be eavesdropped by means of radio-electronic eavesdropping attack. In this scenario, Bob and Eve will have the same instruments to decode the received message. This means that Eve successful intercepts Alice's message. As an anti-eavesdropping measure Alice and Bob can use the following strategy: they select new sub-carriers in Int-OFDM-TCS by changing parameters of $\mathcal{QU}^{\mathbf{b}^0}(\boldsymbol{\theta}^0)$ in the periodical (or random) manner (a priory known for Alice and Bob).

For comparative analysis we use OFDM-TCS, based on one-parameter classical $\mathcal{QF}^{\alpha}(\boldsymbol{\beta},\boldsymbol{\theta}) = \mathcal{QF}^{\alpha}(\mathbf{i})$ and many-parameter Bargmann $\mathcal{QBF}^{\alpha}(\mathbf{i})$ quaternion Fourier transforms in its one-parameter forms

$$\mathcal{QF}^{\alpha}(\mathbf{i}) = \left[\sum_{m=0}^{N-1} e^{\mathbf{i}\cdot s(m)\cdot\alpha}|h_m(k)\rangle\langle h_m(n)|\right],\ \mathcal{QBF}^{\alpha}(\mathbf{i}) = \left[\sum_{m=0}^{N-1} e^{\mathbf{i}\cdot m\cdot\alpha}|h_m(k)\rangle\langle h_m(n)|\right],$$

where $\{h_s(n)\}_{s=0}^{N-1}$ is the set of discrete real-valued Hermite functions, ${}^3\boldsymbol{\mu}_r^p(\beta_r^p,\theta_r^p) \equiv \mathbf{i}$, $\forall r = 1,2,\ldots,n$; $\forall p = 0,1,\ldots,255$ for both transforms.

Simulations were done in MATLAB R2018b. Intelligent OFDM-TCS's parameters are assumed as follows: 256-QAM modulation, the sizes of $\mathcal{QF}^{\alpha}(\mathbf{i})$ and $\mathcal{QBF}^{\boldsymbol{\alpha}}(\mathbf{i})$ are (256×256) (i.e., the number of quaternion-valued subcarriers is 256), every time-slot

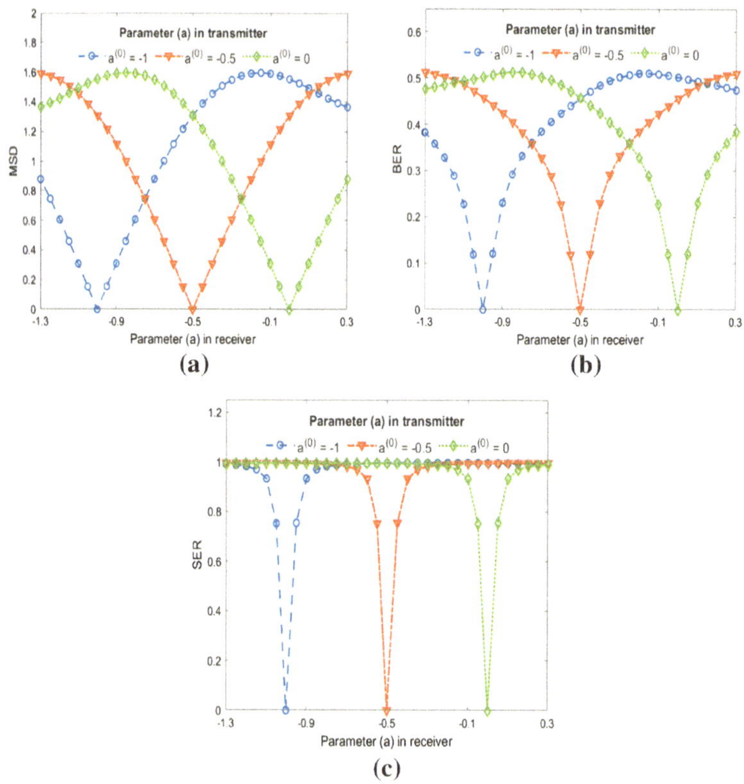

Fig. 6. The average (a) **MSD**, (b) **BER** and (c) **SER** measurements versus a for $\mathcal{QF}^{a}(\mathbf{i})$ with different values of parameter a^0: $a^0 = -1$ (blue curve), $a^0 = -0.5$ (red curve), $a^0 = 0$ (green *curve*). When parameters in transmitter (Alice) and receiver (Eva) are the same ($a = a^0$), we have **MSD** $= 0$, **BER** $= 0$ and **SER** $= 0$.

(OFDM-symbols) is a row from grey-level (256 × 256)-image "Lena", the number of time-slot equal to 256 (*i.e.* equal to the number of "Lena" rows). The length of bit-stream of a single time-slot is equal to 8 × 256 = 2048. Data of 2048 bits are sent in the form of 2568-bit symbols (one symbol is of 8 bits).

Now, we provide some simulation results to substantiate our theoretical claims for $\mathcal{QF}^{a}(\mathbf{i})$ and $\mathcal{QBF}^{a}(\mathbf{i})$ with the following values of parameter $a^{(0)} = \{-1, -0.8, -0.6, -0.4, -0.2, 0\}$. If Eve knows these parameters then she receives the same message as Bob. In order to protect the corporate privacy and the sensitive client information against the threat of electronic eavesdropping Alice and Bob use described above defense mechanism.

It would be interesting to know how MSD, BER and SER are changing with respect to deviation a_1 from initial value a_0. The transmission performances of OFDM system are evaluated by average MSD, BER and SER measurements under 256 time-slot. Figure 6 show the average

$$\mathbf{MSD}(a) = \frac{1}{256}\sum_{l=0}^{255}\mathbf{MSD}[l|a],\ \left\langle \mathbf{BER}^{Bit}_{(A\to B|\xi=0)}[l|a]\right\rangle = \sum_{l=0}^{255}\mathbf{BER}^{Bit}_{(A\to B|\xi=0)}[l|a],$$
$$\left\langle \mathbf{SER}^{Sym}_{(A\to E|\xi=0)}[l|a]\right\rangle = \sum_{l=0}^{255}\mathbf{SER}^{Sym}_{(A\to E|\xi=0)}[l|a]$$

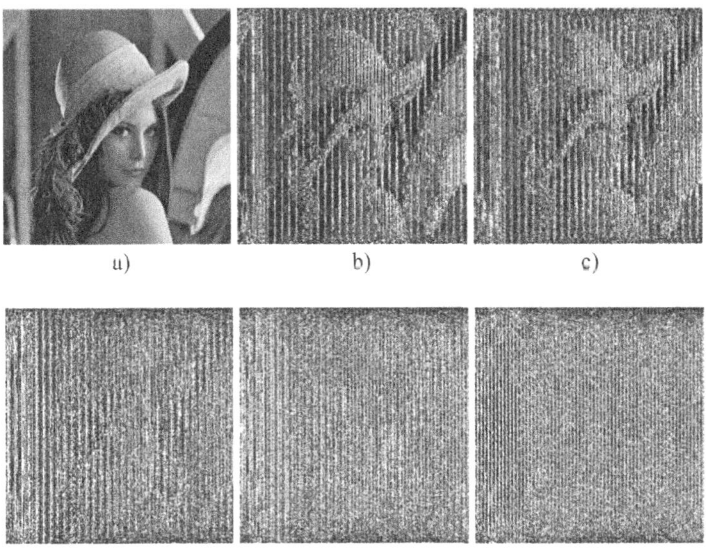

a) b) c) d) e) f)

Fig. 7. Received Eva's messages with different values of parameter a in Alice's OFMD-TCS. Eva continues to work with classical FFT ($a^0 = -1$). Alice uses $\mathcal{QF}^{a}(\mathbf{i})$ with new value of parameter a: (a) $a = -1$, (b) $a = -0.8$, $a = -0.8$, (c) $a = -0.6$, (d) $a = -0.4$, (e) $a = -0.2$, (f) $a = 0$.

measurements versus a_i in noiseless case for $\mathcal{QF}^a(\mathbf{i})$ in the absence of thermal noise ($\xi = 0$) for some types of $\mathcal{QF}^a(\mathbf{i})$ (plotted with different color). When parameters in Alice's and Eva's OFDM-TCS are the same, we have $\mathbf{MSD} = 0$, $\mathbf{BER} = 0$ and $\mathbf{SER} = 0$. This means that Eve successful intercepts Alice's messages.

The changing of parameter a allows to escape eavesdropping. Indeed, all graphics have **V**-like form. It means, that if Alice and Bob change working value of the parameter a ($a^0 \rightarrow a$), but Eve use previous value a^0, then Eve will receive Alice's massage with big mistakes. To illustrate this result, we consider the image "Lena" as Eva's message. Figure 7 shows received Eva's message with different values of a in the Alice's OFDM-TCS, when Eva works with classical DFT.

Similar results we have for OFDM-TCS, based on quaternion Bargmann-Fourier transform $\mathcal{QBF}^a(\mathbf{i})$.

7 Conclusions

In this paper, we have shown a new unified approach to the many-parametric representation of complex and quaternion Fourier transforms. Defined representation of many-parameter quaternion Fourier transforms (MPQFTs) depend on finite set of free parameters, which could be changed independently of one another. For each fixed values of parameter we get the unique orthogonal transform. We develop novel Intelligent OFDM-telecommunication systems based on fractional and multi-parameter Fourier transforms and shown their superiority and practicability from the physical layer security. The new systems use inverse MPQFT (IMPQFT) for modulation at the transmitter and direct MPFQT (DMPFQT) for demodulation at the receiver. Simulation results show that the proposed Intelligent OFDM-TCS have better performance than the conventional OFDM system based on DFT against eavesdropping.

Acknowledgements. The reported study was funded by RFBR, project number 19-29-09022-мк and by the Ural State Forest Engineering's Center of Excellence in «Quantum and Classical Information Technologies for Remote Sensing Systemsю.

References

1. Gupta, M.K., Tiwari, S.: Performance evaluation of conventional and wavelet based OFDM system. Int. J. Electron. Commun. **67**(4), 348–354 (2013)
2. Patchala, S., Sailaja, M.: Analysis of filter bank multi-carrier system for 5G communications. Int. J. Wirel. Microw. Technol. (IJWMT) **9**(4), 39–50 (2019)
3. Kaur, H., Kumar, M., Sharma, A.K., Singh, H.P.: Performance analysis of different Wavelet families over fading environments for mobile WiMax system. Int. J. Future Gener. Commun. Netw. **8**, 87–98 (2015)
4. Davis, J.A., Jedwab, J.: Peak-to-mean power control in OFDM, Golay complementary sequences, and Reed-Muller codes. IEEE Trans. Inf. Theory **45**, 2397–2417 (1999)
5. Michailow, N., Mendes, L., Matthe, M., Festag, I., Fettweis, A., Robust, G.: WHT-GFDM for the next generation of wireless networks. IEEE Commun. Lett. **19**, 106–109 (2015)

6. Manhas, P., Soni, M.K.: Comparison of OFDM system in terms of BER using different transform and channel coding. Int. J. Eng. Manuf. **1**, 28–34 (2016)
7. Hamilton, W.R.: Elements of Quaternions, p. 242. Chelsea Pub. Com., New York (1969)
8. Ward, J.P.: Quaternions and Cayley Numbers: Algebra and Applications, p. 218. Kluwer Academic Publishers, Dordrecht (1997)
9. Labunets, V.G.: Quaternion number–theoretical transform. In: Devices and Methods of Experimental Investigations in Automation, pp. 28–33. Dnepropetrovsk State University Press, Dnepropetrovsk (1981). (In Russian)
10. Sommen, F.: A product and an exponential function in hypercomplex function theory. Appl. Anal. **12**, 13–26 (1981)
11. Sommen, F.: Hypercomplex Fourier and Laplace transforms I. Ill. J. Math. **26**(2), 332–352 (1982)
12. Ernst, R.R., Bodenhausen, G., Wokaun, A.: Principles of Nuclear Magnetic Resonance in One and Two Dimensions. International Series of Monographs on Chemistry, p. 610. Oxford University Press, Oxford (1987)
13. Delsuc, M.A.: Spectral representation of 2D NMR spectra by hypercomplex numbers. J. Magn. Reson. **77**(1), 119–124 (1969)
14. Rundblad–Labunets, E., Labunets, V., Astola, J., Egiazarian, K.: Fast fractional Fourier–Clifford transforms. In: Second International Workshop on Transforms and Filter Banks. TICSP Series, Tampere, Finland, no. 5, pp. 376–405 (1999)
15. Sangwine, S.J., Ell, T.A.: The discrete Fourier transform of a colour image. In: Blackledge, J.M., Turner, M.J. (eds.) Image Processing II Mathematical Methods, Algorithms and Applications, pp. 430–441 (1998)
16. Labunets-Rundblad, E.: Fast Fourier-Clifford transforms design and application in invariant recognition. Ph.D. thesis, p. 26. Tampere University Technology, Tampere, Finland (2000)
17. Rundblad, E., Labunets, V., Egiazarian, K., Astola, J.: Fast invariant recognition of color images based on Fourier–Clifford number theoretical transform. In: EUROPORTO, Conference on Image and Signal Processing for Remote Sensing YI, pp. 284–292 (2000)
18. Labunets, V.G., Kohk, E.V., Ostheimer, E.: Algebraic models and methods of computer image processing. Part 1. Multiplet models of multichannel images. Comput. Opt. **42**(1), 84–96 (2018)
19. Brackx, F., De Schepper, N., Sommen, F.: The 2-D Clifford-Fourier transform. J. Math. Imaging Vis. **26**(1), 5–18 (2006)
20. Brackx, F., Schepper, N., Sommen, F.: The Fourier transform in Clifford analysis. Adv. Imaging Electron Phys. **156**, 55–201 (2009)
21. De Bie, H., De Schepper, N.: The fractional Clifford-Fourier transform. Complex Anal. Oper. Theory **6**(5), 1047–1067 (2012)
22. De Bie, H., De Schepper, N., Sommen, F.: The class of Clifford-Fourier transforms. J. Fourier Anal. Appl. **17**, 1198–1231 (2011)
23. Condon, E.U.: Immersion of the Fourier transform in a continuous group of functional transforms. Proc. Nat. Acad. Sci. **23**, 158–164 (1937)
24. Kober, H.: Wurzeln aus der Hankel- und Fourier und anderen stetigen Transformationen. Quart. J. Math. Oxford Ser. **10**, 45–49 (1939)
25. Bargmann, V.: On a Hilbert space of analytic functions and an associated integral transform. Part 1. Commun. Pure Appl. Math. **14**, 187–214 (1961)
26. Namias, V.: The fractional order Fourier transform and its application to quatum mechanics. J. Inst. Math. Appl. **25**, 241–265 (1980)
27. McBride, A.C., Kerr, F.H.: On Namias' fractional Fourier transforms. IMA J. Appl. Math. **39**, 159–265 (1987)

28. Ozaktas, H.M., Mendlovic, D.: Fourier transform of fractional order and their optical interpretation. Opt. Commun. **110**, 163–169 (1993)
29. Ozaktas, H.M., Zalevsky, Z., Kutay, M.A.: The Fractional Fourier Transform, p. 532. Wiley, Chichester (2001)
30. Carracedo, C.M., Alix, M.S.: The Theory of Fractional Powers of Operators, p. 366. Elsevier, London - New York (2001)
31. Kumar, R., Singh, K., Khanna, R.: Satellite image compression using fractional Fourier transform. Int. J. Comput. Appl. **50**(3), 20–25 (2012)
32. Shannon, C.E.: Communication theory of secrecy systems. Bell Labs Tech. J. **28**(4), 657–671 (1949)
33. Wyner, A.D.: The wiretap channel. Bell Labs Tech. J. **54**(8), 1355–1387 (1975)

New Secure Block Cipher for Critical Applications: Design, Implementation, Speed and Security Analysis

Sergiy Gnatyuk[1,2(✉)], Berik Akhmetov[2], Valeriy Kozlovskyi[1], Vasyl Kinzeryavyy[1], Marek Aleksander[3], and Dmytro Prysiazhnyi[1]

[1] National Aviation University, Kyiv, Ukraine
sergio.gnatyuk@gmail.com, vvkzeos@gmail.com, v.kinzeryavyy@nau.edu.ua
[2] Yessenov University, Aktau, Kazakhstan
berik.akhmetov@ayu.edu.kz
[3] State Higher Vocational School in Nowy Sącz, Nowy Sącz, Poland
aleksandermarek4@gmail.com

Abstract. Most of known methods for confidentiality and privacy ensuring don't provide high-security level against cyberattacks based on linear and differential cryptanalysis and the required cryptographic data processing speed. In view of this, the cryptographic security method for critical infrastructure systems has been developed. On the basis of this method, the Luna-2k17 block cipher was designed. The specifications of this cipher are given in this paper. Also, high bound values of parameters characterizing its practical security against cyberattacks of mentioned categories of cryptanalysis are calculated. Under the same conditions, to evaluate the speed characteristics of ciphers experimental studies were conducted. Results of experiments showed that the Luna-2k17 cipher is faster than GOST 28147-89 cipher approximately in 3.11 times as well as in 1,27 times for the Kalyna and AES.

Keywords: Confidentiality and privacy · Cryptography · Block cipher · Linear cryptoanalysis · Differential cryptoanalysis · Critical applications

1 Introduction

Today critical information infrastructure of the state is defined by fast and impactful changes of its internal environment caused by modern information and communication technologies (ICT) implementation. For example, in civil aviation (sector of critical information infrastructure) so-called critical aviation information systems demand highest level of security [1]. Mentioned systems include air traffic management systems, remote technical service systems, flight plan interphone system etc. [2]. To minimize cyberthreats influence on critical aviation information systems a set of security procedures must be implemented [3]. Among them there is data confidentiality ensuring in accordance to CIA triad. One of the most important areas of data privacy protection is information security using cryptographic methods, the undeniable

Z. Hu et al. (Eds.): AIMEE 2019, AISC 1126, pp. 93–104, 2020.
https://doi.org/10.1007/978-3-030-39162-1_9

advantage of which is protection the data itself rather than access to it. The principal criterion of choosing cryptosystems is the security level against some categories of cyberattacks. But for some specifical targets, the cryptographic data processing speed (as in the case of public key cryptography [4, 5] or some up to date applications based on secret key cryptography) plays a key role.

2 Review of Modern Approaches and Problem Definition

Despite the diversity of modern cryptographic methods and tools, not all of them have required level of efficiency (security and speed) to provide guaranteed data protection [3–5]. Also the development and cost reduction of ICT positively affects the effectiveness of cryptanalysis, one of the most effective methods of which is linear and differential cryptanalysis [6–8]. In [9] two high-performance reliable block encryption algorithms were proposed and its speed and security against linear and differential cryptanalysis was estimated in comparison with most effective modern block ciphers like AES, Kalyna etc. But ICT development as well as cryptanalysis methods and tools enforces to cryptographic security methods improvements and creating new ciphers based on these. From this viewpoint, the main purpose of this paper is to increase the effectiveness (oriented on critical applications with some speed restrictions and requirements for security level) of cryptographic security by new cryptographic data protection comply with high-performance symmetric block cipher.

3 Design and Mathematical Background of Cryptographic Security Method

Suppose t', p', r', q' are natural numbers, $t = 2t', p = 2p', r = 2r' + 1, n = tp, q = pq'$, $w = p, k = 2n, b = 2^{q'}$ (parameter b defines the quantity of different substitution tables (substitutions), which can be used in the method). Then the r-rounded ciphering method $\Im$ with the set of clear text messages (messages to encrypt) $V_n = \{0, 1\}^n$, the set of secret keys V_k and the set of round keys V_{n+q+w} can be described by following sequence of stages:

Stage 1 – Round Keys Producing

At this stage the number of round keys $K_i, K_i \in V_{n+q+w}, i = \overline{1, r}$ is produced from the secret key $K, K \in V_k$.

Step 1. Decomposing secret key $K, K \in V_k$:

$$K = (B_{-4}, B_{-3}, B_{-2}, B_{-1}), \tag{1}$$

$$B_j \in V_{n/2}, \ j = \overline{-4, -1}.$$

Step 2. Producing vectors B_j, $B_j \in V_{n/2}$, $j = \overline{0, c-1}$, $c = \lceil 4r(n+q+w)/n \rceil$:

$$B_j = \begin{cases} \left(S\left(B_{j-4}, W_1, p'\right) \oplus \left(B_{j-2} <<< P_1\right) \oplus B_{j-1} \oplus Q_1\right) <<< B_{j-3}, & j \bmod 4 = 0 \\ \left(S\left(B_{j-3} \oplus B_{j-1}, W_2, p'\right) \oplus \left(B_{j-2} >>> P_2\right) \oplus Q_2\right) >>> B_{j-4}, & j \bmod 4 = 1 \\ \left(B_{j-4} \oplus \left(S\left(B_{j-3}\right) <<< P_3, W_3, p'\right) \oplus B_{j-1} \oplus Q_3\right) >>> B_{j-2}, & j \bmod 4 = 2 \\ \left(S\left(B_{j-4} <<< P_4, W_4, p'\right) \oplus B_{j-3} \oplus B_{j-2} \oplus Q_4\right) <<< B_{j-1}, & j \bmod 4 = 3 \end{cases}, \tag{2}$$

where $\oplus$ is the Boolean operation of binary vectors coordinate wise addition, $X <<< Y$ is the dynamic rotate left of bit sequence X for Y times, and $X >>> Y$ is the dynamic rotate right of the bit sequence X for Y times, W_i, P_i, Q_i are some constants, W_i, P_i, $Q_i \in V_{n/2}$, $i = \overline{1, 4}$.

The substitution S is defined via following formula:

$$S(x, y, z) = \left(s_{y_{z-1}}(x_{z-1}), \ldots, s_{y_0}(x_0)\right), x = (x_{z-1}, \ldots, x_0), y = (y_{z-1}, \ldots, y_0), \tag{3}$$

where $x_j \in V_t$, $y_j \in V_{q'}$, s_{y_j} is the substitution table for the set V_t (one substitution table is chosen among b possible variants by index y_j), $j \in \overline{0, z-1}$.

Step 3. Vectors C_i, $C_i \in V_e$, $i = \overline{1, r}$, $e = e'n/2$, $e' = \lceil 2(n+q+w)/n \rceil$ are formed by concatenation of vectors B_j, $B_j \in V_{n/2}$, $j = \overline{0, c-1}$, $c = \lceil 4r(n+q+w)/n \rceil$ in inverse order (to form one vector C_i it should be used e' number of vectors B_j):

$$C_i = \left(B_{c-1-(i-1)e'} \| B_{c-1-(i-1)e'-1} \| \cdots \| B_{c-1-(i-1)e'-e'+1}\right) \tag{4}$$

Step 4. Calculation of round keys K_i, $K_i \in V_{n+q+w}$, $i = \overline{1, r}$:

$$K_i = (C_i >>> i) \bmod 2^{n+q+w}. \tag{5}$$

The obtained keys K_i, $K_i \in V_{n+q+w}$, $i = \overline{1, r}$ will be used for secret message encryption and decryption.

Stage 2 – Encryption Procedure

At this stage the secret message is encrypted $A = (A_1, A_2, A_3, \ldots, A_u)$, $A \in V_{n \cdot u}$, $A_j \in V_n$, $j = \overline{1, u}$, where u is a natural number.

The encryption function of each $A_j \in V_n$, $j = \overline{1, u}$ is following:

$$F = f_{r, K_r} \circ \ldots \circ f_{1, K_1}. \tag{6}$$

The round function f_{i, K_i} for all $x \in V_n$, $K_i \in V_{n+q+w}$, $i \in \overline{1, r}$ is described as follows:

$$f_{i, K_i}(x) = \begin{cases} \varphi\left(x \oplus k^{(1)}, k^{(2)}, k^{(3)}\right), & \mathit{if} \ \ i < r \\ S\left(x \oplus k^{(1)}, k^{(2)}, p\right), & \mathit{if} \ \ i = r \end{cases}, \tag{7}$$

where $k^{(1)}$, $k^{(2)}$, $k^{(3)}$ are parts of round key K_i ($k = \left(k^{(1)}, k^{(2)}, k^{(3)}\right)$, $k^{(1)} \in V_n$, $k^{(2)} \in V_q$, $k^{(3)} \in V_w$).

The substation S is defined in (3) and the substitution φ is defined by following formula:

$$\varphi(x, y, h) = S(x, y, p)M(h), x \in V_n, y \in V_q, h \in V_w \tag{8}$$

where $x = \left(x_{p-1}, \ldots, x_0\right)$, $y = \left(y_{p-1}, \ldots, y_0\right)$, $x_j \in V_t$, $y_j \in V_{q'}$.

$M(h)$ is the invertible matrix $2p \times 2p$ over Galuis field $GF\left(2^{t'}\right)$, which depends on h, and the multiplication $S(x, y, p) = \left(s_{y_{p-1}}\left(x_{p-1}\right), \ldots, s_{y_0}(x_0)\right)$ on $M(h)$ in (8) is accomplished over this field in the following way:

1. $s_{y_j}\left(x_j\right)$ $\left(j \in \overline{0, p-1}\right)$ is decomposed on 2 t'-bit parts.

$$\left(s_{y_j}\left(x_j\right) = \left(s_{y_j}\left(x_j\right)^{(1)}, s_{y_j}\left(x_j\right)^{(2)}\right), s_{y_j}\left(x_j\right)^{(1)}, s_{y_j}\left(x_j\right)^{(2)} \in V_{t'}\right).$$

2. The vector B is formed from the part $s_{y_j}\left(x_j\right)$ ($j \in \overline{0, p}$):

$$B = s_{y_0}\left(x_0^{(1)}\right) \,||\, s_{y_0}\left(x_0^{(2)}\right) \,||\, \ldots \,||\, s_{y_{p-1}}\left(x_{y_{p-1}}^{(1)}\right) \,||\, s_{y_{p-1}}\left(x_{y_{p-1}}^{(2)}\right).$$

3. Vector B and $M(h)$ multiplying is accomplished over binary vector identification B_j $j \in \overline{0, 2p-1}$ ($B_j \in V_{t'}$) of the matrix $M(p)$ elements.

4 Security Analysis

On the basis of papers [9–11], for the proposed method analytical upper bounds of the parameters that characterize its practical security against cyberattacks of linear and differential cryptanalysis are obtained as following:

$$EDP(\Omega) \le \overset{\approx}{\Delta}_{\oplus}^{r'\lceil B_M/2\rceil+1} \le \Delta_{\oplus}^{r'\lceil B_M/2\rceil+1}, \tag{9}$$

$$EDP(\Omega) \le \overset{\approx}{\Lambda}_{\oplus}^{r'\lceil B_M/2\rceil+1} \le \Lambda_{\oplus}^{r'\lceil B_M/2\rceil+1}, \tag{10}$$

where $EDP(\Omega)$ is the average probability of differential characteristic Ω, $ELP(\Omega)$ is the average probability of linear characteristic Ω, B_M is M matrix branching index, and the parameters $\Delta_{\oplus}$, $\Lambda_{\oplus}$, $\overset{\approx}{\Delta}$, $\overset{\approx}{\Lambda}$ are defined via the following formulas:

$$\Delta_{\oplus} = \max\left\{d_{\oplus}^{(s_j)}(\alpha, \beta) \colon \alpha, \beta \in V_t \backslash \{0\}, j \in \overline{0, b-1}\right\}, \tag{11}$$

$$\Lambda_{\oplus} = \max\left\{l^{(s_j)}(\alpha, \beta) \colon \alpha \in V_t, \beta \in V_t \backslash \{0\}, j \in \overline{0, b-1}\right\}, \tag{12}$$

$$\overset{\approx}{\Delta}_{\oplus} = \max\left\{ b^{-1} \sum_{j=0}^{b-1} d_{\oplus}^{(s_j)}(\alpha,\ \beta) \colon \alpha,\ \beta \in V_t \backslash \{0\} \right\}, \tag{13}$$

$$\overset{\approx}{\Lambda}_{\oplus} = \max\left\{ b^{-1} \sum_{j=0}^{b-1} l^{(s_j)}(\alpha,\ \beta) \colon \alpha \in V_t,\ \beta \in V_t \backslash \{0\} \right\}. \tag{14}$$

In (11)–(14) $d_{\oplus}^{(s_j)}$ is the difference table of the substitution s_j ($j \in \overline{0,\ b-1}$) over the bitwise addition operation modulo 2 and $l^{(s_j)}(\alpha,\ \beta)$ are the tables of linear approximation of the substitution s_j ($j \in \overline{0,\ b-1}$) over this operation.

5 Block Cipher Design

Luna-2k17 block cipher is developed on the basis of proposed method. Such parameters were chosen for he purposes of this cipher: $t' = 8$, $t = 2t' = 16$ (the substitution tables capacity), $p' = 4$, $p = 2p' = 8$, $r' = 4$, $r = 2r' + 1 = 9$ (number of rounds), $n = tp = 128$ (size of data block in bits), $q' = 3$, $q = pq' = 24$, $b = 2^{q'} = 8$ (8 substitution tables are used over the set V_{16}), $w = 8$, $k = 2n = 256$ (the size of secret keys in bits).

The proposed cipher works with 128-bit data blocks and supports a 256 bits length secret key. When expanding a secret key, the required number of 160-bit round keys is generated ($n + q + w = 128 + 24 + 8 = 160$). Data blocks and extended keys are represented as 8×2 byte matrix.

For Luna-2k17 cipher the producing round keys procedure is accomplished via formulas (1)–(5). At the first step one 256-bit round key K is divided into 4 parts: $K = (B_{-4}, B_{-3}, B_{-2}, B_{-1})$ 64 bit length for each. Then at the second step 2 auxiliary 64-bit vectors are calculated B_j, $j = \overline{0, 44}$ ($c = \lceil 4r(n + q + w)/n \rceil = 4 \cdot 9 \cdot (128 + 24 + 8)/128 = 45$).

Here with it is used 64-bit constants W_i, P_i, Q_i, $i = 1, 4$ (see the values of these constants in the Table 1).

Also in step two the substitution S uses 8 substitution tables 16×16 bits. These substitution tables are set up using the calculation of the field inverse element $(C/X)^{-1} \in GF(2^{16})$ with the further execution of Affine transformation over the field $GF(2)$:

$$S(X) = M \cdot (C/X)^{-1} + V, \tag{15}$$

where $X, C, V \in GF(2^{16})$, and M is the invertible square matrix over the field $GF(2)$, size of which is 16×16.

Table 1. The values of 64-bit constants, which are used in the block cipher

Constant	The hexadecimal value of constant
W_1	E702 DDCB BDEC BDD5
W_2	4AB4 2988 9FE1 017A
W_3	7BED 569B 1C78 2259
W_4	15A7 5069 7EF5 439C
P_1	89FC BB23 4DB0 D712
P_2	1970 CB03 479F E7B2
P_3	2FA8 934D EB78 0517
P_4	07C4 20F9 87D1 0ECA
Q_1	8713 DC71 F897 4562
Q_2	D896 3354 4110 80BA
Q_3	5CB1 AE69 0140 DB83
Q_4	A09B E122 36B4 C17A

The parameters C, V and M are presented in hexadecimal values in the Table 2 (each row of matrix M is presented as a single hexadecimal number).

Table 2. Parameters C, V and M, which were used for subs tables setting up over the set V_{16}

Subs table index	M	C	V
0	{0652, 0CA4, 1948, 3290, 6520, CA40, 9481, 2903, 5206, A40C, 4819, 9032, 2065, 40CA, 8194, 0329}	06FB	09F0
1	{32ED, 65DA, CBB4, 9769, 2ED3, 5DA6, BB4C, 7699, ED32, DA65, B4CB, 6997, D32E, vA65D, 4CBB, 9976}	6A8C	760E
2	{32F0, 65E0, CBC0, 9781, 2F03, 5E06, BC0C, 7819, F032, E065, C0CB, 8197, 032F, 065E, 0CBC, 1978}	7992	200B
3	{32FA, 65F4, CBE8, 97D1, 2FA3, 5F46, BE8C, 7D19, FA32, F465, E8CB, D197, A32F, 465F, 8CBE, 197D}	01AC	1E00
4	{3975, 72EA, E5D4, CBA9, 9753, 2EA7, 5D4E, BA9C, 7539, EA72, D4E5, A9CB, 5397, A72E, 4E5D, 9CBA}	7AE3	6EDF
5	{3985, 730A, E614, CC29, 9853, 30A7, 614E, C29C, 8539, 0A73, 14E6, 29CC, 5398, A730, 4E61, 9CC2}	697C	40CD
6	{3B2B, 7656, ECAC, D959, B2B3, 6567, CACE, 959D, 2B3B, 5676, ACEC, 59D9, B3B2, 6765, CECA, 9D95}	4724	68FD
7	{3C54, 78A8, F150, E2A1, C543, 8A87, 150F, 2A1E, 543C, A878, 50F1, A1E2, 43C5, 878A, 0F15, 1E2A}	02EE	75D5

In the third step the 192-bit vectors C_i, $C_i \in V_{192}$, $i = \overline{1,9}$ ($e' = \lceil 2(n+q+w)/n \rceil = 3$, $e = e'n/2 = 192$) are formed. Then 160-bit round keys K_i, $i = \overline{1,9}$ are formed.

For Luna-2k17 cipher <u>*the encryption procedure*</u> is accomplished over the formulas (6)–(8). See the pseudocode of the encryption procedure on the Fig. 1.

The operation $AddKeyMod2\left(state, SubKey^{(1)}\right)$ implies a bitwise modulo 2 addition of 2 corresponding bits of the round key $SubKey^{(1)}$ and the data block *state*.

$MixColumns(state)$ operation represents the linear transformation of matrix *state*. During this operation each 8-byte column of data block *state* is considered as a polynomial over a field $GF(2^8)$ with 8 terms, which is multiplied by fixed polynom $(c(x)$ raised to the power of 7 over the modulo x^8+1. As a polynomial $c(x)$ was chosen the following polinomial:

$$c(x) = 3x^7 + 7x^6 + x^5 + 3x^4 + 7x^3 + 4x^2 + 1Dx + 1$$

(factors are represented in hexadecimal format). As an irreducible polynomial the following polynomial was chosen: $m(x) = x^8 + x^7 + x^5 + x^4 + x + 1$ [12, 13].

Luna-2k17 encryption procedure

<u>Input</u>: 128-bit datab lock $state$, 160-bit round keys

$SubKey_i = \left(SubKey_i^{(1)}, SubKey_i^{(2)}, SubKey_i^{(3)}\right)$, $i = \overline{0, r}$,

$SubKey_i^{(1)} \in V_{128}$, $SubKey_i^{(2)} \in V_{24}$, $SubKey_i^{(3)} \in V_8$.

<u>Output</u>: 128-bit data block $state$.

1. $state = AddKeyMod2\left(state, SubKey_0^{(1)}\right)$;

2. $For\ \ i=1,\ \ i<r,\ \ i{+}{+}\ \ do$

2.1. $state = SubBytes\left(state, SubKey_i^{(2)}, 8\right)$;

2.2. $state = ShiftRows\left(state, SubKey_i^{(3)}\right)$;

2.3. $state - MixColumns\left(state\right)$;

2.4. $state = AddKeyMod2\left(state, SubKey_i^{(1)}\right)$;

3. $state = SubBytes\left(state, SubKey_r^{(2)}, 8\right)$;

4. $state = ShiftRows\left(state, SubKey_r^{(3)}\right)$;

5. $state = AddKeyMod2\left(state, SubKey_r^{(1)}\right)$;

6. $return\ state$.

Fig. 1. The Luna-2k17 block cipher encryprion procedure pseudocode

In $SubBytes\left(state, SubKey^{(2)}, 8\right)$ operation the tabular substitution of every 16 bits of the data block *state* is performed. The Luna-2k17 cipher uses 8 tables over the set V_{16}, where in the choise of the particular table in each round depends on the part of round key $SubKey^{(2)}$ according to the formula (3).

The substitution tables were generated according to the formula (15), the parameters C, V and M, used in generation are represented in Table 2. Substitution tables data is selected so that there are no fixed points, and also that for each substitution table of Luna-2k17 cipher the equality for the parameters is executed: $\Delta_{\oplus} = \Lambda_{\oplus} = 2^{-14}$.

In $ShiftRows(state, SubKey^{(3)})$ operation depending on the part of the round keys $SubKey^{(3)}$ a byte wise shift of elements in the rows of the matrix *state* is performed. The length of the round key $SubKey^{(3)}$ part is 8 bit, so each bit of current vector affects on the shift of corresponding row of the matrix *state*. For instance, if the first bit of the vector $SubKey^{(3)}$ equals 1, then the values of the columns of the first matrix *state* row are swapped, if the first bit of the vector $SubKey^{(3)}$ equals 0, then the values of columns are not changed. See decryption procedure pseudocode at the Fig. 2 [14, 15].

Luna-2k17 decryption procedure
Input: 128-bit data block $state$,160-bit round keys
$SubKey_i = \left(SubKey_i^{(1)}, SubKey_i^{(2)}, SubKey_i^{(3)}\right)$, $i = \overline{0, r}$,
$SubKey_i^{(1)} \in V_{128}$, $SubKey_i^{(2)} \in V_{24}$, $SubKey_i^{(3)} \in V_8$.
Output: 128-bit data-block $state$.

1. $state = AddKeyMod2\left(state, SubKey_r^{(1)}\right)$;
2. $For \quad i = 1, \quad i < r, \quad i{+}{+} \quad do$
   - 2.1. $state = ShiftRows\left(state, SubKey_{r-i+1}^{(3)}\right)$;
   - 2.2. $state = InvSubBytes\left(state, SubKey_{r-i+1}^{(2)}\right)$;
   - 2.3. $state = AddKeyMod2\left(state, SubKey_{r-i}^{(1)}\right)$;
   - 2.4. $state = InvMixColumns\left(state\right)$;
3. $state = ShiftRows\left(state, SubKey_1^{(3)}\right)$;
4. $state = InvSubBytes\left(state, SubKey_1^{(2)}\right)$;
5. $state = AddKeyMod2\left(state, SubKey_0^{(1)}\right)$;
6. *return state*.

Fig. 2. Luna-2k17 block cipher decryption procedure pseudocode

The $InvMixColumns(state)$ operation is a linear transformation of the matrix *state*, which is inverse to $MixColumns(state)$ operation. In this operation every 8-byte column of a data block *state* is considered as a polynomial over a field $GF(2^8)$ with 8 terms, which is multiplied by the fixed polynomial $d(x)$ raised to the power of 7 modulo $x^8 + 1$. The following polynomial is chosen as a polynomial $d(x)$: $d(x) = 7Ax^7 + A1x^6 + F8x^5 + EEx^4 + 29x^3 + 89x^2 + EBx + 51$ (the factors are represented in hexadecimal format).

In the $InvSubBytes\left(state, SubKey^{(2)}\right)$ operation the tabular substitution of each 16 bit of data block $state$. This operation is inverse to $SubBytes\left(state, SubKey^{(2)}\right)$ operation, so it uses the inverse substitution tables in comparing with $SubBytes\left(state, SubKey^{(2)}\right)$ tables.

Figure 3 shows scheme for round fulfilling of Luna-2k17.

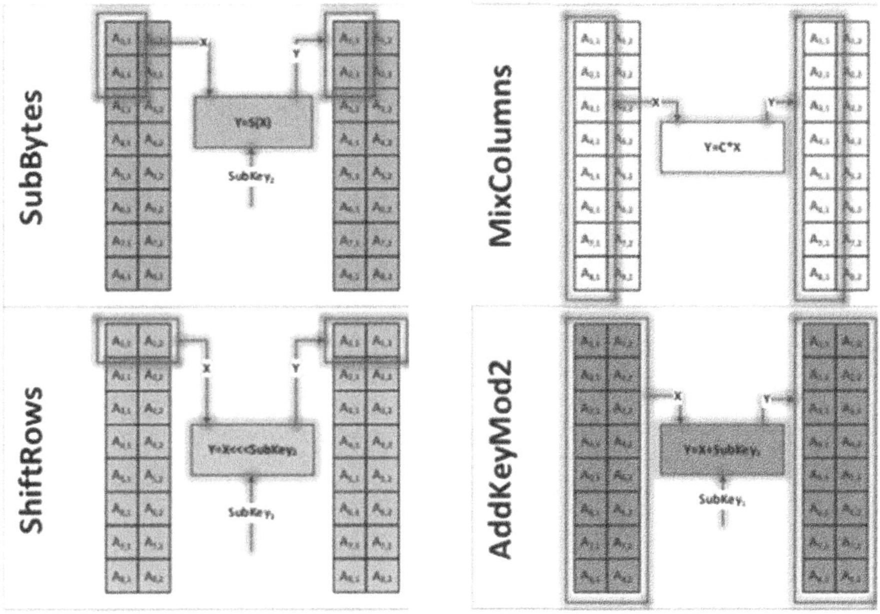

Fig. 3. Scheme of full round fulfilling for Luna-2k17 cipher

6 Experimental Study and Discussion

For experimental study purposes, Luna-2k17 block cipher was implemented as a console application named "Luna-2k17".

The *sequences statistical properties*, created using this application (in the counter mode), were investigated in NIST STS statistical tests environments as well as in Diehard technique. The statistical portraits of the Luna-2k17 block cipher are shown in Fig. 3. For comparison purposes, the results of the sequences testing generated by the Luna-2k17, GOST 28147-89, Kalyna, AES ciphers are given in Table 3. As can be seen from the results, the Luna-2k17 cipher passed a comprehensive control over the NIST STS (Fig. 4) and Diehard (Fig. 5) techniques and showed no worse results than the ciphers above.

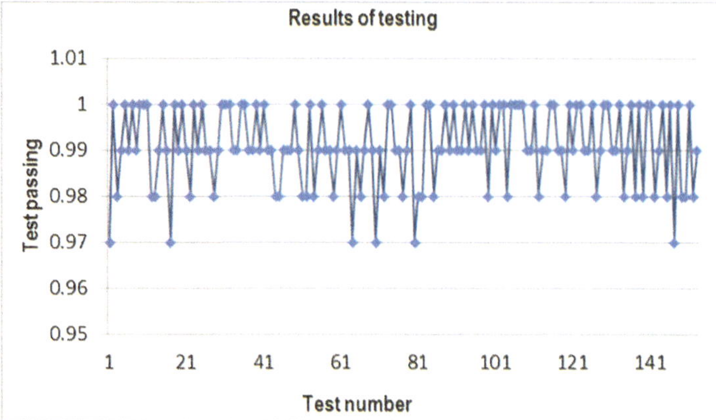

Fig. 4. Luna-2k17 cipher testing results in NIST STS

Test name	ntup	P - value	Assessment
Birthdays	0	0,56788773	Passed
OPERM5	0	0,55416079	Passed
32x32 Binary Rank	0	0,43722351	Passed
6x8 Binary Rank	0	0,76078056	Passed
Bitstream	0	0,22272149	Passed
OPSO	0	0,30707752	Passed
OQSO	0	0,97062324	Passed
DNA	0	0,66572989	Passed
Count the 1s (stream)	0	0,95264882	Passed
Count the 1s (byte)	0	0,94021227	Passed
Parking Lot	0	0,79252967	Passed
Minimum Distance (2d Circle)	2	0,69289211	Passed
Minimum Distance (3d Sphere)	3	0,01723691	Passed

Fig. 5. Luna-2k17 cipher testing results in Diehard tool

Table 3. Sequense testing results using NIST STS

Generator	Number of passed tests	
	99% sequences	96% sequences
BBS	132 (70,2%)	188 (100%)
Kalyna	137 (72,9%)	188 (100%)
GOST 28147-89	130 (69,1%)	186 (98,9%)
AES	133 (70,7%)	188 (100%)
Luna-2k17	141 (75,0%)	188 (100%)

The *speed characteristics* of ciphers are also studied. It has been shown experimentally that the Luna-2k17 cipher is faster than the GOST 28147-89 cipher approximately in 3.11 times, and 1,27 times or the Kalyna and AES ciphers (see Table 4). The research was conducted in the same conditions on Intel (R) Core (TM) i7-2600K CPU 3.4 GHz.

Table 4. The block cipher speed characteristics comparison

Block cipher	Encryption speed (MB/s)
AES-256	64,93
Kalyna-256	71,19
GOST 28147-89	29,02
Luna-2k17	90,48

Also the *security ratings* of the Luna-2k17 cipher over the methods of linear and differential cryptoanalysis are calculated. According to (9)–(14) formulas the parameters upper bounds values, characterizing the Luna-2k17 practical security against linear and differential cryptoanalysis methods are calculated: $\Delta_{\oplus} = \Lambda_{\oplus} = 2^{-14}$, $r' = 4$, $B_M = 9$ – $EDP(\Omega) \leq 2^{-294}$, $ELP(\Omega) \leq 2^{-294}$ and number of round keys $r = 9$. This results shows that if the $r \geq 9$ then the practical security of the Luna-2k17 cipher over the foregoing cryptoanalysis methods is provided.

7 Conclusions

The cryptographic security method was developed, which can increase the effectiveness of cryptographic security by using new procedures sequence of operations in generating round keys and encryption (using substitution tables with increased capacity and randomized linear and non-linear operations). On the basis of this method, the Luna-2k17 symmetric block cipher was developed.

The values of parameters upper bounds characterizing its practical resistibility to cyber attacks of linear and differential cryptanalysis are calculated. Under the same conditions, experimental studies were conducted to evaluate the speed characteristics of ciphers, which showed that the Luna-2k17 cipher is faster than the GOST 28147-89 cipher approximately in 3.11 times, and 1,27 times for the Kalyna and AES ciphers.

Also, the statistical properties of the sequences generated by the Luna-2k17 block cipher were investigated. As a result, it was shown that the Luna-2k17 cipher passed a complex control of the NIST STS and Diehard techniques and showed no worse results than other ciphers.

Future work in this direction can be devoted to security assessment against attacks based on quantum algorithms (Shor's, Grover's etc.) as well as practical implementation of proposed cipher to provide privacy in critical infrastructures (banks, transportation, communications et al.) based on different hardware and software platforms.

Acknowledgments. This scientific work was financially supported as a part of Ukrainian Young Scientists Project of Ministry of Education and Science of Ukraine [№ 0117U006770].

References

1. Gnatyuk, S.: Critical Aviation Information Systems Cybersecurity. Meeting Security Challenges Through Data Analytics and Decision Support. NATO Science for Peace and Security Series – D: Information and Communication Security, vol. 47, № 3, pp. 308–316. IOS Press Ebooks (2016)
2. Gnatyuk, S., Aleksander, M., Sydorenko, V.: Unified data model for defining state critical information infrastructure in civil aviation. In: IEEE 9th International Conference on Dependable Systems, Services and Technologies, pp. 37–42 (2018)
3. Aleksander, M., Dubchak, L., Chyzh, V., Naglik, A., et al.: Implementation technology software-defined networking in wireless sensor networks. In: Proceedings of 2015 IEEE 8th International Conference on Intelligent Data Acquisition and Advanced Computing Systems: Technology and Applications (IDAACS), Warsaw, Poland, 24–26 September (2015)
4. Gnatyuk, S., Okhrimenko, A., Kovtun, M., Gancarczyk, T., Karpinskyi, V.: Method of algorithm building for modular reducing by irreducible polynomial. In: Proceedings of the 16th International Conference on Control, Automation and Systems, Gyeongju, Korea, 16–19 October 2016, pp. 1476–1479 (2016)
5. Hu, Z., Gnatyuk, S., Kovtun, M., Seilova, N.: Method of searching birationally equivalent edwards curves over binary fields. Adv. Intell. Syst. Comput. **754**, 309–319 (2018)
6. Biham, E., Shamir, A.: Differential cryptanalysis of DES-like cryptosystems. J. Cryptol. **4**(1), 3–72 (1991)
7. Lai, X., Massey, J.L., Murphy, S.: Markov ciphers and differential cryptanalysis. In: Proceedings of the Advances in Cryptology – EUROCRYPT 1991, pp. 17–38. Springer (1991)
8. Matsui, M.: Linear cryptanalysis methods for DES cipher. In: Proceedings of the Advances in Cryptology – EUROCRYPT 1993, pp. 386–397. Springer (1994)
9. Gnatyuk, S., Kinzeryavyy, V., Iavich, M., Prysiazhnyi, D., Yubuzova, Kh.: High-performance reliable block encryption algorithms secured against linear and differential cryptanalytic attacks. In: CEUR Workshop Proceedings, vol. 2104, pp. 657–668 (2018)
10. Alekseichuk, A., Kovalchuk, L., Skrynnik, E.: Rating of practical resistance of Kalyna block cipher relative to the difference methods, linear cryptanalysis and algebraic attacks based on homomorphisms. Appl. Radio Electron. **7**(3), 203–209 (2008)
11. Gaeini, A., Mirghadri, A., Jandaghi, G., Keshavarzi, B.: Comparing some pseudo-random number generators and cryptography algorithms using a general evaluation pattern. Int. J. Inf. Technol. Comput. Sci. (IJITCS) **8**(9), 25–31 (2016)
12. Gupta, L.M., Garg, H., Samad, A.: An improved DNA based security model using reduced cipher text technique. Int. J. Comput. Netw. Inf. Secur. (IJCNIS) **11**(7), 13–20 (2019)
13. Gnatyuk, S., Kinzeryavyy, V., Kyrychenko, K., et al.: Secure hash function constructing for future communication systems and networks. In: Advances in Intelligent Systems and Computing, vol. 902, pp. 561–569 (2019)
14. Dychka, I., Tereikovskyi, I., Tereikovska, L., Pogorelov, V., Mussiraliyeva, S.: Deobfuscation of computer virus malware code with value state dependence graph. In: Advances in Intelligent Systems and Computing, vol. 754, pp. 370–379 (2018)
15. Dawood, O.A., Rahma, A.M., Hossen, A.M.: The new block cipher design (tigris cipher). Int. J. Comput. Netw. Inf. Secur. (IJCNIS) **7**(12), 10–18 (2015)

About Direct Linearization Methods for Nonlinearity

Alishir A. Alifov(✉)

Mechanical Engineering Research Institute of Russian Academy of Sciences, Moscow 101990, Russia
a.alifov@yandex.ru

Abstract. All real dynamic systems are nonlinear and potentially oscillatory, despite their separation into various types (physical, chemical, biological, economic, etc.). And the use of linear models that are valid only for small changes in parameters is associated with mathematical difficulties (finding solutions to nonlinear equations). Known methods of analysis and calculation of nonlinear systems have a significant drawback: high labor intensity and time - consuming. In contrast, direct linearization methods reduce these disadvantages by several orders of magnitude. Below are the methods of direct linearization for the calculation of nonlinear systems. Direct linearization of nonlinearity is considered in two cases. In the first case, the nonlinear function depends on one variable, and in the second - on two. Direct linearization methods are compared with a known averaging method. The procedure for applying direct linearization methods is described for calculating oscillatory systems interacting with energy sources.

Keywords: Method · Nonlinearity · Direct linearization · Oscillations

1 Introduction

To put it simply, such a complex process associated with the concept of "artificial intelligence" it is *the ability of a machine* (*device*) to think like a person. As is known, due to the development of electronics in a broader modern definition, a machine is a technical object, which consists of interconnected parts and uses energy to perform the functions assigned to it. Cybernetic machines are able to adapt to environmental changes based on the use of artificial intelligence systems (robots, manipulators, automatic machines, flexible production systems, etc.). The concept of "artificial intelligence" is closely related to the concept of "control system", which in turn includes the technical management structure (a device or a set of devices for manipulating the behavior of other devices or systems). Various variables can be used as a control object in technical objects: mechanical displacements (angular or linear), their speed, electrical variables, temperatures, etc. To calculate the dynamics of machines, mathematical models are used, which in most cases are differential equations in a linear or nonlinear form, describing variables.

As it is known, at present humankind has an acute environmental problem. One of these problems, and very significant, is the reduction of energy consumption. One can say that the theory of oscillatory systems with limited power-supply [1] is directly

Z. Hu et al. (Eds.): AIMEE 2019, AISC 1126, pp. 105–114, 2020.
https://doi.org/10.1007/978-3-030-39162-1_10

connected with the solution of environmental problems because it involves the interaction of an energy source with an oscillatory system.

As shown in [2] and many other works of the author, the fundamental property of matter is oscillatory motion. Oscillations manifest themselves in all real systems, which are characterized by a continuously changing mode of motion in time and space. Under certain conditions, oscillations can occur in many modern technical devices and technological processes, in various kinds of control systems, both with the participation of man and without his participation, etc.

For the analysis of nonlinear oscillatory systems, well-known approximate methods of nonlinear mechanics are widely used: asymptotic averaging method, energy balance, harmonic linearization, etc. [3–17]. All these methods are characterized by laboriousness, which increases with increasing degree of nonlinearity. These drawbacks are significantly reduced using **direct linearization methods (DLM)**, which greatly facilitate the calculation of such systems. The method of direct linearization of nonlinear positional force for the calculation of a conservative system is described in [18]. And in [19], a similar method was used to calculate oscillations for a nonlinear dissipation force (quadratic and intermediate between linear and quadratic). Methods of direct linearization of nonlinearity, in which the linearization coefficients depend on a certain parameter, were proposed in [20–22]. Taking into account this parameter, which determines the accuracy of the method, a direct linearization of the nonlinear resistance force of a fairly general form is considered.

In direct linearization methods, there are no approximations of various orders, and knowledge (specification) of the solution form of the initial nonlinear equation is not required, on the basis of which approximations of different orders are constructed. Such a property of methods ***very significantly reduces the amount of calculations, reduces by several orders cost of time and labor***, which increases their efficiency in application. This is especially valuable when carrying out calculations of real objects regardless of their nature (physical, technical, biological, etc.). The methods of direct linearization are quite simple, which is very important from the standpoint of calculating real technical devices when designing them. Note that as shown in [23] with reference to [24, 25], one of the main problems of nonlinear system dynamics is the high labor costs for analyzing coupled oscillator networks, which play an important role in biology, chemistry, physics, electronics, neural networks and others. The purpose of the article is to describe the methods developed by the author for direct linearization of nonlinearities that depend on one and two variables. First, we consider the direct linearization of nonlinearity, depending on one variable and two variables, then the procedure for applying the DLM to calculate systems with limited excitation.

2 Direct Linearization of Nonlinearity, Depending on One Variable

Let a nonlinear function be given, which is present, for example, in a differential equation. This non-linear function can be replaced by a linear function:

$$F_*(x) = B + kx \tag{1}$$

In (1), the quantities B and k are the linearization coefficients, which are determined from the condition of minimizing the functionals in the form:

$$B = (2r+1)(a^{2r+1})^{-1}\int_0^a F(x)\,x^{2r}dx,\; k = (2r+3)(a^{2r+3})^{-1}\int_0^a F(x)\,x^{2r+1}dx \tag{2}$$

where r is the linearization *accuracy parameter.*

Nonlinear characteristics (observable or obtained from experiments) are in most cases approximated by polynomial functions containing even and odd components (even and odd numbers to the power of x). Note that the even component (components in the more general case) of the function determines the displacement of the center of oscillations during their calculation, and the odd component determines the changes relative to the center. We write the polynomial function in the form:

$$F(x) = \sum_n b_n x^n \tag{3}$$

where $b_n = const$ is approximation coefficients, $n = 0, 1, 2, 3, 4, \ldots$

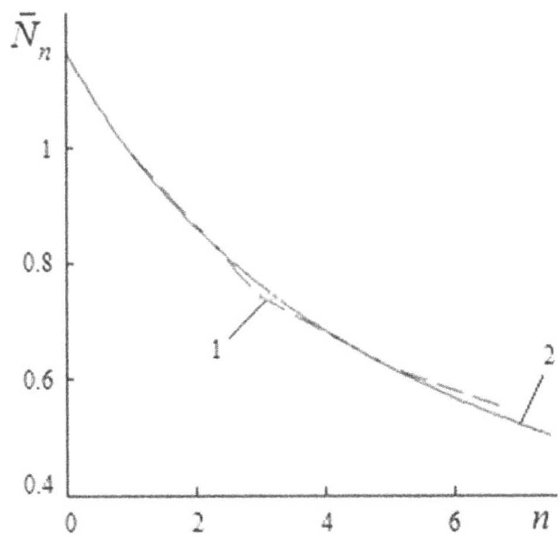

Fig. 1. Comparison of DLM results and asymptotic averaging method.

The calculation of the integrals in (2) for even $(n = 0, 2, 4, \ldots)$ and odd $(n = 1, 3, 5, \ldots)$ components of function (3) leads to the following expressions:

$$B_n = N_n a^n, \qquad \text{n} = 0, 2, 4, \ldots \tag{4a}$$

$$k_n = \bar{N}_n a^{n-1}, \quad \text{n} = 1, 3, 5, \ldots \tag{4b}$$

where $N_n = (2r+1)/(2r+n+1)$, $\bar{N}_n = (2r+3)/(2r+n+2)$.

The linearization coefficients B and k in the linearized characteristic (1) are determined for the function (3) by the expressions

$$B = \sum_n b_n N_n a^n, \; k = \sum_n b_n \bar{N}_n a^{n-1} \tag{5}$$

Regardless of the value of x, in the case of the same importance of the deviations, we can choose $r = 0$, and if the deviations are more significant for large values of x, then we can take $r = 1; 1.5; 2$. The values of the accuracy parameter r in (4a, 4b) can be different for the coefficients N_n and $\bar{N}_n$.

The results obtained by the DLM are compared with known methods, in particular the widely used asymptotic method of averaging nonlinear mechanics. As the results show, the closest results between the DLM and the averaging method are obtained at:

- different values of r for odd n in the interval $r = 1.8 \div 1.85$, and for even n in the interval $r = 0.65 \div 0.7$;
- the same value of r for odd and even n in the interval $r = 1.25 \div 1.3$.

Figure 1 shows a comparison of the coefficient $\bar{N}_n$ at $r = 1.85$ with the numerical coefficients obtained by the asymptotic averaging method for different values of the degree of nonlinearity n. Curve *1* corresponds to the averaging method, curve *2* to the DLM. As you can see, the curves are very close, and in a number of points their complete coincidence takes place. A similar insignificant difference between the curves (coefficient N_n and numerical coefficients of the averaging method), which are not presented for brevity, is also for even n in the interval $r = 0.65 \div 0.7$.

3 Direct Linearization of Mixed Nonlinearity

Consider the method of direct linearization of nonlinearity of the mixed type $F(x, \dot{x}) = f(x)\dot{x}$.

1. Let the nonlinear system described by the equation be given as follows:

$$\ddot{x} + f(x)\dot{x} + x = 0 \tag{6}$$

Introducing a replacement is shown as follows:

$$y = \int x \, dt$$

Equation (6) can be converted to the form:

$$\ddot{y} + G(\dot{y}) + y = 0 \tag{7}$$

where $G(\dot{y}) = \int f(\dot{y}) \, d\dot{y}$.

Now, using the direct linearization method described in p. 2, one can linearize the $G(\dot{y})$ function and then, by solving Eq. (7), we can find the dependence $y(t)$, the time derivative of which gives the solution $x(t)$.

2. Consider the following method of direct linearization of a mixed function $F(x, \dot{x})$. We introduce the ratio as follows:

$$\dot{x}^2 + \upsilon^2 a^{-2} x^2 = \upsilon^2 \tag{8}$$

where a and υ are the maximum values of $|x|$ and $|\dot{x}|$, respectively.
The dependence (8) in another way is the equation of energy and taking into account it, as well as the replacement of $\dot{x} = y$, the linearization coefficients B and k in the linear characteristic of $F_*(\dot{x}) = B + k\dot{x}$ are defined as follows:

$$B = (2r+1)(\upsilon^{2r+1})^{-1} \int_0^{\upsilon} F\left[a\upsilon^{-1}(\upsilon^2 - y^2)^{1/2}, y\right] y^{2r}\, dy \tag{9a}$$

$$k = (2r+3)(\upsilon^{2r+3})^{-1} \int_0^{\upsilon} F\left[a\upsilon^{-1}(\upsilon^2 - y^2)^{1/2}, y\right] y^{2r+1}\, dy \tag{9b}$$

We note that the analytical calculation of integrals (9a, 9b) may in some cases be laborious. Consider, for example, the characteristic $F(x, \dot{x})$ for the function:

$$f(x) = \sum_{i=0}^{n} b_i x^{2i},\ i = 0,\ 1,\ 2, \ldots$$

Based on (7) we have:

$$G(\dot{y}) = \int \left(\sum_{i=0}^{n} b_i \dot{y}^{2i}\right) d\dot{y} = \sum_{i=0}^{n} b_i (2i+1)^{-1} \dot{y}^{2i+1} \tag{10}$$

According to p. 2, taking into account (10) for the coefficient of direct linearization, we get the formula

$$k_1 = \sum_{i=0}^{n} b_i N_i \upsilon^{2i-1} \tag{11}$$

where $N_i = (2r+3)/(2i+1)(2i+2r+3)$.
Using (9b) leads to the following expression for the direct linearization coefficient

$$k_2 = \sum_{i=0}^{n} b_i \tilde{N}_i a^{2i-1} \tag{12}$$

where

$$\tilde{N}_i = (2r+3)\left[\frac{1}{2r+3} - \frac{i}{2r+5} + \frac{i(i-1)}{(2r+7)2!} - \frac{i(i-1)(i-2)}{(2r+9)3!} + \ldots\right].$$

The linearization accuracy parameter r can be selected, as above in p. 2, from the interval (0, 2).

4 The Procedure for Applying the Method of Direct Linearization for the Calculation of Systems with Limited Excitation

In the theory of oscillatory systems interacting with energy sources, at least two equations are considered [1]. One of these equations describes the oscillatory system, the other - the dynamics of the energy source. To illustrate the application of the direct linearization method for calculating such systems, we consider differential equations:

$$\begin{aligned} \ddot{x} + \omega^2 x &= F(x, \dot{x}, \varphi, \dot{\varphi}, \ddot{\varphi}) \\ \ddot{\varphi} &= M(\dot{\varphi}) + H(\varphi, \dot{\varphi}, x, \dot{x}, \ddot{x}) \end{aligned} \tag{13}$$

The first equation in (13) describes the motion of the oscillating system, the second - of the energy source, where $M(\dot{\varphi})$ and $H(\varphi, \dot{\varphi}, x, \dot{x}, \ddot{x})$ reflect the driving force of the energy source and the load on it from the oscillating system, respectively. Equations (13) in the general case are nonlinear and there $F(\cdots)$, $H(\cdots)$ functions can have various specific forms.

Let us make a replacement in the second Eq. (13) and represent it in the form:

$$\dot{\theta} = B_M + k_M \theta + H(\varphi, \theta, x, \dot{x}, \ddot{x}) \tag{14}$$

We use the averaging procedure for the period to Eq. (14). We put $\theta = \Omega + \varepsilon_{osc}$, where Ω is the main part of the solution, and ε_{osc} represents small vibrational components that are not taken into account in the future. This procedure allows you to determine the time-varying value of the velocity Ω of the energy source in the form:

$$\begin{aligned} \dot{\Omega} &= B_M + k_M \Omega + H(a, \Omega) \\ H(a, \Omega) &= \frac{1}{2\pi} \int_0^{2\pi} H(\varphi, \theta, x, \dot{x}, \ddot{x}) \, d\psi \end{aligned} \tag{15}$$

When calculating the integral in (15), the relations

$$x = \upsilon p_\circ^{-1} \cos \psi, \dot{x} = -\upsilon \sin \psi, \ddot{x} = -\upsilon p_\circ \sin \psi, \psi = p_\circ t + \xi$$

where a and $p_\circ$ respectively, the amplitude and frequency of oscillations, $\upsilon = a p_\circ$.

The linearization of the characteristics of the energy source $M(\dot{\varphi})$ can be eliminated, because after the averaging operation we have $M(\Omega)$. Nevertheless, if linearization is necessary, then representing the characteristic of the energy source by a polynomial function

$$M(\dot{\varphi}) = \sum_i \alpha_i \dot{\varphi}^i$$

where $\alpha_i = const$, $i = 0, 1, 2, 3, \ldots$, you can replace it with a linear method of direct linearization.

The linearization coefficients B_M and k_M are determined by expressions of the form (5) provided that a is replaced by Ω_* in the linearization coefficients of the replacement characteristic. Here Ω_* is the maximum value of the speed Ω of the energy source.

5 Calculations and Comparison of Results

To evaluate the results on the DLM, calculations were performed for a number of nonlinear differential equations, including the following equation with a nonlinear function $F(\dot{x})$:

$$m\ddot{x} + k_0\dot{x} + cx = \lambda \sin pt + F(\dot{x}) \tag{16}$$

where $F(\dot{x}) = q\,(1 - \alpha_1 U + \alpha_3 U^3), U = V - \dot{x}$, V, q, α_1 and α_3 are constants.

Equation (16), taking into account the linearization of the function, takes the form

$$m\ddot{x} + k\dot{x} + cx = \lambda \sin pt + q(1 + B_F) \tag{17}$$

where $k = k_0 - qk_F$, B_F and k_F are linearization coefficients for $F(\dot{x})$.

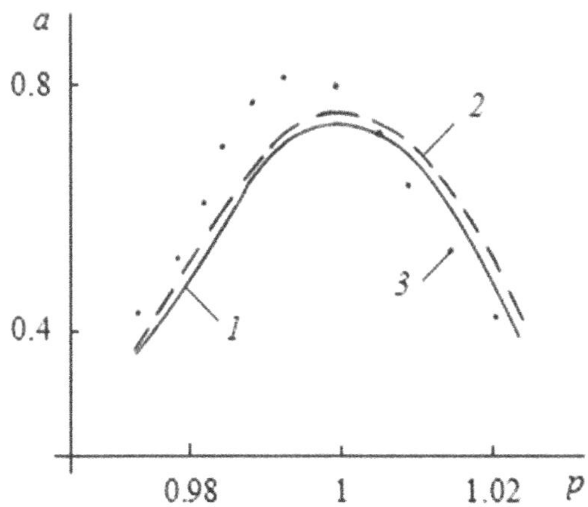

Fig. 2. Results of DLM, asymptotic averaging method and numerical integration.

The solution of Eq. (17) can be constructed by two methods [15], one of which is *the variable replacement method with averaging*. According to this method, for example, the nonlinear functions $F(\dot{x})$ and $f(x)$ is shown in the following equation:

$$\ddot{x} + F(\dot{x}) + f(x) = H(t, x)$$

According to the DLM is reduced to the next equation:

$$\ddot{x} + k\dot{x} + \omega^2 x = H(t, x) \tag{18}$$

where k and ω^2 represent the linearization coefficients for the functions $F(\dot{x})$ and $f(x)$ similarly to (1).

From Eq. (18) using the replacement of variables $x = \upsilon p^{-1} \cos \psi$, $\dot{x} = -\upsilon \sin \psi$, $\psi = pt + \xi$, $\upsilon = \max |\dot{x}|$ and averaging, we obtain the standard form of the equation for determining υ and ξ:

$$\frac{d\upsilon}{dt} = -\frac{k\upsilon}{2} - H_s(\upsilon, \xi), \ \frac{d\xi}{dt} = \frac{\omega^2 - p^2}{2p} - \frac{1}{\upsilon} H_c(\upsilon, \xi) \tag{19}$$

where

$$H_s(\upsilon, \xi) = \frac{1}{2\pi} \int_0^{2\pi} H(\ldots) \sin \psi \, d\psi, \ H_c(\upsilon, \xi) = \frac{1}{2\pi} \int_0^{2\pi} H(\cdots) \cos \psi \, d\psi$$

Equations (19) allow us to study stationary and non-stationary processes. The amplitude of the oscillations is determined by the expression $a = \upsilon p^{-1}$.

For calculations, the following parameters were selected: $\omega = 1\,c^{-1}$, $\lambda = 0.02$ kgf, $k_0 = 0.02$ kgf s cm^{-1}, $q = 0.5$ kgf, $\alpha_1 = 0.84$ s cm^{-1}, $\alpha_3 = 0.18$ s^3 cm^{-3}. The dependences of the amplitude a on the frequency p, which are shown in Fig. 2, were obtained for (17) at V = 1.2 based on equations of type (19). Solid *1* (also represents the results by the asymptotic averaging method) and dashed *2* curves are obtained the DLM with accuracy parameters r = 1.5 and r = 1.3, respectively. The results of numerical integration on the computer of equation (a) are shown by points *3*. Up to the frequency value $p \approx 1$, the results of numerical integration better correspond to curve *2* (r = 1.3), and for frequency values a little more than unity - curve *1* (r = 1.5). Since the numerical coefficients N and $\bar{N}$ in (4a, 4b) depend on the degree of nonlinearity (n), but not on the variable itself, for closer proximity to the results of numerical integration, we can use the ascending and descending parts of curves *1* (r = 1.3) and *2* (r = 1.5). Note that this is not possible in the asymptotic averaging method, in which constant numbers are obtained as a result of averaging a nonlinear function.

6 Conclusions

The considered methods of direct linearization allow linearizing functions with one variable as well as with two variables. A comparison of the results obtained by the DLM and the averaging of nonlinear mechanics showed sufficient proximity. This indicates that instead of the laborious known methods of calculating nonlinear systems, one can use the DLM. The latter are much more effective, because, in comparison with the known methods, their use requires several orders of magnitude less time and labor, which is especially valuable from a practical point of view: for calculations when creating new devices, including technical means of artificial intelligence.

References

1. Alifov, A.A., Frolov, K.V.: Interaction of Non-Linear Oscillatory Systems with Energy Sources. Hemisphere Publishing Corporation and Taylor & Francis Group, New York (1990)
2. Alifov, A.A.: The Fundamental Principle Which Operates the Universe. RCD, Moscow (2012). (in Russian)
3. Bogolyubov, N.N., Mitropolsky, Y.A.: Asymptotic methods in the theory of nonlinear oscillations. Nauka, Moscow (1974). (in Russian)
4. Vibrations in Technology: A Reference Book in 6 Volumes, vol. 2. Oscillations of Nonlinear Mechanical Systems. Mechanical Engineering, Moscow, Russia (1974). (in Russian)
5. Mitropolsky, Y.A., Van Dao, N.: Applied Asymptotic Methods in Nonlinear Oscillations. Springer, Dordrecht (1997)
6. Hayashi, Ch.: Nonlinear Oscillations in Physical Systems. Princeton University Press, Princeton (2014)
7. Esmailzadeh, E., Younesian, D., Askari, H.: Analytical Methods in Nonlinear Oscillations: Approaches and Applications. Springer, Dordrecht (2019)
8. He, J.H.: Some asymptotic methods for strongly nonlinear equations. Int. J. Modern Phys. B **20**(10), 1141–1199 (2006)
9. He, J.H.: Homotopy perturbation method: a new nonlinear analytical technique. Appl. Math. Comput. **135**, 73–79 (2003)
10. Wang, Q., Fu, F.: Variational iteration method for solving differential equations with piecewise constant arguments. Int. J. Eng. Manuf. (IJEM) **2**(2), 36–43 (2012). https://doi.org/10.5815/ijem.2012.02.06
11. Lin, Y., Zhou, L., Bao, L.: A parameter free iterative method for solving projected generalized Lyapunov equations. Int. J. Eng. Manuf. (IJEM) **2**(1), 62 (2012). https://doi.org/10.5815/ijem.2012.01.10
12. Burton, T.D.: A perturbation method for certain nonlinear oscillators. Int. J. Nonl. Mech. **19**(5), 397 (1984)
13. Lu, Q.S.: Special issue on advances of nonlinear dynamics, vibrations and control in China. Int. J. Nonl. Sci. Num. Simul. **6**(1), preface (2005)
14. Chen, D.-X., Liu, G.-H.: Oscillatory behavior of a class of second-order nonlinear dynamic equations on time scales. Int. J. Eng. Manuf. (IJEM) **1**(6) (2011). https://doi.org/10.5815/ijem.2011.06.11

15. Dai, L., Fan, L.: Analytical and numerical approaches to characteristics of linear and nonlinear vibratory systems under piecewise discontinuous disturbances. Comm. Nonl. Sci. Num. Simul. **9**, 417–429 (2004)
16. Panovko, Ya.G.: Fundamentals of Applied Theory of Vibrations and Shock. Mechanical Engineering, Leningrad, Russia (1976). (in Russian)
17. Loytsyansky, L.G.: Free and forced oscillations in the presence of a quadratic and intermediate between the linear and quadratic laws of resistance. Izv. Academy of Sciences of the USSR, Engineering Collection, vol. 18 (1954). (in Russian)
18. Alifov, A.A.: Method of the direct linearization of mixed nonlinearities. J. Mach. Manuf. Reliab. **46**(2), 128–131 (2017). https://doi.org/10.3103/S1052618817020029
19. Alifov, A.A., Farzaliev, M.G., Jafarov, E.N.: Dynamics of a self-oscillatory system with an energy source. Russ. Eng. Res. **38**(4), 260–262 (2018). https://doi.org/10.3103/S1068798X18040032
20. Alifov, A.A.: On the calculation by the method of direct linearization of mixed oscillations in a system with limited power-supply. In: Hu, Z., Petoukhov, S., Dychka, I., He, M. (eds.) Advances in Computer Science for Engineering and Education II, ICCSEEA 2019. Advances in Intelligent Systems and Computing, vol. 938, pp. 23–31. Springer, Cham (2019)
21. Gourary, M.M., Rusakov, S.G.: Analysis of oscillator ensemble with dynamic couplings. In: Hu, Z., Petoukhov, S., He, M. (eds.) Advances in Artificial Systems for Medicine and Education II, AIMEE 2018. Advances in Intelligent Systems and Computing, vol. 902. Springer, Cham (2018)
22. Acebrón, J.A., et al.: The Kuramoto model: a simple paradigm for synchronization phenomena. Rev. Mod. Phys. **77**(1), 137–185 (2005)
23. Bhansali, P., Roychowdhury, J.: Injection locking analysis and simulation of weakly coupled oscillator networks. In: Li, P., et al. (eds.) Simulation and Verification of Electronic and Biological Systems, pp.71–93. Springer (2011)
24. Ashwin, P., Coombes, S., Nicks, R.J.: Mathematical frameworks for oscillatory network dynamics in neuroscience. J. Math. Neurosci. **6**(2), 1–92 (2016)
25. Ziabari, M.T., Sahab, A.R., Fakhari, S.N.S.: Synchronization new 3D chaotic system using brain emotional learning based intelligent controller. Int. J. Inf. Technol. Comput. Sci. (IJITCS) **7**(2), 80–87 (2015). https://doi.org/10.5815/ijitcs.2015.02.10

Models of Information Exchange Between Intelligent Agents

N. Yu. Mutovkina(✉) and V. N. Kuznetsov

Tver State Technical University, Tver, Russia
letter-boxNM@yandex.ru

Abstract. The article discusses the models of information exchange between agents in an abstract intelligent active system. Information exchange is a transfer of messages containing a certain semantic load, from the agent-translator to the agent-recipient. Moreover, in the exchange of information for all agents, the role change is characteristic: the translator becomes the recipient and vice versa. Under the influence of all informational messages, a psychological and business climate is formed in the intellectual active system, which closely correlates with the efficiency of the system. Therefore, it is important to prevent the appearance of information in the system that does not correspond to the truth. To do this, it is necessary to identify the reasons for transmitting false information and to determine managerial influences that would make it possible to minimize the amount of false information transmitted. This article discusses models of the flow of messages transmitted from agent to agent. In these models, parameters are identified, changing which through managerial influences, you can determine which information is true and which information is false. Therefore, to minimize the transmission of false information.

Keywords: Intelligent active system · Intelligent agent · Information exchange · Psycho-behavioral type · Concord · Communication

1 Introduction

The task of information exchange belongs to the class of management tasks since by transferring information of a certain meaning, you can control the agent who receives this information. In the following, the agent transmitting information is called the translator, and the agent receiving the information is called the recipient. Information refers to data about the subject and problem area of the task that is solved, regardless of the form in which this information is presented. The concept of «information» is too capacious to have a universal definition. Therefore, in this paper, information is a collection of data about a complex, non-standard, creative problem Ω, to solve which is the main task of intelligent agents. The joint solution of the problem Ω involves the exchange of information between agents. It should ideally facilitate the joint promotion of agents to the goal [1]. The purpose of agents is to receive remuneration for solving a creative problem Ω. Also, each of the agents seeks to get the greatest part of the reward, not always acting in honest ways. One of such methods is disinformation. However, it should be understood that the transfer of untrustworthy information harms the entire

Z. Hu et al. (Eds.): AIMEE 2019, AISC 1126, pp. 115–125, 2020.
https://doi.org/10.1007/978-3-030-39162-1_11

system, keeping it away from obtaining an acceptable solution to a creative problem. Consequently, an agent who transmits inadequate information in his interests does not receive the desired remuneration either. His interests remain unaccounted for. Besides, a liar agent risks being outside of the intelligent active system (IAS) as unreliable.

Modeling the transfer of reliable and unreliable information helps to understand how the utility function of an intelligent agent is changing and to take the necessary effects to correct its informational behavior. The purpose of this work is to optimize the behavior of agents in the IAS during the information exchange and propose ways to reduce the amount of false information that may appear in the system. The initial data for achieving this goal were observations of the behavior of intelligent agents in solving complex, non-standard tasks in the field of higher education and industry.

Adjusting the behavior of intelligent agents is a process of their learning and contains the idea that the translation of unreliable information will not bring any agent closer to the goal, but, on the contrary, will distance it from it [2].

2 Theoretical Aspects of a Research

In enhancing the efficiency of IAS functioning, an important role is played by mechanisms that make it possible to understand and correctly recognize the behavior of system agents. It is also necessary to develop mechanisms that determine the influence of factors that change the behavior of intelligent agents. It is important to understand when an agent "speaks" the truth or broadcasts a lie. Mistrust and suspicion, as well as unreliable gullibility, are constantly accompanied by any communication processes in the IAS. These intentions play the role of information filters for the recipient. However, it is these filters that often work to the detriment of the recipient himself, letting false information into his consciousness as true or, conversely, blocking and provoking to cast doubt on true information. Naturally, these features of the perception by agents of various kinds of messages significantly impede their activity in the IAS and often become the cause of conflicts.

Since agents, being goal-oriented subjects, can report false information only if they are guided by a specific goal (they are not inclined to "idle" lie), it seems appropriate to use Lipman's psychological theory about lies as a volitional act, aimed at the result [3, 4]. For any volitional act is characterized by the presence of certain internal or external constraints. In the case of lies, the simultaneous presence in the liar's mind of a complex of false and true notions is a brake. An agent lies if, in the struggle between false and true notions, the complex of his false notions wins at the expense of the selfish goals and intentions of the agent himself. If the complex of true notions wins (due to moral notions and notions of consequences), then the agent reports the truth. The agent may report false information consciously (which is dictated by his purpose and desire to thereby secure a win) or unconsciously (if he does not possess the full amount of information or has inaccurate data). If an agent reports incorrect information unconsciously, then he can act as an intermediary between a true liar and the final recipient.

The number of variants for broadcasting true and false messages, as well as messages that only are considered to be true and false, in the IAS is $P_4 = 4! = 24$. Some of these variants are listing in Table 1.

Table 1. Common variants for broadcasting messages in the IAS

Own representations of the translator	Transfer of translator views to the recipient	The real characteristic of the broadcast message	The perception of the message by the recipient
TRUE	TRUE	TRUE	BELIEVE
TRUE	LIE	TRUE	BELIEVE
TRUE	LIE	TRUE	DOES'T BELIEVE
TRUE	TRUE	TRUE	DOES'T BELIEVE
LIE	LIE	TRUE	BELIEVE
LIE	LIE	TRUE	DOES'T BELIEVE
TRUE	TRUE	LIE	BELIEVE

Each informational message of the translator is characterizing by the following parameters:

(1) the translator's own opinion regarding the situation he is considering, the problem area, the task, etc. his subjective assessment of what is happening (may be true, the truth is not fully or far from the truth);
(2) the nature of the transmitted message, i.e. the agent can either communicate his true point of view (as he thinks) or communicate a distorted version (as he wants others to think);
(3) the objective characteristic of the transmitted message ("true" or "false"), i.e. how things are really;
(4) the perception of the broadcast informational message by the recipients: they may believe that this is true, or they may not.

Suppose at the moment t the agent learns something about the problem Ω (gains new knowledge about the subject or problem area, about a possible solution to the problem or the reasons for its occurrence, etc.). If the agent trusts his source, then his views are true. If the agent is sincere and his goals coincide with the interests of the IAS, then he broadcasts this information as it is in his presentation, i.e. like the truth. If the source of the agent is reliable, then this information is true in fact and, if properly fed, it will also be perceived by other agents (see the first variant in Table 1).

It is possible that the agent for some reason wants to hide the true information about the problem. However, he realizes that this information is true. Then, depending on how the agent translates this information into the system, other agents may believe it, although the translator is lying (see the second variant in Table 1) or not (and they will be right). If the translator in the past has established itself as unreliable, then even in the case of the truth of the three first characteristics, other participants of the IAS may not believe him (see the fourth variant in Table 1).

There may be reverse situations where, despite the accuracy of the information, the agent himself thinks that this is a lie, and transmits this information to other agents as a lie in the case of confidence in his veracity. One of the worst variants is that when, with objectively false information, it is perceived by the agent as true, it is broadcast by it as true and the rest of the system believes that it is true.

The art of management in each of these situations consists of the ability to distinguish truth from falsehood and take all possible measures so that all agents act in the IAS based on true premises.

The most frequent manipulations with information messages used in the IAS include [5–7]:

(1) silence (concealment) is the transfer of incomplete true information. This is the best way to hide something, especially when compared to an open-source distortion of facts. Concealment is a passive way of information manipulation;
(2) selection is a selective report to the recipient only of the information that is beneficial to the broadcaster. It is supposed to avoid the truth without the need to resort to outright lies;
(3) distortion (modification of the selection) is such a flow of information when attention is drawn only to the facts most beneficial to the translator. This is a conscious focus only on certain aspects of the problem, which are beneficial to the translator, the presentation of information from a certain point of view;
(4) violation of actual informational proportions (exaggeration, understatement, etc.);
(5) turning over is replacing "white" with "black" and vice versa. This may be a substitution of goals when the translator expresses his interest for the interest of his opponent (an example is a tale of how Brother Rabbit outwitted Brother Fox);
(6) falsification (juggling) is the transfer of false information on the substance of the problem. The translator resorts to this manipulation when one concealment is not enough. In this case, two actions are performing truthful information is concealing and false information is given out for the truth;
(7) disorientation is the transfer of irrelevant true or false information to distract from the essence of the question under consideration. Anything is reported, but not the essence of the problem. In real life, the most common types of disorientation are flattery and slander;
(8) half-truth (shades of truth) is a combination of essential true information with essential false information; one-sided coverage of the facts; inaccurate and vague wording of the provisions under discussion; references to sources with the proviso of the type: "I don't remember who said …"; distortion of a reliable statement using subjective judgments, etc.;
(9) the false conclusion is another trick to avoid open utterances of the lie. It consists of allowing the recipient to conclude the transmits message himself, but at the same time the translator does everything to ensure that this conclusion is false;
(10) false interpretation at the logical level is connecting with the ability to introduce some false premises into the consciousness. For this purpose, such techniques as the "presumption of normality" are used: the message of a large number of true and well-verified judgments, facts, among which only one judgment is false. In the set of all judgments, it is rather difficult to find one false proposition;

(11) the message of truth under the guise of a lie: you can tell the truth, but so that the recipient does not believe it, i.e. to broadcast truth falsely. As the chancellor of the German Empire Otto Eduard Leopold von Bismarck-Schönhausen (1815–1898) said: "If you want to fool the world, tell him the truth" [8].

In any case, the key point in ascertaining whether a given message is true or false is the determination of the motives of the translator, the determination of the reasons for which he considers it necessary to resort to any of the listed manipulation mechanisms.

3 Formalization of Information Exchange in the IAS

Intelligent agents communicate with each other in the IAS in the process of joint work on solving a specific problem Ω. At the same time, they can use both verbal and non-verbal information transfer systems [9, 10]. The message transfer by the agent is carried out according to the scheme shown in Fig. 1.

In this case, the agent can report either all x_i or only a part of them. The set of possible messages with their probabilistic characteristics $\{\overline{X}, \pi(x)\}$ is a message flow (MF). Here $\overline{X} = \{x_1, x_2, \ldots, x_i, \ldots, x_m\}$ is the vector of possible messages distributed in time, $\pi(x_i)$ is the probability of these messages, and, since the probabilities of messages are a complete group of events, their total probability is $\sum_{i=1}^{m} \pi(x_i) = 1$ [11].

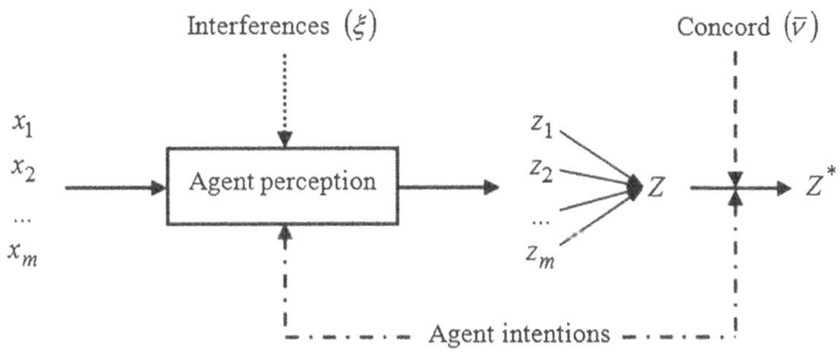

Fig. 1. Message transfer scheme in the IAS

The agent's perception of messages depends on his intentions [12], the level of confidence in the translator, as well as external interference and disturbances. Each message is processed by the recipient, transforming into z_i. Based on the obtained z_i and their generalizations, the final information result Z is forming. Decision making is carried out only after the agreement of Z with other results. The agreed information result Z^* is the starting point for obtaining a final decision in the IAS.

Concord is a procedure that allows you to find a solution that suits all the participants in a system to one degree or another. However, some of the agents who agreed with this decision may experience negative feelings: discomfort, anxiety, anxiety, their insignificance, etc. Reconciling the decision does not guarantee that all agents agree with him. They can only pretend that they are satisfied with the accepted option, but at the same time have internal contradictions. However, negotiation is the most common way out of a crisis [2].

The mathematical model of the MF is a combination of the following distribution functions:

(1) the number of messages transmitted by the translator:

$$C = \left(\left(m \cdot t^{k_{сл.}} \cdot e^{\rho}\right) \cdot y\right)^{\omega_{A_i}} \tag{1}$$

where m is the number of links with other participants of the translator IAS, $m \geq 1$, integral; $k_{сл.}$ is level of difficulty Ω, $0 < k_{сл.} < 1$; ρ is level of goodwill and comfort of a business and psychological climate in the IAS, $0 < \rho < 1$; y is the level of efficiency of a translator's activity depending on its psycho-behavioral type. This level obeys the law of normal distribution; ω_{A_i} is importance of message flow for the translator, $0 < \omega_{A_i} \leq 1$; t is time, $t \to t^*$, where t^* is the moment of finding the solution to the problem, after which its solution will be irrelevant; e is constant, $e \approx 2,718$;

(2) expected usefulness of the message:

$$Y = \left|1 - e^{(\beta \cdot \gamma)^t}\right| \tag{2}$$

where β is degree of novelty of the message for the recipient, $0 \leq \beta \leq 1$; γ is the degree of truthfulness of the message, $0 \leq \gamma \leq 1$. Herewith $\gamma = 0$, if the message is a complete lie, and $\gamma = 1$, if the message is absolute truth;

(3) delivery time of the sequence of messages forming the stream under study:

$$T = |(r_i - r_j)| \cdot \left(\frac{C}{d_{ij}}\right) + (1 - \gamma) \tag{3}$$

where r_i is psycho-behavioral type of translator, $0 < r_i < 1$; r_j is recipient psycho-behavioral type, $0 < r_j < 1$; C is the number of messages, a positive integer greater than or equal to zero ($C = 0$ if the agent does not provide any information); d_{ij} is the tightness of the connection between the i-th and j-th agents, due to their familiarity, joint activities, similar interests, and preferences, etc., $0 \leq d_{ij} \leq 1$;

(4) total time to bring the MF to the recipient, including the time to explain this information:

$$T^* = T + \left|\left(y_{A_i} - y_{A_j}\right)^{\mu \cdot t}\right| \tag{4}$$

where y_{A_i}, y_{A_j} are efficiency levels of the translator and the recipient, respectively, depending on their psycho-behavioral types; μ is level of information perception of the recipient, $0 \leq \mu \leq 1$;

(5) message flow values for an IAS:

$$F_{ц} = C \cdot Y \cdot \left(\omega_{A_i} + \omega_{A_j}\right)^{\varepsilon \cdot \chi \cdot \psi} \tag{5}$$

where ω_{A_i} is the importance of the stream of messages, $0 < \omega_{A_i} \leq 1$; ω_{A_j} is importance of the message flow to the recipient, $0 < \omega_{A_i} \leq 1$; ω_{A_i}, $\omega_{A_j} = 1$ indicate the maximum importance of the MF; ε is indicator of the timely delivery of the informational message. In general, this is a binary parameter, i.e. $\varepsilon = 0$, if the information was not delivered in time, and $\varepsilon = 1$, if timely delivery was performed. The value of 0 or 1 corresponds to a certain range of delivery times: $\varepsilon = \{0, 1 - \sigma\}$, where σ is the assignment, $\sigma \in [0; 1)$. However, depending on the specifics of the situation, the values σ can vary in the range from 0 to 1, if such statements are allowed as "The message was delivered almost on time", "The message, despite the late delivery, have not yet lost its relevance", "Delivery of the message was heavily delaying", etc.; χ is the parameter that shows the intonation coloring of the message (during oral transmission), $0 \leq \chi \leq 1$; ψ is parameter showing emotional coloring of the message (during oral transmission), $0 \leq \psi \leq 1$. Preservation of the emotional and intonational coloring of the informational message is also taken as a unit, if the emotions and intonations are true, and zero otherwise.

In addition to these functions, the characteristics of the MF are influencing by external and internal factors of a stochastic nature.

Based on the performed formalization, it is possible to solve problems of mathematical programming. Changing the parameters, you can get the final characteristics, which are guidelines for certain qualitative results.

4 The Results of the Work of Models of Information Exchange

Let a compromise type translator act in the IAS, entering into an information exchange relationship with three other agents in the process of solving the problem. The initial data describing the parameters of the agents are giving in Table 2. At that, agent A_1 is a translator, and agents A_2, A_3, A_4 are recipients.

The level of the benevolence of the business and psychological atmosphere in the IAS is above average ($\rho = 0,7$).The complexity of the problem solved by A_1 is defining as $k_{сл.} = 0{,}7$. The interaction of agents is carried out within the framework of solving the problem for 20 moments. It is assuming that the parameters $m = 3$; $\rho = 0,7$; $\beta = 0,8$; $\gamma = 0,95$; $\varepsilon = 0,9$; $\chi = 0,5$; $\psi = 0,5$, as well as the characteristics presented in Table 2, do not change for all 20 moments.

Table 2. Initial values of intelligent agent parameters

Descriptions	A_1	A_2	A_3	A_4
r	0,5	0,8	0,3	0,6
y	1,00	0,23	0,52	0,85
d_{ij}	–	0,85	0,45	0,9
ω	0,8	0,8	0,6	0,7
μ	–	0,7	0,5	0,7

It is necessary to find out how the characteristics of the message flow will change with the effects $A_1 \rightarrow A_2$, $A_1 \rightarrow A_3$ и $A_1 \rightarrow A_4$.

The number of messages does not depend on whether the translator intends to lie or report only truthful information, therefore, according to formula (1), A_1 translates the number of messages, shown in Fig. 2.

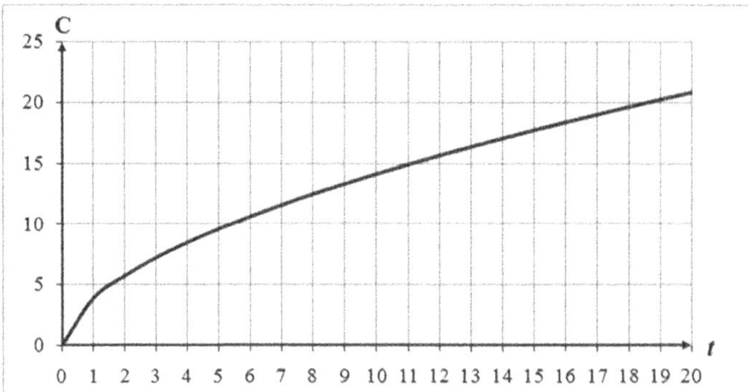

Fig. 2. The number of messages broadcast by A_1

The degree of expected utility of the message, calculated by the formula (2), decreases with time since it is assuming that the agent uses it in solving the problem and then does not return to it. By the last moment, the usefulness of the resource is completely exhausted (Fig. 3).

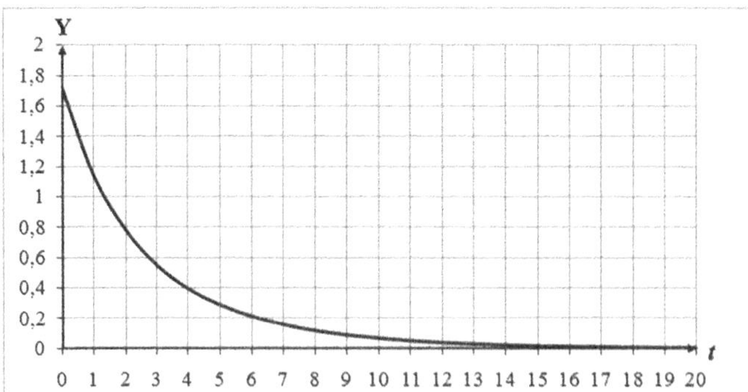

Fig. 3. The dynamics of the usefulness of the flow of messages broadcast A_1

The delivery time of the sequence of messages that form the stream under study, calculated by the formula (3), is distributing as shown in Fig. 4: the greater the discrepancy between the behavioral types of agents, the longer the delivery of messages takes. An important role is also playing by the correlation between agents (d_{ij}): the more "linked" IAS participants are, the easier it is for them to communicate information to each other.

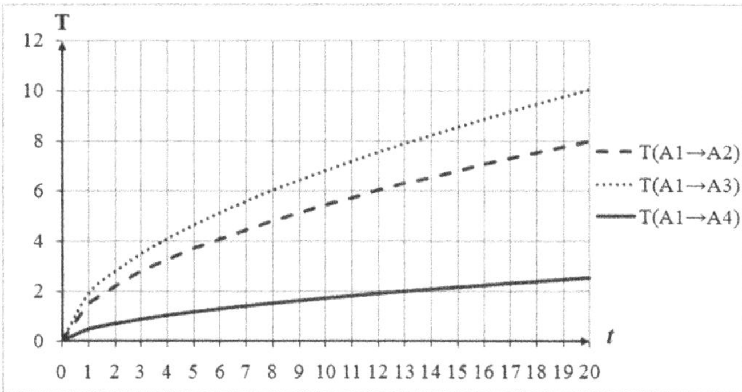

Fig. 4. The distribution of the delivery time of all messages in the stream

The total time for bringing the MF to the recipients, including the time for the translator to explain this information, is distributed as shown in Fig. 5. Obviously, due to the different effectiveness of agents in the system and the level of their informational perception, it is necessary to spend more time on reporting the flow of messages in the effect of $A_1 \to A_3$.

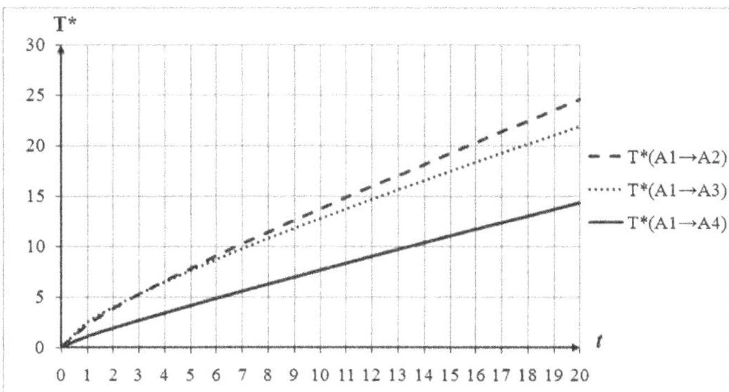

Fig. 5. Total message flow time

The value of the message flow for recipients (Fig. 6), calculated by the formula (5), following the law of diminishing usefulness of operational information decreases with time in all cases of impacts $A_i \to A_j$.

Herewith the highest value of $F_{ц}$ is characteristic of the impact $A_1 \to A_2$.

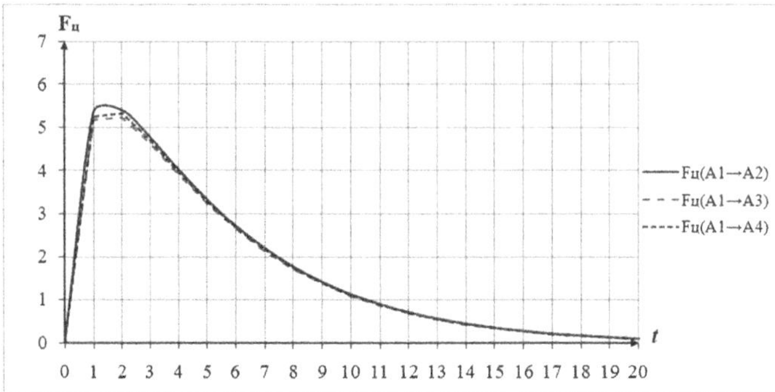

Fig. 6. The value of message flow for recipients

Further experiments showed that with a decrease in the level of the truthfulness of the transmitted information, the expected utility of the MF decreases, and the level of benevolence in the system decreases.

False information has no utility or value for IAS. It, on the contrary, destabilizes the system, creates obstacles and obstacles to achieving the goal. False information is unprofitable even to the agent himself her transmitting. Half-truths can often bring short-term gain, but here the question arises of the subsequent payment for this benefit. The most stable and safe for all agents is the strategy of transmitting true informational messages. The usefulness of such messages decreases naturally because over time they simply lose their relevance.

The degree of usefulness and value of MF with an increasing degree of truthfulness of information increases, and the time for delivery of all messages decreases. This is because broadcasting false information requires more effort than broadcasting the truth.

5 Conclusions

Thus, the article is devoted to the development of messaging models between intelligent agents and the computer implementation of these models. The mathematical model of the message flow is a system consisting of five distribution functions of the following characteristics: the number of transmitted messages; the degree of expected usefulness of the transmit messages delivery time; the total time the agents were convincing of the reliability of the messages; message flow significance for an intelligent active system. The dependencies of these characteristics on the most important anthropocentric parameters of intelligent agents are determined. These parameters include, for example, the level of difficulty of the task, the level of benevolence and comfort of the climate in the system, levels of agent activity depending on their psycho-behavioral type. By changing these and other parameters, you can control the characteristics of the message flow and, therefore, the quality of the circulating information in an intelligent active system.

Acknowledgments. The reported study was funded by RFBR according to the research project No. 17-01-00817A.

References

1. Mutovkina, N.Yu.: The formation of the optimal composition of multi-agent system. In: Hu, Z., Petoukhov, S., He, M. (eds.) Advances in Artificial Systems for Medicine and Education, AIMEE 2017. Advances in Intelligent Systems and Computing, vol. 658, pp. 293–302. Springer, Cham (2017). https://doi.org/10.1007/978-3-319-67349-3_28
2. Mutovkina, N.Yu., Kuznetsov, V.N.: Algorithms for agreement and harmonization the creative solutions of agents in an intelligent active system. In: Hu, Z., Petoukhov, S., He, M. (eds.) Advances in Artificial Systems for Medicine and Education II, AIMEE 2018. Advances in Intelligent Systems and Computing, vol. 902, pp. 651–660. Springer, Cham (2020). https://doi.org/10.1007/978-3-030-12082-5_59
3. Lipman, O.: Falsehood in Truth. O. Lipmann, L. Adam; foreword and trans. of A.E. Brusilovsky, 47 [1] p. Legal Publishing House of Ukraine, Kharkiv (1929)
4. Lipman, O., Romanov, V.V., Romanova, E.V.: Psychology of interrogation of the accused in the criminal process. In: Legal Psychology: Chrestomathy, 352 p. Yurist (2000)
5. Gubanov, D.A., Novikov, D.A., Chkhartishvili, A.G.: Models of reputation and information management in social networks. In: Management of Large Systems: A Collection of Works, vol. 26, no. 1, pp. 209–234 (2009)
6. Dashkova, A.Yu.: Manipulative methods of influencing mass consciousness. Bull. Volga Acad. Public Serv. **1**(22), 74–77 (2010)
7. Mutovkina, N.Yu.: Methods of Coordinated Control in Active Systems: Monograph, 164 p. Tver State Technical University, Tver (2018)
8. Otto von Bismarck: quotes of famous people, aphorisms. https://citaty.info/man/otto-fon-bismark?page=2. Accessed 07 Feb 2019
9. Chouhan, S.S., Niyogi, R.: An analysis of the effect of communication for multi-agent planning in a grid world domain. Int. J. Intell. Syst. Appl. (IJISA) **4**(5), 8–15 (2012). https://doi.org/10.5815/ijisa.2012.05.02
10. Lata, S., Goel, N.: Optimized communication of group mobility in WPAN. Int. J. Comput. Netw. Inf. Secur. (IJCNIS) **8**(3), 10–18 (2016). https://doi.org/10.5815/ijcnis.2016.03.02
11. Wunsch, G.: Systems Theory: Translate from German of T.N. Krenkel, 288 p. Soviet Radio (1978)
12. El Mhouti, A., Nasseh, A., Erradi, M.: Stimulate engagement and motivation in MOOCs using an ontologies based multi-agents system. Int. J. Intell. Syst. Appl. (IJISA) **8**(4), 33–42 (2016). https://doi.org/10.5815/ijisa.2016.04.04

Development of Models of Quantum Biology Based on the Tensor Product of Matrices

Elena Fimmel[1(✉)] and Sergey V. Petoukhov[2]

[1] Mannheim University of Applied Sciences, Paul-Wittsack-Straße 10, 68163 Mannheim, Germany
e.fimmel@hs-mannheim.de

[2] Mechanical Engineering Research Institute, Russian Academy of Sciences, M.Kharitonievsky Pereulok, 4, Moscow, Russia
spetoukhov@gmail.com

Abstract. The Kronecker tensor product of matrices is one of the most important operations in quantum mechanics and quantum informatics. The article is devoted to various applications of this mathematical operation for revealing hidden interrelations among molecular-genetic structures and for developing quantum biology and algebraic biology. Special attention is paid to matrix representations of the set of DNA nucleobases and their hydrogen bonds. These representations reveal, in particular, hidden symmetries in alphabets of DNA *n*-plets and also possibilities of applications of hyperbolic numbers and their extensions for modeling some hidden regularities in long DNA sequences.

Keywords: Genetic code · Kronecker product · Hydrogen bonds · Hyperbolic numbers

1 Introduction

The term "quantum biology" was introduced in 1932 by one of creators of quantum mechanics P. Jordan [1] who analysed the following significant difference between biological and inanimate objects. Jordan correctly pointed out that inanimate objects were governed by the average random motion of millions of particles, such that the motion of a single molecule has no influence whatsoever on the whole object. This insight is usually credited to Erwin Schrödinger, who later claimed that life was different from inorganic chemistry because of its dependence on the dynamics of a small number of molecules. Jordan similarly argued that the few molecules that control the dynamics of living cells within the control center have a dictatorial influence, such that quantum-level events that govern their motion are amplified to influence the entire organism. Jordan believed that living organisms were uniquely able to carry out this amplification in a way that was conspicuously different from inanimate matter. Jordan was convinced he could extend quantum indeterminism from the subatomic world to macroscopic biology. He even made a connection with free will by suggesting a link between quantum mechanics and psychology [1]. At present, the topic of quantum biology and quantum genetics is being developed by many modern authors whose model approaches suppose that living organisms use principles of quantum informatics [2–28].

Z. Hu et al. (Eds.): AIMEE 2019, AISC 1126, pp. 126–135, 2020.
https://doi.org/10.1007/978-3-030-39162-1_12

In quantum mechanics and quantum informatics one of the most important mathematical operations is the Kronecker tensor multiplication of two matrices. By definition, this operation is an operation on two matrices of arbitrary size resulting in a block matrix the way that each of entries of the first matrix is multiplied with the whole second matrix; in the case of symbolic matrices, the multiplication in this context is defined as a concatenation of symbols in the given order [29]. This article describes special applications of the Kronecker tensor multiplication, which allow developing mathematical models and analytical approaches in bioinformatics and algebraic biology, first of all for analyzing structured alphabets of genetic molecules DNA. In DNA, genetic information is written in very long sequential texts using only 4 letters - adenine A, cytosine C, guanine G, thymine T -, which form the alphabet of DNA monoplets. For the genetic system, the combination of these monoplets into doublets, triplets, etc plays an important role; such combinations additionally form DNA alphabets of 16 doublets, 64 triplets, etc., which have their own internal structures (sub-alphabets) because of different genetic coding meaning of their elements. Theoretical physicist Rumer [30, 31] was the first scientist to discover a significant symmetry of the genetic code named after him. He noticed that for 32 codons only the first two bases (he called them roots) decide which amino acid is coded, while for the other 32 the knowledge of the third base is necessary. This important discovery by Rumer can be described very briefly with the help of the Kronecker tensor product, as shown below, although Rumer himself did not explicitly use a tensor product for this purpose. Below other applications of the Kronecker tensor product are considered for analysis of symmetries in the genetic system and for development of new model approaches in bioinformatics.

2 DNA Alphabets as Members of the Single Tensor Family of Matrices

Each of the mentioned DNA alphabets can be represented in a form of appropriate mathematical matrix. Figure 1 shows such representations of the DNA alphabets:

- The alphabet of 4 monoplets in the form of (2 × 2)-matrix;
- The alphabet of 16 doublets in the form of (4 × 4)-matrix;
- The alphabet of 64 triplets in the form of (8 × 8)-matrix.

All these matrices of DNA alphabets can be represented as tensor powers $[\mathrm{C, A; T, G}]^{(n)}$ $n = 1, 2, 3, \ldots$ of the 2 × 2 matrix $[\mathrm{C, A; T, G}]$, which consists of single nucleotide bases. Let us note that in this representation each n-plet, especially each doublet and triplet, automatically occupies its own place within the corresponding n-plet alphabet (see Fig. 1).

In this presentation, some otherwise hidden symmetries of the genetic code are revealed [32–34].

The first remarkable symmetry can be seen when we look at the triplet alphabet thus obtained (see Fig. 1 again) and the corresponding amino acids coded according to the Vertebrate Mitochondrial Code table: the entire set of 8 rows shows itself unexpectedly as a complect of 4 pairs of adjacent rows with identical lists of amino acids and stop-codons in each pair (Fig. 2).

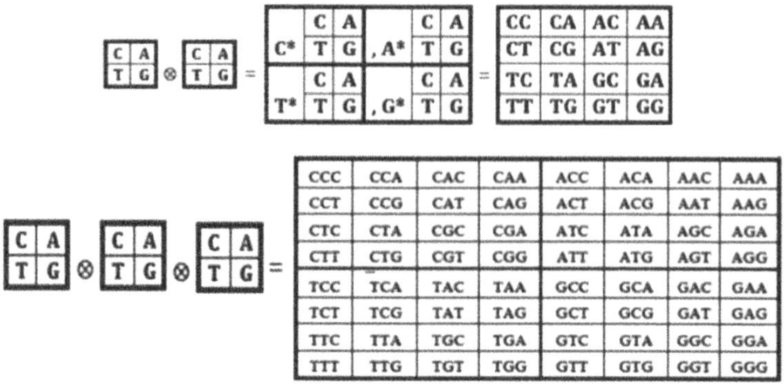

Fig. 1. The tensor family of genetic matrices $[C, A; T, G]^{(n)}$ (here tensor power $n = 1, 2, 3$) of DNA-alphabets of 4 nucleotides, 16 dinucleotides and 64 trinucleotides. The symbol ⊗ means the Kronecker tensor product.

CCC Pro	CCA Pro	CAC His	CAA Gln	ACC Thr	ACA Thr	AAC Asn	AAA Lys
CCT Pro	CCG Pro	CAT His	CAG Gln	ACT Thr	ACG Thr	AAT Asn	AAG Lys
CTC Leu	CTA Leu	CGC Arg	CGA Arg	ATC Ile	ATA Met	AGC Ser	AGA Stop
CTT Leu	CTG Leu	CGT Arg	CGG Arg	ATT Ile	ATG Met	AGT Ser	AGG Stop
TCC Ser	TCA Ser	TAC Tyr	TAA Stop	GCC Ala	GCA Ala	GAC Asp	GAA Glu
TCT Ser	TCG Ser	TAT Tyr	TAG Stop	GCT Ala	GCG Ala	GAT Asp	GAG Glu
TTC Phe	TTA Leu	TGC Cys	TGA Trp	GTC Val	GTA Val	GGC Gly	GGA Gly
TTT Phe	TTG Leu	TGT Cys	TGG Trp	GTT Val	GTG Val	GGT Gly	GGG Gly

Fig. 2. Inside any of the four pairs of rows 1–2, 3–4, 5–6, 7–8, both rows are identical in their lists of encoded amino acids and stop-codons (an appropriate amino acid or stop-codon are shown under each triplet for the case of the Vertebrate Mitochondrial Code, which is the most symmetric among known dialects of the genetic code). Here traditional amino acid designations are used.

Let us now turn to the above mentioned Rumer's dichotomy [30]: the alphabet of 64 triplets is divided by nature into two equal sub-sets on the basis of strong and weak roots, i.e., the first two positions in triplets: (a) 32 triplets with strong roots, i.e., with 8 "strong" doublets AC, CC, CG, CT, GC, GG, GT, TC on their first positions (such triplets are denoted by black color on Fig. 3); (b) 32 triplets with weak roots, i.e., with

8 "weak" doublets AA, AG, AT, GA, TA, TT, TG. Code meanings of triplets with strong roots do not depend on the letters on their third position; code meanings of triplets with weak roots depend on their third letter. Figure 3 shows locations of these "strong" and "weak" doublets and also triplets with strong and weak roots in matrices of 16 doublets and 64 triplets from Fig. 1.

CC	CA	AC	AA
CT	CG	AT	AG
TC	TA	GC	GA
TT	TG	GT	GG

CCC	CCA	CAC	CAA	ACC	ACA	AAC	AAA
CCT	CCG	CAT	CAG	ACT	ACG	AAT	AAG
CTC	CTA	CGC	CGA	ATC	ATA	AGC	AGA
CTT	CTG	CGT	CGG	ATT	ATG	AGT	AGG
TCC	TCA	TAC	TAA	GCC	GCA	GAC	GAA
TCT	TCG	TAT	TAG	GCT	GCG	GAT	GAG
TTC	TTA	TGC	TGA	GTC	GTA	GGC	GGA
TTT	TTG	TGT	TGG	GTT	GTG	GGT	GGG

Fig. 3. The disposition of "strong" doublets and triplets with strong roots (denoted by black) and a "weak" doublets and triplets with weak roots in appropriate matrices from the tensor family of matrices $[C, A; T, G]^{(n)}$ (Fig. 1).

The strong roots and the corresponding triplets are marked as dark in Fig. 3 and the weak ones as light. It is immediately visible that their placement in the tensor product matrices has a very symmetrical character: (a) the left and right halves of the matrix mosaic are mirror-anti-symmetric to each other in its colors: any pair of cells, disposed by mirror-symmetrical manner in these halves, possesses the opposite colors. One can say that each row of this mosaic matrix corresponds to an odd function; (b) both quadrants along each of matric diagonals have identical mosaics; (c) each row of the mosaic matrix $[C\ A;\ T\ G]^{(3)}$ has a meander-line character (the term"meander-line" means here that numbers of black and white fragments are equal to each other along each row). Such structure is typical for Rademacher and Walsh functions, as well as for Hadamard matrices, which are well known in the theory of noise-immunity coding and in quantum mechanics and informatics [24, 32–35].

One can also see from Fig. 3 that the black-and-white mosaic of the (4 $\times$ 4)-matrix of 16 doublets is reproduced in the mosaic of the (8 $\times$ 8)-matrix of 64 triplets on an enlarged scale. It is related with fractal properties of the operation of the Kronecker tensor product [36, Chapter X], which provide an inheritance of mosaic structure of the original matrix under its tensor exponentiation. Figure 4 shows an example of the formation of fractal patterns, the type of which depends on the mosaic of the original matrix.

These fractal properties of the Kronecker tensor product can be used for mathematical modelling many known and genetically inherited fractal-like structures and processes in biological bodies including known relations between cancer and fractals [37].

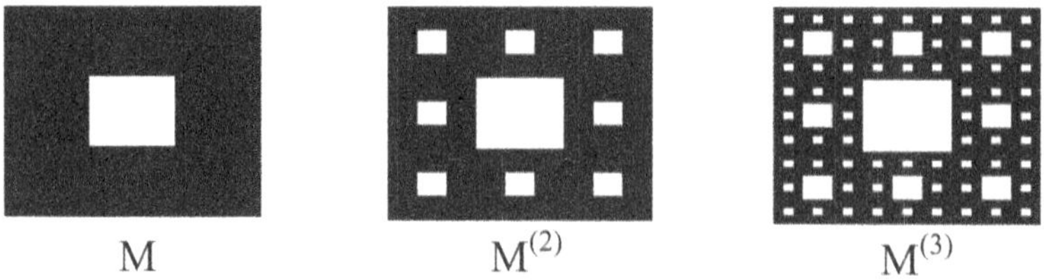

Fig. 4. The example of a fractal carpet of Sierpinski produced in the tensor family of matrices $M^{(n)} = [1, 1, 1; 1, 0, 1; 1, 1, 1]^{(n)}$, where (n) means tensor power. Black and white elements of mosaics correspond to elements 1 and 0.

3 The Tensor Product, Hyperbolic Numbers and Long DNA Sequences

This Section describes using the tensor product of genetic matrices to study some properties of long DNA sequences by means of known algebra of hyperbolic numbers and its 2^n-dimensional extensions. Such research approach is useful for development of algebraic biology; in particular, it allowed revealing hidden numerical regularities of hydrogen bonds in long DNA sequences briefly presented below.

3.1 Short Introduction to Hyperbolic Numbers and Their Tensor Products

Two-dimensional hyperbolic numbers $m_1 = a * \mathbf{1} + b * \mathbf{j}$ (where **1** is the real unit; **j** is the imaginary unit with the property $\mathbf{j} \neq \pm\mathbf{1}$ but $\mathbf{j}^2 = \mathbf{1}$; a, b are real coefficients) are used in physics and mathematics and they have also synonimical names: "split-complex numbers", "perplex numbers" and "double numbers". The collection of all hyperbolic numbers forms algebra over the field of real numbers [38].

Figure 5 shows the matrix representation of these hyperbolic numbers $m_1 = a * \mathbf{1} + b * \mathbf{j}$ and its decomposition into 2 sparse matrices with the multiplication table for these sparse matrices playing the role of basic units of hyperbolic numbers. These sparse matrices are unitary matrices.

$$m_l = a*\mathbf{1} + b*\mathbf{j} \rightarrow \begin{vmatrix} a, b \\ b, a \end{vmatrix} = a* \begin{vmatrix} 1, 0 \\ 0, 1 \end{vmatrix} + b* \begin{vmatrix} 0, 1 \\ 1, 0 \end{vmatrix}$$

*	1	j
1	1	j
j	j	1

Fig. 5. On the left: the matrix representation of hyperbolic numbers $m_1 = a * \mathbf{1} + b * \mathbf{j}$ and its decomposition into two sparse matrices, which represent the real unit 1 (the first sparse matrix) and the imaginary unit j (the second sparse matrix). Here $\mathbf{j}^2 = +1$; a and b are real numbers. On the right: the multiplication table of the basic elements **1** and **j** of hyperbolic numbers.

The second tensor power of the bisymmetric matrix [a, b; b, a] is decomposed into 4 sparse matrices, the set of which is closed relative to multiplication and satisfies to the multiplication table in Fig. 6. The set of these (4 × 4)-matrices corresponds to algebra of 4-dimensional numbers $m_2 = aa * \mathbf{1} + ab * e_1 + ba * e_2 + bb * e_3$.

$$\left|\begin{matrix} a, b \\ b, a \end{matrix}\right|^{(2)} = \left|\begin{matrix} aa, ab, ba, bb \\ ab, aa, bb, ba \\ ba, bb, aa, ab \\ bb, ba, ab, aa \end{matrix}\right| = aa* \left|\begin{matrix} 1\,0\,0\,0 \\ 0\,1\,0\,0 \\ 0\,0\,1\,0 \\ 0\,0\,0\,1 \end{matrix}\right| + ab* \left|\begin{matrix} 0\,1\,0\,0 \\ 1\,0\,0\,0 \\ 0\,0\,0\,1 \\ 0\,0\,1\,0 \end{matrix}\right| + ba* \left|\begin{matrix} 0\,0\,1\,0 \\ 0\,0\,0\,1 \\ 1\,0\,0\,0 \\ 0\,1\,0\,0 \end{matrix}\right| +$$

$$+ bb* \left|\begin{matrix} 0\,0\,0\,1 \\ 0\,0\,1\,0 \\ 0\,1\,0\,0 \\ 1\,0\,0\,0 \end{matrix}\right| = aa*\mathbf{1} + ab*e_1 + ba*e_2 + bb*e_3 .$$

*	**1**	e_1	e_2	e_3
1	**1**	e_1	e_2	e_3
e_1	e_1	**1**	e_3	e_2
e_2	e_2	e_3	**1**	e_1
e_3	e_3	e_2	e_1	**1**

Fig. 6. The decomposition of the matrix [a, b; b, a]$^{(2)}$ into 4 sparse matrices, the set of which is closed relative to multiplication. The multiplication table for this set is shown at the right. The symbol 1 denotes the identity matrix.

The higher tensor powers $n = 3, 4, 5, \ldots$ of the matrix [a, b; b, a] produce matrices, which can be also decomposed into 2^n sparse matrices, the set of which is closed relative to multiplication and which define appropriate algebras of 2^n-dimensional hypercomplex numbers m_n termed as "hyperbolic matrions" of the order n [33, 34].

3.2 An Application for Modelling the Genetic Code

In the DNA double helix, complementary nucleobases A and T are connected by 2 hydrogen bonds and are called *weak bases* (W), while the other two, C and G (*strong bases* (S)), are connected by 3 hydrogen bonds. Taking this into account, the nucleobases can be represented by their numbers of hydrogen bonds with denotations A = T = 2 and C = G = 3. Correspondingly, the genetic matrix [C, A; T, G] is represented as the bisymmetric matrix of numbers of hydrogen bonds [2, 3], which corresponds to hyperbolic number $m_1 = 3 + 2\mathbf{j}$. In long DNA, numbers of weak and strong bases are met with certain percentages denoted by %W and %S, which satisfy the condition %W + %S = 100%. The appropriate matrix of percentages of strong and weak bases is the matrix [%S, %W; %W, %S], which corresponds to hyperbolic number $m_1 = \%S + \%W\mathbf{j}$. Figure 7 shows the second tensor power of the matrices [2, 3] and [%S, %W; %W, %S].

33	32	23	22
32	33	22	23
23	22	33	32
22	23	32	33

%S%S	%S%W	%W%S	%W%W
%S%W	%S%S	%W%W	%W%S
%W%S	%W%W	%S%S	%S%W
%W%W	%W%W	%S%W	%S%S

Fig. 7. The matrices [3, 2; 2, 3]$^{(2)}$ and [%S, %W; %W, %S]$^{(2)}$.

In accordance with Fig. 6, the (numeric!) matrix $[\%S, \%W; \%W, \%S]^{(2)}$ in Fig. 7 corresponds to hyperbolic matrions of the second order m_2. Each (2 × 2)-quadrant of this (4 × 4)-matrix is itself a matrix representation of hyperbolic number because of properties of the Kronecker tensor product. Matrices, received in Fig. 7 by means of the Kronecker tensor product, provokes a hypothesis that percentages of hydrogen doublets SS, SW, WS, WW in long DNA sequences can be modelled as the product of percentages of hydrogen monoplets %S and %W. The testing of this hypothesis on the materials of the eukaryotic and prokaryotic genomes confirmed its validity: the model values of percentages almost coincided with the phenomenological values of percentages in all tested genomes [25, 40].

For example, in the human chromosome №1, phenomenological percentages of hydrogen doublets are the following: %SS = 0.1626, %SW = 0.2546, %WS = 0.2546 and %WW = 0.3282. Using phenomenological percentages of hydrogen monoplets %S = 0.4172 and %W = 0.5828 in this chromosome, calculations of model values of percentages of hydrogen doublets %SS, %SW, %WS and %WW on the basis of the mentioned hypothesis give the following: $\%S * \%S = 0.1741 \approx \%SS$, $\%S * \%W = 0.2431 \approx \%SW$, $\%W * \%S = 0.2431 \approx \%WS$, $\%W * \%W = 0.3397 \approx \%WW$. As a by-product of this modeling, we get that $\%SW \approx \%WS$, since the multiplication of numbers is commutative. One can see that these model percentages almost reproduce phenomenological percentages. Here all values are rounded to four decimal places.

Similar results in long DNA sequences were received for modeling phenomenological percentages of hydrogen triplets, tetraplets and pentaplets by using the exponentiation of the matrix [%S, %W; %W, %S] in the appropriate tensor powers 3, 4 and 5 [25, 40]. Due to this model approach, knowing only percentages %W and %S of hydrogen monoplets in a considered long DNA sequence, one can effectively predict percentages of hydrogen n-plets in this DNA ($n = 2, 3, 4, \ldots$).

These results confirm that - from the standpoint of the stated model approach - regularities of percentages of n-plets of hydrogen bonds in long DNA can be modeled using the algebra of 2-dimensional hyperbolic numbers and its extensions into noted algebras of 2^n-dimensional hypercomplex numbers.

4 Conclusions

Described relations of molecular-genetic structures with the Kronecker tensor product are interesting since this mathematical operation plays an important role in quantum mechanics and quantum computing. The following quotation emphasises this importance: «*This construction is crucial to understanding the quantum mechanics of multiparticle systems ... . The state space of a composite physical systems is the tensor product of the state spaces of the component physical systems*" [41, p. 71, 102]. On the basis of these relations a new class of mathematical models can be created for developing quantum biology and algebraic biology.

In our opinion one of the most significant results concerns the described possibility to use hyperbolic matrions for modelling phenomenological data on percentages of n-plets of hydrogen bonds in long DNA sequences. It is interesting taking into account many other data about the role of hyperbolic numbers in mathematical natural sciences. For example, transformations of hyperbolic rotations, which are particular cases of hyperbolic numbers, are known in the special theory of relativity as Lorentz transformations. In another example, hyperbolic rotations are used to model Weber-Fechner basic psychophysical law [24]. Hyperbolic numbers are represented by symmetric matrices, which always have real eigenvalues and which are significant for the theory of resonances in oscillatory systems with many degrees of freedom [24].

In the whole, the Kronecker tensor product of matrices - with its ability to generate fractal-like matrices and to generate also hyperbolic matrions - seem to be very perspective for their further applications in genetics, bioinformatics, evolutionary biology and biotechnology. In these fields many publications exist [42–47], which are related with materials of our article.

The data on hydrogen bonds in long DNA sequences presented in our article deserve special attention, since hydrogen bonds play an important role in living organisms and in water. For example, L. Pauling wrote that «*the significance of the hydrogen bond for physiology is greater that of any other single structural feature*" [48, p. 450]. Hydrogen is a component of almost all organic matter and is present in all living cells, where the number of hydrogen atoms is almost 63%.

References

1. McFadden, J., Al-Khalili, J.: The origins of quantum biology. Proc. R. Soc. A: Math. Phys. Eng. Sci., 1–13 (2018). https://royalsocietypublishing.org/doi/full/10.1098/rspa.2018.0674
2. Abbott, D., Davies, P.C.W., Pati, A.K. (eds.): Quantum Aspects of Life, foreword by Sir Roger Penrose (2008). ISBN-13: 978-1-84816-253-2
3. Altaisky, M.V., Filatov, F.P.: Genetic information and quantum gas (2001). arXiv:quant-ph/0106123v1. Submitted 22 June 2001
4. Fimmel, E., Petoukhov, S.: Genetic code modeling from the perspective of quantum informatics. In: Hu, Z.B., He, M., Petoukhov S.V. (eds.) Advances in Artificial Systems for Medicine and Education II, pp. 117–126. Springer (2020). ISBN 978-3-030-12081-8
5. Hu, Z.B., Petoukhov, S.V., Petukhova, E.S.: Generalized crystallography, the genetic system and biochemical esthetics. Struct. Chem. **28**(1), 354–368 (2017). https://doi.org/10.1007/s11224-016-0880-0. http://link.springer.com/journal/11224/28/1/page/2
6. Hu, Z.B., Petoukhov, S.V., Petukhova, E.S.: I-Ching, dyadic groups of binary numbers and the geno-logic coding in living bodies. Prog. Biophys. Mol. Biol. **131**, 354–368 (2017). https://doi.org/10.1016/j.pbiomolbio.2017.08.018
7. Hu, Z.B., Petoukhov, S.V., Petukhova, E.S.: On symmetries, resonances and photonic crystals in morphogenesis. Biosystems **173**, 165–173 (2018). https://doi.org/10.1016/j.biosystems.2018.09.004
8. Igamberdiev, A.U.: Quantum mechanical properties of biosystems: a framework for complexity, structural stability, and transformations. Biosystems **31**(1), 65–73 (1993)
9. Igamberdiev, A.I.: Quantum computation, non-demolition measurements, and reflective control in living systems. BioSystems **77**, 47–56 (2004)
10. Igamberdiev, A.I.: Physical limits of computation and emergence of life. BioSystems **90**, 340–349 (2007)
11. Igamberdiev, A.I.: Objective patterns in the evolving network of non-equivalent observers. BioSystems **92**, 122–131 (2008)

12. Igamberdiev, A.I., Shklovskiy-Kordi, N.E.: Computational power and generative capacity of genetic systems. BioSystems **142–143**, 1–8 (2016)
13. Igamberdiev, A.I., Shklovskiy-Kordi, N.E.: The quantum basis of spatiotemporality in perception and consciousness. Prog. Biophys. Mol. Biol. **130**, 15–25 (2017)
14. Josephson, B.D.: The physics of mind and thought. Preprint (2018). https://doi.org/10.13140/rg.2.2.36516.32640/2. https://www.researchgate.net/publication/328968105
15. Matsuno, K.: Cell motility as entangled quantum coherence. BioSystems **51**, 15–19 (1999)
16. Matsuno, K.: Quantum mechanics in first, second and third person descriptions. BioSystems **68**, 107–118 (2003)
17. Matsuno, K., Paton, R.C.: Is there a biology of quantum information? BioSystems **55**, 39–46 (2000)
18. Mikheenko, P.: Possible superconductivity in brain. https://arxiv.org/abs/1812.05602. Submitted 13 December 2018
19. Patel, A.: Quantum algorithms and the genetic code. Pramana – J. Phys. **56**(2–3), 367–381 (2001). arXiv:quant-ph/0002037
20. Patel, A.: Testing quantum dynamics in genetic information processing. J. Genet. **80**(1), 39–43 (2001)
21. Patel, A.: Why genetic information processing could have a quantum basis. J. Biosci. **26**(2), 145–151 (2001)
22. Penrose, R.: Shadows of the Mind: A Search for the Missing Science of Consciousness, 480 p. Oxford University Press, New York (1996)
23. Penrose, R.: Your eyes are not meant for seeing. Community (2019). https://thriveglobal.com/stories/your-eyes-are-not-meant-for-seeing/
24. Petoukhov, S.V.: The system-resonance approach in modeling genetic structures. Biosystems **139**, 1–11 (2016). http://petoukhov.com/PETOUKHOV_ARTICLE_IN_BIOSYSTEMS.pdf
25. Petoukhov, S.V.: The genetic coding system and unitary matrices. Preprints **2018**, 2018040131 (2018). https://doi.org/10.20944/preprints201804.0131.v2
26. Petoukhov, S.V.: The rules of long DNA-sequences and tetra-groups of oligonucleotides, 5th version,, 159 p. 8 October 2018. arXiv:1709.04943v5
27. Petoukhov, S.V.: Structural connections between long genetic and literary texts. Preprints **2018**, 2018120142 (2019). https://doi.org/10.20944/preprints201812.0142.v2. https://www.preprints.org/manuscript/201812.0142/v2
28. Petoukhov, S.V.: Connections between long genetic and literary texts. The quantum-algorithmic modelling. In: Hu, Z., Petoukhov, S., Dychka, I., He, M. (eds.) Advances in Computer Science for Engineering and Education II, ICCSEEA 2019. Advances in Intelligent Systems and Computing, vol. 938, pp. 534–543. Springer, Cham. (2019). https://doi.org/10.1007/978-3-030-16621-2_50. https://link.springer.com/chapter/10.1007/978-3-030-16621-2_50#citeas (online)
29. Bellman, R.: Introduction to Matrix Analysis, 351 p. Mcgraw-Hill Book Company, Inc., New-York (1960)
30. Rumer, Yu.B.: Systematization of the codons of the genetic code. Dokl. Akad. Nauk SSSR **183**(1), 225–226 (1968). (in Russian)
31. Fimmel, E., Strüngmann, L.: Yury Borisovich Rumer and his 'biological papers' on the genetic code. Phil. Trans. R. Soc. A **374**, 20150228 (2016)
32. Petoukhov, S.V.: Hadamard matrices and quint matrices in matrix presentations of molecular genetic systems. Symmetry: Cult. Sci. **16**(3), 247–266 (2005)
33. Petoukhov, S.V.: Matrix genetics, algebras of the genetic code, noise immunity, 316 p. RCD (2008). (in Russian)

34. Petoukhov, S.V., He, M.: Symmetrical Analysis Techniques for Genetic Systems and Bioinformatics: Advanced Patterns and Applications, 271 p. IGI Global, Hershey (2009)
35. Petoukhov, S.V.: Matrix genetics and algebraic properties of the multi-level system of genetic alphabets. Neuroquantology **9**(4), 60–81 (2011)
36. Gazale, M.: Gnomon. From Pharaons to Fractals. Princeton University Press, Princeton (1999)
37. Baish, J.W., Jain, R.K.: Fractals and cancer. Cancer Res. **60**, 3683–3688 (2000)
38. Harkin, A.A., Harkin, J.B.: Geometry of generalized complex numbers. Math. Mag. **77**(2), 118–129 (2004)
39. Fimmel, E., Gumbel, M., Karpuzoglu, A., Petoukhov, S.: On comparing composition principles of long DNA sequences with those of random ones. Biosystems **180**, 101–108 (2019)
40. Petoukhov, S.: Nucleotide epi-chains and new nucleotide probability rules in long DNA sequences. Preprints **2019** (2019). https://doi.org/10.20944/preprints201904.0011.v2
41. Nielsen, M.A., Chuang, I.L.: Quantum Computation and Quantum Information. Cambridge University Press, New York (2010)
42. Mousa, H.M.: DNA-genetic encryption technique. Int. J. Comput. Netw. Inf. Secur. (IJCNIS) **8**(7), 1–9 (2016). https://doi.org/10.5815/ijcnis.2016.07.01
43. Abo-Zahhad, M., Ahmed, S.M., Abd-Elrahman, S.A.: Genomic analysis and classification of exon and intron sequences using DNA numerical mapping techniques. Int. J. Inf. Technol. Comput. Sci. (IJITCS) **4**(8), 22–36 (2012)
44. Hata, R., Akhand, M.A.H., Islam, Md.M, Murase, K.: Simplified real-, complex-, and quaternion-valued neuro-fuzzy learning algorithms. Int. J. Intell. Syst. Appl. (IJISA) **10**(5), 1–13 (2018). https://doi.org/10.5815/ijisa.2018.05.01
45. Awadalla, M.H.A.: Spiking neural network and bull genetic algorithm for active vibration control. Int. J. Intell. Syst. Appl. (IJISA) **10**(2), 17–26 (2018). https://doi.org/10.5815/ijisa.2018.02.02
46. Abuljadayel, A., Wedyan, F.: An approach for the generation of higher order mutants using genetic algorithms. Int. J. Intell. Syst. Appl. (IJISA) **10**(1), 34–35 (2018)
47. Kumar, A., Sharma, R.: A genetic algorithm based fractional fuzzy PID controller for integer and fractional order systems. Int. J. Intell. Syst. Appl. (IJISA) **10**(5), 23–32 (2018)
48. Pauling, L.: The Nature of the Chemical Bond and the Structure of Molecules and Crystals: An introduction to Modern Structural Chemistry, 2nd edn, 664 p. Oxford University Press, London (1940)

A Beautiful Question: Why Symmetric?

Moon Ho Lee[1] and Jeong Su Kim[2(✉)]

[1] Division of Electronics, Chonbuk National University, Jeonju-Si, Korea
moonho@jbnu.ac.kr
[2] Department of ICT Engineering, Soongsil Cyber University, Seoul, Korea
kjs@mail.kcu.ac

Abstract. In this paper we investigate that most of plants have the symmetric property. In addition, the human body is also symmetric and contains the DNA symmetric base complementarity. We can see the logarithm helices in Fibonacci series and helices of plants. The sunflower has a shape of circle. A circle is circular symmetric because the shapes are same when it is shifted on the center. Einstein's spatial relativity is the relation of time and space conversion by the symmetrically generalization of time and space conversion over the spacial. The left and right helices of plants are the symmetric and have element-wise inverse relationships each other. The weight of center weight Hadamard matrix is 2 and is same as the base 2 of natural logarithm. The helix matrices are symmetric and have element-wise inverses matrics base on the finite group theory A B = B A = I_N. Also, we present human a DNA-RNA genetic code with symmetric base complements A = T = U = 30%, C = G = 20% and C + U = G + A. E. Chargaff discovered yeast and octopus DNA base complement [C T; A G] of four componentry A = T = U = 33% and C = G = 17% of the experimental results. This strongly hinted towards the base pair makeup of DNA, although Chargaff did not explicitly state this connection himself. However, it has not been proved in a mathematical analysis view yet. In this paper, we have a simple proof of this problem based on information theory as the doubly stochastic matrix.

Keywords: Symmetry · Helix · Center weighted Hadamard · DNA symmetric base complementarity

1 Introduction

As early as the Italian scientist Galileo Galilei (1564–1662), "is a scholar who made it clear that the laws of nature are mathematical. For example, Helix forms are known to know the providence of nature. There is symmetry in which is Fibonacci sequence and helical pattern. This is found in the spiral of plants and animals and proves that the center weighted Hadamard is based on logarithm with a load of 2 [1–6].

Plants live in a given space, making maximum use of the given spatial information. What is the best way to use space? The first person to comment on the opening of plants was Italian painter Leonardo da Vinci (1452–1519). He was not only the master of Renaissance art, but also a pioneer of science with extraordinary intelligence. If the law of nature is not interrupted, the sixth leaf always appears on the first leaf in many

Z. Hu et al. (Eds.): AIMEE 2019, AISC 1126, pp. 136–147, 2020.
https://doi.org/10.1007/978-3-030-39162-1_13

plants. It plays two roles for plants. The first is to ensure that these buds do not cover each other in the next year in order to receive maximum sunlight, wind, and rain in a given space. Since the five branches go in five different directions, the sixth branch comes out at a certain distance over the first branch. The flower that the sunlight directly touches is mature, but the flower which only reaches the reflection of the sunlight does not mature. There are four ways to leaf on the leaves. The most common way is that the sixth leaf is always above the sixth below. The second is above 2/3 below what is above 2/3. The third method is that the third upper leaf is on the third lower leaf. The fourth method is the method of the fir, and the leaves are layered. The branches come out in pairs of 4, 5 or 6, and at the top, the buds form a pyramid upwards from the stem. In walnut or oak, it forms a hemisphere upwards from the stem.

The second leaf always has its upper part pointed to the sky, allowing it to receive the dew falling in the air. And the leaves are staggered so that they do not hinder each other, which can be seen by observing ivy vines that cover the wall. One is to leave a space between the air and the sun to penetrate them, and the other is to let the water drop on the first leaf fall over the fourth or sixth leaf.

In the occupation of painters, Da Vinci saw that the leaves came out in a spiral pattern along the stem, and found that there was a certain rule here, and most of all, the sixth leaf was found just above the first leaf. When you go to the yard or garden, you can see the most common arrangement of grasses and leaves, such as thistles, umbrellas, horseshoe, mackerel, sweet potato, wormwood, sesame seeds, When they are counted along the stem of any leaf, the first leaf and the sixth leaf (if not exactly) are oriented in the same direction. There is almost no exception if the plant grows vertically straight (grassy plants) or if the leaves grow on the stem. Da Vinci discovered that the angle of the leaf of the plant is close to 144° ($360 \times 2/5$). The composition of this paper is as follows. The composition of this paper is as follows. The spiral of the plant in Sect. 2, Relativity Principal of Einstein in Sect. 3, Center Weighted Hadamard in Sect. 4, DNA Stochastic Entropy Analysis in Sect. 5, Discussion in Sect. 6, Conclusions in Sect. 7.

(a) "The Starry Night" and Sunflower (b) Pine cone helix

Fig. 1. "The Starry Night" and Sunflower, Pine cone helix

2 Spiral of Plants

What is always mentioned with opening is the spirals observed in plants. There is a work by Vincent van Gogh, the painter of Fig. 1(a), called "On a Starry Night." It is a representative work that formed the cloud of the above with two spirals. In addition to this, Van Gogh embodied a vortex-like narration in his work [6]. Van Gogh enjoyed sunflowers when he was in Arles, where he first settled in after arriving in South France. He painted all the houses he lived with yellow paint and called it "the yellow house." Burning sun, sunflower, etc. At that time, the picture was cast in yellow. Then, the flame-like curve and the winding swirling chopsticks slowly began to appear. In pine cones, the yarns form a symmetrical spiral in the form of a vortex. There are 55 and 89, or 89 and 34, spirals of two types rotating in opposite directions. In addition, there are spirals all over the plant, such as dense flowers and arrangement of leaves. The unusual fact here is that the numbers of these spirals are the numbers denominator or numerator of the openings listed above. As an example, the spiral of a pine cone is shown in Fig. 1 (b). The numbers in the pine cones are listed in order of magnitude, and they are Fibonacci sequences as follows. $1, 2, 3, 5, 8, 13, 21, 34, 55, 89, 144, 233, 377\ldots$ It is no coincidence that the number of spirals here is like this. It is because the denominator, the numerator, of the opening that makes the spiral array structure become such a number. That is, the number of spirals is directly related to the opening number.

3 Relativity Principal of Einstein

Einstein, with the help of his friend, mathematician Minkowskii, puts the theory that this world is made up of distorted spaces by using difficult mathematics. At this time, the curved space comes to us as gravity (or gravity) close to us, and in this world we live, in the morning, we open that majestic sun and at night we have a beautiful full moon. Einstein's successful and groundbreaking general relativity began with this symmetry. On the other hand, Herman Wyle, the grandfather of symmetry, thought that if the symmetry of coordinate transformations would allow gravity on moving objects in the sky and earth, there would be no other symmetry inevitably. Herman Weil introduced the concept of phase. What is the concept of phase? The imaginary number is the number of minus squares, and it is written in English. The number that is a mixture of imaginary numbers and real numbers is called a complex number, and a phase is an amount determined by the proportion of a real number and an imaginary number. For example, 2 + 3i means that the number of phases is 3/2, and 5 + 2i corresponds to 2/5. However, according to modern physics, the properties of matter are expressed as wave functions expressed by these complex numbers, and this phase is known to have nothing to do with physical reality. In other words, this world has symmetry about phase shift. By generalizing this phase to be a function of time and space, Einstein's general relativity was derived from the generalization of coordinate transformations. It is known that it is impossible to do so only with matter, and in order to have a phase symmetry, it must have an electromagnetic field. It is the conclusion that light can not have a phase symmetry if it is light because it is the electromagnetic wave. Therefore, when God created this world, light was indispensable in creating a symmetrical beautiful world. In addition, the electromagnetic field plays a role in determining the phase relationship when things move from one place to another. Light (the electromagnetic field) is the basis of everything that makes atoms and crystals and makes TV visible. Einstein has incorporated the following four laws. ① Lavoisier (1743–1794, France): The law of conservation of mass, "The amount of material before and after reaction in all chemical reactions is constant." ② Faraday (1791–1867, England): The law of electronic induction. The electromotive force generated in the circuit by electromagnetic induction (electromagnetic induction) is proportional to the time reduction ratio of the number of times of magnetic flux linkage. Generator and motor principle ③ Maxwell (1831–1879, UK): Electromagnetic wave equation, Symmetry of ∇^2 E and ∇^2 H.

$$v = \frac{1}{\sqrt{\in \mu}}[m/s] = 3 \times 10^8 [m/s],\ \nabla^2 E(1/\frac{\partial^2 E}{\partial t^2}) = \frac{1}{\sqrt{\in \mu}}\ (\text{Constant}),$$

$$\nabla^2 H(1/\frac{\partial^2 H}{\partial t^2}) = \frac{1}{\sqrt{\in \mu}}\ (\text{Constant})$$

④ Newton (1643–1727, United Kingdom): Newton's second law of mechanics, $F = ma$. The mass of an object is not constant but varies with velocity. The mass of the object in the kinematics system B moving at the velocity v with respect to the stationary system A causes a change in the following formula according to the velocity. Thus, energy and mass are inversely proportional and the speed of light is constant. $E = mc^2 \rightarrow E(\frac{1}{m}) = c^2$ (Constant: Refer AB = BA = I_N).

4 Center Weighted Hadamard

The center weighted Hadamard transform (CWHT) is defined. This transform is similar to the Hadamard transform (HT) in that it requires only real operations. The CWHT, however, weights the region of mid-spatial frequencies of the signal more than the HT. A simple factorization of the weighted Hadamard matrix is used to develop a rapid algorithm for the CWHT. The matrix decomposition takes the form of the Kronecker products of fundamental Hadamard matrices and successively lower order weighted Hadamard matrices.

The application of discrete orthogonal transforms for signal and image representation and compression is well known. A thoroughly investigated method, due to the ease and efficiency of its implementation, is based on the Hadamard transform HT, which we call the center weighted Hadamard transform (CWHT). This method retains much of the HT's simplicity, but offers a higher quality of representation than the central region of the image. The scheme was motivated by the fact that the human visual system is most sensitive to the mid-spatial frequencies [9]. The section is organized as follows. In the next section the CWHT is introduced and a recursive relation for the generation of the transform matrix is presented. Next, a fast CWHT algorithm which is similar to a fast HT method is derived for both the forward and inverse transforms that we have proved for examples.

Let the Hadamard and the center weighted Hadamard matrices of order $N = 2^k$ be denoted by $[H]_N$ and $[WH]_N$, respectively. The CWHT of an $N \times 1$ vector $[f]$ and an $N \times N$ (image) matrix $[g]$ are given by

$$[f] = [WH]_N[f], \tag{1}$$

$$[G] = [WH]_N[g][WH]_N. \tag{2}$$

The lowest order WH matrix is of size (4×4) and is defined as follows:

$$[WH]_4 \triangleq \begin{bmatrix} 1 & 1 & 1 & 1 \\ 1 & -2 & 2 & -1 \\ 1 & 2 & -2 & -1 \\ 1 & -1 & -1 & 1 \end{bmatrix}, \tag{3}$$

The inverse of (3) is

$$[WH]_4^{-1} = \frac{1}{4}\begin{bmatrix} 1 & 1 & 1 & 1 \\ 1 & -\frac{1}{2} & \frac{1}{2} & -1 \\ 1 & \frac{1}{2} & -\frac{1}{2} & -1 \\ 1 & -1 & -1 & 1 \end{bmatrix} \tag{4}$$

This choice of weighting was dictated, to a large extent, by the requirement of digital hardware simplicity [6]. As with the Hadamard matrix, a recursive relation governs the generation of higher order WH matrices, i.e.,

$$[WH]_N \triangleq [WH]_{N/2} \otimes [H]_2, [WH]_N[WH]_4^{-1} = [I]_4 \tag{5}$$

Where $\otimes$ is the kronecker product and is the lowest order Hadamard matrix given by (1) to (4), [8]:

$$[H]_2 = \begin{bmatrix} 1 & 1 \\ 1 & -1 \end{bmatrix} \tag{6}$$

We now present a fast algorithm for the CWHT which is related to the fast HT (FHT) algorithm [2–4, 8]: the FHT can be derived by decomposing $[H]_N$ into a product of k sparse matrices, each having rows/columns with only two nonzero elements. In order to develop a similar algorithm for the CWHT, define a coefficient matrix $[WC]_N$ by

$$[WC]_N \triangleq [H]_N[WH]_N. \tag{7}$$

Since $[H]_N^{-1} = 1/N[H]_N$, we have from (7) that

$$[WH]_N = 1/N[H]_N[WC]_N, \tag{8}$$

It is demonstrated that $[WC]_N$ is a sparse matrix with at most two nonzero elements per row and column. Therefore, the fast CWHT (FCWHT) is simply the FHT followed by a sparse matrix $1/N[WC]_N$. To show the sparseness of $[WC]_N$ we start by computing the lowest order $[WC]$, i.e. $[WC]_4$.

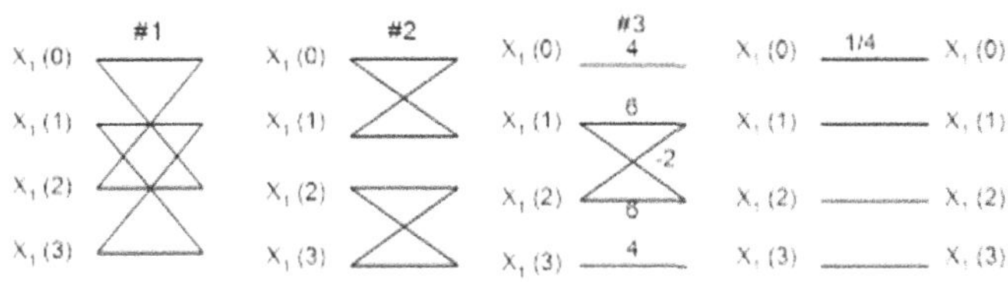

Fig. 2. Fast CWHT flow graph, N = 4

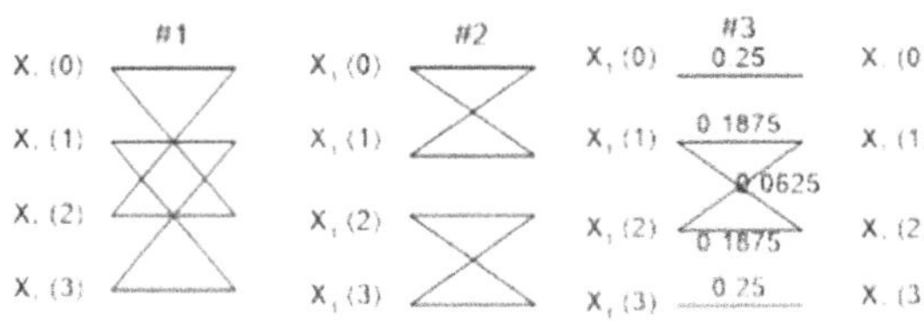

Fig. 3. Fast inverse CWHT flow graph, N = 4

From (1) to (7), we have

$$[WC]_4 = \begin{bmatrix} 1 & 1 & 1 & 1 \\ 1 & -1 & 1 & -1 \\ 1 & 1 & -1 & -1 \\ 1 & -1 & -1 & 1 \end{bmatrix} \begin{bmatrix} 1 & 1 & 1 & 1 \\ 1 & -2 & 2 & -1 \\ 1 & 2 & -2 & -1 \\ 1 & -1 & -1 & 1 \end{bmatrix} = \begin{bmatrix} 4 & 0 & 0 & 0 \\ 0 & 6 & -2 & 0 \\ 0 & -2 & 6 & 0 \\ 0 & 0 & 0 & 4 \end{bmatrix} \tag{9}$$

Clearly $[WC]_4$ is sparse. Using the expansion properties of the Hadamard and weighted Hadamard matrices, (7) can be written as

$$\begin{aligned} [WC]_N &= ([H]_{N/2} \otimes [H]_2)([WH]_{N/2} \otimes [H]_2) \\ &= ([H]_{N/2}[WH]_{N/2}) \otimes ([H]_2[H]_2) \\ &= [WC]_{N/2} \otimes (2[I]_2) \end{aligned} \tag{10}$$

Where $[I]_2$ is the 2×2 identity matrix. Since $[WC]_4$ is symmetric and has at most two non-zero elements in each row, it clearly follows from (1) to (10) that the same is true for $[WC]_8$, and hence, for any $[WC]_N$, $N = 2^k$, $k = 2, 3, 4, \ldots$. Figure 2 shows a flow graph of the 4-point FCWHT algorithm. From this figure it is clear that the first three iterations of the algorithm are those of the FHT. These are followed by the operation of the $[WC]_4$. The $N = 2^k$ point FCWHT algorithm requires $kN + N/2$ real additions and 1.5N real multiplication in contrast with the N-point FHT which requires kN real additions (Fig. 3).

5 DNA Stochastic Entropy Analysis

The genetic code of DNA stochastic base complement is given by [1, 2, 7]. If C = G = 17%, A = T = U = 33%, then the transition channel matrix is

$$P = \begin{bmatrix} C & T \\ A & G \end{bmatrix} = \begin{bmatrix} 0.17 & 0.33 \\ 0.33 & 0.17 \end{bmatrix}. \tag{11}$$

Let the DNA bases [C T; A G] of the Markov process, with the two independent source probabilities, with the 0.17p and 0.33p, to be maximized, then, we have

$$P = \begin{bmatrix} 0.17p & 1-0.17p \\ 1-0.17p & 0.17p \end{bmatrix} = \begin{bmatrix} 0.49987 & 1-0.49987 \\ 1-0.49987 & 0.49987 \end{bmatrix} \approx \begin{bmatrix} 0.5 & 0.5 \\ 0.5 & 0.5 \end{bmatrix}, \tag{12}$$

where is 2.941.

From (12), we have

$$0.17p = 1 - 0.17p, \; p = 2.941. \tag{13}$$

Also, in a similar fashion as (12)

$$P = \begin{bmatrix} 0.33p & 1-0.33p \\ 1-0.33p & 0.33p \end{bmatrix} = \begin{bmatrix} 0.498 & 1-0.498 \\ 1-0.498 & 0.498 \end{bmatrix} = \begin{bmatrix} 0.498 & 0.502 \\ 0.502 & 0.498 \end{bmatrix} \approx \begin{bmatrix} 0.5 & 0.5 \\ 0.5 & 0.5 \end{bmatrix}, \tag{14}$$

where 0.33p = 1 − 0.33p, p is 1.5151.

Therefore, a sum of (12) and (14) is

$$2P = \begin{bmatrix} 0.5 & 0.5 \\ 0.5 & 0.5 \end{bmatrix} + \begin{bmatrix} 0.5 & 0.5 \\ 0.5 & 0.5 \end{bmatrix} = \begin{bmatrix} 1 & 1 \\ 1 & 1 \end{bmatrix}. \tag{15}$$

Finally, (11) is

$$2P = 2\begin{bmatrix} C & T \\ A & G \end{bmatrix} = 2\begin{bmatrix} 0.17 & 0.33 \\ 0.33 & 0.17 \end{bmatrix} = \begin{bmatrix} 0.34 & 0.66 \\ 0.66 & 0.34 \end{bmatrix}. \tag{16}$$

If p is the source probability of the first symbol event, then the entropy function [6] is

$$H_2(p) = p\log_2\left(\frac{1}{p}\right) - (1-p)\log_2\left(\frac{1}{1-p}\right), \tag{17}$$

and is tabulated in the last column of the Table 1. The graph of this function of Shannon and RNA are given by in Fig. 4. Note that at $p = 0$ and $p = 1$ it has a vertical tangent, since

$$\begin{aligned} &\tfrac{d}{dp}\left[p\log_2\left(\tfrac{1}{p}\right) + (1-p)\log_2\left(\tfrac{1}{1-p}\right)\right] \\ &= \left[\log_2\left(\tfrac{1}{p}\right) - 1 - \log_2\left(\tfrac{1}{1-p}\right) + 1\right]\log_2 e \\ &= \log_2\left(\tfrac{1}{p}\right) - \log_2\left(\tfrac{1}{1-p}\right) = 0. \end{aligned} \tag{18}$$

That the maximum occurs at $p = 1/2$ where the derivative is zero.
So,

$$\begin{aligned} &\log_2\left(\tfrac{1}{p}\right) - \log_2\left(\tfrac{1}{1-p}\right) = 0 \\ &\Rightarrow \left(\tfrac{1}{p}\right) - \left(\tfrac{1}{1-p}\right) = 0 \end{aligned} \tag{19}$$

Then, we have

$$p = 1 - p \Rightarrow p = \frac{1}{2}. \tag{20}$$

The RNA base [C U; A G] symmetric entropy is given by

$$\begin{aligned} H_2(p)_{RNA} &= p\log_2\left(\frac{1}{p}\right) - (1-p)\log_2\left(\frac{1}{1-p}\right), \\ &= 0.92482, \; where \;\; p \;\; is \; 0.34 \; or \; 0.66 \end{aligned} \tag{21}$$

and the Shannon entropy is denoted by

$$H_2(p)_{Shannon} = p\log_2\left(\frac{1}{p}\right) - (1-p)\log_2\left(\frac{1}{1-p}\right) = 1, \; where \;\; p \;\; is \; 0.5. \tag{22}$$

The Shannon and RNA Shannon entropy is shown in Table 1 and Fig. 4.

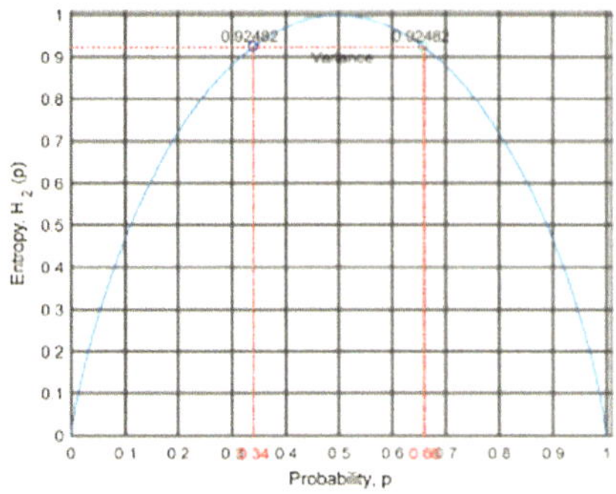

Fig. 4. Shannon and RNA entropy with probability in RNA[C U; A G]

Table 1. Entropy of RNA bases [C U; A G]

p	$\log_2(1/p)$	$p\log_2(1/p)$	$H(p)$
0.40	1.32193	0.52877	0.97095
0.41	1.28630	0.52738	0.97650
0.42	1.25154	0.52565	0.98145
0.43	1.21754	0.52356	0.98582
0.44	1.18442	0.52115	0.98959
0.45	1.15200	0.51840	0.99277
0.46	1.12029	0.51534	0.99538
0.47	1.08927	0.51596	0.99740
0.48	1.05889	0.50827	0.99885
0.49	1.02915	0.50428	0.99971
0.50	1.00000	0.50000	1.00000
0.51	0.97143	0.49543	0.99971
0.52	0.94342	0.49058	0.99885
0.53	0.91594	0.48545	0.99740
0.54	0.88897	0.48004	0.99538
0.55	0.86250	0.47437	0.99277
0.56	0.83650	0.46844	0.98959
0.57	0.81087	0.46225	0.98582
0.58	0.78588	0.45581	0.98145
0.59	0.76121	0.44912	0.97650
0.60	0.73697	0.44218	0.97095

6 Discussion

Why Symmetric? Beautiful when symmetrical. Einstein's theory of relativity proved symmetry and, in the case of symmetric, energy is conserved. Symmetry was proved by linear algebra. We newly proved the symmetric as the weight Hadamard. This paper is an extended paper from References 9. Our approaches can be used in different scientific directions including for example problems studying in [10–13]. Shannon entropy is a popular and very fundamental theory in AI and Education Information Technology. In this paper, we extend to Shannon and RNA entropy analysis in Sect. 5. The Fibonacci sequence, the spiral of plants, and the algebra of animals are symmetrical. In the topological mathematical group definition, Inverse is element $a \in G, \frac{1}{a} \in G$, for only a^{-1}, and then $aa^{-1} = a^{-1}a = e$. Inverse is the element-wise inverse (Appendix). Therefore, the spiral of animals and plants is symmetrical with the left and right spirals, and there is an element-wise inverse in them. The mathematician who discovered this was Emmy Noether (Germany: 1882–1935). Netter is a mathematician who praised Einstein as "the best genius scholar ever since the beginning of women's higher education." The authors confirmed that the symmetric matrix is an element-wise inverse like the inverse symmetry of plants and animals. This is also seen in the Einstein general relativity principle, which revealed the Cosmos Black Hall principle of the universe. That is, astronomical objects of the universe are almost spherical. If the center is turned around the axis, it looks the same before or after turning. It is said to have symmetry at this time. The cosmic law of nature has symmetry (Fig. 5).

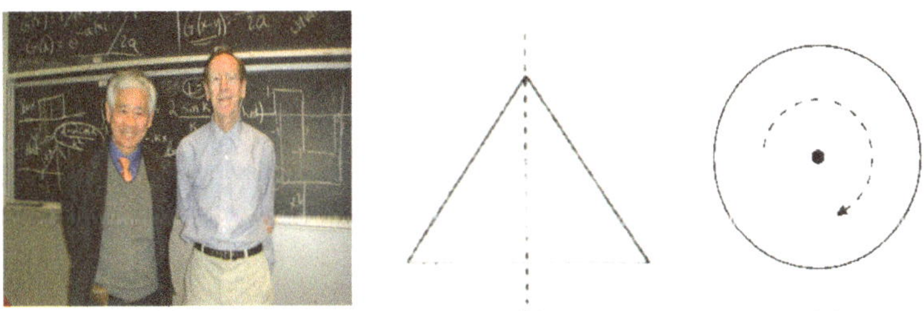

(a) (b) (c)

Fig. 5. (a) MIT Prof. Strang and Moon Ho Lee: Why Symmetric Jacket matrix at MIT Seminar 2012.11.30 (b) Isosceles triangle symmetry (c) Rotational symmetry of circle

7 Conclusions

A beautiful answer is symmetric.

If you look at the tree, the isosceles triangle is symmetrical even if you turn the center axis around. Sunflower flowers are circular. The circle is symmetrical in rotation because the shape is exactly the same even if it rotates about the origin. Beyond the spatial rotation transformation, generalization by the symmetric transformation of time and space, Einstein's special relativity theory is the space-time transformation relation. It is a group theory of symmetry. The possibility was shown by the Center Weighted

Hadamard matrix Moon Ho Lee. That is, we give the middle load value of Hadamard matrix as 2, which is equal to 2 under natural logarithm, the matrix is symmetric and the inverse is element-wise inverse. In 1915, at the University of Göttingen in Germany, Emmy Noether was the most direct and in - depth connection of the abstract world of mechanics (force and motion, natural laws) and symmetry. The theorem of neuron showed that symmetry is the most directly expressed way in nature through the concept of symmetry and energy conservation law. This is why Einstein praised it. More example, E. Chargaff discovered DNA base complement [C T; A G] of four components A = T = U = 33% and C = G = 17% of the experimental results. This strongly hinted towards the base pair makeup of DNA, although Chargaff did not explicitly state this connection himself. However, it has not been proved in a mathematical analysis view yet. In this paper, we have a simple proof of this problem based on symmetric Shannon entropy.

Appendix

Finite group property

1. The transformation of individuals constituting a group is called an element of a group, and each group includes an identical transformation E.
2. The elements of the group can be multiplied with each other. The multiplication of the elements of the group means successive symmetric transformations. A new element made up of the multiplication of elements is also an element of that group.
3. The multiplication of elements satisfies the associative rule and exists in the same group with the inverse of the element of the given group. The symmetric matrix A is $AB = BA = I_N$ in the complex or finite field, and $B = \frac{1}{n}(a_{ij}^{-1})^T$, and satisfies the property of the Jacket matrix.

References

1. Conway, J.H., Guy, R.K.: The Book of Numbers, Springer, New York (1996)
2. Lee, M.H.: The Arithmetic Code of Animal-plant, Youngil (2004)
3. Schneider, M.S.: A Beginners Guide to Constructing the Universe the Mathematical Archetypes of Nature, Art, and Science, Kyeongmunsa (2002)
4. Chargaff, E., Zamenhof, S., Green, C.: Composition of human desoxypentose nucleic acid. Nature **165**(4202), 756–777 (1950). https://en.wikipedia.org/wiki/Erwin_Chargaff
5. Lee, M.H., Hai, H., Lee, S.K.: A mathematical design of Nirenberg RNA standard genetic code and analysis based on the block circulant jacket matrix. Applied to USA patent No. 62/610, 496, 28 December 2017
6. Papoulis, A., Unnikrishna Pillai, S.: Probability, Random Variables and Stochastic Process. International Education (2002)
7. Shannon, C.E.,: A mathematical theory of communication. Bell Syst. Techn. J. **27**, 31–423 and 623–656 (1948)

8. Matthew, H., Petoukhov, S.: Mathematics of Bioinformatics. Wiley, New York (2011)
9. Lee, M.H., Lee, S.K.: A life ecosystem management with base complementary DNA. In: AIMEE 2018, Moscow, 6–8 October 2018 (2018)
10. Wani, A.A., Badshah, V.H.: On the relations between Lucas sequence and fibonacci-like sequence by matrix methods. International Journal of Mathematical Sciences and Computing (IJMSC) 3(4), 20–36 (2017)
11. Bellamkonda, S., Gopalan, N.P.: A facial expression recognition model using support vector machines. Int. J. Math. Sci. Comput. (IJMSC) **4**(4), 42–55 (2018)
12. Hamd, M.H., Ahmed, S.K.: Biometric system design for iris recognition using intelligent algorithms. Int. J. Mod. Educ. Comput. Sci. (IJMECS) **10**(3), 9–16 (2018)
13. Patil, A., Kruthi, R., Gornale, S.: Analysis of multi-modal biometrics system for gender classification using face, iris and fingerprint images. Int. J. Image Graph. Signal Process. (IJIGSP) **11**(5), 34–43 (2019)

Advances in Medical Approaches

Modelling of Bimorph Piezoelectric Elements for Biomedical Devices

Constantine Bazilo(✉)

Cherkasy State Technological University,
Shevchenko blvd, 460, Cherkasy 18006, Ukraine
b_constantine@ukr.net

Abstract. The relevance of the use of various functional elements of piezoelectronics in biomedical informational and measuring systems is explained by their high reliability, small dimensions and weight. These aspects greatly facilitate the solution of the problem of miniaturization of such systems. Because of the absence of reliable and valid methods of bimorph piezoelectric transducer's mathematical model constructing the need is to solve the problem of the excitation of transverse bending oscillations in bimorph piezoelectric element. Construction and features of mathematical description of axisymmetric transverse bending oscillations in bimorph piezoelectric element are considered. The solution of the problem of transverse bending oscillations excitation in bimorph piezoelectric element by the difference of electric potentials is obtained.

Keywords: Piezoelectric disk · Bimorph element · Physical processes · Mathematical description

1 Introduction

According to the results of the analysis of the world market of micro devices that are based on the piezoelectric effect the consulting company "Yole Développement" (France) found that, for today, this market has almost reached the mark of $ 85 million and by 2023 it will be about $ 310 million (while the Compound Annual Growth Rate will be 30.3%) [1]. The unconditional advantages of piezoelectric ceramics PZT (lead zirconate titanate) using are low cost of raw materials and technology for manufacture, as well as high inertness to the influence of climatic factors, electrical and mechanical strength, high sensitivity to external mechanical influences and others.

Piezoelectric transducers and transformers of various shapes and sizes are used in control and measuring devices, microtechnology and biomedical devices. As transformers, for example, they can be used in high voltage secondary power sources, in devices that need a large voltage conversion ratio, particularly in medical technology. Piezoceramic transformers are used in ionizers and ozonizers, which can be used for disinfection of premises, water purification and other applications [2].

Piezoelectric disks with surfaces partially covered by electrodes are often used to create various functional piezoelectronic devices. It should be especially noted that this piezoelectric element has compatibility with microsystem technology, so it can be made as biomedical microelectromechanical structures (bioMEMS) [3, 4].

Z. Hu et al. (Eds.): AIMEE 2019, AISC 1126, pp. 151–160, 2020.
https://doi.org/10.1007/978-3-030-39162-1_14

Many scientists made a significant scientific contribution to the development of the theoretical, mathematical, methodological and experimental basis for the development of new devices with piezoelectric ceramics. For solving problems related to the improvement of manufacturing technologies for piezoceramic elements have been devoted, for example, works [5, 6]. The aim of the paper is to solve the problem of the excitation of oscillations of transverse bending in bimorph piezoelectric element.

2 Formal Problem Statement

Let us consider the principle of operation of piezoelectric devices on the example of a piezoelectric voltage transformer, which is well known [7].

Applying the electric potential difference U_1 is an amplitude value of the electric potential difference on a pair of electrodes that partially cover the bottom and front surfaces of the piezoelectric plate, harmonic oscillations of material particles are disturbed in the volume of the plate. The oscillations of material particles are accompanied by dynamic deformations of the infinitely small elements of the volume of a piezoelectric. Due to the direct piezoelectric effect in the piezoelectric, which is deformed, polarization charges arise. Some of these charges are collected by the second pair of electrodes, which, like the first pair, partially covers the surfaces of the piezoelectric plate. The polarization charge on the second pair of electrodes causes an electric current in the conductor that connects one of the electrodes of the second pair to the load resistance Z_l. The voltage $U_2 = Z_l I$ is an output signal of piezoelectric transformer. Obviously, the transformation ratio $K(\omega, \Pi)$ (symbol Π defines a set of geometrical, and physical and mechanical parameters of piezoelectric transformer; ω is a circular frequency), is equal to the ratio of the output signal to the input impact, i.e.

$$K(\omega, \Pi) = \frac{U_2}{U_1} = \frac{Z_l I}{U_1}$$

and can be considered as mathematical model of piezoelectric transformer.

With the help of a personal computer it is possible to work out several combinations of geometrical, physical and mechanical parameters of piezoelectric transformer within a few hours, and find a combination of ones that ensures the implementation of the specified parameters of the device. A qualitative mathematical model can significantly reduce the time and cost of developing new piezoelectric transformers [8].

3 Literature Review

Many publications have been devoted to the construction and research of mathematical models of piezoelectric transducers. The basics of the calculation of piezoelectric transducers' transfer characteristics are considered, for example, in [9–13].

However, in many papers only processes occurring in a piezoelectric disk with a surface that is fully covered by electrodes are described. The works devoted to calculating the parameters and characteristics of piezoelectric transducers can be divided into two groups. The first group includes the works that are based on the field model of piezoceramic disk [9–11]. The second group includes works based on the use of equivalent circuits [12, 13]. Mathematical models of piezoelectric transducers based on the analysis of the so-called equivalent circuits do not take into account the obvious fact that the movements of the material particles of the piezoelectric disk must satisfy the second and third Newton's laws. Ignoring this fundamental position takes these models beyond the mechanics of a deformable solid, and, as a result, determine their inadequacy to real objects and physical processes occurring in them [14].

Based on the above, it can be argued that currently there are no reliable and valid methods of constructing mathematical models of piezoelectric transformers. Therefore, an urgent task is to develop a mathematical model of bimorph piezoelectric element for informational and measuring systems.

4 Materials and Methods

Let consider the construction, which is shown in Fig. 1*a*. Position 1 in Fig. 1*a* denotes a metal disk, the thickness of which $2h$ is substantially less than its diameter $2R$. Along the contour $\rho = R$ the metal disk is rigidly fixed on an absolutely fixed support (position 3). Wherein, the elements of the cylindrical surface ($\rho = R$; $0 \leq \phi \leq 2\pi$; $-h \leq z \leq h$) (ρ, ϕ, z are coordinate axes of the cylindrical coordinate system) do not have the ability to move along the coordinate axes ρ, ϕ, z. Two identical electroded piezoceramic disks are attached to the surfaces $z = \pm h$ of a metal disk by conductive glue (position 2 in Fig. 1*a*). Research of conducting electrodes of piezoelectric elements modified by low-energy ribbon-shaped electron stream is presented in [15]. Electroded surfaces of piezoceramic disks, which are in electrical contact with a metal plate through a thin layer of conductive adhesive, always have the same zero (Fig. 1*a*) electric potential. Piezoceramic disks are not glued to the metal plate in an arbitrary manner, but in such a way that the direction of electrical polarization of these disks, which is shown in Fig. 1*a*, as bold arrows were the same.

If on the top and bottom electroded surface of the piezoceramic disks are applied the electric potential difference as shown in Fig. 1*a*, the physical state of each of the disks will be determined by the following rations:

$$\sigma_{ij}^{(\pm)} = c_{ijk\ell}^{E}\varepsilon_{k\ell}^{(\pm)} - e_{kij}E_k^{(\pm)},\ D_m^{(\pm)} = e_{mij}\varepsilon_{ij}^{(\pm)} + \chi_{mn}^{\varepsilon}E_n^{(\pm)} \tag{1}$$

where the plus sign determines the physical state of the upper piezoceramic disk, and the minus sign determines the lower; $c_{ijk\ell}^{E}$, e_{kij}, χ_{mn}^{ε} are material constants or elements of the matrices of elastic modulus, piezoelectric modules and dielectric constants; $\sigma_{ij}^{(\pm)}$, $\varepsilon_{k\ell}^{(\pm)}$, $E_k^{(\pm)}$ and $D_m^{(\pm)}$ are amplitude values of the components of elastic stress and strain tensors, electric field strength and electrical induction vectors, respectively.

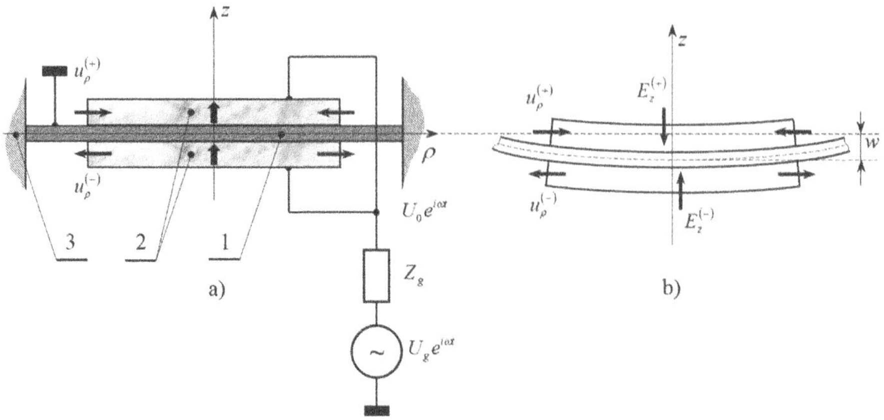

Fig. 1. The main components of bimorph piezoelectric element (a) and transverse bending formation circuit (b)

Let us say that at an arbitrarily fixed moment of time a positive electrical potential U_0 is on the electroded surfaces $z = \pm(\alpha + h)$, where α is a thickness of piezoceramic elements of equal size. It is obvious that the axial components $E_z^{(\pm)}$ of electric field intensity vector in the upper and lower disks have opposite directions, i.e. $E_z^{(\pm)} = \mp E_z$ (Fig. 1*b*). If the disks are glued to the metal plate strictly coaxially, then the stress-strain state of the piezoceramic disks and of all (Fig. 1*a*) construction has an axial symmetry. This means that the shear stresses $\sigma_{\phi\beta}^{(\pm)}$ $(\beta = \rho, z)$ and strains $\varepsilon_{\phi\beta}^{(\pm)}$ are zero.

Bimorph piezoceramic element (Fig. 1) is completely naturally divided into two parts such as active zone $(0 \le \rho \le R_0)$, where R_0 is piezoceramic disk radius, and passive zone of metal ring $R_0 \le \rho \le R$.

To construct the equation of harmonic oscillations in the active zone of bimorph element let us consider the integral characteristics of the stressed state of this area.

Firstly, let us write the expressions for calculating the deformations in an active zone. The equation of harmonic oscillations of the material particles of the active zone follows from the conditions of dynamic equilibrium of an element of the volume of the active zone of bimorph piezoelectric element:

$$\nabla^4 w_0 - \lambda_0^4 w_0 = 0 \qquad (2)$$

where $\nabla^4 w = \frac{\partial^4 w}{\partial \rho^4} + \frac{2}{\rho}\frac{\partial^3 w}{\partial \rho^3} - \frac{1}{\rho^2}\frac{\partial^2 w}{\partial \rho^2} + \frac{1}{\rho^3}\frac{\partial w}{\partial \rho}$; w_0 is a deflection of the active zone of bimorph piezoelectric element depending on the values of the radial coordinate ρ; $\lambda_0 = \sqrt{4}\left(2h\rho_m + 2\alpha\rho_{pe}\right)\omega^2/D_1$ is wave number of harmonic oscillations of transverse bending of the active zone of bimorph piezoelectric element; ρ_m and ρ_{pe} are densities of metal plate and piezoceramics. In deriving Eq. (2) it was taken into account that $\sigma_z^* \equiv 0$.

Axial movement w_1 of the material particles of the passive zone, i.e., the metal ring $R_0 \leq \rho \leq R$, of bimorph piezoelectric element are solutions of the following equation:

$$\nabla^4 w_1 - \lambda_1^4 w_1 = 0 \tag{3}$$

where $\lambda_1 = \sqrt{4}2h\rho_m\omega^2/D_m$ is a wave number of harmonic oscillations of material particles of metal ring; D_m is a bending stiffness of a metal disk.

Let us consider the features of joint solution of differential equations (2) and (3).

Since the active zone of the bimorph piezoelectric element contains a point $\rho = 0$, so the general solution of Eq. (2) is the function $w_0(\rho)$, which is defined as:

$$w_0(\rho) = A_1 J_0(\lambda_0\rho) + A_2 I_0(\lambda_0\rho) \tag{4}$$

where A_1 and A_2 are constants to be defined; $J_0(\lambda_0\rho)$ and $I_0(\lambda_0\rho)$ are Bessel function and modified zero order Bessel function [16].

The general solution of Eq. (3) is

$$w_1(\rho) = A_3 J_0(\lambda_1\rho) + A_4 N_0(\lambda_1\rho) + A_5 I_0(\lambda_1\rho) + A_6 K_0(\lambda_1\rho) \tag{5}$$

where $A_3, \ldots, A_6$ are constants to be defined; $N_0(\lambda_0\rho)$ and $K_0(\lambda_0\rho)$ are Neumann function and zero order Macdonald function [16].

Common solutions $w_0(\rho)$ and $w_1(\rho)$ should ensure continuity of stress-strain state of a bimorph on conditional boundary $\rho = R_0$ between active and passive zones of bimorph piezoelectric element. This is achieved by stitching on this border the kinematic and dynamic characteristics of stress-strain state of active and passive zones:

$$w_0(\rho)|_{\rho=R_0} = w_1(\rho)|_{\rho=R_0}, \left.\frac{\partial w_0(\rho)}{\partial \rho}\right|_{\rho=R_0} = \left.\frac{\partial w_1(\rho)}{\partial \rho}\right|_{\rho=R_0}, \tag{6}$$

$$M_\rho(\rho)|_{\rho=R_0} = \left.M_\rho^{(m)}(\rho)\right|_{\rho=R_0}, Q(\rho)|_{\rho=R_0} = Q^{(m)}(\rho)|_{\rho=R_0} \tag{7}$$

where $M_\rho^{(m)}(\rho)$ and $Q^{(m)}(\rho)$ are linear densities of bending moments and shear forces in metal ring of passive zone of bimorph element. The numerical values are:

$$\begin{aligned} M_\rho^{(m)}(\rho) &= -D_m\left(\frac{\partial^2 w_1}{\partial \rho^2} + \frac{\nu}{\rho}\frac{\partial w_1}{\partial \rho}\right), \; Q^{(m)}(\rho) \\ &= D_m\left(\frac{\partial^3 w_1}{\partial \rho^3} + \frac{1}{\rho}\frac{\partial^2 w_1}{\partial \rho^2} - \frac{1}{\rho^2}\frac{\partial w_1}{\partial \rho}\right). \end{aligned} \tag{8}$$

After substituting determines (4) and (5) into matching conditions for solutions (6)–(7), we obtain a non-uniform system of four algebraic equations containing six unknown constants $A_1, \ldots, A_6$. Missing two equations are delivered by contour fixing conditions $\rho = R$. In practice, three methods of fixing are most simply implemented. They are rigid, hinged and free fixing.

Conditions (6)–(7) and conditions for fixing the contour $\rho = R$ form a heterogeneous system of six algebraic equations, which contains exactly six constants $A_1,\ldots,A_6$. This system of equations has the following form:

$$m_{jk}A_k = \delta_{j3}\frac{M_0}{D_1\lambda_0^2}, \; j,k = 1,\ldots,6 \tag{9}$$

where coefficients m_{jk} for rigid, hinged and free fixing are defined in [17]; δ_{j3} is Kronecker symbol equal to one when $j = 3$ and equal to zero for all $j \neq 3$. The right part of the third equation in the system of equations (9) can be presented as follows: $M_0/\left(D_1\lambda_0^2\right) = W_0U_0$, where $W_0/2(h+\alpha/2)/\left(D_1\lambda_0^2\right)$ is an absolute sensitivity of the active zone of bimorph piezoelectric element (the dimension is meter divided by volt).

The solution of the system of equations (9) can be presented as follows:

$$A_k = (-1)^{k+3}W_0U_0\frac{\Delta_{3k}}{\Delta_0}, \; k = 1,\ldots,6 \tag{10}$$

where Δ_{3k} is an algebraic complement with unknown coefficient; A_k is a determinant of matrix sized 5×5, which is obtained from the matrix of coefficients m_{jk} system of equations (9) by crossing out the third line and k-th column; Δ_0 is a determinant of matrix 6×6 which is obtained from the coefficients m_{jk} of system of equations (9).

5 Experiments and Results

Obviously, by manipulating the dimensions of the piezoceramic disks, one can control the numerical values of resonant frequencies of bimorph piezoelectric element. The numerical values of dimensionless frequencies $\Omega = \lambda_1 R$ of the first three resonances were determined for various types of contour $\rho = R$ fixing of metal disk and are shown in Table 1. The symbols r_0 and α_0 indicate the dimensionless radius and thickness of piezoceramic disks, where $r_0 = R_0/R$ and $\alpha_0 = \alpha/h$. In the process of calculating were used the following material constants: metal (steel) plate: Young's modulus $E = 200\ GPa$, Poisson's ratio $v = 0,28$, density $\rho_m = 7800\ \text{kg}/\text{m}^3$, plate half thickness $h = 5 \cdot 10^{-4}$ m, radius $R = 5 \cdot 10^{-2}$ m; piezoceramic (PZT type ceramics) disks: elastic modulus $c_{11}^E = 110\ GPa$, $c_{12}^E = 62\ GPa$, $c_{33}^E = 100\ GPa$, piezoelectric modulus $e_{31} = -9\ \text{C}/\text{m}^2$, $e_{33} = 18\ \text{C}/\text{m}^2$, dielectric permittivity $\chi_{33}^{\varepsilon} = 1300\chi_0$, dielectric constant $\chi_0 = 8,85 \cdot 10^{-12}\ \text{F}/\text{m}$, density $\rho_{pe} = 7400\ \text{kg}/\text{m}^3$, disk thickness $\alpha = 10^{-3}$ m, disk radius $R_0 = 1,5 \cdot 10^{-2}$ m. The cyclic frequency $f = 9694$ Hz corresponds to the dimensionless frequency $\Omega = 10$. The method of determination of physical and mechanical constants of piezoceramic materials is described in [18, 19].

Calculations show that the first three dimensionless resonance frequencies of bimorph element with contour $\rho = R$ rigid fixing have the highest values in comparison with the frequencies of resonances for free and hinged fixing of the edge of metal plate. This can be explained by the fact that the integral rigidity of the structure of bimorph

Table 1. The numerical values of dimensionless frequencies Ω_n of the first three resonances of bimorph piezoelectric element with contour $\rho = R$ for $\alpha_0 = 2$

r_0	Rigid fixing			Hinged fixing		
	Ω_1	Ω_2	Ω_3	Ω_1	Ω_2	Ω_3
0,001	3,196215	6,306421	9,439463	2,214820	5,449477	8,610093
0,05	3,184003	6,277701	9,391830	2,210765	5,425611	8,566268
0,10	3,155050	6,255022	9,427965	2,200993	5,396206	8,580123
0,15	3,122535	6,299843	9,616115	2,190085	5,410380	8,727195
0,20	3,096476	6,418144	9,893565	2,182346	5,482482	8,967366
0,25	3,082320	6,593766	10,182627	2,181023	5,607511	9,242890
0,30	3,082214	6,804526	10,380713	2,188314	5,773709	9,480346
0,35	3,096416	7,020831	10,395259	2,205664	5,965299	9,587031
0,40	3,124117	7,199459	10,305584	2,234067	6,158825	9,539750
0,45	3,163734	7,290985	10,318486	2,274255	6,320354	9,465294
0,50	3,212845	7,279736	10,551852	2,326769	6,412789	9,518145
0,55	3,267952	7,208348	10,979425	2,391881	6,422981	9,770863
0,60	3,324275	7,140299	11,444410	2,469337	6,378194	10,191795
0,65	3,375984	7,125115	11,672058	2,557860	6,325803	10,627492
0,70	3,417348	7,192587	11,569838	2,654375	6,311404	10,829371
0,75	3,444991	7,349094	11,431646	2,753188	6,375197	10,762664
0,80	3,460677	7,560465	11,412258	2,845851	6,550967	10,635236
0,85	3,474202	7,741617	11,652274	2,923016	6,844382	10,659116
0,90	3,509730	7,812019	11,995156	2,978466	7,171252	11,032349
0,95	3,638117	7,819491	12,090464	3,012738	7,365918	11,561329
0,999	4,300487	8,487432	12,707134	3,032881	7,404109	11,689884

element with contour $\rho = R$ rigid fixing of the metal plate is the highest one in comparison with the total (integral) rigidity of the whole structure in the case of other methods of fixing. The frequency of the first resonance with rigid and hinged fixing firstly decreases and then begins to increase with increasing the parameter r_0. This is because of the rigidity of bimorph element for small values of r_0 increases at a slower rate in comparison with the increase of the mass of oscillating element. Upon reaching a certain value of the parameter r_0 its further increase is accompanied by a leading increase in the rigidity of bimorph piezoelectric element. These dependences become more pronounced as the thickness α_0 of piezoceramic disk increases. For the second and third resonances, two (second resonance) and three (third resonance) intervals are observed for the numerical values of the parameter r_0, where there is a decrease and increase in numerical values of dimensionless frequencies of resonances.

To complete the mathematical description of bimorph piezoelectric element it is necessary to determine the potential U_0 on the electroded surfaces (Fig. 1).

From the definition of axial component $D_z^{(\pm)}$ of electric induction vector it follows that electric charges $Q^{(\pm)}$ on the electroded surfaces of the upper and lower piezoceramic disk are equal in absolute value, but have opposite signs:

$$Q^{(\pm)} = 2\pi \int_0^{R_0} \rho D_z^{(\pm)} d\rho = \mp 2\pi (h + \alpha/2) e_{31}^* R_0 \left.\frac{\partial w_0(\rho)}{\partial \rho}\right|_{\rho = R_0} \mp C_0^\sigma U_0 \tag{11}$$

where $C_0^\sigma = \pi R_0^2 \chi_{33}^\sigma / \alpha$ is dynamic electric capacitance of piezoceramic disk as a part of bimorph piezoelectric element.

Taking into account the definition (4) of the deflection $w_0(\rho)$ of the neutral layer in the active zone of bimorph element and the general solution (10) of the system of algebraic equations (9), we can write the following expression:

$$\left.\frac{\partial w_0(\rho)}{\partial \rho}\right|_{\rho = R_0} = -\lambda_0 W_0 U_0 \left[\frac{\Delta_{31}}{\Delta_0} J_1(\lambda_0 R_0) + \frac{\Delta_{32}}{\Delta_0} I_1(\lambda_0 R_0)\right]$$

where λ_0 is a wave number of bending oscillations in the active zone of bimorph element; $W_0 = 2(h + \alpha/2)\, e_{31}^* / \left(\lambda_0^2 D_1\right)$ is an absolute sensitivity of the active zone.

Substituting the last expression into formula (11), we get the following result:

$$Q^{(\pm)} = \pm C_0^\sigma U_0 [\Psi_0(\omega, \Pi) - 1] \tag{12}$$

where $\Psi_0(\omega, \Pi)$ is a function that depends on the frequency and set of parameters (symbol Π) of the components of active zone of bimorph piezoelectric element:

$$\Psi_0(\omega, \Pi) = 2K_{31}^2 \frac{c_{11}^E (h + \alpha/2)^2 \alpha}{\Delta_0 D_1} \times \left[\Delta_{31} \frac{2 J_1(\lambda_0 R_0)}{\lambda_0 R_0} + \Delta_{32} \frac{2 I_1(\lambda_0 R_0)}{\lambda_0 R_0}\right]$$

where $K_{31}^2 = \left(e_{31}^*\right)^2 / \left(c_{11}^E \chi_{33}^\sigma\right)$ is squared electromechanical coupling coefficient of piezoceramics in the mode of planar oscillations.

Let us rewrite expression (12) as $Q^{(\pm)} = C_0^\sigma U^{(\pm)} [\Psi_0(\omega, \Pi) - 1]$, where $U^{(\pm)} = \pm U_0$. From the last expression, it follows that:

$$I^{(\pm)} = -i\omega C_0^\sigma U^{(\pm)} [\Psi_0(\omega, \Pi) - 1].$$

The electrical impedance $Z_{pe}^{(\pm)}$ of the disks in the composition of bimorph piezoelement is determined from Ohm's law for a part of the electrical circuit as follows:

$$Z_{pe}^{(\pm)} = \frac{U^{(\pm)}}{I^{(\pm)}} = -\frac{1}{i\omega C_0^\sigma [\Psi_0(\omega, \Pi) - 1]} = Z_{pe}. \tag{13}$$

It is clear that equal electrical impedances $Z_{pe}^{(\pm)}$ are connected in parallel. For this reason, the electrical impedance of bimorph element is $Z_{BE} = Z_{pe}^{(\pm)}/2 = Z_{pe}/2$.

The electric potential U_0 with the presence of the final output resistance Z_g in the source of the difference of electric potentials is determined as:

$$U_0 = \frac{U_g Z_{BE}}{Z_{BE} + Z_g} = \frac{U_g Z_{pe}}{Z_{pe} + 2Z_g}.$$

Substituting relation (13) into the last expression and obtained result into the general solution (10), we obtain the final form of the expression for calculating the numerical values of the coefficients A_k:

$$A_k = \frac{(-1)^{k+3} W_0 U_g \Delta_{3k}}{\{1 - 2i\omega C_0^\sigma Z_g [\Psi_0(\omega, \Pi) - 1]\}\Delta_0}. \quad (14)$$

Expression (14) completes the solution of the problem of the excitation of oscillations of transverse bending in bimorph piezoelectric element by the difference of electric potentials, which is produced by a real generator with an output electrical impedance Z_g.

6 Conclusions

Piezoelectric bimorph transducers are widely used in control and measuring devices, microtechnology and biomedical devices because of their high reliability, small dimensions and weight. The features of mathematical description of transverse bending axisymmetric oscillations of bimorph piezoelectric element are considered in the article. The problem of the excitation of transverse bending vibrations in a bimorph piezoelectric element by the difference of electric potentials produced by a real generator with an output electrical impedance is solved.

Acknowledgements. The research leading to these results was made within the framework of a state budgetary research topic "Development of highly efficient intellectual complex for creation and research of piezoelectric components for instrumentation, medicine and robotics".

References

1. Report "Status of the MEMS Industry 2018". Yole Development, Lyon, France (2018)
2. Sharapov, V.: Piezoceramic Sensors. Springer, Heidelberg (2011)
3. Varadan, V., Vinoy, K., Jose, K.: RF MEMS and Their Application. Technosphere, Moscow (2004). (in Russian)
4. Jivani, R.R., et al.: Biomedical microelectromechanical systems (BioMEMS): revolution in drug delivery and analytical techniques. Saudi Pharm J. **24**(1), 1–20 (2016). https://doi.org/10.1016/j.jsps.2013.12.003

5. Jian, L.: Design of active vibration control system for piezoelectric intelligent structures. Int. J. Educ. Manag. Eng. (IJEME) **2**(7), 22–28 (2012). https://doi.org/10.5815/ijeme.2012.07.04
6. Nakamura, K. (ed.): Ultrasonic Transducers: Materials and Design for Sensors, Actuators and Medical Applications. Woodhead Publishing, Philadelphia (2012)
7. Vijaya, M.S.: Piezoelectric Materials and Devices: Applications in Engineering and Medical Sciences. CRC Press LLC, Boca Raton (2017)
8. Petrishchev, O.N., Bazilo, C.V.: Principles of mathematical modeling of transformers that operate on planar axisymmetric vibrations of a piezoceramic disk. Herald Cherkasy State Technol. Univ. **3**, 10–20 (2015). (In Russian)
9. Wu, L., Chure, M.-C., Chen, Y.-C., et al.: Electrode size and dimensional ratio effect on the resonant characteristics of piezoelectric ceramic disk. In: Ceramic Materials – Progress in Modern Ceramics, pp. 25–40. InTech (2012)
10. Calas, H., Moreno, E., Eiras, J.A., et al.: Model for radial modes in a thin piezoelectric annular array. Jpn. J. Appl. Phys. **47**(10), 8057–8064 (2008)
11. Peerasaksophol, M., Srilomsak, S., Laoratanakul, P., Kulworawanichpong, T.: Design and implementation of ring-dot piezo-electric ballasts for 36-W fluorescent lamps. Eur. J. Sci. Res. **64**(2), 189–205 (2011)
12. Buchacz, A., Placzek, M., Wrobel, A.: Modelling of passive vibration damping using piezoelectric transducers – the mathematical model. Maint. Reliab. **16**(2), 301–306 (2014)
13. Livingston, D., Kumar, K.P., Venugopal, N.: Modelling and simulation of multiple piezoelectric transformer converters. Int. J. Emerg. Technol. Adv. Eng. **3**(8), 237–245 (2013)
14. Petrishchev, O.N., Bazilo, C.V.: Principles and methods of the calculation of transfer characteristics of disk piezoelectric transformers. Part 2: the procedure of calculation of parameters and characteristics of the simplest disk piezoelectric transformer. Herald of Cherkasy State Technological University, no. 4, pp. 10–23 (2015). (in Russian)
15. Medianyk, V.V., Bondarenko, Yu.Yu., Bazilo, C.V., Bondarenko, M.O.: Research of current-conducting electrodes of elements from piezoelectric ceramics modified by the low-energy ribbon-shaped electron stream. J. Nano-Electron. Phys. **10**(6), 06012-1–06012-6 (2017). https://doi.org/10.21272/jnep.10(6).06012
16. Abramowitz, M., Stegun, I.A.: Handbook of Mathematical Functions with Formulas, Graphs, and Mathematical Tables. Science, Moscow (1979). (in Russian)
17. Petrishchev, O.N., Bazilo, C.V.: Principles and Methods of Mathematical Modelling of Oscillating Piezoelectric Elements. Gordienko Publ., Cherkasy (2019). (in Russian)
18. Bazilo, C., Zagorskis, A., Petrishchev, O., Bondarenko, Y., Zaika, V., Petrushko, Y.: Modelling of piezoelectric transducers for environmental monitoring. In: Proceedings of 10th International Conference "Environmental Engineering", Vilnius Gediminas Technical University, Lithuania (2017). https://doi.org/10.3846/enviro.2017.008
19. Petrishchev, O.N., Bazilo, C.V.: Methodology of determination of physical and mechanical parameters of piezoelectric ceramics. J. Nano-Electron. Phys. **9**(3), 03022-1–03022-6 (2017). https://doi.org/10.21272/jnep.9(3).03022

Robust Operational-Space Motion Control of a Sitting-Type Lower Limb Rehabilitation Robot

Santhakumar Mohan[1,2], Jayant Kumar Mohanta[3], Laxmidhar Behera[3], Larisa Rybak[2], and Dmitry Malyshev[2](✉)

[1] Indian Institute of Technology Palakkad, Palakkad, India
[2] Belgorod State Technological University named after V.G. Shukhov, Belgorod, Russian Federation
rlbgtu@gmail.com
[3] Indian Institute of Technology Kanpur, Kanpur 208016, India

Abstract. This paper presents a robust motion control of a sitting-type lower limb rehabilitation robot (LLRR) in its operational-space. The mathematical background of the proposed robot is discussed and its motion control design in the task-space based on a double-loop control approach is derived herein along with its closed-loop system stability analysis. The motion tracking performance analysis of the proposed scheme is demonstrated using computer based numerical simulations. For numerical simulations and to validate the effectiveness of the motion control strategy, the clinically obtained test gait data is used for the desired motion trajectory of the lower limb rehabilitation robot.

Keywords: Limb rehabilitation · Manipulator workspace · Design of manipulators · Parallel robot

1 Introduction

Currently, in practical health care, there are several applied problems, the best way to solve which is to use robotic tools. These tasks concern not only the treatment and rehabilitation of patients with the impaired musculoskeletal system, but also the implementation of their self-care functions, social adaptation, and replenishment of lost motor and communication functions. It should be noted that the need to create robotic means of helping people with disabilities is steadily growing. This is due to the high frequent disability of the population as a result of vertebral injuries, stroke and other neurological diseases. According to studies conducted in 2018 [1] in 2014 alone, 795 thousand people with stroke were registered in the USA, in the Russian Federation - 419 thousand. According to the Ministry of Health of the Russian Federation, in 2017, 427,963 people were registered, for whom acute cerebrovascular disorders were first detected. Patients who have experienced acute attacks of this type of disease either cannot do without outside help or are completely deprived of the opportunity to move around on their own. Loss of self-care imposes special obligations on family members

Z. Hu et al. (Eds.): AIMEE 2019, AISC 1126, pp. 161–172, 2020.
https://doi.org/10.1007/978-3-030-39162-1_15

of the patient. A decrease in the labor potential of citizens ultimately leads to a decrease in the country's gross domestic product. At the same time, state losses from one patient who received a disability amount to about 1,200,000 rubles per year. According to the registration list of a brain stroke of the Research Institute of Neurology of the Russian Academy of Medical Sciences, by the end of the acute period (3 weeks from the onset of the stroke), impaired motor activity was observed in 80% of 100 patients who survived. 31% of stroke patients require outside help, 20% cannot walk on their own. Currently, over 1 million stroke survivors live in Russia. Of these, a third are people of working age, but only every fourth patient returns to work [2].

According to epidemiological data from the World Disability Report published by the World Health Organization, more than 1 billion people, about 15% of the world's population, have some form of disability. From 110 million (2.2%) to 190 million (3.8%) people 15 years and older experience significant difficulties in functioning. Moreover, disability rates are rising due to an aging population and an increasing burden of chronic health problems. The number of patients with severe disabling consequences of neurological diseases requiring special conditions for survival and special rehabilitation methods is about 4% among them. In Russia, 70% of patients in need of rehabilitation are neurological patients. According to the World Health Organization, more than 10 million people suffer from traumatic brain injury each year and make up 31–50% of the total structure of injuries, traumatic damage to the central nervous system in the general structure of the primary disability of the adult population is 0.7–4%, in the structure of post-traumatic primary disability of the adult population – 44%, or 3.6 per 10,000 of the population annually. The frequency of cerebral strokes is 23% in the structure of neurological pathology.

Disorders of the limbs must undergo a rehabilitation process to restore the normal functioning of the limbs. However, rehabilitation therapeutic treatment depends on the limbs and differs from the upper to lower limbs. Treatment of the upper limbs is aimed at restoring the nervous system, muscle ability, and strength of the hands, while treatment of the lower extremities is focused on various joint movements of the legs and their synchronized movements.

Lower limb rehabilitation or therapeutic treatment has been a hot topic in recent years, as mechanized systems promised effective results and led to significant improvements in the recovery of patients using robotic physiotherapy [3]. In the treatment of lower limbs: each lower limb has a certain movement (specific to the body and limbs) and requires special settings for a separate limb. These problems make the development of the system quite complex and require modular and reliable mechanical systems for the treatment of lower extremities. However, treatment in the lower extremities is very common in the field of rehabilitation and covers a wide range of patients. Robots/mechanisms for the rehabilitation of the lower extremities undergo human-machine interaction and must meet the criteria of human safety standards. Thus, the control problem has two levels: a higher level and a lower level for a safe therapy. Higher-level control determines strategies for using various modes in various situations, the necessary transformation of limb movement data into device status outputs, failover, etc. While lower-level control focuses on minimizing errors in trajectory tracking tasks.

It is widely recognized that a person's locomotion depends both on the basic patterns generated at the level of the spine and on the prognosis and reflex-dependent precise control of these patterns at different levels [4–6]. These physiological movements recorded in healthy people are carried out in patients with disorders of the lower extremities. These data sequences for the joints of the lower extremities are called gait data (walking pattern). In addition to gait data, the lower extremities have movements such as hip flexion and extension, knee flexion and extension, ankle flexion, and back flexion. Gait training restores the synchronization of muscle action in the lower limb and is processed separately to strengthen each motor joint to strengthen each joint of the leg.

The problems of using parallel robot architectures in medical applications, such as rehabilitation and physiotherapeutic exercises, were considered in [7, 8]. An important characteristic of rehabilitation robots is the work area. To determine the working area, various numerical methods of interval analysis, as well as grid approximation, are used [9]. The application of these approaches to the solution is associated with significant computational difficulties since the problem has a large dimension. In [10], the method of non-uniform coverings for approximating the set of solutions of a system of nonlinear inequalities was considered, and in [11–13], the application of this method to determine the working area of some types of planar robots was considered. In the article, these methods are developed for constructing the working area of a rehabilitation robot for the required ranges of rotation angles in the hip, knee and ankle joints while walking simulating: 1st step phase - flexion in the hip joint from 0° to 20°, flexion in the knee joint from 0° to 60°, extension in the ankle joint from 0° to 10°; 2 phase - 0°, 0° and 90° respectively; 3 phase - extension in the hip joint from 0° to 10°, in the knee joint 0°, flexion in the ankle joint from 0° to 20°.

The output from the mid-level controller is the desired state of the device for robot control [14–16]. This state may consist of a combination of joint positions, velocities, and torques. The design of the device and the actuators it contains will have a strong influence on how well the desired state can be achieved, and thus must be considered during the development of the mid and low-level controllers. Commanding joint positions or velocities is a straightforward approach to robot control, for which an abundance of theory exists. In this case, the robot is tasked with the precise reproduction of a pre-defined trajectory. This type of controller works best when the output mechanical impedance (as defined below) of the actuator is high relative to the load, thus enabling the device to reject perturbations from the user or the environment. Controlling the position or velocity of the device is useful when the desired trajectory and the interaction forces are well-characterized; but both of these may be difficult to predict given the dynamic nature of locomotion. Interaction with stiff objects (e.g. the environment) can lead to instability and the generation of high forces under this type of control [17]. Presumably, it is because of these issues that very few examples of assistive devices for lower-limbs were found that use position control. Torque control is possible when the output mechanical impedance of the actuator is low relative to the load, and is useful for providing assistive forces when the desired position is ill-defined or unimportant. Problems can arise, however, when torques are applied without regard to the position of the joint.

The motion control strategy of the robotic system varies case-to-case and it control structure should be flexible. For example, in some therapeutic treatments, the patient's efforts have no effect, but in some treatments, the patient provides a significant effort to the system. Therefore, the motion control scheme should be adaptive accordingly to the therapeutic treatments and patient's conditions. i.e., it depends on the type of therapy. To control the proposed rehabilitation robot with minimal tracking errors and provide an effective performance (that is, at a satisfactory designed level) on the repetitive task-specific lower-limb rehabilitation therapeutic motions, the motion controller should be synthesized with all constraints and found the optimum controller design. The optimum controller should overcome the system uncertainties, process and measurement noises, and further reject the unknown disturbances from the external factors and patient's responses. Considering the proposed system as a nonlinear system and the actuator dynamics is much faster than the desired motion trajectory reference, a robust controller namely, a double-loop motion control based on the integral backstepping design is proposed along with a nonlinear disturbance observer and employed for a sitting-type rehabilitation robot.

The rest of the paper is organized as follows: Sect. 2 discusses the mathematical background of the sitting-type lower limb rehabilitation robot. Section 3 describes the motion control design and its stability analysis, Sect. 4 discusses the performance analysis of the proposed scheme during a clinical gait tracking and its results and finally, Sect. 5 concludes the summary of the paper.

2 Mathematical Background

In this paper, a sitting/lying type lower limb rehabilitation robot is considered for the performance analysis which is providing an improved rehabilitation robotic system. The conceptual design of the system is given in Fig. 1. The device consists of two mechanisms namely a planar 2PRP-2PPR parallel manipulator and a planar RRR serial manipulator as a passive orthosis system. Figure 2 shows the kinematic arrangement of these two mechanisms in side-vertical plane.

The kinematics of the system is required to map the position from the joint space to task space and vice-versa. The control motion can only be given to the joint space to move the end-effector in task space. In order to accomplish the task, the manipulator is operated with a control scheme and the schemes based on the law need the joint space position and velocity.

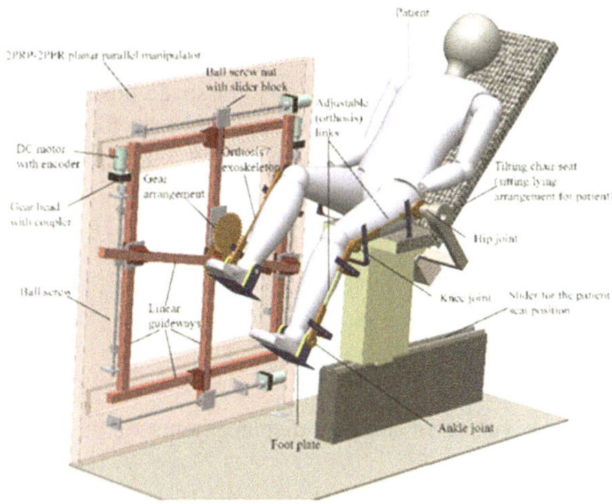

Fig. 1. Conceptual design of the lower limb rehabilitation system

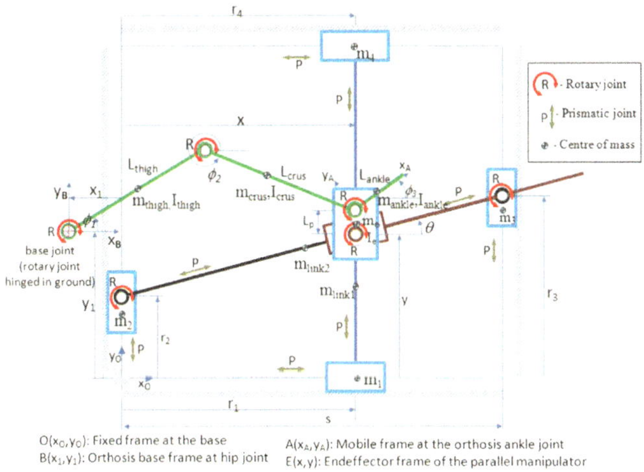

Fig. 2. Kinematic arrangement and locations of center of masses of the lower limb rehabilitation robotic system

The end-effector position is obtained from the forward kinematic relation of the parallel manipulator. The forward kinematic relation is given in Eq. (1).

$$
\begin{aligned}
\boldsymbol{\mu} &= [x \quad y \quad \theta]^T = f(\boldsymbol{\eta}) \\
\boldsymbol{\mu} &= \left[r_1 \quad r_2 + \left(\frac{r_1(r_3 - r_2)}{s}\right) \quad \tan^{-1}\left(\frac{(r_3 - r_2)}{s}\right) \right]^T,
\end{aligned} \tag{1}
$$

where, $\boldsymbol{\mu} \in \Re^{3\times 1}$ is the vector of operational-space positions, i.e., end-effector positions and orientation. $\boldsymbol{\eta} \in \Re^{3\times 1}$ is the vector of linear displacements of the actuated linear joints i.e., $\boldsymbol{\eta} = [r_1 \quad r_2 \quad r_3]^T$(configuration or joint space positions or displacements).

Similarly, the end- effector position is obtained from the forward kinematic relation of the serial manipulator (i.e., a passive orthosis) and is given in Eq. (2).

$$\boldsymbol{\mu} = [x \quad y \quad \theta]^T = f(\boldsymbol{\Phi})$$
$$\boldsymbol{\mu} = \begin{bmatrix} x_1 + L_{thigh}\cos\phi_1 + L_{crus}\cos(\phi_1+\phi_2) \\ y_1 + L_{thigh}\sin\phi_1 + L_{crus}\sin(\phi_1+\phi_2) + L_p \\ G(\phi_1+\phi_2+\phi_3) \end{bmatrix}^T \tag{2}$$

where $\boldsymbol{\Phi} \in \Re^{3\times 1}$ is the vector of angular displacements of the passive joints of the orthosis i.e., $\boldsymbol{\Phi} = [\phi_1 \quad \phi_2 \quad \phi_3]^T$.

The forward velocity relation is derived using the Jacobian matrix which is function of joint space feedbacks. Inverse of this matrix can be used for calculating the joint space or configuration space velocity vector which is used to control the manipulator. The Jacobian matrix can be calculated with the help forward kinematic relation shown in Eqs. (1) and (2) by partially differentiating the operational-space pose vector with respect to the configuration-space variables of the parallel manipulator and serial orthosis. This gives the forward velocity relationships are given as follows:

$$\dot{\boldsymbol{\mu}} = \begin{pmatrix} \dot{x} \\ \dot{y} \\ \dot{\theta} \end{pmatrix} = \mathbf{J}_m(\boldsymbol{\eta})\dot{\boldsymbol{\eta}}$$
$$\dot{\boldsymbol{\mu}} = \begin{bmatrix} 1 & 0 & 0 \\ \frac{r_3-r_2}{s} & 1-\frac{r_1}{s} & \frac{r_1}{s} \\ 0 & -\frac{s}{s^2+(r_3-r_2)^2} & \frac{s}{s^2+(r_3-r_2)^2} \end{bmatrix} \begin{pmatrix} \dot{r}_1 \\ \dot{r}_2 \\ \dot{r}_3 \end{pmatrix} \tag{3}$$

$$\dot{\boldsymbol{\mu}} = \begin{pmatrix} \dot{x} \\ \dot{y} \\ \dot{\theta} \end{pmatrix} = \mathbf{J}_o(\boldsymbol{\Phi})\dot{\boldsymbol{\Phi}}$$
$$\dot{\boldsymbol{\mu}} = \begin{bmatrix} -L_{thigh}S_1 - L_{crus}S_{12} & -L_{crus}S_{12} & 0 \\ L_{thigh}C_1 + L_{crus}C_{12} & L_{crus}C_{12} & 0 \\ G & G & G \end{bmatrix} \begin{pmatrix} \dot{\phi}_1 \\ \dot{\phi}_2 \\ \dot{\phi}_3 \end{pmatrix} \tag{4}$$

where $\mathbf{J}(\boldsymbol{\eta})$ is the Jacobian (velocity transformation) matrix which maps the joint space velocities to the task space velocities of the parallel manipulator. $\mathbf{J}(\boldsymbol{\Phi})$ is the Jacobian matrix which maps the joint space velocities to the task space velocities of the serial orthosis. $S_1 = \sin\phi_1, S_{12} = \sin(\phi_1+\phi_2)$, $C_1 = \cos\phi_1$ and $C_{12} = \cos(\phi_1+\phi_2)$.

The dynamic equation of motion of the robotic system is obtained with the help of Euler-Lagrangian method, and it can be represented in matrix and vector form as follows:

$$\mathbf{M}(\boldsymbol{\eta})\ddot{\boldsymbol{\eta}} + \mathbf{C}(\boldsymbol{\eta}, \dot{\boldsymbol{\eta}})\dot{\boldsymbol{\eta}} + \mathbf{F}(\dot{\boldsymbol{\eta}}) + \mathbf{G}(\boldsymbol{\eta}) = \boldsymbol{\tau}_c + \boldsymbol{\tau}_d, \tag{5}$$

where, $\ddot{\boldsymbol{\eta}} \in \Re^{3\times 1}$ is the vector of linear accelerations of the actuated linear joints (configuration or joint space accelerations). $\mathbf{M}(\boldsymbol{\eta}) \in \Re^{3\times 3}$ is the symmetric positive definite mass (inertia) matrix; $\dot{\boldsymbol{\eta}} \in \Re^{3\times 1}$ is the vector of linear velocities of the actuated linear joints (configuration or joint space velocities). $\mathbf{C}(\boldsymbol{\eta}, \dot{\boldsymbol{\eta}}) \in \Re^{3\times 3}$ is the centripetal and Coriolis force matrix; $\mathbf{F}(\dot{\boldsymbol{\eta}}) \in \Re^{3\times 1}$ is the frictional force vector; $\mathbf{G}(\boldsymbol{\eta}) \in \Re^{3\times 1}$ is the gravitational force vector; $\boldsymbol{\tau}_c \in \Re^{3\times 1}$ is the control (actuator) input vector; $\boldsymbol{\tau}_d \in \Re^{3\times 1}$ is the disturbance vector (which includes external disturbances and internal uncertainties namely parameter variations, noises, etc.).

3 Motion Control Design

The equation of motion of the rehabilitation robotic system can be represented in two first-order subsystems, in other words, kinematic and dynamic equations are as follows:

$$\begin{aligned} \dot{\boldsymbol{\Phi}} &= \mathbf{J}_o(\boldsymbol{\Phi})^{-1}\mathbf{J}_m(\boldsymbol{\eta})\dot{\boldsymbol{\eta}} = \mathbf{J}(\boldsymbol{\eta}, \boldsymbol{\Phi})\dot{\boldsymbol{\eta}} \\ \ddot{\boldsymbol{\eta}} &= \mathbf{M}(\boldsymbol{\eta})^{-1}(\boldsymbol{\tau}_c + \boldsymbol{\tau}_d - (\mathbf{C}(\boldsymbol{\eta}, \dot{\boldsymbol{\eta}})\dot{\boldsymbol{\eta}} + \mathbf{F}(\dot{\boldsymbol{\eta}}) + \mathbf{G}(\boldsymbol{\eta}))) \end{aligned} \tag{6}$$

$$\begin{aligned} &\dot{\mathbf{x}}_1 = \mathbf{g}_1(\mathbf{x}_1)\mathbf{x}_2, \\ &\dot{\mathbf{x}}_2 = \mathbf{f}_2(\mathbf{x}_1, \mathbf{x}_2) + \mathbf{g}_2(\mathbf{x}_1, \mathbf{x}_2)\mathbf{u} \\ &\mathbf{x}_1 = \boldsymbol{\Phi}, \mathbf{g}_1(\mathbf{x}_1) = \mathbf{J}(\boldsymbol{\eta}, \boldsymbol{\Phi}), \mathbf{x}_2 = \dot{\boldsymbol{\eta}}, \mathbf{g}_2(\mathbf{x}_1, \mathbf{x}_2) = \mathbf{M}(\boldsymbol{\eta})^{-1} \\ &\mathbf{f}_2(\mathbf{x}_1, \mathbf{x}_2) = \mathbf{M}(\boldsymbol{\eta})^{-1}(\boldsymbol{\tau}_d - (\mathbf{C}(\boldsymbol{\eta}, \dot{\boldsymbol{\eta}})\dot{\boldsymbol{\eta}} + \mathbf{F}(\dot{\boldsymbol{\eta}}) + \mathbf{G}(\boldsymbol{\eta}))) \end{aligned} \tag{7}$$

The proposed motion control design is as follows:

$$\mathbf{u} = \mathbf{g}_2(\mathbf{x}_1, \mathbf{x}_2)^{-1}\left\{\begin{array}{l} \mathbf{g}_1(\mathbf{x}_1)^{-1}(\ddot{\mathbf{x}}_{1d} + \mathbf{K}_1\dot{\tilde{\mathbf{x}}}_1) + \\ \dot{\mathbf{g}}_1(\mathbf{x}_1)^{-1}(\dot{\mathbf{x}}_{1d} + \mathbf{K}_1\tilde{\mathbf{x}}_1) + \\ \mathbf{K}_2\left(\mathbf{g}_1(\mathbf{x}_1)^{-1}(\dot{\mathbf{x}}_{1d} + \mathbf{K}_1\tilde{\mathbf{x}}_1) - \mathbf{x}_2\right) + \\ \mathbf{g}_1(\mathbf{x}_1)^T\tilde{\mathbf{x}}_1 + \mathbf{f}_2(\mathbf{x}_1, \mathbf{x}_2) \end{array}\right\}, \tag{8}$$

where $\tilde{\mathbf{x}}_1 = \mathbf{x}_{1d} - \mathbf{x}_1$, $\mathbf{x}_{1d} = \boldsymbol{\Phi}_d$ is the desired lower limb joint angles (the desired joint angles of the orthosis, based on clinical gait data). $\tilde{\mathbf{x}}_2 = \mathbf{x}_{2d} - \mathbf{x}_2, \mathbf{x}_{2d} = \mathbf{g}_1(\mathbf{x}_1)^{-1}(\dot{\mathbf{x}}_{1d} + \mathbf{K}_1\tilde{\mathbf{x}}_1)$. $\mathbf{K}_1 = \mathbf{K}_1^{\mathrm{T}} > 0, \mathbf{K}_2 = \mathbf{K}_2^{\mathrm{T}} > 0$ are the controller gain matrices.

To validate the stability of the control scheme, a candidate Lyapunov function is defined as follows:

$$V(\tilde{\mathbf{x}}_1, \tilde{\mathbf{x}}_2) = \frac{1}{2}\left(\tilde{\mathbf{x}}_1^{\mathrm{T}}\tilde{\mathbf{x}}_1 + \tilde{\mathbf{x}}_2^{T}\tilde{\mathbf{x}}_2\right) \tag{9}$$

The selected candidate Lyapunov function is always positive being sum of squares of real numbers. Its time derivative along with trajectories is given as follows:

$$\dot{V}(\tilde{\mathbf{x}}_1, \tilde{\mathbf{x}}_2) = \tilde{\mathbf{x}}_1^{\mathrm{T}}\dot{\tilde{\mathbf{x}}}_1 + \tilde{\mathbf{x}}_2^{\mathrm{T}}\dot{\tilde{\mathbf{x}}}_2 \tag{10}$$

Based on the definitions given above, the error derivatives of $\tilde{\mathbf{x}}_1, \tilde{\mathbf{x}}_2$ are given as below:

$$\begin{aligned} \dot{\tilde{\mathbf{x}}}_1 &= \dot{\mathbf{x}}_{1d} - \dot{\mathbf{x}}_1 \\ \dot{\tilde{\mathbf{x}}}_2 &= \mathbf{g}_1(\mathbf{x}_1)^{-1}\left(\ddot{\mathbf{x}}_{1d} + \mathbf{K}_1\dot{\tilde{\mathbf{x}}}_1\right) + \dot{\mathbf{g}}_1(\mathbf{x}_1)^{-1}(\dot{\mathbf{x}}_{1d} + \mathbf{K}_1\tilde{\mathbf{x}}_1) - \dot{\mathbf{x}}_2 \end{aligned} \tag{11}$$

Substituting the variables from Eqs. (7) and (8) in (11) and rearranging the Eq. (10) based on 11, it becomes,

$$\dot{V}(\tilde{\mathbf{x}}_1, \tilde{\mathbf{x}}_2) = -\left(\tilde{\mathbf{x}}_1^{\mathrm{T}}\mathbf{K}_1\tilde{\mathbf{x}}_1 + \tilde{\mathbf{x}}_2^{\mathrm{T}}\mathbf{K}_2\tilde{\mathbf{x}}_2\right) \tag{12}$$

The time derivative of the selected candidate Lyapunov function is always negative, therefore the closed-loop system is asymptotically stable and the tracking errors converge to zero asymptotically. In other words, the lower limb rehabilitation robot will follows the given desired joint angles.

4 Simulation and Results

To show the effectiveness of the motion control strategy numerical simulations are conducted. A clinically obtained gait data [18] shown in Fig. 3a is used to validate the trajectory tracking operation. The effectiveness is checked by plotting the error norms for the trajectory tracking operation. Modeling of the work area according to Eq. (2) was performed using interval analysis [19, 20] in accordance with the parameters for catch change required for rehabilitation: $\varnothing_1 = [-20°, 10°]$, $\varnothing_2 = [-60°, 0°]$, $\varnothing_3 = [-10°, 20°]$. The simulation results are presented in Fig. 3b.

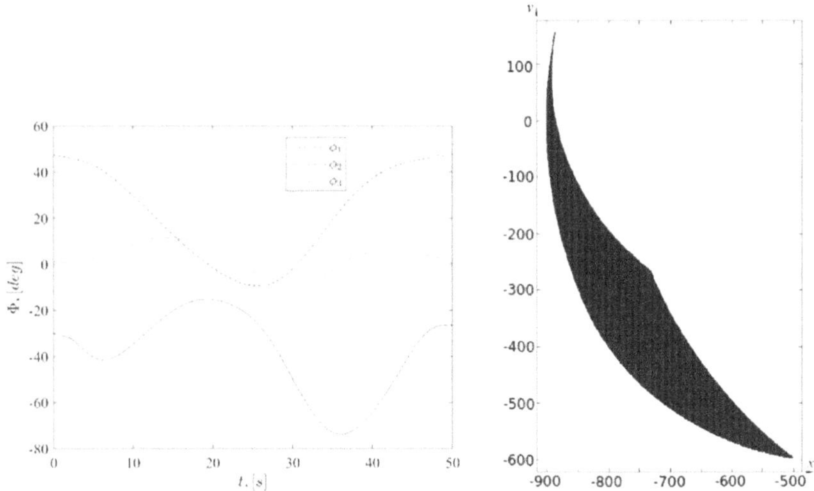

(a) (b)

Fig. 3. Simulation results: (a) Clinical gait data for hip, knee and ankle joint, (b) RRR mechanism workspace.

The calculation time for the approximation accuracy δ = 1 mm on a personal computer was 21 s. It can be seen from Fig. 3b that the overall dimensions of the working area are 401 × 754 mm.

Figure 4 shows the transformation of the clinical gait data into manipulator task space which also determines the manipulators travel and span required for performing the operation.

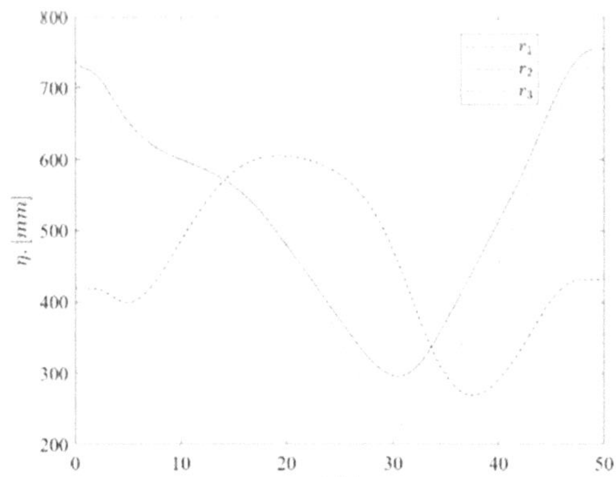

Fig. 4. Required task space for the manipulator

The stroke length of actuators are taken as 0.6 m for each of the actuator leg length of patient is taken as $L_{thigh} = 0.52$ m, $L_{crus} = 0.48$ m, mass of the dynamic components $m_1 = m_4 = 2.5$ kg, $m_2 = 1$ kg, $m_3 = 1$ kg, $m_{link1} = 2$ kg, $m_{link2} = 2$ kg, $m_{crus} = 5.5$ kg, $m_{thigh} = 8$ kg, G = 0.25, s = 1.2 m.

Figure 5 shows the results obtained from the dynamic simulation of the manipulator. Error in the joints are recorded and plotted. Results show that the error values are within $\pm 1^{o}$ which is safe for this kind of application (Fig. 6).

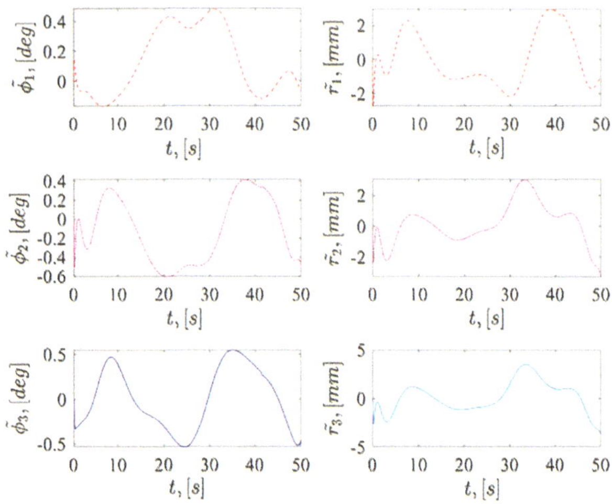

Fig. 5. Error plots for operational space i.e. hip, knee and ankle joint movements

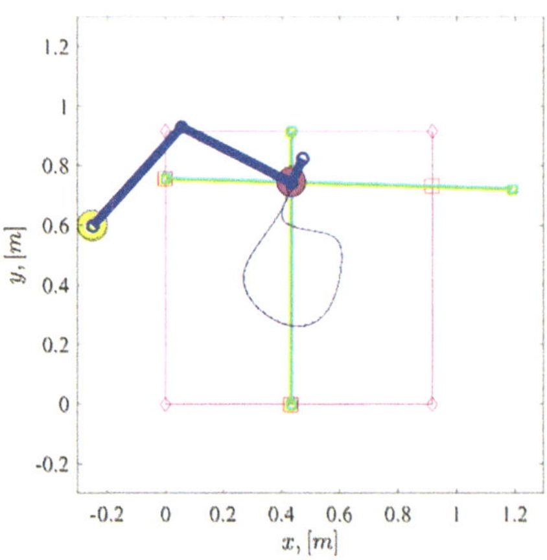

Fig. 6. Simulation results of trajectory tracking operation

5 Conclusions

Robust operational space motion control strategy is designed for the kind of manipulator system where the manipulators joints are not directly connected to the required output motion. In this case the 2PRP-2PPR parallel manipulator is able to control its end-effector within the required workspace for rehabilitation but the control variable is in another system which is of RRR configuration. So for controlling the error on one system from another system need some intermediate transformation which is addressed using this architecture of control strategy.

For control as well to stabilize the output intermediate states are required to stabilize final output which is the basis of the control scheme. Simulation results clearly show the stable behavior of the control scheme and make it a reliable control strategy for high precision operations.

Acknowledgements. This work was supported by the Russian Science Foundation, the agreement number 19-19-00692.

References

1. Samorodskaya, I.V., Zayratyants, O.V., Perkhov, V.I., Andreev, E.M., Vaisman, D.Sh.: Trends in stroke mortality rates in Russia and the USA over a 15-year period. Arch. Pathol. **80**(2), 30–37 (2018)
2. Early rehabilitation of patients after acute ischemic stroke. Science and healthcare, 2 (2014)
3. Vashisht, N., Puliyel, J.: Polio programme: let us declare victory and move on. Indian J. Med. Ethics **9**(2), 114–117 (2012)
4. Truelsen, T., Bonita, R.: The worldwide burden of stroke: current status and future projections. In: Handbook of Clinical Neurology, 3rd edn, vol. 92, pp. 327–336 (2009)
5. Loeb, G.E.: Neural control of locomotion. BioSciences **39**, 800–804 (1989)
6. Stein, P.S.G., Stuart, D.G., Grillner, S., Selverston, A.I.: Neurons, Networks, and Motor Behavior. MIT Press, Cambridge (1999)
7. Huang, G., et al.: Design and feasibility study of a leg-exoskeleton assistive wheel-chair robot with tests on gluteus medius muscles. Sensors **19**(3), 548 (2019)
8. Cafolla, D., Russo, M., Carbone, G.: CUBE, a cable-driven device for limb rehabilitation. J. Bionic Eng. **16**, 492–502 (2019)
9. Merlet, J., et al.: Determination of 6D workspaces of Gough-type parallel manipulator and comparison between different geometries. Int. J. Robot. Res. **18**(9), 902–916 (1999)
10. Evtushenko, Y.: Numerical methods for finding global extreme (case of a non-uniform mesh). U.S.S.R. Comput. Maths. Math. Phys. **11**, 1390–1403 (1971)
11. Evtushenko, Y., Posypkin, M., Rybak, L., Turkin, A.: Approximating a solution set of nonlinear inequalities. J. Global Optim. **7**, 129–145 (2018)
12. Evtushenko, Y., Posypkin, M., Turkin, A., Rybak, L.: The non-uniform covering approach to manipulator workspace assessment. In: Proceedings of the 2017 IEEE Russia Section Young Researchers in Electrical and Electronic Engineering Conference, El-ConRus 2017, pp. 386–389 (2017)
13. Rybak, L.A., Posypkin, M.A., Turkin, A.V.: Method for approximating the workspace of the parallel robot. Int. J. Pharmacy Technol. **8**(4), 25045–25055 (2016)

14. Joumah, A.A., Albitar, C.: Design optimization of 6-RUS parallel manipulator using hybrid algorithm. Int. J. Inf. Technol. Comput. Sci. (IJITCS) **10**(2), 83–95 (2018)
15. Mohammed, R.H., Elnaghi, B.E., Bendary, F.A., Elserfi, K.: Trajectory tracking control and robustness analysis of a robotic manipulator using advanced control techniques. Int. J. Eng. Manuf. (IJEM) **8**(6), 42–54 (2018)
16. Sahraoui, M., Khelfi, M.F., Salem, M.: Sequential adaptive fuzzy inference system based intelligent control of robot manipulators. Int. J. Intell. Syst. Appl. (IJISA) **6**(11), 49–56 (2014)
17. Colgate, J.E., Hogan, N.: Robust control of dynamically interacting systems. Int. J. Control **48**(1), 65–88 (1988)
18. Stansfield, B.W., Hillman, S.J., Hazlewood, M.E., Robb, J.E.: Regression analysis of gait parameters with speed in normal children walking at self-selected speeds. Gait Posture **23**, 288–294 (2006)
19. Jaulin, L.: Applied Interval Analysis: With Examples in Parameter and State Estimation, Robust Control and Robotics. Springer, New York (2001)
20. Moore, R.E.: Interval Analysis. Prentice Hall, Englewood Cliffs (1966)

Design and Practice of Training System for Sports Broadcasting and Hosting Talents Based on OBE Concept in the Medium Age

Ziye Wang[1](✉), Mengya Zhang[2], and Yao Zhang[2]

[1] College of Journalism and Communication, Wuhan Sports University, Wuhan 430079, China
kathy8899@126.com

[2] School of Logistics Engineering, Wuhan University of Technology, Wuhan 430063, China

Abstract. In the era of media convergence, excellent sports broadcasting and hosting professionals are the important force to enrich public sports cultural life and promote the construction of "healthy China". Under the condition that the traditional education methods cannot meet the modern requirements, this paper takes the training of sports broadcasting and hosting professionals as the research object, under the guidance of OBE concept, analyses the role orientation and requirements of sports broadcasting hosts, probes into the ability of professionals to apply the theoretical knowledge they have learned to practice, and puts forward some suggestions. The modes of "coming in & walking out & sinking down", "professional mentor & enterprise mentor", "course design & research subject selection", "order training & elite education", and "Project Cooperative Research + Experimental Training Practice" are designed. The practical training system of sports broadcasting and hosting talents and the concrete implementation scheme are also designed. The system analysis is carried out through the results of practical application. The research shows that the training system can play an important role in the training of sports broadcasting and hosting talents.

Keywords: Practice training system · OBE · Sports broadcasting and hosting

1 Introduction

Sports broadcasting and hosting professionals play an important role in the construction of "healthy China" [1, 2]. In the era of media integration, sports program hosts and commentators have more advanced and higher requirements. Not only need to have a comprehensive personal quality, professional quality and professional skills, but also need to have Internet thinking and information capture capabilities, while completing the "acquisition, editing and broadcasting" of a series of tasks, as well as the practical operation of modern broadcasting equipment [3, 4]. Only in this way can the students be well qualified for their own work in the future.

Guided by the concept of OBE (Outcome-based education), with the students' learning output as the guidance, and emphasizing the ability that students need to

Z. Hu et al. (Eds.): AIMEE 2019, AISC 1126, pp. 173–185, 2020.
https://doi.org/10.1007/978-3-030-39162-1_16

obtain to design and complete teaching, it should done that closely combine theoretical learning with practice, train more excellent talents through the design and implementation of training system, and promote the construction of "healthy country" [5].

2 The Role and Requirements of Sports Broadcasting Host

2.1 Role Orientation of Sports Broadcasting Host

Sports program hosts and commentators play an important role in the mass media and make great contributions to the promotion of sports activities and social life [6]. Many scholars have studied and discussed this topic from different perspectives. In sports competitions, commentary activities are reported live or relayed according to signals from other sources [7]. Competition commentators need to make detailed and vivid descriptions or comments on the athletes' movements, performances, key scenes, atmosphere on the spot, and the specific circumstances of athletes and coaches, as well as the relevant background materials such as competition rules [8]. By listening to the commentary and commentary, the audience can get the details of the competition.

2.2 Professional Requirements of Sports Broadcasting and Hosting Talents

The future professional role of sports broadcasting and hosting talents has special professional requirements [9, 10]. They need to understand and love sports. They need to have a comprehensive mastery and understanding of sports activities and events. They also need to have a wise mind, a keen perspective, rapid and accurate judgment, accurate and reasonable expression, warm and moderate interaction with the audience. At the same time, they need to master new technologies and skills.

Sports broadcasting hosts need to have comprehensive personal qualities, professional qualities and professional skills. Individual literacy mainly includes social responsibility, humanistic literacy, new media thinking and innovative spirit [11]. Their professional quality mainly refers to: knowing and knowing sports knowledge and event knowledge, explaining it from a professional perspective, guiding the public to better appreciate sports and actively participate in sports activities. Language skills are their basic skill. Professional skills mainly include: the ability to control the scene, the ability to interact with the audience, the ability to complete multiple tasks such as editing and broadcasting at the same time, and skilled use of new media tools.

In the era of media convergence, TV, radio, television, newspapers, the Internet, mobile Internet (mobile terminal) and other media means are constantly integrated, which provides many very good communication platforms for some major sports events and beautiful sports moments, and also puts forward the higher quality for sports commentators' requirement [12–14]. A good sports commentator must be a journalist at the same time, with the ability of news reporting and news sensitivity, as well as the ability to control the scene and capture details and broad international horizons. At the same time, they must master or learn the use of modern editing and broadcasting equipment in the shortest possible time.

3 Training of Sports Broadcasting and Hosting Personnel Under the OBE Concept

OBE (Outcome-based education) is a goal-oriented (output-oriented) education model developed in the 1990s [15, 16]. It is guided by students' learning output and emphasizes the implementation of teaching around the abilities that students need to acquire [17, 18].

OBE concept pays attention to the learning achievement of the learner, emphasizes the degree of knowledge and ability of the learner, and holds that the educator must have a clear expectation of the ability level that the learner will achieve when he graduates, and organizes the teaching process with the expected learning output as the central goal, so as to ensure that the learner can achieve the desired goals [19, 20].

Sports broadcasting and hosting is a profession with obvious practicality, which needs to attach great importance to the combination of theory and practice, and use the knowledge learned to complete the work of information dissemination. Therefore, in the design of the curriculum system, the weight of experiment and practice should be emphasized, and maintain good cooperation with employers, so that students can get in touch with the actual work as soon as possible, and get practical exercise and continuous growth [21].

Based on the new requirements of sports broadcasting and hosting in the era of media convergence, this paper focuses on the ability of sports commentary and hosts to apply the theoretical knowledge to practice, designs the training system, and systematically analyses the results of practice [22].

4 Design and Practice of Training System for Sports Broadcasting and Hosting Talents

Educational innovation is the requirement of the times [23, 24]. Based on the urgent need of the reform of the practical training curriculum of sports broadcasting and hosting specialty, this paper puts forward such modes as "invite in + go out + sink down", "professional tutor + enterprise tutor", "course design + research subject selection", "order training + elite education" and so on. The training system is planned and the concrete practice is designed which takes the cultivation of sports broadcasting host as the object and focuses on the cooperation of the employing unit. The overall design framework is shown in Fig. 1.

4.1 The Mode of "Coming in and Walking out and Sinking Down"

In the design of the practical training system for the training of sports broadcasting and hosting talents, we have designed the teaching mode of "coming in & walking out & sinking down" shown in Table 1. Among them, "coming in" teaching: invite industry tutors to enhance students' practical professional skills; "walking out" to visit: in-depth employment enterprises, understand the reality of the industry, enhance college students' professional awareness; "sinking down" to study: comprehensively improve the quality of teachers and the level of integrating theory with practice, in order to strengthen the professional operation ability and comprehensive quality of college students, since the quality and level of teachers play a direct and important role in the cultivation of students [25].

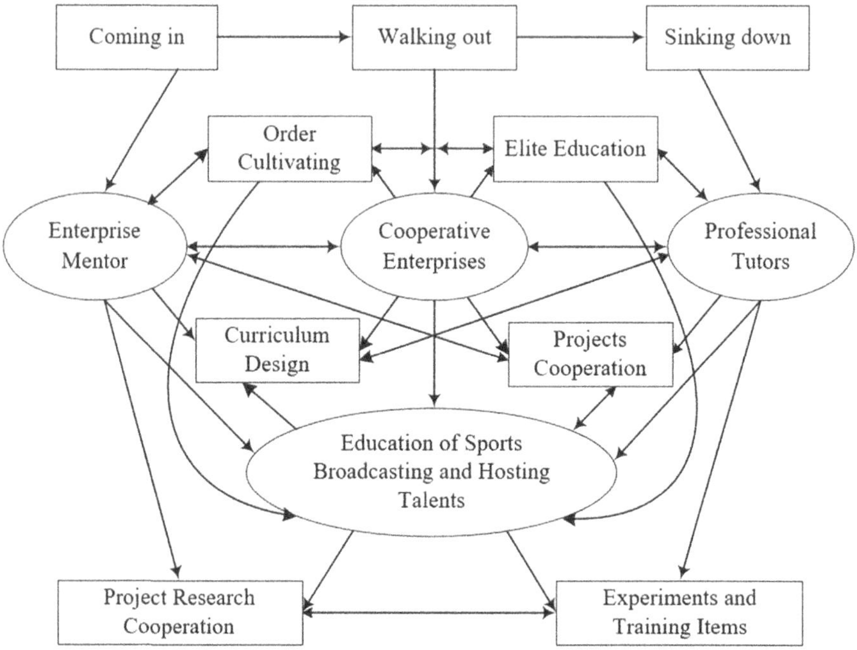

Fig. 1. The implementation framework of sports broadcasting and hosting training course

4.1.1 Curriculum Reform Needs "Coming in"

The training of sports broadcasting and hosting talents needs to learn many theoretical courses, including mainstay courses, such as Chinese Language and Literature, Journalism and Communication, Drama and Film and Television, and some major courses, which includes Introduction to Broadcasting and Hosting Art, Putonghua Speech and Broadcasting Pronunciation, Foundation of Broadcasting Creation, Broadcasting and Hosting of Broadcasting Programs, Broadcasting and Hosting of Television Programs, Introduction to Linguistics, Principles of Journalism, Principles of Communication, Comments on Sports Interpretation.

But if the study of the course is limited to the teaching of theory, students will not have a profound understanding, and the ability to use theoretical knowledge to solve practical problems will be inadequate [26]. Therefore, it is necessary to invite the relevant personnel of the employing unit into the school to participate in the reform and design of the curriculum (especially the training courses).

Table 1. Implementation plan of the mode of "coming in & walking out & sinking down"

Mode	Implementation plan
Coming-in	• Invite famous TV hosts and technicians from We-Media companies to come in the campus and participate in the reform and evaluation of the undergraduate training courses of sports broadcasting and hosting specialty • Invite professionals such as radio, television and all-media presenters to join the teachers and teach students practical courses such as cognitive practice, professional practice and graduation practice
Walking-out	• Arranges out-of-school training to enable students to go out of the campus and into the television stations and other media companies to carry out training courses • 2. Encourage teachers to go out of the campus for social practice learning, and guide students' cognitive practice and graduation practice together with outside tutors
Sinking-down	• Encourage teachers to sink down and devote themselves to the study of teaching, and enriching the courses with practical cases • Arouse students to set their heart to study, to make use of what they have seen and heard in enterprises, make up for their professional knowledge and consolidate their professional skills

4.1.2 Practice Class Requires to "Walk out"

At present, professional theory courses and training courses in research universities are often carried out on campus, and traditional practice teaching is carried out through laboratory simulation. This kind of teaching environment can stimulate students' enthusiasm and passion for learning to a certain extent. However, if the training class is moved to the actual scene, so that students can truly feel the atmosphere and needs of the relevant jobs, and learn in practice, it will be more effective to promote students' absorption and application of theoretical knowledge.

4.1.3 Teachers Must "Sink down"

Talents cannot be trained without excellent teachers. Teachers in schools need not only a high degree and a strong scientific research ability, but also practical experience. Therefore, social practice should be included in the annual assessment criteria of teachers. On the one hand, it can open up cooperative units; on the other hand, it can guide the students who practice in cooperative units. The more prominent benefit is that it can enrich the cases of teachers' teaching and build excellent teachers from social practice and research and teaching reform. In a sense, only teachers who can "sink" and study hard can bring out excellent students who "sink" and study hard.

4.2 "Professional Mentor + Enterprise Mentor" Model

It has been proved that the teaching practice mode of combining theory with practice by introducing "double tutorial system" has been effective [27, 28]. Teachers who cooperate with schools, professionals or technical experts are invited to serve as out-of-school tutors to train students' ability to integrate theory with practice. To guide students in professional learning and practical training by "double tutors" mode, so as to improve students' innovative ability and practical ability. As shown in Table 2.

Table 2. Implementation plan of the mode of "Professional mentor + enterprise mentor"

Subject	Implementation plan
Professional mentor	• Employees: Teachers with excellent ideological and moral character (such as those who have won the title of "the most favorite teacher of students"), school teachers with outstanding professional abilities • Responsibilities: Ideological and moral education, professional courses, competition activities inside and outside the school guidance
Enterprise mentor	• Employees: cooperative radio and television stations or other enterprises in the industry, managers or technicians with years of experience in the enterprise • Job Duties: Arrangement and guidance of internship, classroom teaching in school, guidance of graduation thesis, guidance of text practice and subject competition activities, etc.

4.2.1 Professional Tutor Function

The employment of professional tutors needs ideological and psychological guidance for students, and also needs to provide support for students from the aspects of professional ethics training, professional knowledge learning, learning interest training, professional knowledge innovation, etc. Encourage and guide students to actively participate in various professional competitions inside and outside the school, and improve students' practical ability and transformation ability of scientific research achievements.

4.2.2 Enterprise Tutor Function

Actively "invite in" the enterprise mentor, on the one hand, as the guidance of the training class, on the other hand, it also helps students to complete the arrangement and guidance of graduation internship work. Their duties include: cultivating professional ethics, training professional skills, shaping professional ability, and creating professional value. What should to be done is to encourage students not only to achieve self-worth, but also to help enterprises create value.

4.3 "Course Design + Research Subject" Model

The curriculum design of undergraduates is an important link to check whether the teaching objectives have been achieved or not. It is also a comprehensive practical link in the teaching plan. The mode of "curriculum design + consulting project" is to arrange practical curriculum design topics, promote cooperation between cooperative enterprises and professional tutors in consulting projects, and integrate consulting projects with students' curriculum design. On the one hand, we have completed the course design from the perspective of enterprise cognition, on the other hand, we have helped professional tutors to actively complete the consulting project of cooperative enterprises. As shown in Table 3.

Table 3. The plan of "course design + research subject"

Project content	Implementation plan
Implementation of consulting projects	• Teachers in schools should cooperate with enterprises to develop research topics of school-enterprise cooperation • Encourage school students and personnel of enterprises and institutions to participate in project research • The research subject should be feasible, and the implementation effect of the project should be assessed according to the application of the enterprise
Topics for curriculum design	• Tightly combine the practical problems of enterprises to select topics, or tightly integrate the hot issues of the current society • Enterprise tutors and professional tutors work together to guide and evaluate

4.3.1 Implementation of Consultation Project

Teachers in schools are encouraged to cooperate with TV stations and media companies in establishing consulting projects and integrating them with curriculum design, so as to make curriculum design more practical and innovative by combining theory with practice and writing high-level curriculum design reports.

4.3.2 Topic Selection for Course Design

For the assessment of curriculum design, the evaluation criteria can be determined by enterprise and professional tutors from the aspects of feasibility, innovation and practicability [29], so that the content of the selected topic can be closely combined with the solution of practical problems, and the practicability of curriculum design can be highlighted.

4.4 "Order Training + Elite Education" Model

The training of "order + elite" talents mainly includes the setting of modular curriculum system, the implementation of flexible teaching forms, the adoption of diversified teaching methods and the evaluation of Diversified Assessment methods. As shown in Table 4.

Table 4. The plan of "order training + elite education"

Project	Contents of the project
Curriculum system setting	• Basic quality module, professional compulsory module, order module, elective module, practice module and so on
Implementation of teaching forms	• School-enterprise cooperation, teacher-student interaction, work-study alternation and other forms to form learning, learning to do
Adoption of teaching methods	• Modern teaching, interactive teaching, diversified teaching, learning in thinking, thinking in learning
Evaluation of assessment method	• Written and oral examinations, open and closed papers, examinations and examinations, graduation design and graduation practice, in-school and out-of-school training, corporate and professional tutors and other diversified assessment

What needs to be emphasized in particular is the improvement of teaching methods. Diversified teaching according to one's aptitude, focusing on a variety of advanced teaching methods and means, using modern teaching methods such as micro-courses, flipping classes, mobile classes and intelligent classes, interactive teaching such as case analysis, discussion, task-driven, demonstration simulation, process-oriented, role-playing, brainstorming, situational teaching, etc. By means of multi-media teaching, network teaching, video teaching, on-site demonstration teaching and other diversified means, students are guided to study in thinking and thinking in learning, to promote students' interest in learning, to cultivate students' learning methods, and to stimulate students' knowledge absorption.

Under the circumstances of strengthening the training link, the examination methods also need to be improved accordingly. We can also choose a variety of written and oral examinations, open and closed papers, examinations and examinations, graduation design and graduation practice, on-campus and off-campus training and other assessment methods to comprehensively examine the students' ability to use knowledge and solve practical problems. Because it is impossible to objectively examine students' comprehensive abilities by simply relying on the way of examinations, therefore, the effect of students' learning can be assessed in many ways.

4.5 "Project Cooperative Research + Experimental Training Practice" Model

Practice and innovation are important directions and ways of personnel training. The implementation of this project model is shown in Table 5.

Table 5. Content of "Project Cooperative Research + Experimental Training Practice" model

Scheme design	Implementation plan
Collaborative project research	• College teachers and cooperative enterprises jointly plan research topics • To determine the research topics and contents according to the actual situation of enterprises; Absorbing competent students to participate in research • According to the research expertise of teachers, find out the problems existing in enterprises and determine the research topics
Experiments, training and innovative activities	• To cultivate and organize students to participate academic competitions and innovative entrepreneurship projects at different levels • To provide tutors support for guiding students, and give them certain awards according to the award-winning situation • To set standards for the employment, requirements and assessment of tutors and award in innovative entrepreneurship training

4.5.1 Workload Calculation of Teacher Project Research

Teachers participating in school-enterprise cooperative research projects can get some workload calculation according to the content and results of the research in the year-end assessment, and award those teachers.

4.5.2 Establishment of Students' Innovative Credit

According to the different levels of innovative activities and competitions students participating in (e.g. national, provincial and school levels), some innovative credits are awarded. At the same time, innovative credits are included in the graduation conditions of students. Those who participate in research projects of in-school tutors and corporate tutors can also be recorded with innovation credits.

5 Evaluation of the Training System of Sports Broadcasting and Hosting Talents

Whether the design of system scheme and training platform is reasonable and effective requires a set of scientific methods to evaluate [30, 31]. For the training of sports media professionals, including sports broadcasting and hosting, it can be seen from Fig. 1 that the design of the training system deals with different groups and levels. The evaluation of the training program also includes a number of indicators, which have the relationships of parallel and subordinate.

The first group is about the basic ideas and schemes. Whether those aspects, i.e. "Coming in", "walking out" and "sinking down" are well done or not determines the correctness and rationality of the training system.

Name the first group of index as A_1, A_2 and A_3 respectively, which show that the group of indicators occupies a relatively large weight in the evaluation system. According to whether the invited experts have high level and quality, and whether the role of personnel training is obvious, A_1 can also be classified as A_{11}, A_{12}, A_{13}, ...

Similarly, other two indexes can be divided into A_{21}, A_{22}, A_{23}, ... and A_{31}, A_{32}, A_{33}, ...

The second group is "enterprise mentor" and "professional mentor", whose candidates have a direct impact on the effect of training, which are expressed in B_1 and B_2 respectively.

Their source and level can also be used in the next level of indicators, expressed as B_{11}, B_{12}, B_{13}, ... and B_{21}, B_{22}, B_{23}, ...

"Curriculum design" and "research topic" are the third group of indicators, expressed in C_1 and C_2. The former mainly solves the design and adjustment of school curriculum system, which needs to be carried out according to the needs of employers. The latter is directly facing the practical problems of enterprises, refining the highlights of research, and at the same time, improving the research ability of teachers and students.

The fourth group of "order education" and "elite education" index are designed according to actual needs, both of which cannot be separated from practical training education. D_1 and D_2 are used to indicate the accuracy of their positioning. More specifically, it is available to use the next level of indicators D_{11}, D_{12}, D_{13}, ..., to express and measure.

The fifth part of indicators is "cooperative project research" and "experimental, practical and innovative activities", expressed as E_1 and E_2. Colleges and enterprises can not only focus on solving the practical problems of enterprises, but also jointly carry out other research project declarations. E_1 can be decomposed into E_{11}, E_{12}, E_{13}, ... For students, E_2 is to score all levels of competitions and experiments, practical training and innovative activities; for teachers who guide students' practical training and innovation, and schools should also support special funds and award them at the end of the year. E_2 is formed by E_{21}, E_{22}, E_{23}, ...

There are five first-level indicators, A, B, C, D, E respectively. Their corresponding secondary indicators are: A - A_1, A_2, A_3; secondary indicators are: A_{11}, A_{12}, A_{13}, ...; A_{21}, A_{22}, A_{23}, ...; A_{31}, A_{32}, A_{33}, ...;

B - B_1, B_2, the secondary indicators are: B_{11}, B_{12}, B_{13}, ... B_{21}, B_22, B_{23}, ...;

Similarly, there are:

C - C_1, C_2, secondarily: C_{11}, C_{12}, C_{13}, ... C_{21}, C_{22}, C_{23}, ...;
D - D_1, D_2: D_{11}, D_{12}, D_{13}, ... D_{21}, D_{22}, D_{23}, ...;
E - E_1, E_2: E_{11}, E_{12}, E_{13}, ... E_{21}, E_{22}, E_{23}....

The evaluation index set $U = (U_1, U_2, U_3, U_4, U_5) = (A, B, C, D, E)$ is established.

Using analytic hierarchy process, the weight set corresponding to the evaluation factor set U is determined to be W.

$W = (w_1, w_2, w_3, w_4, w_5)$
Among them, w_i is the weight,
$0 < w_k < 1, k - 1, 2, \ldots n$

Establish a collection of reviews.

$V = (v_1, v_2, v_3, v_4, v_5)$

Among them, V_J represents the ranking result of evaluation content, $j - 1, 2, \ldots M$.

Then, a comprehensive evaluation matrix is built. According to the general fuzzy comprehensive evaluation method, the evaluation results of the training program can be obtained.

6 The Training Characteristics and Teaching Reform of Interpretation Talents in Wuhan Sport University

In the training of sports commentators, the broadcasting and hosting specialty of Wuhan Sport University adheres to the principles of systematization of personnel training, individualization of direction selection and omnipotence of actual operation.

In the course design, we pay attention to the coherence of knowledge and skills and the relevance of theory and practice in the training of talents in the direction of sports explanation. In the third semester of undergraduate students, the work focuses on the cultivation of students' interests and the completion of sports program appreciation. By the fourth semester, students will be led to understand the game, sports viewing and live reporting. The fifth semester focuses on skills acquisition, sports commentary and commentary, as well as sports exhibition and live broadcast. By the sixth semester, the students are required to use their knowledge skillfully to comment on new sports media, and simulate sports events in the form of experiments.

Practice is definitely the key to explain the cultivation of talents. The college has constructed a perfect classroom experiment and built a provincial experimental demonstration center "all-media live broadcast laboratory" to meet all the requirements of the simulation experiment of sports events broadcast, such as shooting, recording, broadcasting, explanation and uploading of video signal.

The project team devoted itself to the design and implementation of the training system and achieved remarkable results. The implementation of the project is mainly carried out from two aspects: in-class practice and out-of-class practice. In-class practice is to use the laboratory platform to introduce practical explanation into the classroom where students directly use the live broadcast platform for the practice of event commentary, and the teaching is carried out according to the actual situation as the feedback.

The content of extracurricular practice is to absorb undergraduate and graduate students under the guidance of professional and part-time tutors, and jointly build an interactive platform of all-media sports in Hubei province, Han Dynamics Sports. Through the whole course of tracking, in-depth report of the game, the students are capable to familiarize with the process of sports broadcasting and interpretation. Significant results have been achieved. Many students stand out in various competitions at the national, provincial and school levels. Before graduation, many of them were recruited by TV stations and other media companies as employees.

7 Conclusions

Focusing on the learning achievement of the learner, emphasizing the degree of knowledge and ability of the learner, it is believable that the educator must have a clear expectation of the ability level that the learner will achieve when he graduates, and organize the teaching process with the expected learning output as the central goal. The OBE concept can guarantee the learner's ability, and achieve the desired goal.

For sports program hosts and commentators in the media environment, the ability to integrate theory with practice is essential. This requires our education to strengthen the design and practice of practical training. Some special practice modes such as "come in & go out & sink down", "professional mentor & enterprise mentor", "course design & research subject selection", "order training & elite education", so on and so forth, are proved to be useful ways to cultivate sport broadcasting and host talent.

In the era of media convergence, only the integrative talents with theory and practice, comprehensive quality and innovative spirit can play their due role in the construction of "healthy China".

Acknowledgements. This paper is supported by —(1) Hubei Teaching Research Project "Research on the Reform of Sports Media Professional Training Model Based on OBE Concept"; (2) Wuhan Sports University Young Teachers' Scientific Research Fund Project "The MECE study under the background of Sports Events media convergence".

References

1. Rowe, D.: Media studies and sport. Soc. Sci. Sport 135–161 (2014)
2. Abeza, G., O'Reilly, N., Reid, I.: Relationship marketing and social media in sport. Int. J. Sport Commun. **6**(2), 120–142 (2013)
3. Zhang, K.: Assessment of media professionals with four dimensions. Youth Reporters **25**, 68–69 (2015). (in Chinese)
4. Li, C., Wang, X.D.: The role function and quality structure of sports commentators. News Res. **6**, 65–66 (2014). (in Chinese)
5. Xu, X.F.: Value development of sports media. Sci. Technol. Inf. **15**, 107 (2014). (in Chinese)
6. Ren, Y., Gu, X.: Educational technology: questions and answers to discipline development. Educ. Res. **40**(1), 141–152 (2019). (in Chinese)
7. Zhang, D.S., Zhang, G.H., Li, F., et al.: Essential characteristics and subject role of sports commentary. J. Wuhan Sports Univ. **51**(5), 5–9 (2017). (in Chinese)
8. Zhuo, W.: Language rhetoric ability of the host in the age of medium integration. Radio TV J. (8), 79–80. (in Chinese)
9. Zhang, D.S., Zhang, G.H., Li, F.: Intrinsic characteristics and roles of sports commentary. J. Wuhan Sport Univ. **5**, 5–9 (2017). (in Chinese)
10. Chen, Y.G., Li, H.: Main problems and Countermeasures of commentary on sports commentary in China. Southeast Commun. **4**, 138–140 (2014). (in Chinese)
11. Meese, J., Podkalicka, A.: Practices of media sport: everyday experience and audience innovation. Media Int. Aust. **155**(1), 89–98 (2015)
12. Li, L.: Sports news communication in the age of fusion media. Journalistic Front **04**, 137–138 (2016). (in Chinese)

13. Li, T.T.: The impact of new media on the spread of sports events. Sci. Technol. Commun. **05**, 118–119 (2015). (in Chinese)
14. Zhong, P., Li, M.H.: Analysis on the development of sports event communication under the new media environment. J. Chongqing Univ. (Soc. Sci. Edn.) **06**, 143–147 (2013). (in Chinese)
15. Castillo, R.C.: A paradigm shift to outcomes-based higher education: policies, principles and preparations. Int. J. Sci.: Basic Appl. Res. **14**(1), 174–186 (2014)
16. Tshai, K.Y., Ho, J.H., Yap, E.H., et al.: Outcome-based education–the assessment of programme educational objectives for an engineering undergraduate degree. Eng. Educ. **9**(1), 74–85 (2014)
17. Yusoff, Y.M., Fuaad, N.F.A., Yasin, R.B.M., et al.: Achievement of program outcomes in outcome based education implementation—a meta analysis. In: Proceedings of the 2014 International Conference on Industrial Engineering and Operational Management, Bali, Indonesia, pp. 7–9 (2014)
18. Kumar, M.K.P., Agrawal, S., Bhat, H.: Effectiveness of project based learning on outcome based education-a case study. J. Eng. Educ. Transf. **29**(3), 132–137 (2016)
19. Tam, M.: Outcomes-based approach to quality assessment and curriculum improvement in higher education. Qual. Assurance Educ. **22**(2), 158–168 (2014)
20. Khan, A.A., Madden, J.: Speed learning: maximizing student learning and engagement in a limited amount of time. Int. J. Modern Educ. Comput. Sci. (IJMECS) **8**(7), 22–30 (2016)
21. Xu, J., Wang, G., Li, B.: Improvement of practical ability of sports journalism majors. J. Wuhan Inst. Phys. Educ. **41**(5), 74 (2007). (in Chinese)
22. Wang, Z., Zhang, Q., Zhang, M.: Training platform construction of omni-media live broadcast of sport event. In: Hu, Z., Petoukhov, S., Dychka, I., He, M. (eds.) ICCSEEA 2019. AISC, vol. 938, pp. 205–216. Springer, Cham (2020)
23. Ellis, A.K.: Research on Educational Innovations. Routledge, Abingdon (2014)
24. Robles, A.C.M.O.: Blended learning for lifelong learning: an innovation for college education students. Int. J. Modern Educ. Comput. Sci. (IJMECS) **4**(6), 1–8 (2012)
25. Bansal, S., Gaffar, A., Dalrymple, O.: Building faculty expertise in outcome-based education curriculum design. In: 2015 IEEE Frontiers in Education Conference (FIE), pp. 1–8. IEEE (2015)
26. Barman, L., Bolander-Laksov, K., Silén, C.: Policy enacted–teachers' approaches to an outcome-based framework for course design. Teach. High. Educ. **19**(7), 735–746 (2014)
27. Zhen, H.: The vocational school-enterprise cooperation: problems and policy analysis. China High. Educ. Res. **1**, 90–93 (2013). (in Chinese)
28. Qiu, X., Wang, B.: Discussion on school-enterprise cooperation of secondary occupation education strategy. J. Gannan Normal Univ. **4**, 027 (2013). (in Chinese)
29. Kotevsk, Z., Tasevska, I.: Evaluating the potentials of educational systems to advance implementing multimedia technologies. Int. J. Modern Educ. Comput. Sci. (IJMECS) **9**(1), 26–35 (2017)
30. Haji, E.E., Azmani, A., Harzli, M.E.: Using AHP method for educational and vocational guidance. Int. J. Inf. Technol. Comput. Sci. (IJITCS) **9**(1), 9–17 (2017)
31. Liu, S., Chen, P.: Research on fuzzy comprehensive evaluation in practice teaching assessment of computer majors. Int. J. Modern Educ. Comput. Sci. (IJMECS) **7**(11), 12–19 (2015)

Analysis of the Structure and Workspace of the Isoglide-Type Robot for Rehabilitation Tasks

Gagik Rashoyan, Konstantin Shalyukhin, Anton Antonov(✉), Aleksandr Aleshin, and Sergey Skvortsov

Blagonravov Mechanical Engineering Research Institute of the Russian Academy of Sciences, Moscow 101000, Russia
ant.ant.rk@gmail.com

Abstract. This article provides a comparative analysis of the various structural schemes of Isoglide-type mechanisms. Several variants of the kinematic chains with the different number and position of the links and joints are presented. Next, the paper presents a methodology for constructing a maximum workspace for such mechanisms, which is based on the use of the chord method. The working area for one selected type of mechanisms is modeled by means of the suggested approach for different values of the links' lengths.

Keywords: Parallel manipulator · Isoglide · Robotic rehabilitation system · Workspace · Chord method

1 Introduction

It is known that a disability can be caused by the diseases associated with the belonging to an elderly age group, as well as the various disorders (stroke, trauma, paralysis, etc.). A rehabilitation process is necessary to restore the impaired functional activity of the limbs. Rehabilitation practice in the treatment of lower extremities requires focusing on various movements of the legs joints.

Based on the foregoing, the choice of the rational manipulator mechanical system with respect to ergonomics and structure becomes the actual problem for the rehabilitation of the lower limbs. The efficiency of the recovery process largely depends on it. It should be noted that each of the lower limb characteristics (weight, type, size) imposes requirements on the parameters of the mechanical system. A solution to this problem is required for the design of reliable and effective therapeutic rehabilitation systems based on special mechanisms [1].

2 Analysis of Isoglide-Type Mechanisms

According to the analysis, it was found that the human foot performs spatial curvilinear movements. The 3-DOF mechanism, suggested by Kong and Gosselin [2] and also known as "Isoglide", is considered as the basic scheme.

Z. Hu et al. (Eds.): AIMEE 2019, AISC 1126, pp. 186–194, 2020.
https://doi.org/10.1007/978-3-030-39162-1_17

The mechanism (Fig. 1) consists of three kinematic chains with orthogonally located translational pairs.

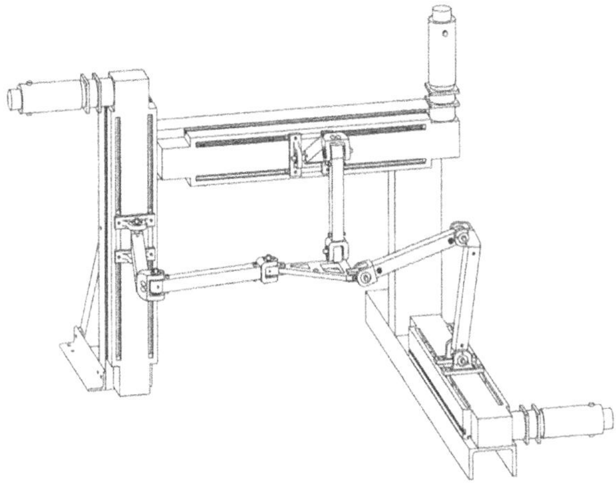

Fig. 1. Isoglide mechanism, suggested by Kong and Gosselin

This mechanism is kinematically decoupled and has three degrees of freedom, that can be calculated by means of screw theory [3–5]. Decoupling is ensured by the presence of a structural group of three rotational pairs with the parallel axes in each kinematic chain. Such a scheme provides a constant gear ratio between the drives and the output link and a complete kinematic decoupling between three linear coordinates.

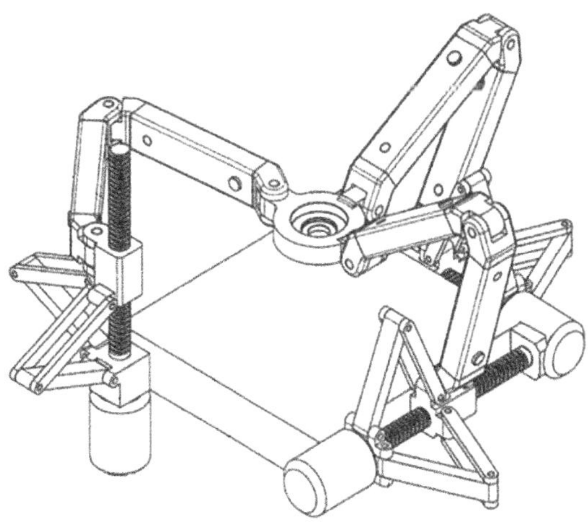

Fig. 2. Isoglide mechanism with additional kinematic chains

An analysis of this scheme was conducted in IMASH RAN and a prototype with slight modification was created (Fig. 2). The mechanism was experimentally studied for stiffness [6]. The modification involves the presence of additional kinematic chains, "elbows", that allows eliminating translational kinematic pairs (guides).

A number of new modifications of these structural schemes were proposed. The first one is known as a translational-directing parallelogram mechanism [4, 5, 7]. Each kinematic chain consists of one driven translational pair, located on the base, and two translational pairs, designed as a parallel link mechanism (Fig. 3).

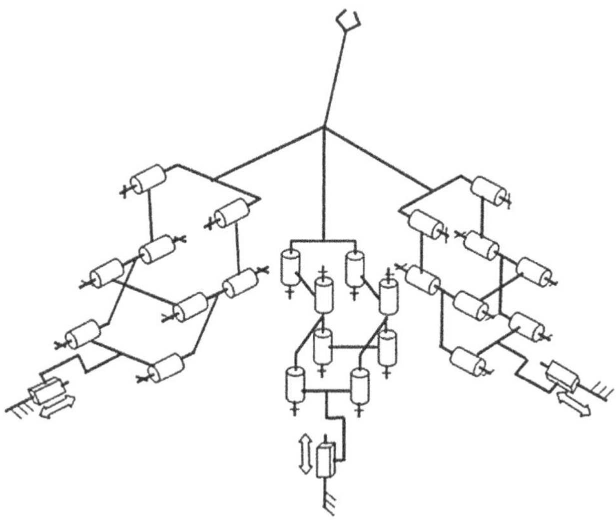

Fig. 3. Translational mechanism with parallelograms

The second modification of the mechanism with three translational degrees of freedom and kinematic decoupling is shown in Fig. 4 [8]. Gear blocks perform the functions of the parallel link mechanisms.

Both modifications have the isomorphism property, that is, each motor corresponds to the movement of the output link in only one coordinate.

The scheme presented in Fig. 1 is optimal for rehabilitation tasks since the kinematic chains consist of a minimum number of movable joints and links. The mechanism based on this scheme can be used for a rehabilitation robot (Fig. 5).

One of the most important criteria for the mechanism, used for rehabilitation purposes, is the geometry of its workspace. Let's consider the working area analysis for the Isoglide-type mechanism.

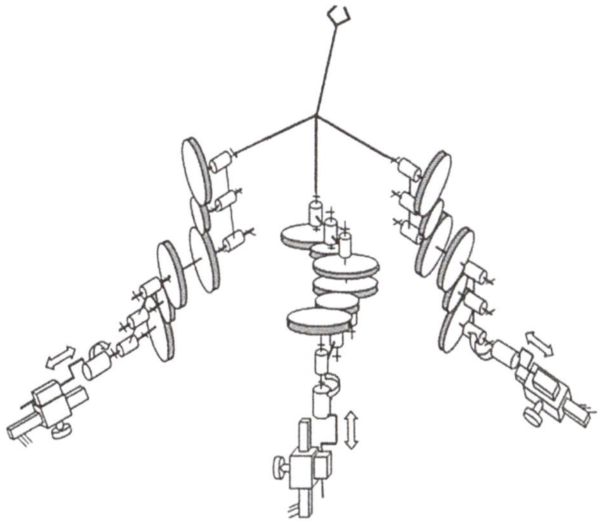

Fig. 4. Mechanism with gear blocks

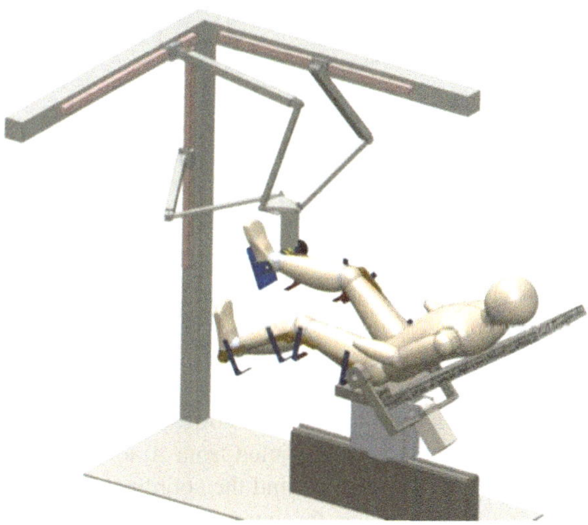

Fig. 5. Rehabilitation system based on Isoglide mechanism

3 Workspace Analysis

Here we present an approach for maximum workspace calculation, which combines features of two different methods: a discrete method and an optimization chord method. The idea of this method is as follows. First, the whole space around the mechanism output link is discretized along one axis. Then, the optimization chord method is used in each such workspace slice to find the boundary of a planar 2-DOF mechanism maximum workspace. This optimization method provides a fast and effective way to calculate the workspace boundary for planar mechanisms with respect to different constraints. The chord method with examples of its application is described in [9–11], and we will not focus on it here.

The first step to analyze mechanism workspace is to examine its kinematic model. This model connects the output link coordinates with the driven coordinates. For considered spatial 3-DOF mechanism (Fig. 6), the output link coordinates can be described as a position vector $\mathbf{p} = (x\ y\ z)^T$ of the point P (Fig. 6), and the driven coordinates are the linear displacements $\mathbf{q} = (q_1\ q_2\ q_3)^T$.

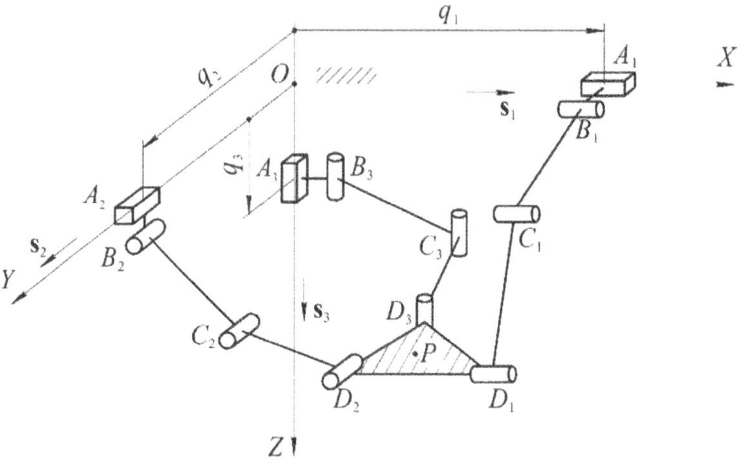

Fig. 6. Kinematic model of the considered mechanism

One can write down the following equation that connects **p** and **q** [2]:

$$q_i = \mathbf{s}_i^T(\mathbf{p} + \mathbf{PD}_i - \mathbf{A}_i\mathbf{B}_i), i = 1\ldots3, \tag{1}$$

where $\mathbf{s}_i$ is the unit vector of the coordinate axis corresponding to q_i; $\mathbf{PD}_i$ is the vector pointed from P to D_i; $\mathbf{A}_i\mathbf{B}_i$ is the vector pointed from A_i to B_i (Fig. 6).

With the known coordinate q_i, one can find the coordinate of the point B_i and then find the coordinates of the point C_i as the intersection point of two circles with the centers in B_i and D_i and the radii of B_iC_i and D_iC_i respectively. All these calculated coordinates allow finding angles in mechanism joints. Thus, the kinematic model is completed.

In the process of workspace analysis, one should deal with the mechanical constraints. For the considered mechanism we can specify the following:

- constraints on the driven coordinates **q**;
- constraints in the passive rotational joints B_i, C_i, and D_i.

All these constraints can be taken into account in the chord optimization method as inequalities [9], and the values of the linear displacements and joint angles are directly calculated using the kinematic model.

Let's consider an example of the workspace analysis. Suppose, the mechanism has the following parameters (all in the relative units): for all kinematic chains $A_iB_i = 5$, $B_iC_i = C_iD_i = 50$, the output link is an equilateral triangle with a side equals 20. The point P is in the center of this triangle. Let each driven coordinate q_i vary from 0 to 100. The minimum angle between A_iB_i and B_iC_i will be equal to 15° and the minimum angle between B_iC_i and C_iD_i will be the same. Also, let's constrain the maximum value of the angle between B_iC_i and C_iD_i to 170° to avoid the aligning of these links: this position corresponds to the mechanism singular position.

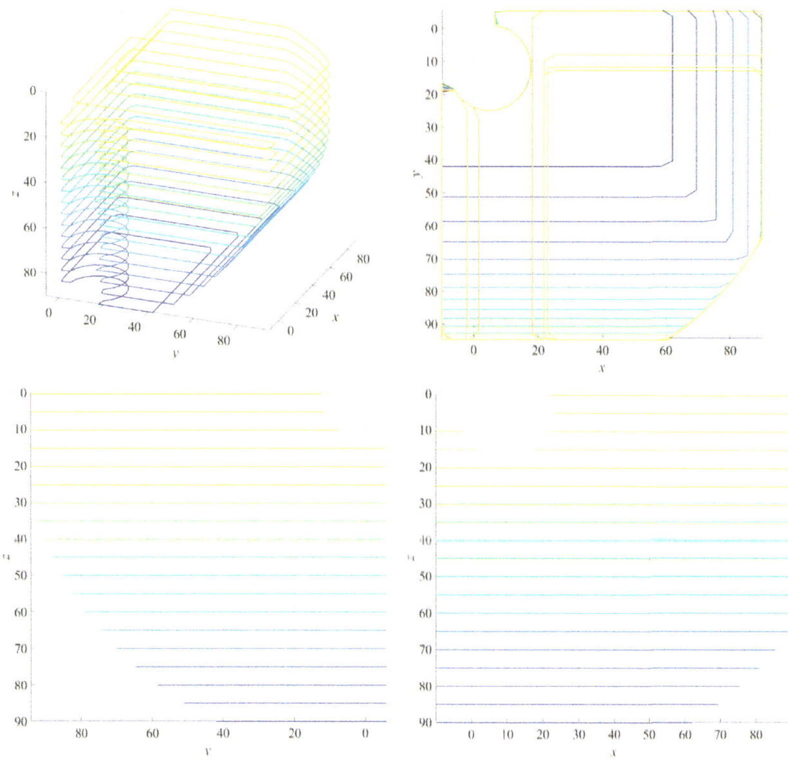

Fig. 7. Maximum workspace of the mechanism with $A_iB_i = 5$ and $B_iC_i = C_iD_i = 50$

MATLAB program using the standard function "fmincon" for solving optimization problem was written to build the maximum workspace. The slices of the working area were performed along the Z axis every 5 units from 0 to 100. Figure 7 demonstrates the obtained workspace and its projections on different coordinate planes.

Another example was performed for $B_iC_i = 60$ and $C_iD_i = 40$ (note that their sum stays the same). Figure 8 shows the results.

Comparing the results in Figs. 7 and 8, one can see that the volume of the workspace becomes smaller due to an unachievable area for small values of x and y. The greater the difference between the lengths of the links B_iC_i and C_iD_i, the smaller will be the workspace.

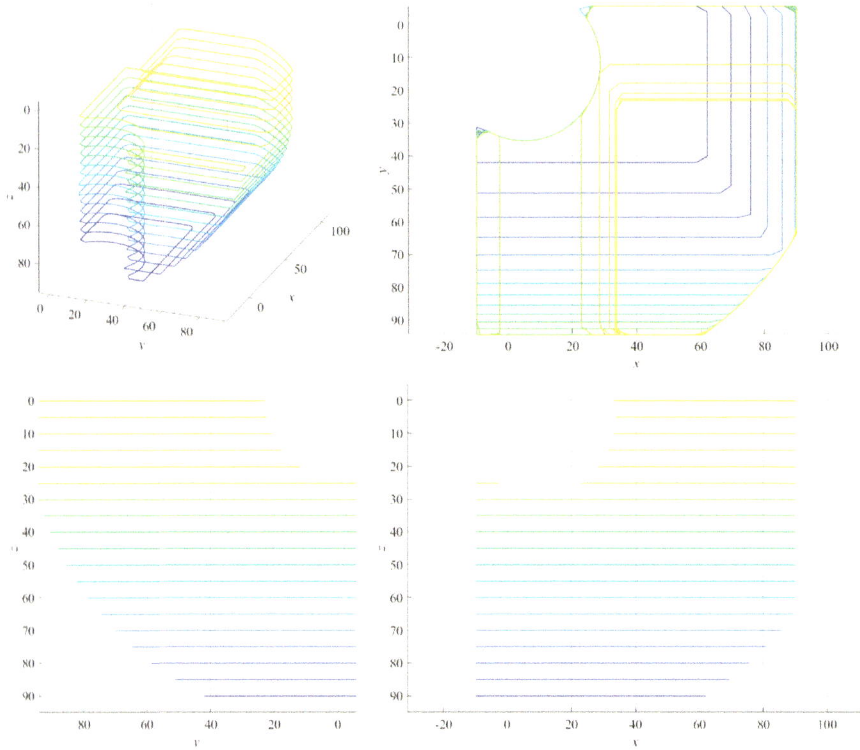

Fig. 8. Maximum workspace of the mechanism with A_iB_i = 5, B_iC_i = 60, and C_iD_i = 40

The final example was performed for the initial values of B_iC_i and C_iD_i, but the value of A_iB_i was increased from 5 to 10. The obtained working area is presented in Fig. 9.

As we can see from the plots in Figs. 7 and 9, the mechanism unachievable area displaced due to the change of A_iB_i, but the radius of this area did not change. The further increase of this parameter will only displace the unachievable zone.

There are also several other well-known methods for analyzing the mechanism's workspace. For example, Kong [2] used the geometrical approach to find the maximum working area. This method is intuitive and very illustrative when accounting only the limits on the generalized coordinates, however it becomes difficult to use it concerning the other constraints, e.g. the passive joints limits. Another approach [12, 13] is to use the discretization along each of the coordinate axes. This method allows to take into account the above-mentioned constraints with no trouble, but the main drawback of this method is a long computational time for precise results. There is also a method based on an interval analysis [14, 15], but it also takes a long time.

The suggested approach has the advantages of both of these methods: it can deal with almost any of the mechanism's constraints and takes far less time for computation comparing to the discretization method. The performed modeling showed the effectiveness of this approach.

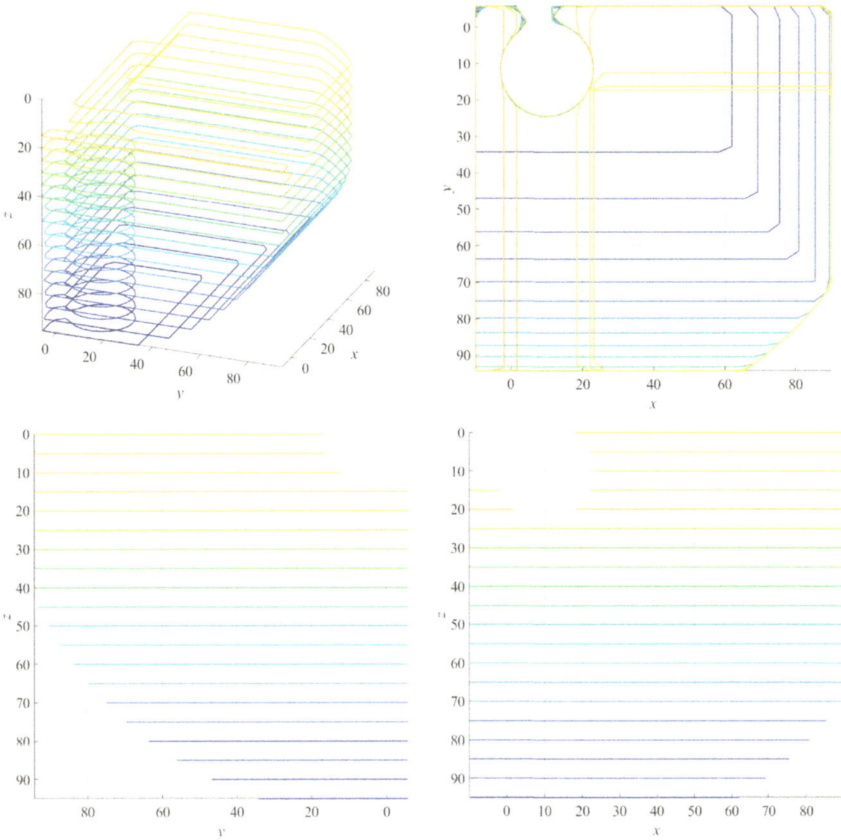

Fig. 9. Maximum workspace of the mechanism with A_iB_i = 10, B_iC_i = 50, and C_iD_i = 50

4 Conclusions

The paper presents Isoglide-type mechanisms and the workspace analysis for one of them. The maximum working area was obtained using the optimization chord method, which was adapted to analyze a spatial mechanism. The suggested approach can be used as an efficient alternative for such traditional methods as geometric and discrete.

The workspace analysis is essential for the mechanism's dimensional synthesis. It is also important for motion planning because all of the robot's trajectories have to stay within the obtained working area. This feature should be considered by the robot's control system, and it will be the subject of future research.

Acknowledgment. This work was supported by Russian Science Foundation, grant no. 19-19-00692.

References

1. Plitan, F., Badri, A., Meigolinedjad, J., Keshavarz, M.: Adaptive artificial intelligence based model base controller: applied to surgical endoscopy telemanipulator. Int. J. Intell. Syst. Appl. **5**(9), 113–115 (2013)
2. Kong, X., Gosselin, C.M.: Kinematics and singularity analysis of a novel type of 3-CRR 3-DOF translational parallel manipulator. Int. J. Robot. Res. **21**(9), 791–798 (2002)
3. Kong, X., Gosselin, C.M.: Generation of parallel manipulators with three translational degrees of freedom based on screw theory. In: Proceedings of 2001 CCToMM Symposium on Mechanisms, Machines and Mechatronics, Montreal, Canada, p. M3–01–012 (2001)
4. Glazunov, V., Kheylo, S.: Dynamics and control of planar, translational, and spherical parallel manipulators. In: Zhang, D., Wei, B. (eds.) Dynamic Balancing of Mechanisms and Synthesizing of Parallel Robots, pp. 365–402. Springer, Cham (2016)
5. Rashoyan, G.V., Shalyukhin, K.A., Gaponenko, E.V.: Development of structural schemes of parallel structure manipulators using screw calculus. In: IOP Conference Series: Materials Science and Engineering vol. 327, p. 042090 (2018)
6. Glazunov, V.A., Kasilov, V.P., Kozyrev, A.V., Levin, S.V., Shalyukhin, K.A.: The parallel structure manipulator with three orthogonal translational degrees of freedom and its rigidity analysis. Eng. Autom. Probl. **3**, 48–56 (2015)
7. Glazunov, V.: Design of decoupled parallel manipulators by means of the theory of screws. Mech. Mach. Theory **45**(2), 239–250 (2010)
8. Shalyukhin, K.A., Rashoyan, G.V., Aleshin, A.K., Skvortsov, S.A., Levin, S.V., Antonov, A.V.: Problems of kinematic analysis and special positions of mechanisms of robots with parallel structure. J. Mach. Manuf. Reliab. **47**(4), 310–316 (2018)
9. Snyman, J.A., du Plessis, L.J., Duffy, J.: An optimization approach to the determination of the boundaries of manipulator workspaces. J. Mech. Des. **122**(4), 447–456 (2000)
10. Hay, A.M., Snyman, J.A.: The chord method for the determination of nonconvex workspaces of planar parallel manipulator. Comput. Math Appl. **43**(8–9), 1135–1151 (2002)
11. Hay, A.M., Snyman, J.A.: A multi-level optimization methodology for determining the dextrous workspaces of planar parallel manipulators. Struct. Multidisciplinary Optim. **30**(6), 422–427 (2005)
12. Joumah, A.A., Albitar, C.: Design optimization of 6-RUS parallel manipulator using hybrid algorithm. Int. J. Inf. Technol. Comput. Sci. (IJITCS) **10**(2), 83–95 (2018)
13. Nanda, S.K., Panda, S., Subudhi, P.R.S., Das, R.K.: A novel application of artificial neural network for the solution on inverse kinematics controls of robotic manipulators. Int. J. Intell. Syst. Appl. (IJISA) **4**(9), 81–91 (2012)
14. Chablat, D., Wenger, Ph., Merlet, J.: Workspace analysis of the Orthoglide using interval analysis. In: Lenarčič, J., Thomas, F. (eds.) Advances in Robot Kinematics, pp. 397–406. Springer, Dordrecht (2002)
15. Kumar, V., Sen, S., Roy, S.S., Das, S.K., Shome, S.N.: Inverse kinematics of redundant manipulator using interval Newton method. Int. J. Eng. Manuf. (IJEM) **5**(2), 19–29 (2015)

Hyperbolic Numbers, Genetics and Musicology

Sergey V. Petoukhov(✉)

Mechanical Engineering Research Institute, Russian Academy of Sciences, M.Kharitonievsky pereulok, 4, Moscow, Russia
spetoukhov@gmail.com

Abstract. The article is devoted to applications of 2-dimensional hyperbolic numbers and their algebraic extensions in the form of 2^n-dimensional hyperbolic numbers in bioinformatics, algebraic biology and musicology. These applications reveal hidden interconnections between structures of different biological phenomena. It helps to understand living bodies as holistic essences, structural organisation of which has relations with musical harmony, first of all, with Pythagorean musical scales on the basis of the quint ratio 3/2. Presented results are connected with a problem of genetic fundamentals of aesthetics and inborn feeling of harmony. On the basis of these and some other results, the author believes that 2^n-dimensional hyperbolic numbers and their matrix representations are a key element for effective algebraic modeling many aspects of inherited structural organisation of biological bodies and for developing algebraic biology. The received results lead to new approaches in bioinformatics, musicological analysis, acoustic biotechnologies and artificial intelligence.

Keywords: Hyperbolic numbers · Genetics · DNA · Tensor product · Musical scales · Quint ratio

1 Introduction

The main task of the mathematical natural sciences is the creation of mathematical models of natural systems. Development of models and formalized theories depends highly on those mathematical notions and instruments, on which they are based. Modern science knows that different natural systems could possess their own individual geometries and their own individual arithmetic [1].

Various kinds of multi-dimensional numbers – complex numbers, hyperbolic numbers, dual numbers, quaternions and other hypercomplex numbers – are used in different branches of modern science. They have played the role of the magic tool for development of theories and calculations in problems of heat, light, sounds, fluctuations, elasticity, gravitation, magnetism, electricity, current of liquids, quantum-mechanical phenomena, special theory of relativity, nuclear physics, etc. For example, in physics thousands of works - only in XX century – were devoted to quaternions of Hamilton (their bibliography is in [2]).

The idea about special mathematical peculiarities of living matter exists long ago. For example Vernadsky [3] put forward the hypothesis on a non-Euclidean geometry of living nature. It seems an important task to investigate what systems of multi-dimensional numbers are connected or can be connected with ensembles of parameters

Z. Hu et al. (Eds.): AIMEE 2019, AISC 1126, pp. 195–207, 2020.
https://doi.org/10.1007/978-3-030-39162-1_18

of the genetic code and inherited biological peculiarities. Some results of such investigation are presented in this article. They are connected with hyperbolic numbers and their algebraic extensions, matrix forms of which give a new class of mathematical models in genetics, musicology and some other scientific fields.

2 Musical Harmony and the Quint Ratio 3/2

As known, thoughts about the key significance of musical harmony in the organization of the world exist from ancient time. For example, one can quote here a classical work of Chinese literature "Spring and Autumn" by Lu Bu We about the fundamental role of music and numbers 3 and 2 as numbers of Heaven and Earth: "*The origins of music lie far back in the past. Music arises from Measure and is rooted in the great Oneness. ... Music is founded on the harmony between Heaven and Earth*" (this citation is taken from the book [4]). In Ancient China the ratio 3/2, traditionally termed as the quint ratio (or the pure perfect fifth), was used as the fundament of quint music scales. After Ancient Chinese, Pythagoreans also considered numbers 2 and 3 as the female and male numbers (or Yin and Yang numbers), which can give birth to new musical tones in their interconnection. Ancient Greeks attached an extraordinary significance to search of the quint 3:2 in natural systems because of their thoughts about musical harmony in the organization of the world. For example, Archimedes considered as the best result of his life a detection of the quint 3/2 between volumes and surfaces of a cylinder and a sphere entered in it.

For Europeans the idea of musical harmony is connected basically with the name Pythagoras. The Pythagorean musical scales, which are based on the quint ratio 3/2, played the main role in the Pythagorean's doctrine about a cosmic meaning of musical harmony. Figure 1 shows the known interconnection of sound frequencies of notes of Pythagorean 7-stages scale (a heptatonic scale) on the basis of the ratio 3/2 when notes are spaced in the appropriate octaves.

fa (F)	do (C)	sol (G)	re (D^1)	la (A^1)	mi (E^2)	si (B^2)
87	130	196	293	440	660	990
$\mathbf{(3/2)^{-3}}$	$\mathbf{(3/2)^{-2}}$	$\mathbf{(3/2)^{-1}}$	$\mathbf{(3/2)^{0}}$	$\mathbf{(3/2)^{1}}$	$\mathbf{(3/2)^{2}}$	$\mathbf{(3/2)^{3}}$

Fig. 1. The quint sequence of the 7 notes of the Pythagorean musical scale. The upper row shows the notes. The second row shows their frequencies. The third row shows the ratios between the frequencies of these notes to the frequency 293 Hz of the note re (D^1). The designation of notes is given on Helmholtz system. Values of frequencies are approximated to integers.

Pythagoras created the mathematical foundations of ancient Greek music, borrowing in a certain degree some ancient knowledge on musical harmony. His theory used the discovery that the frequency of a vibrating string is inversely proportional to its length and that musical consonances can be represented by the ratios of small integer numbers, first of all the octave ratio 2:1 and the quint ratio 3:2. These ideas became the basic fundamental ones of all music theory from antiquity to even modern times. For most Europeans from antiquity, quint scales in music are connected with this Pythagorean mathematical theory of musical harmony and with divisions of vibrating strings in the quint ratio 3:2.

In a general case, the Pythagorean scale is any scale, which can be constructed from only quint ratios 3:2 and octaves 2:1 [5, p. 163]. One of known Pythagorean scales is a pentatonical scale, which is a five-stages music scale, all the sounds of which can be arranged in quint ratios. Its example is the set of the following 5 notes with their sound frequencies from Fig. 1: do(C)-sol(G)-re(D^1)-la(A^1)-mi(E^2) or respectedly 130–196–293–440–660 Hz. Other examples of Pythagorean scales are tetratonic and tritonic scales, which are correspondingly 4-stages and 3-stages music scales, all the sounds of which can be arranged by the quint ratio, for instance, 130–196–293–440 Hz for the tetratonic scale and 130–196–293 Hz for the tritonic scale.

The remarkable historical fact is that these Pythagorean musical scales on the basis of the quint ratio were used by different civilisations around the world long before Pythagoras without knowledge of any mathematical laws [6–10]. For example, the pentatonical scale is the foundation of traditional music of the Chinese, Vietnamese, Mongols, Turkic peoples (Bashkirs, Tatars, Chuvashes, etc.), the Inca Empire and the peoples of the South Andes in general. Pentatonics is also found in European musical folklore and in the oldest layers of the Russian folk song (especially in the so-called calendar ritual songs). Tetratonic music was noted as common in Polynesia and Melanesia. Tetratonic scales were known for example among the Plains Indians, the Arapaho, Blackfoot, Crow, Omaha, Kiowa, Pawnee, Sioux, some Plateau tribes, the Creek Indians, and in the Great Basin region among the Washo, Ute, Paiute, and Shoshone. In the Southwest, the Navajo people also largely used the pentatonic and tetratonic, occasionally also tritonic scales. Tetratonic, as well as tritonic scales, were commonly used by the tribal peoples of India, such as the Juang and Bhuyan of Orissa state [11]. Tetratonic scales are generally associated with prehistoric music [12].

Leibniz declared that music is arithmetic of soul, which computes without being aware of it. But what is there in living organisms that determines the special attraction of musical scales on the basis of the quint ratio 3/2 for representatives of various civilizations and epochs? A possible answer lies in the structural features of DNA molecules that are carriers of genetic information in humans and other living organisms. The author draws attention to the fact that the parametric structure of DNA molecules is connected in many ways with the quint ratio 3/2 at various levels of their parametric organization [13, 14]. The author has paid attention that numbers 2 and 3 and the quint ratio 3/2 are implemented in parameters of molecular-genetic systems.

Molecules of heredity - DNA and RNA – contain sequences of 4 "letters" or nucleobases: adenine (A), cytosine (C), guanine (G), thymine (T) (or uracil U in RNA). Letters A-T(U) and C-G form complementary pairs with 2 and 3 hydrogen bonds in them, respectively. From the standpoint of its sequence of two and three hydrogen bonds, each DNA molecule is a long chain of numbers 2 and 3 of type 32232332....

The genetic code encodes sequences of 20 amino acids in proteins by means of 64 triplets (three-letter words) that represent all possible combinations of these four letters

(ATC, TTA,...). Since A = T = 2, C = G = 3, each triplet has a numeric representation as a product of number of hydrogen bonds of its constituent letters. For example, the triplet ACT is represented by number 2*3*2 = 12. Each of 64 triplets is represented by one of such numbers of hydrogen bonds $2^3 = 8$, $2^{2*}3 = 12$, $2^*3^2 = 18$, $3^3 = 27$, the pairwise relations between which are equal to the quint 3/2 in varying integer degrees (by analogy with music tetratonic scales), for example, $27/8 = (3/2)^3$, $18/8 = (3/2)^2$, etc.

Under considering pairs of adjacent triplets, then DNA molecule appears as a quint sequence of 7 kinds of numbers of hydrogen bonds with the following numeric representation: $2^6 = 64$, $2^{5*}3 = 96$, $2^{4*}3^2 = 144$, $2^{3*}3^3 = 216$, $2^{2*}3^4 = 324$, $2^*3^5 = 486$, $3^6 = 729$. Pairwise ratios in this series of numbers are equal to the quint 3/2 in the same powers as in the Pythagoras 7-stage scale in Fig. 1. If, for example, the frequency of 87 Hz of the note "F" is compared with the first number 64 of this series, then all other numbers of this series will correspond precisely to the other frequencies of the Pythagoras scale. Then any sequence of triplets (e.g., insulin gene GGC-ATC-GTT-GAA-CAG-TGT-...) can be associated uniquely with a sequence of notes of Pythagoras 7-stages scale (figuratively speaking, we have "music of genes in the Pythagoras scale").

Accordingly, each DNA molecule as a chain of hydrogen bonds is characterized by its own sequences of the quint 3/2 in different integer degrees. By analogy with quint musical scales, depending on the chosen lengths of nucleobase fragments of DNA, we have - on the basis of hydrogen bonds - various systems for transmitting information signals with quint-power relations between signals.

The quint ratios are realized in DNA not only for the hydrogen bonds of complementary nucleobases, but also for several other parameters, such as sums of atoms in the rings of purines and pyrimidines (numbers 9 and 6 with the ratio 3/2), or sums of protons in the rings of complementary nitrogenous bases (numbers 60 and 40 with the ratio 3/2), and others. Chains of these parameters in DNA form their own sequences of quint ratios, which correspond to sequences of note frequencies in quint music scales. In other words, Nature created DNA as a plexus of various sequences of quint ratios ("a quint polyphony of DNA"). The harmony of the parametric organization of the genetic system is akin to the musical harmony of the Pythagorean quint scales.

As it was reminded above, over the centuries from Ancient China to antiquity, the numbers 2 and 3 were considered respectively as female and male numbers (that is as Yin and Yang numbers) forming the important pair. The author proposes their consideration not as separate one-dimensional numbers but as two separate parts of two-dimensional number. Mathematics knows 3 main kinds of two-dimensional numbers: complex numbers, hyperbolic (or double) numbers and dual numbers [15]. By a few reasons, we choice hyperbolic numbers for a presentation of numbers 3 and 2 as two parts of single two-dimensional number $G_2 = 3 + 2j$, where j is imaginary unit with its feature $j^2 = + 1$; the index 2 refers 2-dimensionality of the number G_2. This hyperbolic number can be expressed as a point or a vector on a two-dimensional plane with Cartesian coordinates, in which the axis of abcissus is considered the axis of Yang-numbers, and the axis of ordinates is considered the axis of Yin-numbers. Figure 2 shows this coordinate system and also the matrix form of presentation of hyperbolic numbers with its decomposition into 2 sparse matrices playing the role of real and imaginary basis units of hyperbolic numbers. This matrix [3, 2; 2, 3] is conditionally termed "quint matrix" since its components 3 and 2 give the ratio 3/2. (The same quint matrix [3, 2; 2, 3] appears under a consideration of DNA alphabet C, A, T, G and its three binary sub-alphabets [13, Chapter 2; 14, Chapter 4]).

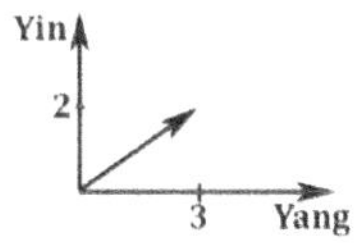

$$G_2 = \begin{vmatrix} 3, 2 \\ 2, 3 \end{vmatrix} = 3* \begin{vmatrix} 1, 0 \\ 0, 1 \end{vmatrix} + 2* \begin{vmatrix} 0, 1 \\ 1, 0 \end{vmatrix};$$

*	1	j
1	1	j
j	j	1

Fig. 2. The matrix presentation of the 2-dimensional hyperbolic numbers $G_2 = 3 + 2j$. The first sparse matrix [1, 0; 0, 1] is the identity matrix, the second sparse matrix [0, 1; 1, 0] presents imaginary unit j having the property $[0, 1; 1, 0]^2 = [1, 0; 0, 1]$. The multiplication table of these sparse matrices, where 1 refers the matrix [1, 0; 0, 1], is also shown at right.

3 Applications of Algebra of Hyperbolic Numbers and Its Extensions

This Section shows relations of Pythagorean musical scales with hyperbolic numbers and their algebraic extensions on the basis of the tensor product of matrices.

The collection of all hyperbolic numbers (their synonimical names: "double numbers", "split-complex numbers", "perplex numbers", etc.) forms algebra over the field of real numbers [15]. Hyperbolic numbers are well known in mathematics and theoretical physics and they have matrix form of their general presentation: $G_2 = [a, b; b, a]$, where a and b are real numbers.

The second tensor power of the (2*2)-matrix $[a, b; b, a]$ gives the (4*4)-matrix (Fig. 3) representing 4-dimensional hyperbolic numbers G_4 (in his previous publications the author used the term "hyperbolic matrions" for such multidimensional numbers [13, 14], but now for simplicity the term "2^n-dimensional hyperbolic number" will be used): $G_4 = x_0e_0 + x_1e_1 + x_2e_2 + x_3e_3$. Figure 3 shows a decomposition of this (4*4)-matrix into 4 sparse matrices playing the role of one real ($e_0 = 1$) and three imaginary basis units e_1, e_2 and e_3 of these 4-dimensional hyperbolic numbers. The set of these 4 sparse matrices is closed relative to multiplication and defines the multiplication table (Fig. 3) of algebra of 4-dimensional hyperbolic numbers.

By analogy, n-th tensor powers of (2*2)-matrix presentation $[a, b; b, a]$ of 2-dimensional hyperbolic numbers, give 2^n-dimensional hyperbolic numbers, which can be decomposed into 2^n sparse matrices playing the role of real and imaginary basis units of such hyperbolic numbers. The set of these 2^n sparse matrices is closed relative to multiplication and define an appropriate multiplication table of algebra of 2^n-dimensional hyperbolic numbers [13, 14].

	x_0 x_1 x_2 x_3		1 0 0 0		0 1 0 0		0 0 1 0		0 0 0 1
$G_4 =$	x_1 x_0 x_3 x_2	$= x_0$	0 1 0 0	$+ x_1$	1 0 0 0	$+ x_2$	0 0 0 1	$+ x_3$	0 0 1 0
	x_2 x_3 x_0 x_1		0 0 1 0		0 0 0 1		1 0 0 0		0 1 0 0
	x_3 x_2 x_1 x_0		0 0 0 1		0 0 1 0		0 1 0 0		1 0 0 0

*	e_0	e_1	e_2	e_3
e_0	e_0	e_1	e_2	e_3
e_1	e_1	e_0	e_3	e_2
e_2	e_2	e_3	e_0	e_1
e_3	e_3	e_2	e_1	e_0

Fig. 3. The matrix presentation of 4-dimensional hyperbolic numbers $G_4 = x_0e_0 + x_1e_1 + x_2e_2 + x_3e_3$ and its decomposition into 4 sparse matrices. The first sparse matrix e_0 is the identity matrix; other three sparse matrices represent imaginary units e_1, e_2 and e_3. The multiplication table of these 4 sparse matrices is also shown (at bottom).

Let us show now that exponentiation of the quint matrix [3, 2; 2, 3] into tensor powers $n = 2, 3, 4,\ldots$ generate 2^n-dimensional hyperbolic numbers, whose components form sets similar to the sets of sound frequencies of the Pythagorean quint scales in the following sense: ratios between any pair of their components are equal to the ratio 3/2 in integer powers. The tensor product of matrices [16] is widely applied in mathematics, physics, informatics, etc. It is used for algorithmic generation of higher dimensional spaces on the basis of spaces with smaller dimensions. By definition, the tensor product of two square matrices V and W of the orders m and n respectively is the matrix $Q = V \otimes W = \|v_{ij}{*}W\|$ with the order $m{*}n$. For example, the second tensor power of the initial (2*2)-matrix $[3, 2; 2, 3]^{(2)}$ gives the (4*4)-matrix [9, 6, 6, 4; 6, 9, 4, 6; 6, 4, 9, 6; 4, 6, 6, 9], representing the 4-dimensional hyperbolic number $9e_0 + 6e_1 + 6e_2 + 4e_3$ where e_0, e_1, e_2, e_3 are basis units from Fig. 3. The set of components of this hyperbolic number consists of numbers 4, 6, 9 with the following ratios between them: 9/6 = 3/2, $9/4 = (3/2)^2$, 6/4 = 3/2. The same ratios characterize the above mentioned tritonic musical scale 130–196–293 Hz: 293/196 = 3/2, $293/130 = (3/2)^2$, 196/130 = 3/2.

The third tensor power of (2*2)-matrix $[3, 2; 2, 3]^{(3)}$ gives an appropriate (8*8)-matrix representing the 8-dimensional hyperbolic number $27 + 18s_1 + 18s_2 + 12s_3 + 18s_4 + 12s_5 + 12s_6 + 8s_7$, where $s_1, s_2, \ldots, s_7$ are imaginary units. The set of components of this hyperbolic number consists of numbers 8, 12, 18, 27; pairwise ratios between them are the same as pairwise ratios between sound frequencies in the above mentioned Pythagorean tetratonic scale 130–196–293–440 Hz. By analogy the sixth tensor power of (2*2)-matrix $[3, 2; 2, 3]^{(6)}$ leads to 64-dimensional hyperbolic number, whose components form the set with the same values of pairwise ratios as in Pythagorean 7-stages scale in Fig. 1.

The matrix form of presentation of 2^n-dimensional hyperbolic numbers deserves special attention by the following reasons:

(1) This presentation form is based on symmetric matrices, which are closely related with the theory of resonances of oscillatory systems, having many degrees of

freedom, and also with Punnett squares from Mendelian genetics of inheritance of traits in living organisms [17–19];

(2) These symmetric matrices can be interpreted as metric tensors, which are main invariants in Riemanian geometry and which can be used in the theory of morpho-resonance morphogenesis [17, 18];

(3) These symmetric matrices are related with hyperbolic rotations [ch x, sh x; sh x, ch x], which are particular cases of hyperbolic numbers and are connected with the theory of biological phyllotaxis laws [20–22], with problems of locomotion control [23], with the main psychological law of Weber-Fechner [17], with Lorenz transformations in the special theory of relativity;

(4) These symmetric matrices are related with the theory of solitons of sine-Gordon equation [13, 14, 24]. Such solitons are the only relativistic type of solitons; they were put forward for the role of the fundamental type of solitons of living matter in the book [24];

(5) In long DNA sequences of nucleobases, where complementary nucleobases C and G (A and T) are linked by 3 (2) hydrogen bonds, tensor family of hyperbolic numbers $[3\%, 2\%; 2\%, 3\%]^{(n)}$ (n = 1, 2, 3, 4, 5; 3% and 2% denote percentages of numbers 3 and 2 of hydrogen bonds in the analyzed DNA sequence) can effectively models percentages of monoplets, doublets, triplets, tetraplets and pentaplets of these numbers 3 and 2 of hydrogen bonds [25].

Symmetric matrices possess a wonderful property to express resonances [16, 26]. The expression y = A*S models the transmission of a signal S via an acoustic system A, represented by a relevant matrix A. If an input signal is a resonant tone, then the output signal will repeat it with a precision up to a scale factor y = λ*S by analogy with a situation when a musical string sounds in unison with the neighboring vibrating string. In the case of a matrix A, its number of resonant tones S_i corresponds to its size. They are termed its eigenvectors, and the scale factors λ_i with them are termed its eigenvalues or, briefly, spectrum A. One of the main tasks of the theory of oscillations is a determination of natural frequencies (mathematically, eigenvalues of operators) and the natural forms of oscillations of bodies. To find all the eigenvalues λ_i and eigenvectors of the matrix A, which are defined by the matrix equation A*s = λ*s, the "characteristic equation" of the matrix A is analyzed: det(A − E) = 0, where E – the identity matrix (see more in [17]). Matrices, which are relevant to the various problems of the theory of oscillations, are usually symmetric real matrices [27]. Such matrices have real eigenvalues and their eigenvectors are orthogonal.

Taking into account the described relations of Pythagorean musical scales with some 2^n-dimensional hyperbolic numbers, the author proposes using hyperbolic numbers as a new mathematical tool in musicology for a possible revealing hidden regularities in products of musical creativity. The speech is that each of Pythagorean k-stages scales (k = 2, 3, 4, 5, 6, 7,...) can be formally connected with an appropriate multi-dimensional hyperbolic number and its matrix representation. For example, the Pythagorean tetratonic scale with sound frequencies of its notes 130–196–293–440 Hz can be formally expressed by a matrix form of 4-dimensional hyperbolic number $130e_0 + 196e_1 + 293e_2 + 440e_3$ where e_0, e_1, e_2 and e_3 are basis units in its matrix representation from Fig. 3. In such case, musical notes of this Pythagorean scale have the following presentations using separate basis units of 4-dimensional hyperbolic numbers: the note do(C) with its frequency 130 Hz is represented by the square matrix $130e_0$; the note sol(G) is represented by the square matrix $196e_1$; the note re (D^1) - by the square matrix $293e_2$; the note la (A^1) - by the square matrix $440e_3$. In such presentation each note of the considered musical scale has its own set of 4 eigenvectors and eigenvalues in multi-dimensional configurational space of this scale. Let us note a special role of eigenvectors in the proposed approach.

Each basis unit of 2^n-hyperbolic numbers is represented by a corresponding symmetric $(2^{n*}2^n)$-matrix, which is an orthogonal matrix and has its own set of orthogonal eigenvectors. This orthogonal set is a corresponding vector basis of 2^n-dimensional space. For example in the case of any 2-dimensional hyperbolic number $a*j_0 + b*j_1$ (Fig. 2.4) its real component aj_0 is presented by the matrix a*[1, 0; 0 1], which has two orthogonal eigenvectors [1, 0] and [0, 1] independently on value of the coefficient a ($a \neq 0$). This pair of eigenvectors defines the first vector basis of the 2-dimensional space of existence of hyperbolic numbers. The imaginary term bj_1 is presented by the matrix b*[0, 1; 1, 0], which has another pair of orthogonal eigenvectors $[-2^{-0.5}, 2^{-0.5}]$, $[2^{-0.5}, 2^{-0.5}]$ independently on value of the coefficient b ($b \neq 0$). This pair of eigenvectors defines the second vector basis of the considered 2-dimensional space. In other words, the pairs of eigenvectors are determined only by basis units j_0 and j_1. These two pairs of eigenvector bases can be considered as a two-term vector alphabet of basis units of hyperbolic numbers in case of 2-dimensional space. By analogy in the case of 4-dimensional hyperbolic numbers and their space, the 4-term eigenvector alphabet of their 4 basis units exists. In a general case of 2^n-dimensional hyperbolic numbers, the 2^n-term eigenvector alphabet of their 2^n basis units exists. Each member of such alphabet is a set of 2^n orthogonal vectors. The author briefly calls such alphabets of eigenvector bases of matrix representations of basis units of 2^n-dimensional hyperbolic numbers as «hyperbolic eigenvector alphabets» or simply «hyper-alphabets». Here the prefix "hyper" is the beginning of the word "hyperbolic" and its use is additionally justified by the fact that each member of such hyper-alphabet contains in itself another alphabet of a set of eigenvectors of the corresponding basis unit.

Any transformation of one such eigenvector basis into another (that is a transformation of one member of a hyper-alphabet into another) is provided by means of an orthogonal matrix (orthogonal operator). Orthogonal operators preserve the space metric and define transformations of proper and improper rotations. Any sequence of basis units (or their sums) of 2^n-dimensional hyperbolic numbers corresponds to a certain sequence of eigenvector bases of these units, and also to a sequence of orthogonal matrices transforming successively these bases. Such algebraic sequences can be used for transmitting information. Taking into account some results of his previous published studies, the author supposes that genetic sequences are related with such algebraic sequences.

Moreover, the author puts forward the hypothesis that alphabets of eigenvectors of matrix representations of basis units of 2^n-dimensional hyperbolic numbers play a key role in transmitting biological information and that they can be considered as a foundation of coding information at different levels of biological organization. The corresponding languages using such alphabets define many inherited phenomenological structures in biology including molecular-genetic structures.

As known, the principle of transmitting information in the form of certain texts composed on the basis of certain "alphabets" is widely used in living organisms: genetic information is recorded in DNA molecules in the form of texts based on the DNA alphabet; music is a sequence of sound frequencies of one or another musical scale (that is, the "alphabet" of note sound frequencies of one octave); literary texts are written on the basis of literary alphabets, etc. The author believes that various alphabets and texts in these bioinformational fields can be effectively modeled and studied on the basis of the presented hidden algebraic alphabets as their joint algebraic foundation. This approach is connected with the theme of a «grammar of biology», which term was introduced by E. Chargaff in 1971.

Since alphabets are used as foundations of corresponding languages, each algebraic hyper-alphabet in 2^n-dimensional spaces with a concrete number n can be considered as a foundation of a corresponding algebraic language. From this model standpoint, many biological languages can be considered as using these algberaic hyper-alphabets. In particular, with this algebraic approach researchers get new mathematical opportunities to study hidden aspects of harmony of musical works.

Let us use that each kind of 2^n-dimensional hyperbolic numbers under fixed n has its own hyper-alphabet of 2^n orthogonal systems of its 2^n basis units. Transitions from one member of such hyper-alphabet to other members are determined by transformations of mentioned orthogonal rotations. From the proposed algebraic point of view, any sequence of musical notes is related with a sequence of the mentioned rotations of orthogonal systems of eigenvectors inside appropriate multi-dimensional vector space. But musical works are based not only on the sound frequencies of notes of a particular musical system but also on a system of note durations: 1, 1/2, 1/4, 1/8, 1/16, 1/32, 1/64, 1/128. In musical pieces each note is a symbiosis of a sound frequency and one of these durations. Correspondingly in the proposed approach on the foundation of 2^n-dimensional hyperbolic numbers, such symbiosis can be expressed as a sum of appropriate basis units of these numbers (see details in [28]).

The traditional use of one-dimensional numbers in musicology made it possible to study the relationship of musical sound frequencies in musical systems and chords in line with the Pythagorean theory of the relationship of musical frequencies with dividing strings into parts. But it is obvious that knowledge of only the harmonious interrelations of sound frequencies is not enough to reveal the harmony of the musical work as a whole. For example, you can take two pieces of music, which use the same set of sound frequencies, but – because of different sequences of the same musical sounds in them - one piece will provides a charming effect on the listener, and the other will leave him indifferent or will cause a negative emotion. This indicates that in musical works there is some other - additional - type of harmony, reflected in the transitions between sounds and providing harmonic development of the theme of a musical work (musical plasticity) [28].

The author supposes that applications of 2^n-dimensional hyperbolic numbers can be useful in many topics (including for example topics in [29–34]).

The final part of this Section is devoted to the use of 2^n-dimensional hyperbolic numbers for modeling inherited fractal-like forms of biological bodies. Living bodies in a course of their ontogenesis from the embryonic state to the mature state gradually increase the number of body parts. Accordingly, the number of parameters, characterizing the developing body, increases. This leads to appropriate phased increasing a dimensionality of a configuration space of parameters of the body. In many cases of such ontogenetic development one can see the following iterative process: body structural elements, which exist at a previous stage of ontogenesis, produce - at the next step of ontogenesis - new elements with similar structures. In the result, after some repetitions of this ontogenetic procedure, a complex fractal-like structure of the multi-level body appears. A multidimensional configuration space of parameters of such body has a fractal-like set of its different subspaces having similar patterns of parametric states. One of many examples of such phased producing a fractal-like structure of multi-level body is ontogenetic producing new and new dichotomic branches in some plants.

A great number of publications are devoted to algorithmic creation of fractal-like geometric figures in spaces of fixed dimensionality, first of all, on 2-dimensional complex plane. There are also known works devoted to constructions of fractal geometric patterns on the plane of hyperbolic (or double) numbers [35, 36]. In contrast to these works, the author proposes to study an algorithmic generation of patterns, which are similar each other, not in a space of fixed (!) dimensionality but in different subspaces of multidimensional configurational spaces of parameters of multi-level bodies under their phased ontogenetic development.

The author notes the following possibility of modelling of such multi-step development of biological systems receiving - from time to time - new and new parameters for their configurational spaces. Let us take hyperbolic number $[f_1(t), f_2(t); f_2(t), f_1(t)]$ whose components $f_1(t)$ and $f_2(t)$ are functions of time. If this (2*2)-matrix is tensor multiplied on the left by a hyperbolic number [1] acting as a generator of additional dimensions of the configurational space, the result is (4*4)-matrix representing 4-dimensional hyperbolic number $f_1(t)*e_0 + f_2(t)*e_1 + f_1(t)*e_2 + f_2(t)*e_3$ (here e_0, e_1, e_2 and e_3 are basis units from Fig. 3). This 4-dimensional configurational space repeats in its subspaces (namely in the first plane on the basis vectors e_0 and e_1, and in the second plane on the basis vectors e_2 and e_3) the same functions $f_1(t)$ and $f_2(t)$, which were in the initial 2-dimensional space.

Repeating the required number of times this operation of the tensor multiplication on the left using the generator [1], we obtain a hierarchical tree of 2^n-dimensional hyperbolic numbers and their corresponding 2^n-dimensional configurational spaces of parameters for algorithmic modelling a multi-step ontogeneis of a fractal-like morphogenetic construction. Different levels of this tree of configurational spaces have subspaces with the same functions $f_1(t)$ and $f_2(t)$, which were in the initial 2-dimensional space; in this sense one can speech about a fractal-like structure of this hierarchy of multi-dimensional configurational spaces of parameters.

4 Some Concluding Remarks

Hyperbolic numbers and their algebraic extensions in the form of 2^n-dimensional hyperbolic numbers can be the basis of a new wide class of models in algebraic biology and also in musicology. From their point of view many different biological phenomena happen to be structurally interconnected. New results are obtained for understanding any living body as a holistic essence and also for revealing hidden structural relations of musical harmony with genetic systems and inherited physiological systems. Author's results are connected with a problem of genetic basis of aesthetics and inborn feeling of harmony. According to words of famous physicist Nobel prize winner R. Feynman about feeling of musical harmony, "*Whether far we stand from Pythagoras in understanding of why only some sounds are pleasant for hearing? The general theory of aesthetics, apparently, has been moved forward not significantly since Pythagorean times*" [37].

As far as we know, till now all works in fields of musicology and theory of aesthetics were traditionally based on one-dimensional numbers in contrast to wide usage of multidimensional numbers in many fields of modern mathematical natural sciences. On the basis of his research results, the author proposes using the concrete type of multidimensional numbers - 2^n-dimensional hyperbolic numbers - to study genetic basis of aesthetics and inborn feeling of musical harmony and to develop algebraic biology in close relations with other heritable biological structures.

The notion of "number" is the main notion of mathematics and mathematical natural sciences. Pythagoras has formulated the famous idea: "*Numbers rule the world*" or "*All things are numbers*". This Pythagorean slogan arised not because that the number can express a quantity of objects. Pythagoras was engaged in figured numbers associated with geometric figures: triangular, square, 5-angled, 12-angled, etc. Seeing that different numbers can dictate different geometric shapes, he came up with the idea that numbers have an internal structure and able to organize the outside world according to their properties. In view of this idea, natural phenomena should be explained by means of systems of numbers; the systems of numbers play a role of the beginning for uniting all things and for expressing the harmony of nature [1, p. 21, 24]. For the Pythagoreans, the number expressed the "essence" of everything, and therefore the phenomena should be explained only with the help of numbers; it was numerical relations that served as the unifying principle of all things and expressed the harmony and order of nature.

Many prominent scientists and thinkers were supporters of this Pythagorean standpoint or of one similar to it. As W. Heisenberg noted, modern physics, where matrices are used as a higher form of numbers, is moving along the same path along which the Pythagoreans walked [38]. Not without reason Russell [39] noted that he did not know any other person who could exert such influence on the thinking of people as Pythagoras. Taking this into account, one can believe there is no more fundamental scientific idea in the world, than this idea about a basic meaning of numbers.

Our research results and the proposed approach can be considered as a further development of this fundamental idea of Pythagoras in connection with the structural organization of the genetic system and inherited biological phenomena. The author recommends the further wide development of this Pythagorean approach and applications of 2^n-dimensional hyperbolic numbers in fields of the genetic and other genetically heritable biological systems. He hopes that on this way biology will be included - step by step - in the field of mathematical natural sciences. As a result, in particular, new acoustic biotechnologies and new approaches to the creation of artificial intelligence will arise.

Acknowledgments. Some results of this paper have been possible due to a long-term cooperation between Russian and Hungarian Academies of Sciences on the topic "Non-linear models and symmetrologic analysis in biomechanics, bioinformatics, and the theory of self-organizing systems", where S.V. Petoukhov was a scientific chief from the Russian Academy of Sciences. The author is grateful to participants of this international cooperation and to all the colleagues who discussed the materials of this article with him.

References

1. Kline, M.: Mathematics. The Loss of Certainty. Oxford University Press, New York (1980)
2. Gsponer, A., Hurni, J.-P.: Quaternions in mathematical physics (2): Analytical bibliography (2008). http://arxiv.org/abs/math-ph/0511092, 3th version from 6 July 2008
3. Vernadsky, V.I.: The chemical structure of the biosphere and its environment. Moscow, Nauka, 205 p. (1965). (in Russian)
4. Hesse, H.: The Glass Bead Game. St Martins Press (2002). ISBN 0312278497, 9780312278496
5. Sethares, W.A.: Tuning, Timbre, Spectrum, Scale, p. 163. Springer, London (2005). ISBN 1-85233-797-4
6. Day-O'Connell, J.: Pentatonicism from the Eighteenth Century to Debussy. University Rochester Press, Rochester (2007). ISBN 978-1-58046-248-8
7. Christidis, A.-F., Arapopoulou, M., Christi, M.: A History of Ancient Greek: From the Beginnings to Late Antiquity. Cambridge University Press, Cambridge (2007). ISBN 978-0-521-83307-3
8. Apel, W.: Harvard Dictionary of Music. Harvard University Press, Cambridge (1969). ISBN 978-0-674-37501-7
9. Olsen, D.A., Sheehy, D.E. (eds.) The Garland Encyclopedia of World Music, Volume 2: South America, Mexico, Central America, and the Caribbean. Taylor & Francis, New York (1998). ISBN 0824060407
10. Todd Titon, J.T.: Worlds of Music: An Introduction to the Music of the World's Peoples, Shorter Version. Cengage Learning, Boston (1996). ISBN 9780028726120
11. Bhattacharya, S.: Ethno-musicology and India, p. 54. Indian Publications (1968). Accessed 22 June 2012
12. Baines, A.: Woodwind Instruments and Their History. Courier Dover Publications, New York (1991). ISBN 978-0-486-26885-9
13. Petoukhov, S.V.: Matrix genetics, algebras of the genetic code, noise immunity. Moscow, RCD, 316 p. (2008). (in Russian)
14. Petoukhov, S.V., He, M.: Symmetrical Analysis Techniques for Genetic Systems and Bioinformatics: Advanced Patterns and Applications. IGI Global, Hershey (2009)
15. Kantor, I.L., Solodovnikov, A.S.: Hypercomplex Numbers. Springer, Berlin (1989). ISBN 978-0-387-96980-0
16. Bellman, R.: Introduction to Matrix Analysis. Mcgraw-Hill Book Company, New York (1960)
17. Petoukhov, S.V.: The system-resonance approach in modeling genetic structures. Biosystems **139**, 1–11 (2016)
18. Petoukhov, S.V.: The concept of resonances in genetics ("resonant bioinformatics"). Selected Proceedings of the VII International Congress "Weak and Super-Weak Fields and Radiations in Biology and Medicine", St. Petersburg, vol. 7, pp. 1–41 (2015). http://www.biophys.ru/lib/sci/resonance/460-resonance-00001

19. Petoukhov, S.V.: Matrix genetics and algebraic properties of the multi-level system of genetic alphabets. Neuroquantology **9**(4), 60–81 (2011)
20. Bodnar, O.Ya.: Geometry of phyllotaxis. Reports of the Academy of Sciences of Ukraine, №9, pp. 9–15 (1992)
21. Bodnar, O.Ya.: Golden Ratio and Non-Euclidean Geometry in Nature and Art. Publishing House "Sweet", Lviv (1994)
22. Stakhov, A., Olsen, Sc. (collaborator): Mathematics of Harmony: From Euclid to Contemporary Mathematics and Computer Science, 694 p. World Scientific Pub Co Inc., Singapore (2009). ISBN-10: 981277582X, ISBN-13: 978-9812775825
23. Smolyaninov, V.V.: Spatio-temporal problems of locomotion control. Uspekhi Fizicheskikh Nauk **170**(10), 1063–1128 (2000). https://doi.org/10.3367/UFNr.0170.200010b.1063
24. Petoukhov, S.V.: Biosolitons, 288p. GP Kimrskaya tipographia, Moscow (1999)
25. Petoukhov, S.: The Genetic Coding System and Unitary Matrices. Preprints 2018, 2018040131. https://doi.org/10.20944/preprints201804.0131.v2. Accessed 27 Sept 2018
26. Balonin, N.A.: New Course on the Theory of Motion Control (Novyi kurs teorii upravleniia dvizheniem). Saint Petersburg State University, Saint Petersburg (2000). (in Russian)
27. Gladwell, G.M.L.: Inverse Problems in Vibration, 452 p. Kluwer Academic Publishers, London (2004)
28. Rashevskij, P.K.: Tensor Analysis and Riemannian Geometry. Nauka, Moscow (1964)
29. Angadi, S.A., Hatture, S.M.: Biometric person identification system: a multimodal approach employing spectral graph characteristics of hand geometry and palmprint. Int. J. Intell. Syst. Appl. (IJISA) **3**, 48–58 (2016)
30. Sahana, S.K., AL-Fayoumi, M., Mahanti, P.K.: Application of modified ant colony optimization (MACO) for multicast routing problem. Int. J. Intell. Syst. Appl. (IJISA) **4**, 43–48 (2016)
31. Algur, S.P., Bhat, P.: Web video object mining. Int. J. Intell. Syst. Appl. (IJISA) **4**, 67–75 (2016)
32. Hata, R., Akhand, M.A.H., Islam, M.M., Murase, K.: Simplified real-, complex-, and quaternion-valued neuro-fuzzy learning algorithms. Int. J. Intell. Syst. Appl. (IJISA) **10**(5), 1–13 (2018). https://doi.org/10.5815/ijisa.2018.05.01
33. Awadalla, M.H.A.: Spiking neural network and bull genetic algorithm for active vibration control. Int. J. Intell. Syst. Appl. (IJISA) **10**(2), 17–26 (2018). https://doi.org/10.5815/ijisa.2018.02.02
34. Abuljadayel, A., Wedyan, F.: An approach for the generation of higher order mutants using genetic algorithms. Int. J. Intell. Syst. Appl. (IJISA) **10**(1), 34–35 (2018)
35. Pavlov, D.G., Panchelyuga, M.S., Panchelyuga, V.A.: On the form of the analogue of the Julia set at zero value of the parameter on the plane of the double variable. Hypercomplex Numbers Geom. Phys. **6**(11), 146–151 (2009)
36. Pavlov, D.G., Panchelyuga, M.S., Panchelyuga, V.A.: On the form of analogues of the Julia set on the plane of a double variable. Hypercomplex Numbers Geom. Phys. **6**(12), 161–175 (2009)
37. Feynman, R.: The Feynman Lectures on Physics, vol. 4. Perseus Publishing, New York (1998)
38. Heisenberg, W.: Physics and Philosophy: The Revolution in Modern Science. Penguin Classics, New York (2000)
39. Russel, B.: A History of Western Philosophy. Book One, Part I, Chapter III. Simon & Schuster/Touchstone, New York (1967)

Metric Properties of Visual Perception of Mirror Symmetry

T. Rakcheeva(✉)

Mechanical Engineering Research Institute of the Russian Academy of Sciences, 4 Maly Kharitonievskiy Pereulok, Moscow 101990, Russia
rta_ra@list.ru

Abstract. The work is devoted to an experimental investigation of visual perception of symmetry. Symmetry participates in the mechanism of the evolutionary adaptability of a living organism, providing the form that is optimal for existence. Not only all living beings, but also objects created by their creativity, one way or another, correspond to the laws of symmetry. Functional expediency was transformed into such aesthetic categories as structural proportionality, harmony. The main developments of psychophysical research are carried out in areas in which qualitative characteristics of the perception of symmetry are used, such as emotional reaction, aesthetic experience, imaginative association, and even taste sensations. The goal of this work is an experimental study of the metric properties of perception and reconstruction of elementary symmetries as a manifestation of symmetry models of human intelligence. The experimental research was implemented on a computer in the format of the reconstruction of a partially given symmetry composition. The subject had to finish the composition in accordance with the transformation of the given symmetry. Independent or regulated factors are symmetry elements, symmetry objects and symmetry transformations. The dependent or resulting factors are functions of the regulated factors and reflect the accuracy of the reconstruction of a given symmetry composition. The visual perception of mirror symmetry in a vertical configuration for a "point" object is characterized by statistically significant stable correctness of the average value on the population and the horizontal-vertical anisotropy of errors for individuals; axis symmetry direction is a matter of principle. This work represents the first and necessary stage of research related to the perception of arbitrary forms.

Keywords: Psychophysics · Symmetry · Visual perception · Metric properties · Mirror symmetry · Experimental research · Engineering psychology

1 Introduction

Symmetry is present in the life of mankind throughout its history. This is evidenced by the preserved artifacts, ancient drawings, household items et al. Symmetry participates in the mechanism of the evolutionary adaptability of a living organism, providing the form that is optimal for existence. The more complex the organism, the more complex its symmetry [1–4]. Functional expediency was transformed into such aesthetic categories as structural proportionality, harmony. "Symmetry - in a broad or narrow

Z. Hu et al. (Eds.): AIMEE 2019, AISC 1126, pp. 208–217, 2020.
https://doi.org/10.1007/978-3-030-39162-1_19

sense - is the idea by which people throughout the centuries tried to comprehend and create order, beauty and perfection!" [1]. Not only all living beings, but also objects created by their creativity, one way or another, correspond to the laws of symmetry [1, 2]. Together with Plato, Weil believed that "the mathematical laws governing nature are a source of symmetry in nature, and the intuitive implementation of this idea in the creative spirit of the artist is a source of symmetry in art".

Thus, the aesthetic sense of structural harmony is a manifestation of the human *model* of intelligence, formed both evolutionally and in the process of education. It contains both the basic mechanisms of interaction with the world, inherent in all healthy people, and the individual characteristics of a particular complexity. In accordance with this model, a person interacts with the world around him and responds to its incentives [3, 4].

In times of antiquity and the revival of the concepts of symmetry and asymmetry, great importance was attached from the special mathematical idea of Plato to the expedient coherence of Aristotle. The modern era, filling this concept with the content of functional necessity, is marked, as is well known, by outstanding scientific breakthroughs in various fields of science. All this is accompanied by a rise in interest in the study of symmetry in general, and in the visual perception of symmetries in particular. Comprehensive studies of the perception of symmetries are carried out in engineering psychology [5–7]. Many features and characteristics of the process of perception at different levels of the mental processing of visual information by a person are revealed [6]. Various models of this process are also formulated: from simulation models to mathematical ones [8]. However, in one research it is impossible to take into account all the properties of a complex multifactorial process of perception of symmetries.

The main developments of psychophysical research are carried out in areas in which qualitative characteristics of the symmetry perception are used, such as emotional reaction, aesthetic experience, imaginative association, and even taste sensations [9, 10]. Thus, experimental investigations of the visual perception of symmetric and random patterns indicate that symmetrical patterns are associated: with a verbal reaction of positive valence and a high level of arousal, and random patterns – with a reaction of negative valence and a low level of arousal [11]. In well-known studies related to the perception of the "Zen garden of stones", a person who is in a particular observation point experiences a pleasant feeling of harmony. The results are formulated as a hypothesis about the "resonant mechanism" of the perceptual effect between the sensory perception of the real visual image (stones) and the sense of the virtual image (treelike) of the symmetrical relations intrinsic to man [9, 12].

Significantly fewer works contain quantitative studies of the properties of visual perception of symmetries and conformity to natural laws that determine their correlation [13–18]. Of particular interest are the attempts of a mathematical interpretation of the visual perception of an object from the point of view of the correlation-resonance relations of its form with symmetry [9, 14].

The process of visual perception of symmetries is one of the basic mechanisms of the visual system. As is known, about 50 ms is required to detect symmetric structures in conditions of complex background. In addition, the human visual system, as a rule, provides a certain structural and semantic organization in the perception of even random images (blots, clouds, coffee residues). This all testifies to the presence in the human analyzer (and not only man) of ontogenetically formed symmetry models. In this connection, the researches of the nature and features of the perception of symmetries from the standpoint of the metric ratios of their geometric structures seem to be quite important.

In geometry, the symmetry of a structural object can be defined as the invariance of the transformations of its self-alignments in space [1]. Geometric invariance in this case means either compatible equality of objects (congruence) or mirror equality (combination with an overturn). There are symmetries of transfer, rotation, reflection, inversion. Each of them is also characterized by an element of symmetry - a center, an axis, or other varieties (in general). Variants of symmetry structures on the plane are limited to mathematically admissible combinations of transformations and symmetry elements. So, for example, various linear, border compositions, as is known, are possible seven (Fig. 1a), and the number of flat ornamental configurations is limited to seventeen. The artistic richness and diversity of symmetry samples, examples of which are shown in this figure, are provided by the additional arbitrary variability of the form of the generating elements [1, 2].

a)

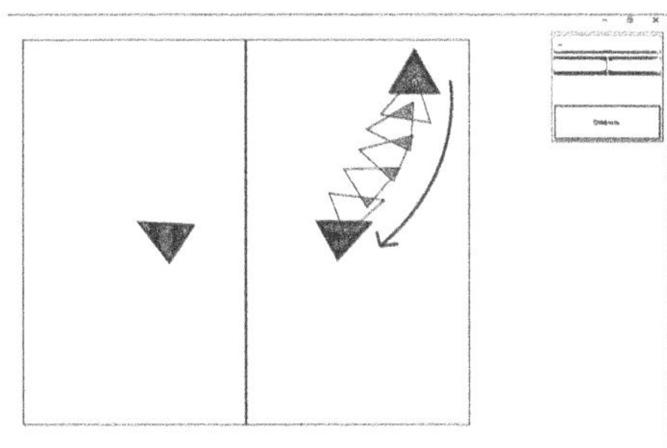

b)

Fig. 1. Artistic symmetry patterns - borders (*a*). Scheme of experiment (*b*).

The goal of this work is an experimental study of the metric properties of perception and reconstruction of elementary symmetries on the plane as a manifestation of symmetry structure models of human intelligence.

2 Formulation of the Problem

The experimental research was implemented in the format of the reconstruction of a partially given symmetry composition. The subject had to finish the composition in accordance with the transformation of the given symmetry.

The investigation of the whole variety of symmetry structures is a significant amount and goes beyond the scope of one work. In addition, the set of compositional symmetries has a hierarchical structure. Therefore, it is necessary first to investigate the properties of simple symmetry configurations that form the basis of the entire set of symmetry compositions.

Arbitrary symmetry composition includes the following three components:

- symmetry object;
- symmetry element;
- symmetry transformation.

Each of them is represented by variants of its geometrical characteristics, namely:

- object may be of different dimensions, shape, size and orientation;
- element may be of different dimensions, shape and orientation;
- transformation may be of transfer along the axis, rotation around the center, reflection from the axis or center.

In addition to variations in the individual characteristics of each of the components of symmetry, possibly affecting perception, it is necessary to take into account the influence and variations of the compositional interactions of these components in the overall structure of this symmetry. For example, an object can be located at a different distance from the symmetry element or have a different mutual orientation.

Thus, even the lowest level of the compositional structure of symmetry includes a sufficiently large number of parameters, each of which requires some of its values. This paper presents the results of experiments performed under the conditions of the following configuration:

- symmetry object: point, segment, triangle;
- symmetry element: center-point, axis-line;
- transformations: central inversion, reflection from the axis, central rotation.

The point ***P***, being an object of null dimension, has neither shape, nor size, nor orientation, but only has the distance to the element of symmetry. The segment ***S*** has a linear dimension and does not have, therefore, a shape, but has a size and orientation. The triangle ***T*** varies in all the above parameters of the shape, size, orientation. Center-point ***C***, as an element of symmetry, does not have its own variation parameters. In contrast, the axis of symmetry ***m*** has one orientation parameter. The symmetry transformations of the central inversion and axial reflection each have one parameter of the distance ***d*** between the object and the symmetry element, and the central rotational one has an additional parameter - the angle of rotation of the object relative to the fixed center of symmetry.

The listed parameters are independent regulated factors. The dependent or resulting factors are functions of the regulated factors and reflect the accuracy of the reconstruction of a given symmetry composition. In quantitative terms, the resulting factors are the coordination errors ***x***, ***y*** and orientation α of the test symmetry object, the statistical analysis of which is the results of this research.

As a result of the experimental research, it is expected to receive answers to such questions, as: whether there are systematic statistically significant errors of the symmetry perception; whether such factors as spatial orientation of symmetry elements and objects have a significant effect on errors; what is the nature of errors.

3 Description of the Experiment

The experimental studies under discussion were implemented on a computer. To perform this work, special software has been written, which allows investigating a wide range of quantitative estimates of symmetry reconstruction in the interactive mode. In the course of the experiment, the computer automatically performed all the registrations required for each test subject: personal data, the results of his actions, calculations and estimates, both the coordination object errors of symmetry and its orientation, providing the possibility of an exact quantitative estimation of the values.

The scheme of the experiment in each test was as follows.

After entering personal data and settings of the experiment parameters performed on the computer screen in the registration mode, the interface of the working screen appears (Fig. 1a). A small area on the right is reserved for the controls of the experiment. In the center of the screen is the working area of the square form (160 × 160 mm). In this area, the subject is presented with an incomplete symmetry composition, where there is an element of symmetry and one of the objects of symmetry for given values of parameters. In this figure, the vertical axis of symmetry is set as an element of symmetry, which is installed in the center and divides the working area into two equal parts. In the left part the symmetry object is placed in a certain orientation and at a given distance from the axis of symmetry. In this case, it is a triangle.

Outside this composition, on the service panel in the upper corner of the screen, there is also a copy of the object in an arbitrary orientation. The task of the subject is to symmetrize the composition by placing a copy of the object in the place corresponding to the given composition and in the correct orientation if the object is not a point. To do this, the subject, in accordance with the instructions, must "take" a copy of the object with the mouse and, moving it around the screen and rotating it, achieve the correct completion of the symmetric composition *from the point of view of its visual perception*. After fixing the solution of this test, the following composition automatically appears.

The time of the symmetrization procedure was not limited, so that the subject could concentrate on the accuracy of the task. The program recorded the test results: in the form of coordinates of the object location and its orientation angle in the screen coordinate system. Knowing the correct coordinates, the computer calculated the subject's errors, both metric (x - in the horizontal direction and y - in the vertical) and orientational (α - angle in the positive direction), which were recorded in the database for further statistical processing.

The experiments involved groups of 15–23 subjects aged 20 to 25 years old with normal vision. For each subject, the complete experiment consisted of a series of tests with a grid of parameters for each adjusted factor. The execution time of the complete test set screen was about 30 min, so the work was carried out intermittently.

4 Results of the Experiment

Consider the results of tests performed for central inversion and mirror symmetry in the following configurations:

- elements: center *C*, axis m: *V*(0°), *G*(90°), *L*(−45°), *R*(45°);
- objects: point *P*, segment *S*, triangle *T*;
- transformations: d = 20, 40, 60 mm.

The combinations of these parameters make up a set of 45 experiment tests for each subject.

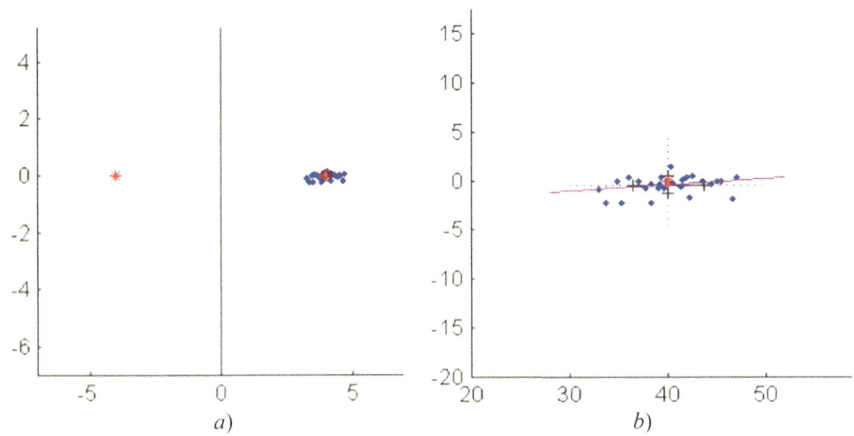

Fig. 2. Sample data for the **PmV**-experiment: (*a*) d = 40 mm; (*b*) on an enlarged scale (Color figure online)

Figure 2a presents the results of: with mirror symmetries and a "point" object of symmetry. The axis of symmetry is set in a vertical orientation. The object of symmetry - "point" is a circle of minimum size (≈1 mm). It is located in the center of the left half at a distance of 40 mm from the axis of symmetry (marked by the characters "*" in red). To the right in the mirror-symmetrical position is the same symbol, which marks the mathematically "ideal" position, completing given composition. The set of blue dots is a sample of the results of performing this task in the process of testing by a group of subjects (30 tests).

The same set is shown on an enlarged scale in Fig. 2b. The dashed lines indicate the average values of x and y, and the symbol "+" fixes standard deviations. Red indicates the line of the regression - it has an almost horizontal direction. As can be seen from the figure, the set has a special form allowing drawing certain conclusions. Namely: (1) there is no significant shift in the center of the sample - the correct solution is close to the mathematical center of the sample; (2) there is a significant elongation of the sample in the horizontal direction, which indicates the absence of isotropy of the spatial distribution of these errors. Pearson's χ^2-criterion confirms the normal distribution of sample data. Therefore, criteria for parametric statistical analysis were used to obtain quantitative estimates. Estimation of the sample mean shift by the Student's criterion gives the actual value $t_f = \{0.15(x), 2.00(y)\}$ which confirms the absence of a statistically significant shift of the mean value at the 1% and 5% significance levels. The horizontal-vertical asymmetry of the sample distribution can be estimated by the Fisher parametric criterion for comparing the variances. For this sample, this criterion gives the actual value $F_f = 11.09$, which significantly exceeds the critical value $F_{crit} \approx 2.5$ for 1% significance level. The principal component method gives horizontal and vertical vector components approximately.

Figure 3a allows comparing the results of the same **PmV**-experiment for the three distances mentioned above between the object and the axis of symmetry. It can be seen from the figure that the mean stability and the shape of the sample distribution also hold for other distances. Indeed, the statistical estimates suggest that the lack of systematic bias error and the horizontal-vertical anisotropy of the distribution of perceptual errors characterize the mirror symmetry regardless of the distance. But, as follows from the figure and the given estimates, the degree of their manifestation depends on the distance.

Thus, from this experiment we can conclude: the visual perception of mirror symmetry in a vertical configuration for a "point" object is characterized by statistically significant stable correctness of the average value on the population and the horizontal-vertical anisotropy of errors for individuals.

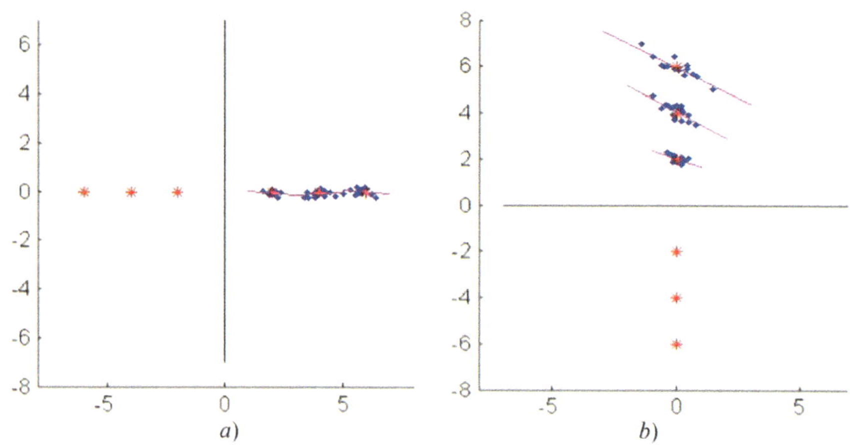

Fig. 3. Sample data for three distances: (*a*) PmV-experiment; (*b*) PmG-experiment

In Fig. 3b presents the results of a similar "point" **PmG**-experiment with the axis of mirror symmetry, located horizontally. The figure shows sample data for three distances. Here, statistically insignificant displacements of sample means are also observed (Student's test confirms this even at the 1% significance level). The form of the sample set also demonstrates a high degree of anisotropy, but has a different trend. Perhaps this is due to our natural asymmetry of the right and left. Each sample in the figure shows straight-line linear regression segments. Angular regression parameters - the correlation coefficients are equal: $-0.35(\approx -20°)$, $-0.56(\approx -30°)$, $-0.53(\approx -28°)$, respectively.

In Fig. 4 shows the results of a "point" experiment with oblique axes of mirror symmetry to the left: **PmL** (Fig. 4a) and to the right: **PmR** (Fig. 4b).

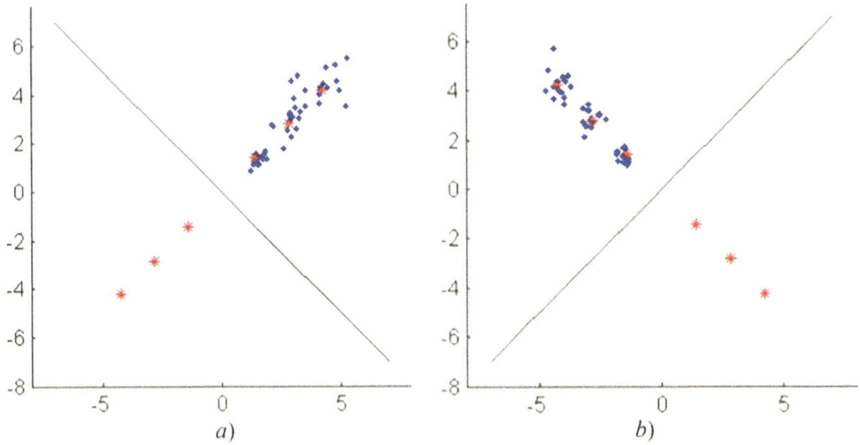

Fig. 4. Sample data for three distances: (*a*) PmL-experiment; (*b*) PmR-experiment

Statistically significant stability of the reconstruction on average is also observed in these cases. However, there is a noticeable difference in the dispersions - for PmL variations are significantly larger than for PmR, and they do not have, in contrast to PmV and PmG, a pronounced trend ($F_f = 1.35 \div 3.39$). Finally, PmR has the most compact isotropic and unbiased form ($F_f = 1.35 \div 2.14 < F_{5\%}$). Quantitative comparisons of variants related to the orientation of the axis of symmetry show that with increasing distance, this difference appears more and becomes statistically significant. Confirmation is received by Fisher's criterion for comparing variances of distances from centers of the samples on 5% significance level.

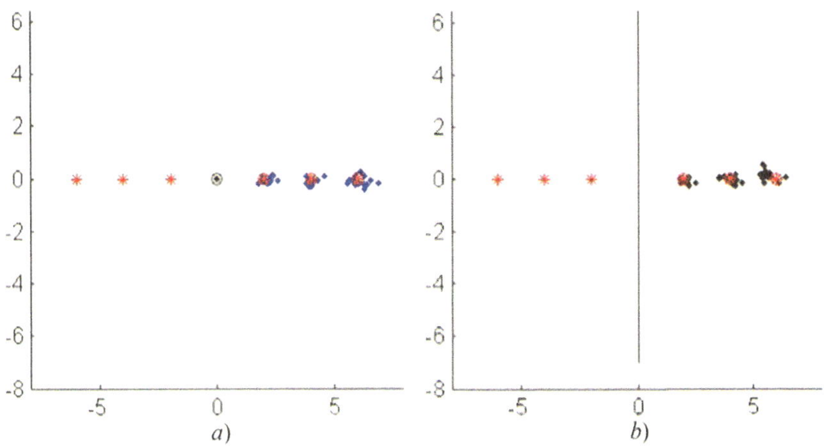

Fig. 5. Sample data for three distances: (*a*) PC-experiment; (*b*) SmV-experiment

It is also of interest to comparisons the symmetries of the "point": (1) for mirror symmetry with the vertical axis (PmV on Fig. 3a), and (2) for central symmetry with the horizontal level of the segment between the object and the center (PC on Fig. 5a). Although there is no significant difference in average values, the difference in form is still noticeable. The anisotropy coefficient is much lower.

Comparisons of mirror-vertical symmetry in the context of other objects (the segment of Fig. 5b, rather like data were obtained for triangle) give a similar result: the stability of the means and the differences in the distributions, which in this case are also due to the orientation of the object, as well as its form.

In general, the volume of experiments performed is much wider, but a discussion of their results is beyond the scope of this work.

5 Conclusions

The mirror symmetry of a point is characterized by the stability of an average value close to "ideal" (also for other objects). The error of the mean is more often larger in the orthogonal direction to the axis of symmetry than in the parallel one. Dispersion strongly depends on both the distance to the axis of symmetry and its direction. The orientation distribution of the dispersion has a trend, the severity of which also depends on the direction of the symmetry axis.

The laws of symmetry, as is known, occupy a large place in architecture, design, decorative and applied art, therefore the experimental studies described and their results may be useful here. Recently, there has been a need for visual representation of a complex of a large number of heterogeneous data for the description, recognition and control of objects or processes [19, 20]. The presentation of such multifactorial data by symmetric compositions of borders or ornaments seems promising.

The investigation of the perception of symmetries plays a small role in engineering psychology and medicine so far [10, 20, 21]. But, the experiments showed that the results obtained even with such an elementary object as a "point" can have practical applications associated with the identification of diagnostic signs of a one or another pathology of spatial orientation. Along with the use of optical-geometric illusions, it seems appropriate and effective to use the individual characteristics of the perception of symmetries in social and medical psych diagnostics. The results obtained may also have practical application in other areas of professional activity for example, in education [9, 21–23]. It seems to be a pressing issue of introducing relevant subjects in the learning process, since it is difficult to overestimate their role in aesthetic education.

References

1. Weil, H.: Symmetry, p. 192c. Nauka, Moscow (1968)
2. Shubnikov, A.V., Koptsik, V.A.: Symmetry in Science and Art, 560p. Computer Research Institute, Moscow-Ijevsk (2004)
3. Gregory, R.L.: A reasonable eye, 232 p. Editorial URSS (2003)
4. Fress, P., Piaget, J.: Experimental Psychology, Issue 4, Progress, Moscow, 344 p. (1978)

5. Hamada, J., Amano, K., Fukuda, S.T., Uchiumi, C., Fukushi, K., van der Helm, P.A.: A group theoretical model of symmetry cognition. Acta Psychol. **171**, 128–137 (2016). https://doi.org/10.1016/j.actpsy.2016.10.002
6. Treder, M.S.: Behind the looking-glass: a review on human symmetry perception. Symmetry **2**, 1510–1543 (2010). https://doi.org/10.3390/sym2031510
7. Van der Helm, P.A.: Symmetry perception. In: Wagemans, J. (ed.) Oxford Handbook of Perceptual Organization, pp. 108–128. Oxford University Press, Oxford (2015). https://doi.org/10.1093/oxfordhb/9780199686858.013.056
8. Poirier, F.J.A.M., Wilson, H.R.: A biologically plausible model of human shape symmetry perception. J. Vis. **10**, 1–16 (2010)
9. Artemenkov, S.L., Shukova, G.V., Mironova, K.V.: Visual perception of symmetry as a factor of aesthetic experience. Exp. Psychol. **11**(1), 166–177 (2018). https://doi.org/10.17759/exppsy.2018110110
10. Turoman, N., Spence, Ch.: Cross-modal correspondence between visual symmetry and taste. Perception **45**(2), 329 (2016). Suppl, 39th European Conference on Visual Perception (ECVP) 2016 Barcelona
11. Pecchinenda, A., Bertamini, M., Makin, A.D.J., Ruta, N.: The pleasantness of visual symmetry: always, never or sometimes. PLoS ONE **9**(3), e92685 (2014). https://doi.org/10.1371/journal.pone.0092685
12. Van Tonder, G.J., Lyons, M.J.: Visual perception in Japanese rock garden design. Axiomathes **15**, 353–371 (2005). https://doi.org/10.1007/s10516-004-5448-8
13. Van der Helm, P.A.: Weber-Fechner behavior in symmetry perception? Attention Percept. Psychophysics **72**, 1854–1864 (2010)
14. Zhukova, K.V., Reier, I.A.: Basic skeleton connectivity and a parametric shape descriptor. Mach. Learn. Data Anal. **1**(10), 1354–1368 (2014)
15. Rakcheeva, T.A.: The factor of symmetrization in the identification of forms of curves. In: Seventh Kurdyumov readings. Synergetics in the Natural Sciences: Materials intern. interdisciplinary Scientific Conf. Tver, pp. 156–158 (2011)
16. Rakcheeva, T.A.: The influence of form on the perception of distance. In: Eighths Anniversary Kurdyumov Readings: Synergetics in the Natural Sciences, Proceedings of the International Interdisciplinary Conference, pp. 117–119 (2012)
17. Rakcheeva, T.A., Nikolaeva, E.C., et al.: Orthogonal illusion of visual perception. Modern Med. Theory Practice **1**, 42–51 (2004)
18. Rakcheeva, T.A.: Metric invariants of the illusion of intersection. In: Proceedings of the International Conference “Machines, Technologies and Materials for Modern Engineering, dedicated to the 80th Anniversary of IMASH RAS, p. 160 (2018)
19. Al Walid, M.H., Anisuzzaman, D.M., Saif, A.S.: Data analysis and visualization of continental cancer situation by Twitter scraping. Int. J. Modern Educ. Comput. Sci. (IJMECS) **11**(7), 23–31 (2019). https://doi.org/10.5815/ijmecs.2019.07.03
20. Hamd, M.H., Ahmed, S.K.: Biometric system design for iris recognition using intelligent algorithms. Int. J. Modern Educ. Comput. Sci. (IJMECS) **10**(3), 9–16 (2018). https://doi.org/10.5815/ijmecs.2018.03.02
21. Marouf, A.A., Ashrafi, A.F., Ahmed, T., Emon, T.: A machine learning based approach for mapping personality traits and perceived stress scale of undergraduate students. Int. J. Modern Educ. Comput. Sci. (IJMECS) **11**(8), 42–47 (2019). https://doi.org/10.5815/ijmecs.2019.08.05
22. Isong, B.: A methodology for teaching computer programming: first year students perspective. Int. J. Modern Educ. Comput. Sci. (IJMECS) **6**(9), 15–21 (2014)
23. Dominic, M., Francis, S.: An adaptable e-learning architecture based on learners’ profiling. Int. J. Modern Educ. Comput. Sci. (IJMECS) **7**(3), 26–31 (2015)

Comparative Analysis of Human Adaptation to the Growth of Visual Information in the Problems of Recognition of Formal Symbols and Meaningful Images

A. V. Koganov[1(✉)] and T. A. Rakcheeva[2]

[1] FGU FNC NIISI RAN, 36 Nakhimovsky st., corp. 1, Moscow 117218, Russia
akoganov@yandex.ru
[2] IMash RAN, 4, Bardina st., Moscow 117334, Russia

Abstract. We describe the engineering-psychological experiment, continuing the study of the ways of human adaptation to increasing complexity of logical problems with the method of presentation of the series of tasks of increasing complexity. Tasks require calculations in associative or nonassociative system of operations. By the nature of the change in the time of solving the problem, depending on the number of necessary operations, it can be concluded that a purely sequential method of solving problems or connecting additional brain resources to the solution in parallel. In a previously published experimental work, a person in the process of solving an associative problem recognized color pictures with meaningful images. In the new study, a similar problem is solved for abstract monochrome geometric shapes. The analysis of the result showed that for the second case the probability of the test subject's transition to a parallel method of processing visual information is significantly reduced. As a control series of problems (to separate parallel work from the acceleration of the sequential algorithm), as in the previous experiment, a non-associative comparison problem in cyclic arithmetic, presented in the visual form of the game "stone, scissors, paper" is used.

Keywords: Parallel counting · Engineering psychology · Testing · Algebra · Associativity · Recognition of visual images

1 Introduction

This work continues the research described in [1–7], the strategy of changing the ways of solving computational problems by a human being under the condition of increasing complexity of calculations. The main direction of research was to determine the ability of a person to move to parallel calculations with increasing complexity of calculations, when the task allowed it. We have set engineering psychological experiment where used the method of presenting a series of tasks of increasing complexity. These tasks require calculations in an associative or nonassociative system of operations. By the nature of the change in the time of solving the problem, depending on the number of necessary operations and their algebraic associativity, it can be concluded that a purely

Z. Hu et al. (Eds.): AIMEE 2019, AISC 1126, pp. 218–230, 2020.
https://doi.org/10.1007/978-3-030-39162-1_20

sequential method of solving problems or connecting additional brain resources to the solution in parallel. This research method is described in detail in [2–5]. In the previously published experimental work [3, 4] people in the process of solving the associative problem recognized color icons with meaningful images.

The research of the change of algorithm for the solving of like tasks, when the input dates complexity changes, is very important for development of article intellect system. In the work of that direct it require to teach the computer to solve anything task class as well as to convergence the algorithm work to the human method of solve, for example [8–15]. To the effect it is necessary to match the development of technical systems with researches in human engineering.

In the new study, a similar problem is solved for abstract monochrome geometric shapes. In mathematical and logical sense both problems are equivalent. A person remembers a set of standard objects that are easily distinguishable visually. Then he gets a table with a variable number of cells, each of which is one of the standard objects and must determine which of these objects in the table is missing, or to establish that all objects are available. There is a guarantee that there is no more than one object. This problem allows an effective solution with parallel analysis of different parts of the table and with parallel recognition of different objects.

As a control series of problems (to separate parallel work from the acceleration of the sequential algorithm), as in the previous experiment, a non-associative comparison problem in cyclic arithmetic, presented in the visual form of the famous game “stone, scissors, paper” is used. The person received a table in which cells are drawn randomly stone, scissors or a sheet of paper. This table should be interpreted as a single line with a line break, as in a book. Consistently applying the rules of “stone stronger than scissors”, “scissors stronger than paper”, “paper stronger than stone”, it was necessary to determine the “winner” of the first pair of cells, the winner between him and the next cell, and so on to the last cell. The final winner is the solution to the problem.

In both tasks, the answer was given in the form of a number assigned to a standard object in a separate table that is in the field of view of the person. A person could stop the countdown when he found the answer in graphical form. The problem disappeared. After that, you could search for the answer number in the table without changing the recorded time for solving the problem.

Tasks were presented on the computer screen in a standard form. The necessary commands “start”, “stop”, “remember answer” were formed by mouse click on the corresponding virtual button. The set of the answer in each task and the setting of the table size for a series of tasks were made by the keyboard.

Tasks in the form of a table of a certain size were presented in series of 10 tasks. The number of the task in the series was displayed on the screen. There were four series for each task with the size of tables 8, 12, 16, 20. Next, the size of the table will be called the complexity of the problem. The number of recognition and comparison operations required for both tasks is proportional to the complexity. The first presented a non-associative problem “rock, paper, scissors”.screen in the special field in format of several strings within interface window. That window contains the button for driving the new task and stop-watch (start), the button for pause stop-watch (stop), the window for input the result of solution and the button for writing that results in the result file, the window for input the task parameter (the number of symbols in input dates), the window for indicating the number of the task in series with last value of parameter. The interface window scheme shows on Fig. 1.

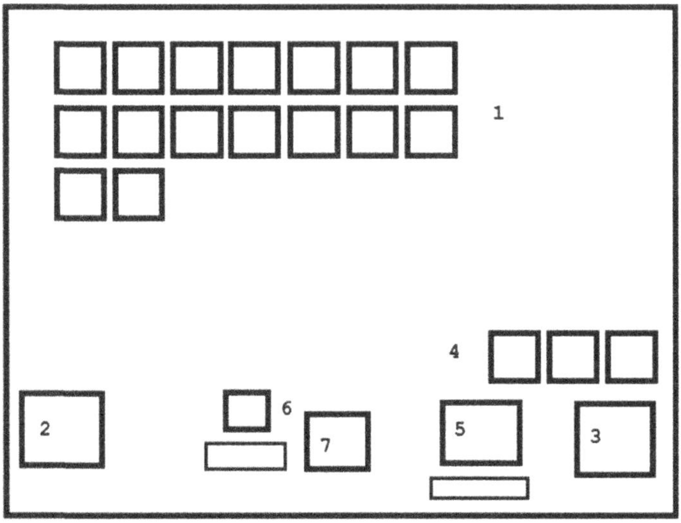

Fig. 1. The scheme of Interface Window in experiment. 1. The Field for showing the input task dates. 2. The bottom "START". 3. The bottom "STOP". 4. The field for presentation of the alphabet at the task with the numbers of symbols for the result record with bottom "write". 5. The window for writing of result. 6. The window for task parameter specify with bottom "write". 7. The counter of the tasks in currant series after the last parameter specifying.

The authors are grateful employees of National Russian University named after A. N. Kosygin for their help in setting up the experiment. The work performed on the topic of state assignment for research work number 065-2029-0007.

2 Method of Processing the Results of the Experiment

During the experiment, the values of the problem solution time were memorized. The average solution time and standard deviation of this mean were calculated for each series of problems. Next, we will call the two values of the mean time of the solution strictly different, if their intervals if their intervals (plus or minus the standard deviation) do not intersect, and conditionally different, if each of them is outside this interval of the partner.

If average times are growing with the growth of complexity in both tasks, we recorded a uniform sequential way of solutions in the associative task. If in a nonassociative task time increases, and in the associative task, there is no reliable (strongly or relatively) the growth time, it registers the parallel method of solution of the associative task. If there is no reliable growth in both problems, then the sequential solution of both problems is recorded with the acceleration of calculations as the complexity increases. If there is no reliable growth in the nonassociative problem, but there is an increase in the average time in the associative problem, then the presence of significant interference factors in the experiment (inadequate result) is recorded.

The mathematical motivation for such a system of data processing was given in [2–4]. If the human brain performs a standard operation at constantly performance then the proofing time grows in proportion to number of the operation which is necessary for proofing (the task complexity).

If the increase in the number of operations leads to reduced time of performance one operation (the calculate speed grows) then increase the time of solving the test task decreases with increasing of the task complexity. In that case, the curve of the proof time is convex bottom as the function of task complexity. The properties is independent the structure of the task.

If the task allowed effectively parallel performing of average operations simultaneously then it is possible to monitor a convex bottom function of depending the proof time an task complexity off increasing the speed of one operation. In that case, if the task don't allowed the effectively parallel execution then the time increase remains linear an the task complexity.

In article [1, 2] we show that the calculation in associative algebra may by realized in parallel regime with big boot of all processors. The parallel calculations in nonassociative algebra require using big number of processors with low coefficient of useful processor time. Therefore the realization of execution of several simultaneous operations is energetically uneconomic. That mathematical apparatus is laid in basis of experiments series.

$$(((x_1 \circ x_2) \circ x_3) \circ x_4) = ((x_1 \circ x_2) \circ (x_3 \circ x_4)) \tag{2.1}$$

Formula (2.1) showed as in associative algebra three steps of calculation transform to two steps. Each step corresponds to the level of bracket nesting. In nonassociative algebra such decomposition is not possible.

In real experiment a probability factors interfere to the time of task solution. Those factors are connected with unavoidable inputting extern signals in the human brain. The sources of that activity may to lie in around environment or in human organism. Furs more the probability forming of input dates for the task induces different difficult of sensing. For example the tasks may to have difference variety of symbols in input dates. Consequently it is necessary to produce the serious of the task same complexity and to calculate the average solution time and its standard derivation.

3 The Analysis of the Results

Currently, we have conducted a preliminary experiment, the purpose of which was to test a new mathematical software and identify the features of human perception of a new type of problems. At this stage, four people took part in the experiment. However, the analysis of their results showed that for the new variant of objects (abstract monochromatic figures) the probability of transition of the subject to a parallel method of processing visual information is significantly reduced in comparison with the previous experience, where the subjects recognized color meaningful images. In the first experiment, all subjects (8 people) showed a parallel way to solve the associative problem. In the new experiment, two (IS1, IS2) showed a sequential solution to both problems. One of them (IS2) showed a possible acceleration in a nonassociative problem. Two people (IS3, IS4) showed a parallel method of solving an associative problem with a uniform sequential solution of a nonassociative test. The results of the experiment are presented in Table 1 and Fig. 2.

Table 1. Average times of solving a series of problems.

Associative task			Nonassociative task		Persona
The string length	t_{cp}: The average time (s)	sig: The standard deviation of the mean (s)	t_{cp}: The average time (s)	sig: The standard deviation of the mean (s)	Personal code
8	6,1	1,02	4,8	0,51	IS1
12	6,4	0,79	10,5	1,07	
16	7,9	0,79	10,9	0,94	
20	8,2	1,03	16,5	1,24	
8	5,5	1,09	5,3	0,45	IS2
12	7,5	0,64	5,5	0,83	
16	7,8	1,12	11,1	1,57	
20	8,2	0,54	9,0	1,26	
8	9,5	0,67	7,5	1,09	IS3
12	9,3	0,65	10,4	1,47	
16	10,6	0,60	12,1	0,98	
20	10,4	1,44	14,0	0,92	
8	5,4	0,86	11,9	1,26	IS4
12	5,0	1,10	12,0	1,28	
16	5,9	0,74	15,8	2,26	
20	6,9	0,84	17,3	1,63	

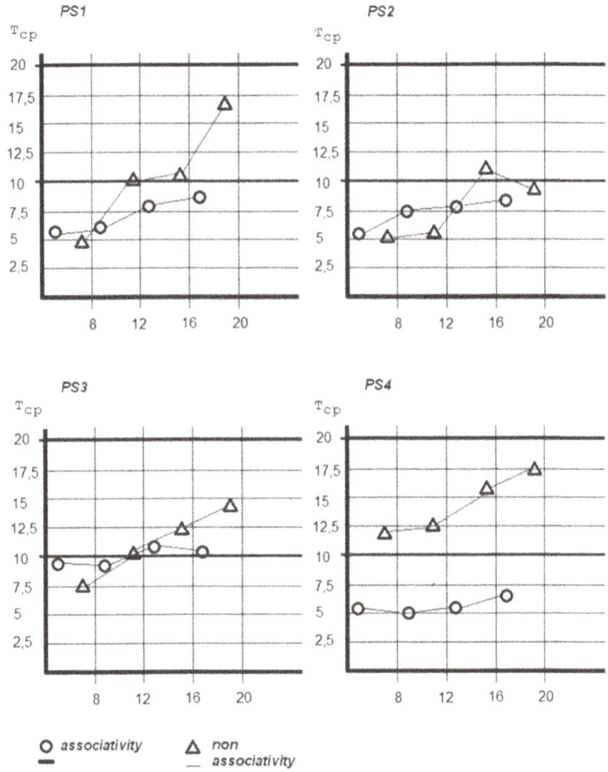

Fig. 2. The dependence of the average time to solve the problem on the complexity of the problem. Each graph summarizes data on two (associative and non-associative) problems. Graphs IS1 and IS2 correspond to the sequential method of solving both problems. The graphs IS3 and IS4 correspond to the parallel solution of the associative problem.

In that experiment the dates are coding by pictures of stone, scissors and paper for nonassociative task (Fig. 3), and by collection of geometric figures in associative task (Fig. 4)

Fig. 3. The coding of date for the task "stone, scissors, paper". It are using realistic images of correspond objects.

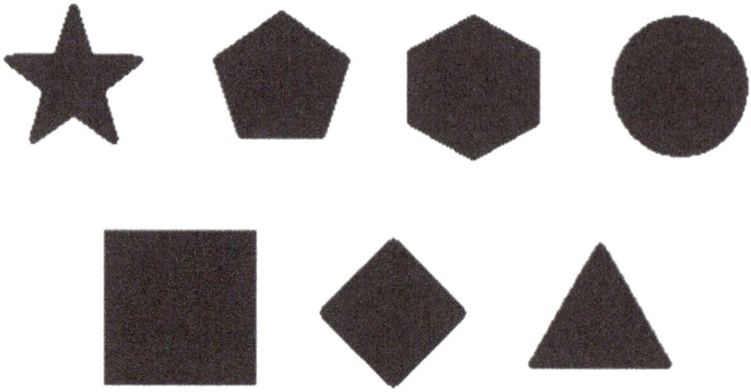

Fig. 4. The coding of date for the task "The find of the lacking object" in new experiment.

Nonassociative task exactly repeat those task in previous experiment. In both experiments the encoding uses realistic pictures of the stone, scissors and paper. For associative task in first experiment it was used the recognition images well-known real objects (locomotive, tree, house, horse, ship, car). Those icons are shown on Fig. 5. Therefore the refusal of several testing persons to parallel strategy adaptation to increase of task complexity connects, probability, with different encoding of information for recognition.

Fig. 5. Encoding date in the task "The find of the lacking object" with using realistic objects.

For comparison of two experiments, we show several diagrams which correspond to recognition of familiar things (Fig. 6).

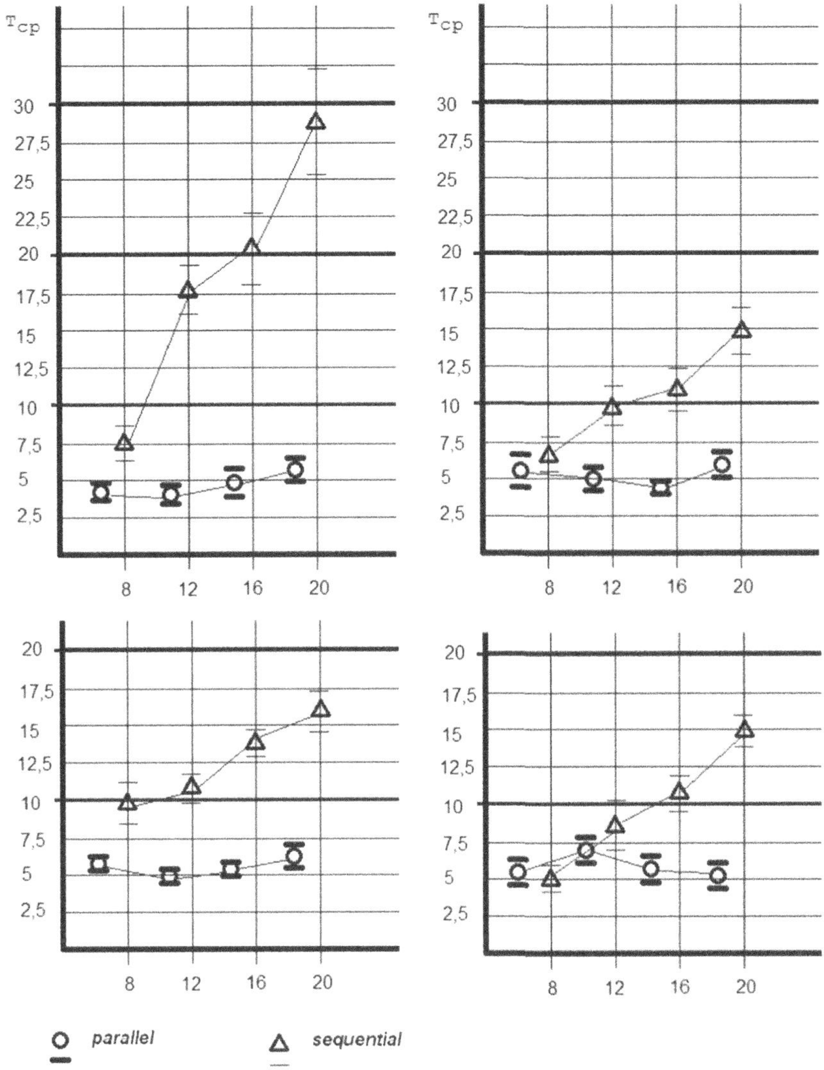

Fig. 6. The diagrams of average time to solve problems with recognition of realistic images. All testing persons used the strategy of parallel calculations for adaptation to growing complexity of tasks.

4 Mathematical Model of the Compute Strategy Change

The analysis above allows us to suggest that the presence of a figurative interpretation of the input information facilitates the transition of a person to parallel computing. At this man does not understand this process. Apparently there are congenital structures of the central nervous system that create parallel information processing mechanisms for working with recognizable real objects.

We will try to build a mathematical model of such a structure to solve the problems described above. Primary information enters the human visual receptors (bloc A). Then it excites the work of primary brain analyzers, probably working in the subcortical areas (block B). The processed signal enters the cortex, where the symbols of the initial data of of the incoming task are recognized (block D). These data are transferred to the structure of the cortex, which arose in the process of learning and training to solve the problem (processor unit E). The output signal of this structure enters the region of output neurons that cause the work of effectors that produce a result in the external environment (block F). This diagram is shown in Fig. 7. Since processor unit E has a purely logical structure, it is likely that the sequential organization of computations is characteristic of it.

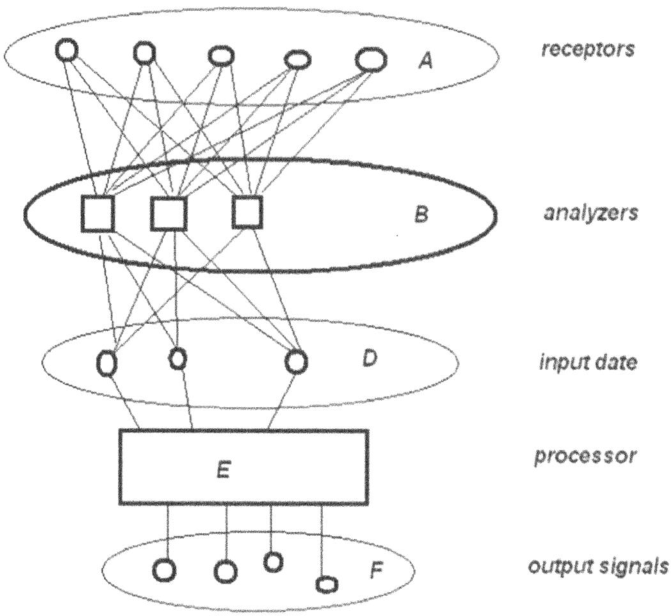

Fig. 7. The scheme of uniprocessor organization of calculations in the human brain. A—signal from the visual receptors. B—primary analyzers. D—character recognition of the source data of the task. E—a processor for solving a problem in the cerebral cortex. F—output neurons for re-recording the result.

In the case when the input information has an additional interpretation in the form of recognition of objects known to a person, additional data processing, which performs this recognition, is simultaneously activated (block C). This system does not depend on what task the person with the source data solves. When the complexity of the task becomes large, the brain can use this parallel process to speed up data processing. Then the objects previously identified in recognition begin to be perceived as a duplicate of the source data. To process them, a duplicate processor is created that solves the problem (block G). Each of processors E and G can now process its own part of the data and give it to the third processor, which forms the final result from the parts (synthesis block S). Thus, a parallel computational scheme is implemented. This process is shown in Fig. 8.

The formation of blocks is not necessary. Therefore, parallel computations for figurative data encoding are not always observed.

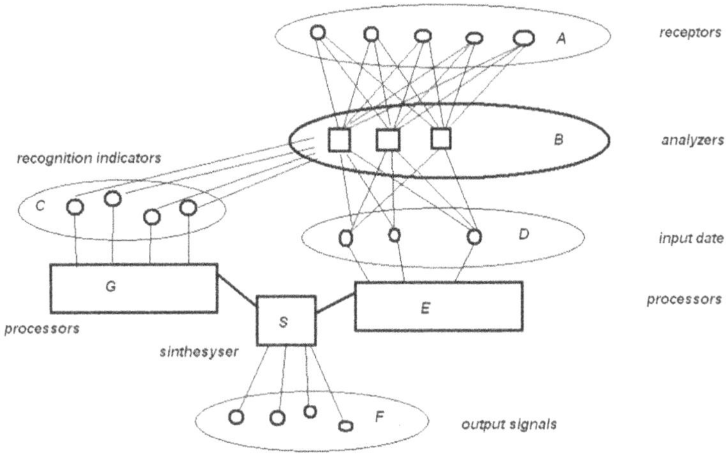

Fig. 8. Scheme of multiprocessor implementation of calculations using recognition of objects depicted on input characters. Additional function blocks arise. C—recognition of depicted objects. G—processors for the task solved on a recognition data. S—a synthesizer of the results of all the processors to produce a common result. The remaining blocks coincide with the description in Fig. 7.

We introduce compact schemes for the above strategies of organizing computations and brief notation for them by the names of the last blocs.

Sequential strategy (EF)

$$A \to B \to D \to E \to F$$

Parallel strategy (SF)

$$\begin{array}{ccccccc} A & \to & B & \to & D & \to & E \\ & & \downarrow & & & & \downarrow \\ & & C & \to & G & \to & S \to F \end{array}$$

On the base of such model we consider when the formation of parallel algorithm is effectively. On the scheme of Fig. 8 it shows that for small volume of input date the conversion to parallel regime is not profitable. The using two steps for processing in that case retard the task solving. Additional, the work of three processors requires more energy.

When the computing is nonassociative the multiprocessor strategy is no effective even for big volume of dates. For reduction of the solution time it is require very many processors with very low coefficient of using time [1, 2]. Therefore the acceleration of solution requires big expenditures of energy. That fact causes the sequential computing for nonassociative task which was observed in all experiments.

If it is possible sizable to decrease the time of solution with few expenditures energy and resource then parallel strategy is gainful and observed in experiments.

That considering allows building of the model of intuitive human choice of computing algorithm. Let and are the average time of solution in schemes and respectively; and are the correspond expenditure of recourse for algorithm realization. The Recourse may measure as energy. In those model the persona characterizes by two coefficients of sensitivity to parameters the time and resource and respectively. The function of target which the persona tries minimized is

$$\Phi(T,R) = a_T T + a_R R \tag{4.1}$$

Decision to using sequential algorithm corresponds to unequation

$$\Delta\Phi = a_T(T_{EF} - T_{SF}) + a_R(R_{EF} - R_{SF}) \leq 0. \tag{4.2}$$

In that model a different persons choice of different strategy because they has different coefficient of sensitivity. When it is use the imaginative encoding of input date the probability of parallel algorithm augments seeing the difference $R_{EF} - R_{SF}$ descends. The block C is built narrow.

5 Conclusions

The experiment showed that when solving logical problems with abstract geometric monochromatic figures, a person uses a sequential method of solving much more often than when solving similar problems with meaningful color images.

The nature of this phenomenon is not entirely clear. Perhaps for realistic images in the brain there is a special innate recognition mechanism that is physiologically adapted to parallel work. But for abstract images, this mechanism is not inducted in the work. Recognition is carried out by a universal logical system, which is more difficult to translate into parallel computing. This phenomenon requires additional research.

Perhaps this phenomenon is associated with the preferred use of drawings on road signs and indicative, as well as in advertising and computer icons.

The method of identifying parallel ways of solving the problem was also effective for the analysis of preferred forms of encoding of operational information.

The nature of this phenomenon is not entirely clear. Perhaps for realistic images in the brain there is a special innate recognition mechanism that is physiologically adapted to parallel work. But for abstract images, this mechanism is not inducted in the work. Recognition is carried out by a universal logical system, which is more difficult to translate into parallel computing. This phenomenon requires additional research.

A mathematical model of the process of choosing an algorithm for solving the problem is constructed. The model explained the fact that three are differences in strategies for adapting to the complexity of a task for different people and the effect of increasing the percentage of people who choose a parallel algorithm when the source data is figuratively encoded.

This allows us to continue the cycle of work in the direction of variation of forms of presentation of information. Particular attention should be paid to the visual arrangement of the data on the monitor screen, the size of the symbol image, the color scheme of the representation and the associations that symbols cause in humans. Those parameters that will ensure maximum performance of information processing can be effectively used in the development of human—machine dialogue systems. This is an important area of development of artificial intelligence systems.

References

1. Koganov, A.V., Rakcheeva, T.A.: Tests of parallel information processing on the basis of algebra and formal automata. In: Hu, Z., Petoukhov, S., He, M. (eds.) Advances in Artificial Systems for Medicine and Education. AISC, vol. 658, pp. 68–78. Springer, Cham (2018). ISBN 978-3-319-67348-6
2. Koganov, A.V., Rakcheeva, T.A.: Tests for checking the parallel organization of the logical calculation on the basis of algebra and automata. In: The Computer Research and Modeling, vol. 9, no. 4, pp. 621–638. The Computer Research Institute (UGU), The Machine research institute at RAN in the name of A.A. Blagonravov (2017). (ISSN 2076-7633, russian)
3. Koganov, A.V., Rakcheeva, T.A.: Experimental detection of the parallel organization of mental calculations of a person on the basis of two algebras having different associativity. In: Hu, Z., Petoukhov, S.V., He, M. (eds.) AIMEE2018 2018. AISC, vol. 902, pp. 139–149. Springer, Cham (2020). https://doi.org/10.1007/978-3-030-12082-5_13. (Print ISBN 978-3-030-12081-8; Online ISBN978-3-030-12082-5; Series Print ISSN2194-5357; Series Online ISSN2194-5365)
4. Koganov, A.V., Rakcheeva, T.A., Prikhodko, D.I.: Experimental identification of the organization of mental calculations of the per-son on the basis of algebras of different associativity. In: The Computer Research and Modeling, vol. 11, no. 2, pp. 311–327. The Computer Research Institute (UGU), The Machine research institute at RAN in the name of A.A.Blagonravov (2019). (ISSN 2076-7633). https://doi.org/10.20537/2076-11-2-311-327
5. Koganov, A.V.: The growing inductor spaces and the analysis of the parallel algorithms. Program. Prod. Syst. Appl. Int. J. Probl. Theory Practicum Control **2**, 33–38 (2010)
6. Koganov, A.V., Zlobin, A.I., Rakcheeva, T.A.: The research of the possibility of the parallel processing of the information by man in the series of the tasks of the high complexity. In: The Computer Research and Modeling, vol. 5, no. 5, pp. 845–861. The Computer Research Institute (UGU), The Machine research institute at RAN in the name of A.A.Blagonravov (2013)
7. Koganov, A.V., Zlobin, A.I., Rakcheeva, T.A.: The task of the calculation of the trajectory with homogenous the distribution of solutions. In: The Computer research and modeling, vol. 6, no. 5, pp. 803–828. The Computer Research Institute (UGU), The Machine research institute at RAN in the name of A.A.Blagonravov (2014)

8. Popov, G., Mastorakis, N., Mladenov, V.: Calculation of the acceleration of parallel programs as a function of the number of threads, January 2010. https://www.researchgate.net/publication/228569958
9. Fischer, R., Plessow, F.: Efficient multitasking: parallel versus serial processing of multiple tasks. Front. Psychol. **6**, 1366 (2015). https://doi.org/10.3389/fpsyg.2015.01366)
10. Marouf, A.A., Ashrafi, A.F., Ahmed, T., Emon, T.: A machine learning based approach for mapping personality traits and perceived stress scale of undergraduate students. Int. J. Modern Educ. Comput. Sci. (IJMECS) **11**(8), 42–47 (2019). https://doi.org/10.5815/ijmecs.2019.08.05
11. Nandi, D., Saif, A.S., Prottoy, P., Zubair, K.M., Shubho, S.A.: Traffic sign detection based on color segmentation of obscure image candidates: a comprehensive study. Int. J. Modern Educ. Comput. Sci. (IJMECS) **10**(6), 35–46 (2018). https://doi.org/10.5815/ijmecs.2018.06.05
12. Ajay, A., Singh, P.K.: Novel digital image water marking technique against geometric attacks. Int. J. Modern Educ. Comput. Sci. (IJMECS) **7**(8), 61–68 (2015). https://doi.org/10.5815/ijmecs.2015.08.07
13. Kant, R.: A study of creativity of secondary school children as a correlate of some television viewing habits. Int. J. Modern Educ. Comput. Sci. (IJMECS) **4**(10), 33–39 (2012)
14. Ali, H.K., George Amalarethinam, D.I.: Activity recognition with multi-tape fuzzy finite automata. Int. J. Modern Educ. Comput. Sci. (IJMECS) **5**(5), 60–65 (2013). https://doi.org/10.5815/ijmecs.2013.05.07
15. Tran, L.B., Le, T.H.: Person authentication using relevance vector machine (RVM) for face and fingerprint. Int. J. Modern Educ. Comput. Sci. (IJMECS) **7**(5), 8–15 (2015). https://doi.org/10.5815/ijmecs.2015.05.02

Chaotic Algorithms of Analysis of Cardiovascular Systems and Artificial Intelligence

Ivan V. Stepanyan[1(✉)] and Alexey A. Mekler[2]

[1] Mechanical Engineering Research Institute of the Russian Academy of Sciences, 4, Malyi Kharitonievsky pereulok, 101990 Moscow, Russian Federation
neurocomp.pro@gmail.com
[2] Saint Petersburg State Pediatric Medical University, 2 Litovskaya str., 194100 St. Petersburg, Russian Federation
mekler@yandex.ru

Abstract. Despite the intensive development of the dynamical systems theory and artificial intelligence, which is quite a powerful theoretical apparatus, an adequate description of chaotic processes at cardiovascular systems is a rather complicated problem. In this paper, the dynamical systems theory is applied to cardiovascular studies by processing the recorded signals with stochastic neural networks as well as dynamic chaos methods. The method of investigation is the reconstruction of dynamic systems attractor. Phase-temporal characteristics of human pulse waves were discussed, the new concept of the stochastic-graph of the pulse wave was shown. The attractor of heart pulse waves was reconstructed and its correlation dimension was estimated.

Keywords: Chaotic dynamics · Quasi-neural network · Clusterization · Chaotic attractor · Sphygmograph · Pulse waves

1 Introduction and Formulation of the Problem

Physiologists and clinicians are increasingly paying attention to the problem of the real-time pulse wave analysis. The development of pulse wave analysis technology is the know-how of developer firms. Since the problem is very relevant and multifaceted, it is proposed to use mathematical methods related to the theory of chaos.

The dynamic chaos theory is a mathematical apparatus that describes the behavior of nonlinear dynamical systems. The pioneers of the theory are Henri Poincaré, mathematicians Kolmogorov and Arnold, and Moser, who built the chaos theory called Kolmogorov-Arnold-Moser theory [1]. The theory introduces concepts of the stable orbits of the system and attractors (including strange attractors as attracting Cantor structures).

Historically, the use of the "chaos" concept is based on putting forward hypotheses, which refers to ancient Greek philosophers. A chaotic process is a process, which behavior is deterministic, but the subsequent state of such a system is described both by quantities that can be predicted and also can be random. However, according to [2, 3],

Z. Hu et al. (Eds.): AIMEE 2019, AISC 1126, pp. 231–240, 2020.
https://doi.org/10.1007/978-3-030-39162-1_21

the development of any process over time (whether deterministic or probabilistic) will be random. A special interpretation of chaotic processes can put it in order.

An illustrative process with chaotic dynamics is the change in the characteristics of a human's pulse wave and the variability of heart rate. The time between the beats of a heart is not constant but forms an attractor when processed by chaotic methods.

Pulse wave velocity is the biomechanical parameter, which indicates the velocity of arterial pulse propagation at the cardiovascular system. It is used clinically as it relates to arterial stiffness, biological aging, and arteriosclerosis [4]. The high social significance of cardiovascular risks leads to need in improved methods of analysis of biomechanical parameters of the heart and human cardiovascular system [5, 6].

The pulse wave sensor based on fiber-optical transducer [7] developed in the laboratory of biomechanical systems IMASH RAS allows us to obtain highly accurate sphygmographic data of high resolution. This data was analyzed using described below chaotic models and preliminary stored as a digital row of amplitudes at a time scale. For the analysis of this data, artificial intelligence [8] methods, the methods of dynamic chaos [9] and stochastic clusterization [10] were used.

This article consists of related works review; methods with the description of the used algorithms; stochastic clusterization part with visualization of graph structure, chaotic dynamics, visualization of reconstructed attractors of pulse waves and conclusions.

2 Related Works

The cardiovascular system has a non-linear dynamics of heart rate variability. There are many works about analysis of heart rate dynamics by methods derived from nonlinear mathematics and chaos theory (some of them are [11–13]). Work [14] shows that the paradox regarding the classic power spectral analysis of heart rate variability (HRV) is whether the characteristic high- (HF) and low-frequency (LF) spectral peaks represent stochastic or chaotic phenomena. The resolution of this fundamental issue is key to unraveling the mechanisms of HRV, which is critical to its proper use as a noninvasive marker for cardiac mortality risk assessment and stratification in congestive heart failure (CHF) and other cardiac dysfunctions.

Work [15] revealed that the nonlinear dynamic methods could have clinical and prognostic applicability also in short-time ECG series. Dynamic analysis based on chaos theory during the exercise ECG test points out the multifractal time series in CHD patients who loss normal fractal characteristics and regularity in HRV.

Optimal levels of chaos and fractality can be associated with physiological health and function in natural systems. Paper [16] explains mathematical concepts of fractals and chaos, such as fractal dimension, criticality, bifurcation, and iteration, and how they are related to biology. Interesting results and visualizations were presented at [17] In the analysis of the dynamic of heart rate, the complexity of visibility graphs is seen as a sign of short term variability in signals. The dynamics of the signals were studied both before and during meditation by examining the complexity of visual graphs using graph index complexity. Generally, the obtained results showed that the heart rate signals were more complex during meditation.

It is known that the theory of dynamic chaos applies to problems of artificial intelligence and digital signal processing. In [18] the 3-dimensional chaotic system is firstly investigated and then an intelligent controller, based on brain emotional learning was implemented. This new chaotic system is synchronized. In [19] chaos-based digital image watermarking algorithm based on redundant discrete wavelet transform and singular value decomposition is proposed. The method is shown to be robust against both the geometrical and image processing attacks and to provide a better watermark concealment via computer simulations. Paper [20] presents fuzzy-model-based designs for the control and synchronization of a chaotic system. Paper [21] presents the solution of the multi-objective optimal dispatch problem of a solar-wind-thermal system by the improved stochastic fractal search algorithm. In [22], the observation equation is converted to the new observation equation with white noise. By the observation equation for the white observation noise, this work proposes new Wiener estimation algorithms for the fixed-point smoothing and filtering estimates in linear discrete-time wide-sense stationary stochastic systems.

3 Methods

Stochastic clusterization is a quasi-neural network algorithm based on the frequency-structured graph representation of big data. The analyzed data have to be a one-dimensional sequence of values. With regards to the problem under consideration, such data is a dynamics of high-resolution sphygmograph. Each edge of the graph is a quasi-neuron related to each semantic unit. In studying of high-resolution sphygmographic data every unit is dealt with equal-size part of dynamic sphygmographic data that is a sequence of amplitudes of length L. Every quasi-neuron related to the corresponding semantic unit have a step value Ω of activation function f:

$$
\begin{aligned}
f(n) &= 0 \; if \, \Omega < n \\
f(n) & \quad 1 \; if \, \Omega \geq n
\end{aligned}
$$

Ω is a resolution (scaling) factor that plays the role of a lag parameter equal for all quasi neurons. Cycles in graphs represent and identify repeating patterns in sequences of units for specific Ω, L, and n. By finding out proper parameters informative hidden processes are becoming visualized.

4 Stochastic Clusterization

Cardiovascular events in the relatively low-frequency vibrations directly related to the movements of cardiac valves and pump units (that lays not only in the traditional area of tone and heart noises) were researched with an application of a sensor based on the fiber-optic converter.

The stochastic process of pulse waves can be represented as a quasi-neural network structure constructed using methods of stochastic artificial neural networks (clusterization) [10] and corresponding adjacency matrix. To build a graph model the digital flow of pulse wave amplitudes was analyzed. Structurally-frequency characteristics of pulse waves are fixed in the structure of a weighted graph constructed using the following algorithm: each vertex of the graph corresponds to the fragment of pulse wave data flow and graph arcs connect adjacent fragments. All fragments have a fixed length, which is a free parameter of the algorithm. Taking into account the frequency characteristics of each fragment, the constructed graph shows the internal structure of a pulse wave and can be used for farther analysis such as recognition of biomedical data related to cardiovascular system parameters.

The characteristics of human cardiovascular system parameters were algorithmically constructed and displayed in the graph (Fig. 1). In clusters (separate fragments of graph), the repeatable structure chains of biological information can be distinguished. This algorithm provides a structural filter for subsequent detection and analysis of patterns in the characteristics of the individual human cardiovascular system.

Individual human cardiovascular system characteristic is a personal biometric data that can be researched for person identification, monitoring of the biological aging process and other purposes with the methods of artificial intelligence.

Fig. 1. Stochastic clusterization of pulse wave data with a quasi-neural network graph representation. Main structural-frequency clusters were detected in the analyzed specimen of a person's pulse wave.

5 Chaotic Dynamics

For biomechanical information processing and research of the human cardiovascular system based on the heart rate monitoring, a mathematical algorithm that involves the assessment of the dynamic (phase and temporal) characteristics of the received signal as a noise signal has been implemented. This was accomplished by filtration with the definition narrow-band of selected harmonic component of the signal relative to the reference phase signal phase shift dynamics with further analysis based on the phase and time characteristics. Useful information is extracted from the nature of the change thus determined phase shift in time.

For further research, the reconstructed attractor of the human's heartbeat was built in a lag parameter space with verification [23]. Attractor analysis showed, even though it looks quite simple - the dynamics in it has a distinct multi-dimensional character. Figure 2 shows the image of the attractor reconstructed in 3-dimensional lag space (unit lag and equal to the first minimum of mutual information function).

Each variant was constructed in three projections. We have tried to estimate the dimension of the reconstructed attractor. Calculations were made using the Grassberger-Procaccia algorithm [24]. This method has been widely applied to signals of biological nature, e.g. electroencephalogram [25]. According to this algorithm, for each x_i we calculate the proportion of points that are located on the distance not exceeding some value e from this point in the reconstructed attractor. Then this value is averaged over all points in the reconstructed attractor. We can write the following relation that is called correlation integral:

$$C(\varepsilon) = \frac{1}{N^2} \sum\nolimits_{\substack{i,j=1 \\ i \neq j}}^{N} \Theta\left(\varepsilon - \left|\vec{x}_i - \vec{x}_j\right|\right),$$

where $\Theta(x) = \begin{cases} 0, & x < 0 \\ 1, & x > 0 \end{cases}$ – Heaviside's function and $\left|\vec{x}_i - \vec{x}_j\right|$ – the distance between x_i and x_j in reconstructed attractor. When e is small correlation integral behaves as a power of e:

$$C(\varepsilon) \propto \varepsilon^{D_2} \text{and } D_2 \approx \frac{logC(\varepsilon)}{log\varepsilon},$$

where D_2 is correlation dimension of reconstructed attractor and it may be estimated as a slope of linear region of $C(\varepsilon)$, plotted in the double logarithmic scale.

This algorithm was modified for use with the signals of biological nature, which are too short and noisy to apply this algorithm in its original form [26]. Implementing of this method for time series study requires process stationarity and a sufficiently large observation time.

For example, according to the criteria [27], the time series should be of at least $D^{2+0.4D_2}$ points. Here, D_2 – is the estimated value of the correlation dimension of reconstructed attractor. We have had 802 points only. Registering the longer signal seems impossible due to violation of the process stationarity requirements. As a result, time series length is not sufficient for a precise D_2 estimation.

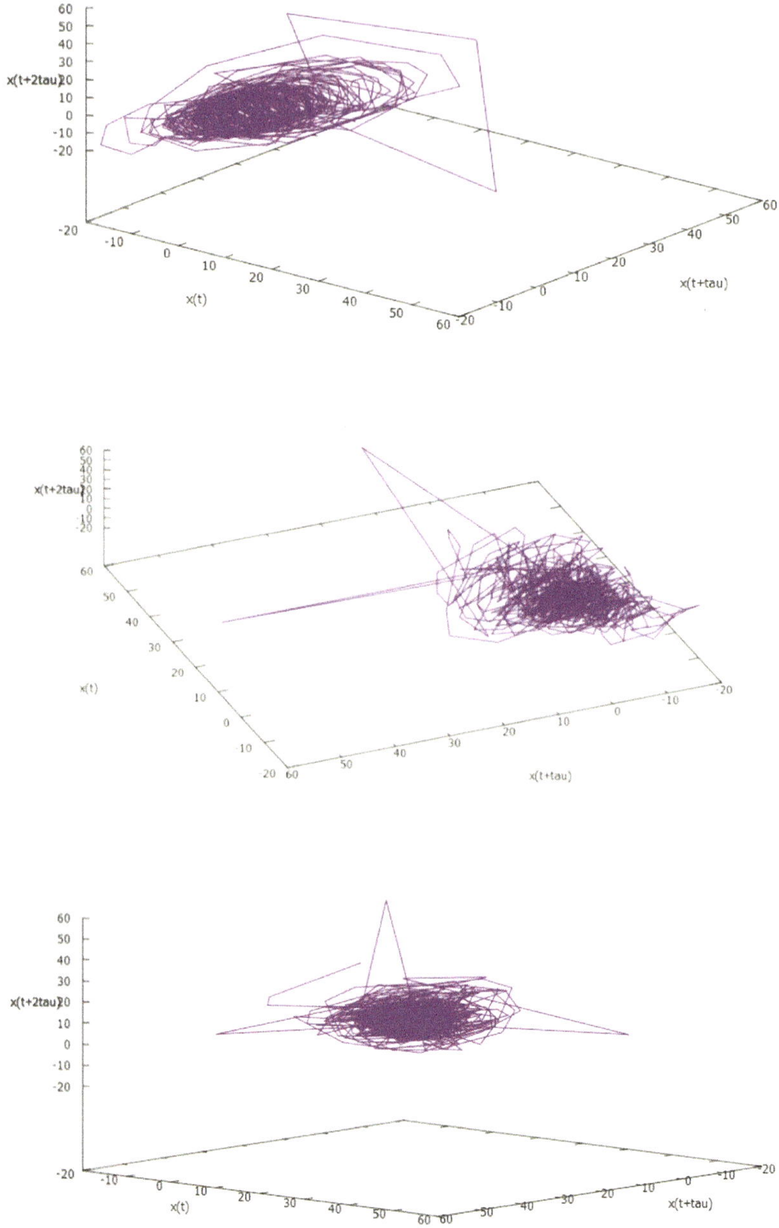

Fig. 2. Reconstructed attractors of heart pulse waves in a 3-dimensional lag space (unit lag and equal to the first minimum mutual information function).

To estimate the embedding dimension D_{emb}, that is minimally necessary to reconstruct the attractor we used the method of false nearest neighbors [28] that was realized in TISEAN software package [29]. However, due to too small length of the time series under study the program stopped the calculation when D_{emb} reached 5, due to the lack of points for statistics. When values $D_{emb} < 5$ the relative false nearest neighbors number was more than 20% (FNN ratio > 0,2), that is comparable with the values obtained in the studies of stochastic or multidimensional dynamical processes. Also, with D_{emb} increasing from 3 to 4 FNN ration increased too while normally it should decrease with D_{emb} increasing (Fig. 3). These circumstances suggest the multidimensional dynamics, combined with a lack of long-time series of the test. Also, a graph of $D_2(D_{emb})$ was plotted (Fig. 4). D_{emb} can be estimated from this graph as the dimension in which the relationship of D_2 (D_{emb}) enters to saturation.

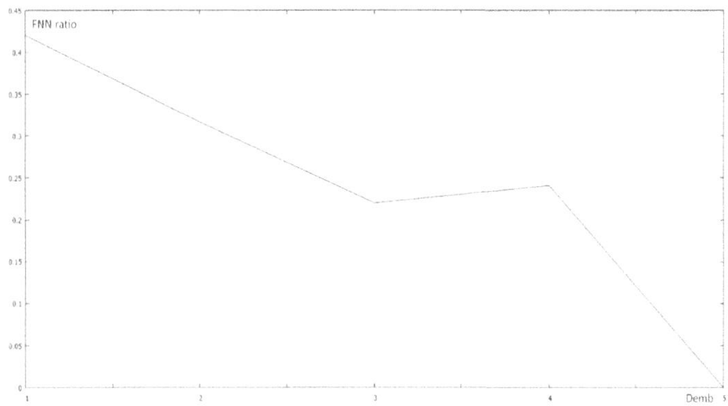

Fig. 3. The relative number of false nearest neighbors for different dimensions of the embedding space.

Here one can see that with values of $D_{emb} < 6$ no saturation occurs. Further increasing of the value of D_{emb} is not reasonable due to the insufficient length of the time series. Thus, there is a reason to approve that the analyzed signal has a multidimensional dynamics and dimension of the chaotic attractor of the human cardiovascular system (if it exists) is no less than 5. To continue investigations and answer the question if there exist deterministic chaotic dynamics in the signal, we need longer time series that at the same time meet the requirements of stationarity.

Chaotic algorithms allow getting additional information about the dynamic processes of any complexity and length. Particularly, proven algorithms are applicable to recognition problems. To do this, the structural representation of data in the form of a graph or trajectory in the phase space can be used. Thus, the processed sequences with the described methods can be used in artificial intelligence systems to form complex reasoning.

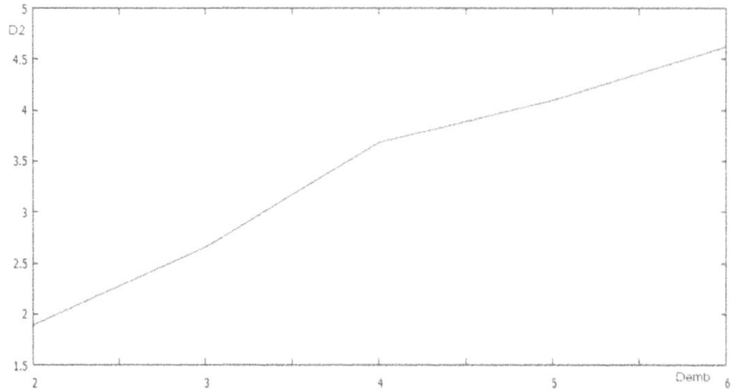

Fig. 4. The dependence of the correlation dimension of the reconstructed attractor on the embedding dimension.

The developed algorithms and the obtained results can be used for further studies of analysis of high-resolution dynamic data such as sphygmograph as well as for signal processing theory.

For a modeling agent, a deterministic environment is simpler than a stochastic one. The sequences of data that were processed with chaotic algorithms can consist of hundreds and thousands of additional units that can be used in artificial intelligence systems to form reasoning concerning complex causal relationships between objects or within complex objects.

6 Conclusions

The possibility of using the methods of chaotic dynamics and clusterization of biomedical data and the presence of specific regularities due to the nature of the characteristics of the researched processes are shown.

As a result of research, we found that the human pulse wave contains a distinct characteristic of the state of the human circulatory system with multidimensional chaotic dynamics, the parameters of which were quantified. Dynamic characteristics were analyzed using chaotic models and quasi-neural networks. For these purposes, stochastic clustering and chaotic attractor methods were chosen, which are related to each other by the lag parameter, which plays the role of a viewing window. The attractor of dynamic human heart characteristics with an estimate of its dimension was reconstructed.

Acknowledgments. The authors wish to sincerely thank Igor Yavelov for devices used in this research.

References

1. Dumas, H.S.: The KAM Story – A Friendly Introduction to the Content, History, and Significance of Classical Kolmogorov–Arnold–Moser Theory. World Scientific Publishing (2014). ISBN 978-981-4556-58-3. Chapter 1: Introduction
2. Kac, M., Logan, J., Montroll, E.W., Lebowitz, J.L. (eds.): Fluctuation Phenomena. North-Holland, Amsterdam (1976)
3. Nelson, E.: Quantum Fluctuations. Princeton University Press, Princeton (1985)
4. Nichols, W.W.: Clinical measurement of arterial stiffness obtained from noninvasive pressure waveforms. Am. J. Hypertens. **18**(1 Pt 2), 3S–10S (2005). https://doi.org/10.1016/j.amjhyper.2004.10.009. PMID 15683725
5. Khazaee, A.: Heart beat classification using particle swarm optimization. Int. J. Intell. Syst. Appl. (IJISA) **5**(6), 25–33 (2013)
6. Queyam, A.B., Pahuja, S.K., Singh, D.: Doppler ultrasound based non-invasive heart rate telemonitoring system for wellbeing assessment. Int. J. Intell. Syst. Appl. (IJISA) **10**(12), 69–79 (2018)
7. Yavelov, I.S, Rochagov, A.V.: "Pulse" pulse diagnosis and computer analyzer. Moscow-Izhevsk: Research Center "Regular and chaotic dynamics" (2006)
8. Kurama, V., Alla, S., Rohith, V.K.: Image semantic segmentation using deep learning. Int. J. Image Graph. Sig. Process. (IJIGSP) **10**(12), 1–10 (2018). https://doi.org/10.5815/ijigsp.2018.12.01
9. Anthony, M., Wang, K., Hu, B.: Qualitative Theory of Dynamical Systems. Taylor & Francis (2001). ISBN 978-0-8247-0526-8. OCLC 45873628
10. Stepanyan, I.V., Yavelov, I.S., Saveliev, A.V., Do, O.K., Svirin, V.I., Pleshakov, K.V.: Phase-impulse analysis of the pulse wave and biopotentials of the human brain. Biomed. Radio Electron. **4**, 81–83 (2015)
11. Sassi, R., Cerutti, S., Lombardi, F., Malik, M., Huikuri, H.V., Peng, C.K., Schmidt, G., Yamamoto, Y., Document Reviewers, Gorenek, B., Lip, G.Y., Grassi, G., Kudaiberdieva, G., Fisher, J.P., Zabel, M., Macfadyen, R.: Advances in heart rate variability signal analysis: joint position statement by the e-Cardiology and the ESC Working Group Europ Heart Rhythm Association co-endorsed by the Asia Pacific Heart Rhythm Society. Europace. http://dx.doi.org/10.1093/europace/euv015. First published online: 15 July 2015
12. Huikuri, H.V., Makikällio, T.H., Perkiomaki, J.: Measurement of heart rate variability by methods based on nonlinear dynamics. J. Electrocardiol. **36**(Suppl), 95–99 (2003)
13. Melillo, P., Bracale, M., Pecchia, L.: Nonlinear Heart Rate Variability features for real-life stress detection. Case study: students under stress due to university examination Biomed. Eng. Online **10**, 96 (2011). Published online 7 Nov 2011. https://doi.org/10.1186/1475-925x-10-96
14. Wu, G.Q., Arzeno, N.M., Shen, L.L., et al.: Chaotic signatures of heart rate variability and its power spectrum in health, aging and heart failure. PLoS ONE **4**(2), e4323 (2009). https://doi.org/10.1371/journal.pone.0004323
15. Krstacic, G., Krstacic, A., Smalcelj, A., Milicic, D., Jembrek-gostovic, M.: The, "Chaos Theory" and nonlinear dynamics in heart rate variability analysis: does it work in short-time series in patients with coronary heart disease? Ann. Noninvasive Electrocardiol. **12**(2), 130–136 (2007)
16. Korolj, A., Wu, H.T., Radisic, M.: A healthy dose of chaos: using fractal frameworks for engineering higher-fidelity biomedical systems. Biomaterials **219**, 119363 (2019)
17. Nasrolahzadeh, M., Mohammadpoory, Z., Haddadnia, J.: Analysis of heart rate signals during meditation using visibility graph complexity. Cogn. Neurodyn. **13**(1), 45–52 (2019)

18. Ziabari, M.T., Sahab, A.R., Fakhari, S.N.S.: Synchronization new 3D chaotic system using brain emotional learning based intelligent controller. Int. J. Inf. Technol. Comput. Sci. (IJITCS) **7**(2), 80–87 (2015). https://doi.org/10.5815/ijitcs.2015.02.10
19. Çetinel, G., Çerkezi, L.L.: Robust chaotic digital image watermarking scheme based on RDWT and SVD. Int. J. Image Graph. Sig. Process. (IJIGSP) **8**(8), 58–67 (2016). https://doi.org/10.5815/ijigsp.2016.08.08
20. Ziabari, M.T., Moarefianpur, A., Morvarid, M.: Fuzzy stability and synchronization of new 3D chaotic systems. Int. J. Inf. Eng. Electron. Bus. (IJIEEB) **6**(5), 53–62 (2014). https://doi.org/10.5815/ijieeb.2014.05.08
21. Tyagi, T., Dubey, H.M., Pandit, M.: Multi-objective optimal dispatch solution of solar-wind-thermal system using improved stochastic fractal search algorithm. Int. J. Inf. Technol. Comput. Sci. (IJITCS) **8**(11), 61–73 (2016). https://doi.org/10.5815/ijitcs.2016.11.08
22. Nakamori, S.: New RLS wiener smoother for colored observation noise in linear discrete-time stochastic systems. Int. J. Inf. Technol. Comput. Sci. (IJITCS) **6**(1), 13–24 (2014). https://doi.org/10.5815/ijitcs.2014.01.02
23. Akimov, N.N., Buhnin, A.V., Milov, V.R., Koltsov, V.A.: Kuranov AAA method of verification models based technical systems mining processes. Inf. Measur. Oper. Syst. **11**, 19–25 (2015)
24. Grassberger, P., Procaccia, I.: Characterization of strange attractors. Phys. Rev. Lett. **50**, 346–349 (1983)
25. Goshvarpour, A., Goshvarpour, A.: Classification of electroencephalographic changes in meditation and rest: using correlation dimension and wavelet coefficients. Int. J. Inf. Technol. Comput. Sci. (IJITCS) **4**(3), 24–30 (2012)
26. Mekler, A.: Calculation of EEG correlation dimension: Large massifs of experimental data. Comput. Methods Programs Biomed. **92**, 154–160 (2008)
27. Tsonis, A.A., Elsner, J.B.: The weather attractor over short timescales. Nature **333**, 545–547 (1988)
28. Kennel, M.B., Brown, R., Abarbanel, H.D.I.: Determining embedding dimension for phase-space reconstruction using a geometrical construction. Phys. Rev. A **45**, 3403 (1992)
29. Hegger, R., Kantz, H., Schreiber, T.: Practical implementation of nonlinear time series methods: the TISEAN package. CHAOS **9**, 413 (1999)

Synchronization of Neural Ensembles in the Formation of Attention in the Brain

M. Mazurov[✉]

Russian Economic University G.V. Plekhanova, Moscow, Russia
mazurov37@mail.ru

Abstract. The method of studying the synchronization of relaxation self-oscillations, based on a modified axiomatic method and using the properties of uniform almost-periodic functions is used. A computational algorithm is used to study the synchronization of relaxation self-oscillations, using axiomatic algebraic models and properties of the theory of uniform almost periodic functions. It is shown that synchronization is a flexible and efficient process for shaping the attention of other cognitive processes to certain external informational influences. The five synchronization modes of neural ensembles of 100 peripheral neurons were investigated: asynchronous mode, full synchronization, partial synchronization, "incorrect" synchronization mode, transient phase-dynamic process. The complex synchronization regimes of relaxation self-oscillations are considered: "incorrect" synchronization, the presence of specific and "phase-dynamic" transient processes caused by the properties of uniform almost-periodic functions. Discussed the adequacy of the used mathematical computer model for the formation of attention.

Keywords: Synchronization · Relaxation self-oscillations · Axiomatic theory · Uniform almost periodic functions · Kronecker inequalities · Computer algorithms

1 Introduction

The task of forming attention when synchronizing a neural ensemble from a central oscillator is considered. This task is of great practical importance for understanding the mechanism of cognitive processes [1–9, 16, 26–28, 31]. The architecture of the model for studying attention formation, including interneuron connections, is presented in Fig. 1. The elements of the first layer are peripheral neurons (PN). They are feature detectors in the primary areas of the new cortex, activated by external stimuli. It is assumed that the external input to the PN is large enough to put the neuron in a pacemaker mode with a certain pulse generation frequency. In the case of visual stimuli, the input signals are formed by an image located on a flat grid, PN s are located on the grid with the same dimensions, with each PN receiving a signal from a pixel having the same coordinates on the grid as the PN. Generally speaking, PN must be connected by exciting lateral connections, but to simplify the model in this work, these connections are eliminated so that all interaction between PN goes through the central control element CN2.

Z. Hu et al. (Eds.): AIMEE 2019, AISC 1126, pp. 241–252, 2020.
https://doi.org/10.1007/978-3-030-39162-1_22

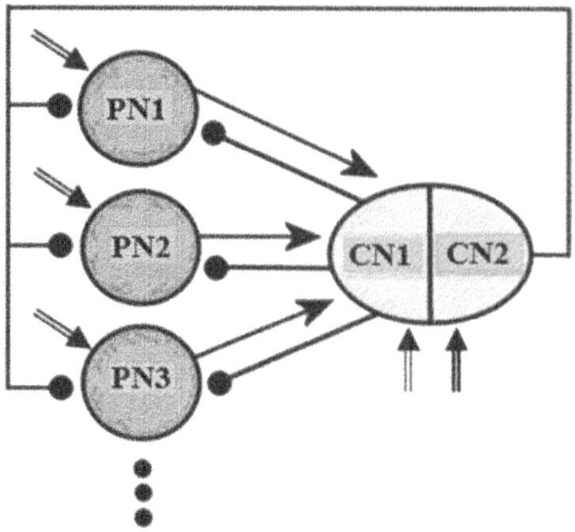

Fig. 1. The architecture of connections in the attentional formation model with the central element and peripheral neurons PN, encoding the signs of external stimuli CN1 and CN2 - central neurons that control the focus of attention. Exciting connections are shown by black arrows, lines with black circles at the ends show brake connections, double gray arrows show external input signals.

The central control element is extremely simplistic and consists of two neurons, CN1 and CN2. CN1 is used to synchronize a certain ensemble of PN. A visual object represented by a PN ensemble working in phase with CN1. CN2 controls the sequential synchronization of CN1 with different PN ensembles, which is interpreted as a sequential selection of different objects. Both neurons, CN1 and CN2, send brake signals to all monitors. Communications from PN come only on CN1 and are exciting. The power of synaptic transmission is assumed to be fixed and the same for all such connections. There are no links between CN1 and CN2. Thus, CN2 does not receive any signals from other neurons of the system. Only a constant external signal arrives at the input of CN2, which converts the CN2 into an oscillatory mode.

While the PN braking force from the TsN1 side is fixed, the brake signals arriving at the PN from the CN2 change with time. The latter is due to the short-term synaptic plasticity of bonds from CN2 to PN. If both neurons CN2 and PN are simultaneously active for a sufficiently long time, the force of the inhibitory synaptic connection between them increases. If at least one of these neurons has low level of activity, the strength of the synaptic connection decreases [27, 34]. This method of adapting the brake coupling from CN2 to PN is used to simulate an experimental fact, according to which, when successively viewing objects, the probability of re-selecting an object that has already been viewed falls [32, 33]. To describe the neurons of the system, the Hodgkin-Huxley model is usually used [29].

2 Computational Methods for Studying Synchronization of Self-oscillations During the Formation of Attention, Using Axiomatic Algebraic Models and Methods of the Theory of Uniform Almost-Periodic Functions

The axiomatic model of a self-oscillatory relaxation system is defined as two different functions on two consecutive temporal regions [18–22, 35].

$$f(t) = \begin{cases} f_1(t) & 0 \leq t \leq t_1 \\ f_2(t) & t_1 < t \leq T \end{cases}$$

$$f_{\partial}(t) = \begin{cases} \infty & \\ f_3(t) & \end{cases}$$

where $f(t)$ is the function characterizing the form of relaxation self-oscillation; $f_1(t)$, $f_2(t)$ - functions characterizing relaxation self-oscillation in the range of "fast" and "slow" changes in relaxation self-oscillation; $f_{\partial}(t)$- dynamic excitation threshold of the self-oscillatory system; $f_3(t)$- a function characterizing the dynamic excitation threshold in the interval of the "slow" phase.

A qualitative approximation of the functions $f(t)$ and $f_{\partial}(t)$- potential and dynamic excitation threshold for a neuron is shown in Fig. 2.

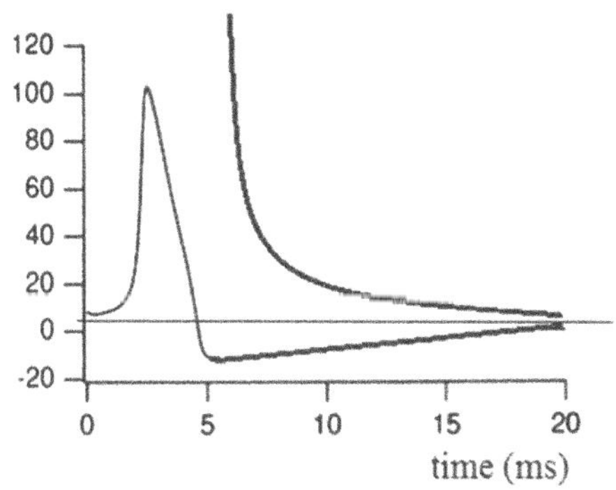

Fig. 2. The shape of the potential and the dynamic threshold in the modified axiomatic algebraic model of a neuron [18, 22].

Using this algorithm, a number of problems of synchronization of electrical processes in the heart, which are relaxation self-oscillations, were investigated [18–22]. Consider the numerical computer implementation of the axiomatic model. The algebraic approximation of the "fast" and "slow" phases of the relaxation oscillation, the dynamic threshold, and the external approximating signal in the form of short pulses are used. From an applied point of view, such an approximation of the synchronizing signal is quite adequate, since in real applications (for example, when synchronizing neural ensembles), the synchronization of relaxation self-oscillations is also produced by a relaxation oscillation in the form of short nerve impulses.

A computational algorithm for the study of synchronization in the algebraic realization of the form of relaxation self-oscillations and the form of a dynamic threshold can be represented in the form of the formulas mentioned above. In the numerical implementation, the allocation of almost-periods of a uniform almost-periodic function obtained as a result of synchronization is performed by extracting the remainder R from dividing the current process time by the synchronizing oscillation period using the built-in Matlab 7 procedure

$$R = \mathrm{mod}\frac{t}{T_c} = \mathrm{mod}\frac{n\Delta t}{T_c},$$

where: t is the current time, Δt is the time step, n is the number of steps. The remainder R is equal to the argument of the algebraic approximation of two functions of the "fast" and "slow" phases of the relaxation oscillation and the dynamic excitation threshold corresponding to the "slow" phase of motion. Next, we use a condition that describes the extraordinary excitation of a relaxation self-oscillation when the dynamic threshold is exceeded by an instantaneous amount equal to the value of the relaxation self-oscillation and the amplitude value of the external clock signal, that is, the condition for extraordinary excitation is

$$\mathrm{mod}\frac{n\Delta t}{T_c} + U_c \geq f_\partial(\mathrm{mod}\frac{n\Delta t}{T_c}).$$

Part of the program code that implements the last algorithm is given below.

```
j=0;
y1=0.1;
y2=-0.4;
 for kf=1:1:numPoints
koeff(kf)=k1+0.0002*(kf-1);
end
 T_zapazd = zeros(numPoints,1);
T_zapazd(:,:) = 0;
 curT = zeros(numPoints);
for num=1:1:numPoints
    for j=1:1:N
       curT(num) = curT(num) + 1;
       if curT(num) < 40
          yrez(j,num)=yi(curT(num));
       else
          yrez(j,num) = y2+koeff(num)*curT(num);
          if mod(j,period) == 0 && (yrez(j,num) + amp) >= 0.1
             curT(num) = 0;
          end
          if yrez(j,num) >= 0.1
             curT(num) = 0;
          end
       end
    end
end
figure;
 subplot(11,1,1);
 hLog=plot([1:1:N],yrez(:,1),'k'); set(hLog,'LineWidth',5);
 subplot(11,1,2);
 hLog=plot([1:1:N],yrez(:,2),'k'); set(hLog,'LineWidth',5);
 subplot(11,1,3);
 hLog=plot([1:1:N],yrez(:,3),'k'); set(hLog,'LineWidth',5);
subplot(11,1,4);
 hLog=plot([1:1:N],yrez(:,4),'k'); set(hLog,'LineWidth',5);
 subplot(11,1,5);
 hLog=plot([1:1:N],yrez(:,5),'k'); set(hLog,'LineWidth',5);
 subplot(11,1,6);
 hLog=plot([1:1:N],yrez(:,6),'k'); set(hLog,'LineWidth',5);
subplot(11,1,7);
 hLog=plot([1:1:N],yrez(:,7),'k'); set(hLog,'LineWidth',5);
 subplot(11,1,8);
 hLog=plot([1:1:N],yrez(:,8),'k'); set(hLog,'LineWidth',5);
 subplot(11,1,9);
 hLog=plot([1:1:N],yrez(:,9),'k'); set(hLog,'LineWidth',5);
subplot(11,1,10);
 hLog=plot([1:1:N],yrez(:,10),'k');set(hLog,'LineWidth',5);
 subplot(11,1,11);
 hLog=plot([1:1:N],ysinx(:,1),'k'); set(hLog,'LineWidth',5);
```

The use of axiomatic algebraic models with approximation of the "fast" and "slow" phases of relaxation self-oscillation and dynamic threshold by algebraic functions is effective in studying the synchronization of neural ensembles [14–18]. In the future, to study the synchronization modes of neural ensembles in the formation of attention, we will use the above software.

3 Study of Synchronization Modes, Possible When Forming Attention

The mathematical model of the synchronization of an ensemble of neurons in the formation of attention is represented by an ensemble of peripheral neurons that generate pulses simultaneously with the CN1. Thus, the study of synchronization modes in terms of the model is reduced to the study of modes that may occur depending on the selected parameters. Simulation modeling allowed us to identify the following five types of model dynamics: (1) asynchronous mode; (2) full synchronization; (3) partial synchronization; (4) the mode of "incorrect" synchronization; (5) transitional phase-dynamic process. The following examples illustrate the conditions under which these types of dynamics arise. The CN2 neuron does not play a significant role in shaping the dynamics of the model, it is only important for switching attention from one object to another. In this regard, in this section, this neuron is excluded from the model. Numerical calculations were carried out mainly for the model containing CN1 and 200 PI, which were divided into two groups of 100 PN in each group. The neurons in the first group received a larger input signal and had a higher eigenfrequency of pulse generation. For convenience of illustration of the results, we limit ourselves to the case of a model with 10 PN. Qualitatively, the behavior of a model with such a reduced number of PN is no different from a model containing 200 PN.

In Fig. 3 shows the dynamics of the membrane potential of peripheral neurons in the case when all connections are zeroed. Such dynamics of neural activity, in which neurons operate independently of each other, is called asynchronous mode. Note that in the asynchronous mode, the CN1 does not generate pulses, since it receives subthreshold external excitation. Asynchronous mode, that is, the lack of synchronization, calculated using the program implemented in Matlab 7, is shown in Fig. 3

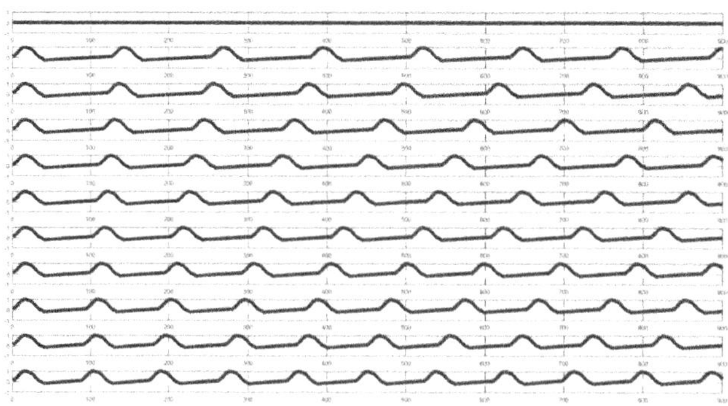

Fig. 3. Asynchronous mode, that is, lack of synchronization in the absence of communication between neurons.

In Fig. 4 shows the model dynamics corresponding to the full synchronization mode, when all peripheral neurons have the same frequency, equal to the frequency of CN1. From the content point of view, this means setting the process of attention to one specific input object.

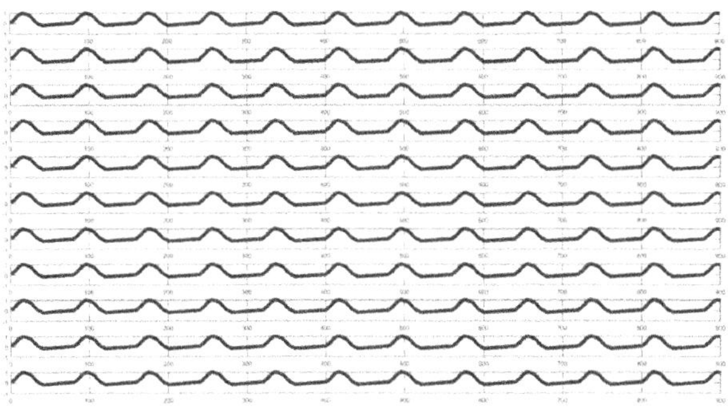

Fig. 4. The dynamics of the model corresponding to the mode of complete synchronization of peripheral neurons. The graph above shows the rhythm of the synchronizing neuron.

In this mode, all PNs generate pulses simultaneously with CN1, and the ratio of the number of pulses of CN1 and PN is 1 : 1. The generation of CN1 pulses is due to sufficiently high excitation signals.

The partial synchronization mode occupies, as it were, an intermediate position between the asynchronous mode and the full synchronization mode. In this mode, the neurons of one group trigger in phase with CN1, while the neurons of another group of pulses have different generation frequencies. The mode of partial synchronization of neurons of group A is presented in Fig. 5.

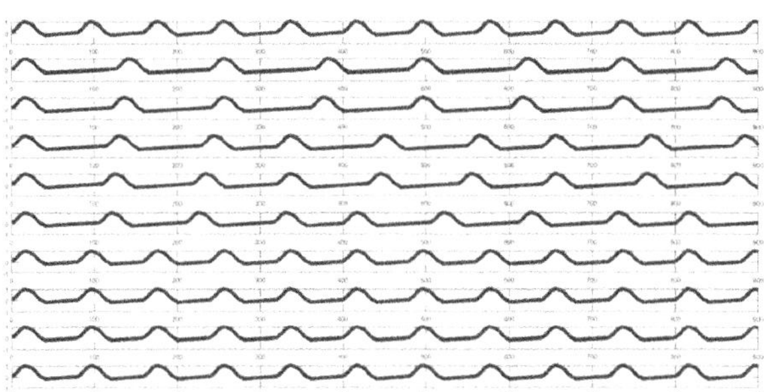

Fig. 5. The mode of partial synchronization of peripheral neurons. Peripheral neurons with numbers from 7 to 10 (the last four neurons) are synchronized with CN1, the generation frequency of which is illustrated by the first graph of the figure. Peripheral neurons 2 through 6 are not synchronized.

In the structure of possible types of synchronization, a significant place is occupied by synchronization modes other than the 1/1 type synchronization mode, that is, the synchronization modes of the type (m/n ($n \neq 1$). These modes are called "wrong" synchronization modes. An example of such modes of various types is shown in Fig. 6.

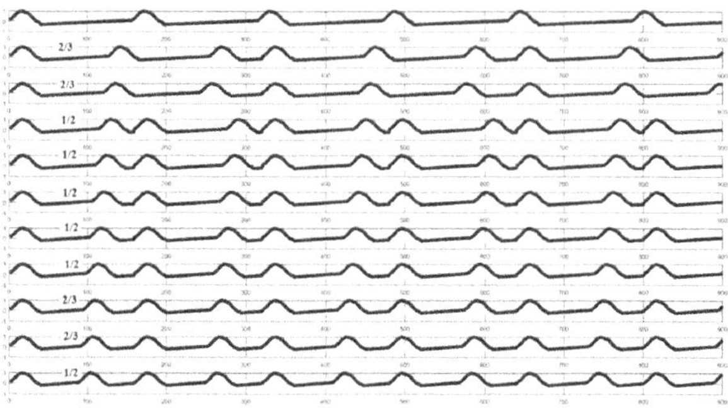

Fig. 6. Examples of incorrect synchronization modes. The type of synchronization mode is indicated at the beginning of the mode graphics. The top chart is the sync signal.

Let us consider in more detail transients occurring from the beginning of synchronization to the establishment of a stationary one. Experimental studies show that with variation of parameters, the transition from one mode to another does not occur instantaneously, but passes through the so-called transition state. Let us consider in more detail transient processes during synchronization of relaxation self-oscillations.

4 Transient Processes During Synchronization of Relaxation Self-oscillations - Phase-Dynamic Transient Processes

Stationary synchronization is preceded by transients. It is established that these transients can be divided into two types. These are, firstly, systemic transient processes and, secondly, some specific transient processes, which will be called phase-dynamic ones.

In linear systems, transients are determined by the dynamic properties of the system. Consider the differential equation of the non-linear van der Pol oscillator, which has the form [14, 15, 17, 23, 24]

$$y'' + \varepsilon(1 - y^2)y' + y = 0.$$

The transients in the van der Pol oscillator with different values of the parameter are illustrated in Fig. 7, a–d.

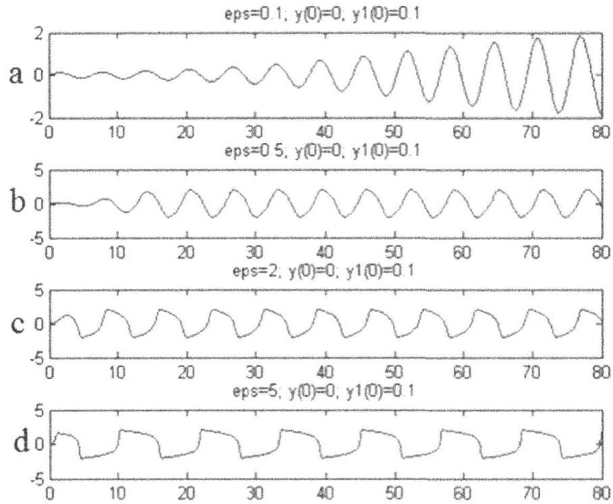

Fig. 7. System transients in the establishment of stationary oscillations in the Van der Pol autogenerator for the following parameter values ε : (a) $\varepsilon = 0.1$; (b) $\varepsilon = 0.5$; (c) $\varepsilon = 2$; (d) $\varepsilon = 5$. Graphs of system transients were obtained in [18].

Consider transient processes in synchronization of a relaxation oscillator by a sequence of short pulses. Let be T_0 the initial phase shift of the second synchronization pulse relative to the relaxator pulse. If the second and subsequent synchronization pulses do not cause an extraordinary excitation of an oscillator, then a transient process takes place that lasts until the first extraordinary excitation of the relaxator. This specific phase-dynamic transient is not related to the system properties of the relaxer itself, but is due to the properties of the synchronization signals and the width of the oscillator's sensitivity zone to the synchronization signal. The properties of phase-dynamic transient processes are closely related to the properties of almost periodic functions, their properties directly follow from the properties of almost periodic functions. The sequence of synchronizing functions and the sequence of sensitive intervals of the relaxator can be considered as two periodic functions with incommensurable periods; thus, their sum is a uniform almost-periodic function. From here it is possible to immediately establish the main property of phase-dynamic transients—their maximum duration is equal to the duration ε- almost the period of the function, which is the sum of two periodic functions—the synchronizing effect and the sequence of sensitive zones.

At the initial moment of time, the phases of the sequences of pulses of the pacemaker of the relaxation self-oscillating system and the synchronizing signal are arbitrary; without loss of generality, they can be considered equal T_0. The condition of the first synchronization by short pulses can be found as the minimal solution by the Chebyshev inequality [18–22, 25]

$$-\varepsilon < (T_1 n_1 - T_2 n_2 + T_0) \leq 0.$$

An experimental study of phase-dynamic transient processes using computer calculations was carried out. An axiomatic model of a neuron was used as a relaxation self-oscillatory system. An example of one of these calculations is shown in Fig. 8.

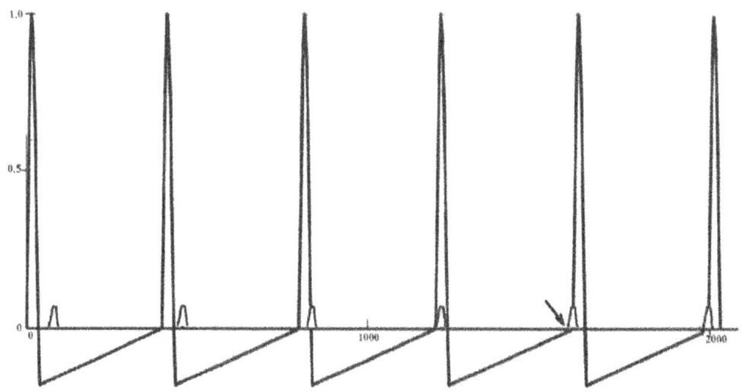

Fig. 8. Phase-dynamic transients of the artificial neuron. This is a phase-dynamic transient process lasts for the first four pulses of relaxation oscillations. The beginning of the stationary synchronization is shown in the figure by the arrow. The period of the self-oscillatory activity of the neuron is T = 400 (arbitrary units), the synchronization period = 380.

In the above computational experiment, stable synchronization occurred after 4 external clock pulses. In the general case, synchronization occurs after a different number of synchronizing pulses.

5 Conclusions

The study of the synchronization of peripheral neurons from the central neuron during the formation of attention allowed us to identify the following five types of model dynamics: (1) asynchronous mode; (2) full synchronization; (3) partial synchronization; (4) the mode of "incorrect" synchronization; (5) transitional phase-dynamic process.

It should be noted that the parameter area corresponding to the transition state is rather wide. Thus, the transition from full synchronization to partial synchronization or from partial synchronization with one group to partial synchronization with another group takes place within a certain time, which affects the speed of processes in the central nervous system. The synchronization bandwidth during the implementation of the attention process increases with an increase in the number of connections between neurons and the amplitude of the synchronizing effect. From the point of view of neurobiology, this fact is important because a higher visibility of the stimulus usually leads to a higher activity of the neurons representing this stimulus in the cerebral cortex, and to a greater likelihood of choosing this stimulus in FV [1–9, 16, 26–28, 31]. This model better fits this data.

References

1. Abarbanel, G.D.I., Rabinovich, M.I., Selverston, A.: Synchronization in neural ensembles. Successes phys. sci. **166**(4), 363–390 (1996)
2. Borisyuk, G.N., Borisyuk, R.M., Kazanovich, Ya.B., Luzyanina, T.B., Turova, T.S., Tsymbalyuk, G.S.: Oscillatory neural networks. Math. Results Appl. Mat. Simul. **4**(1), 3–43 (1992)
3. Borisyuk, G.N., Borisyuk, R.M., Kazanovich, Ya.B., Ivanitsky, G.R.: Modeling the dynamics of neural activity and information processing in the brain - the results of the "decade". Successes Phys. Sci. **172**(10), 1189–1214 (2002)
4. Borisyuk, G.N., Borisyuk, R.M., Kazanovich, Ya.B., Ivanitsky, G.R.: UFN, 172: 10 (2002), 1189–11214; Phys. Usp., 45:10, 1073-1095 (2002)
5. Vinogradova, O.S.: Hippocampus and memory. M: Science (1975)
6. Ivanitsky, G.R., Medvinsky, A.B., Tsyganov, M.A.: From the dynamics of population autowaves formed by living cells to neuroinformatics. Adv. Phys. Sci. **164**(10), 1041–1072 (1994)
7. Kazanovich, Ya.B., Borisyuk, R.M.: Synchronization in the neural network of phase oscillators with the central element. Math. Model. **6**(8), 45–60 (1994)
8. Kazanovich, Ya.B.: The theory of temporal correlation and segmentation models of visual information in the brain. Math. Biol. Bioinform. **5**(1), 43–97 (2010)
9. Kazanovich, Ya.B.: Oscillatory neural network models of cognitive functions of the brain. Thesis for the degree of Dr.Sc. Pushchino, p. 463 (2018)
10. Saleh, B.J., Qasim al-Aqbi, A.T., Falih Saedi, A.Y., Salman, L.: Comparative study of inspired algorithms for trajectory-following control in mobile robot. Int. J. Mod. Educ. Comput. Sci. (IJMECS) **10**(9), 1–10 (2018)
11. Moshref, M., Al-Sayyad, R.: Developing ontology approach using software tool to improve data visualization (case study: computer network). Int. J. Mod. Educ. Comput. Sci. (IJMECS) **11**(4), 32–39 (2019). https://doi.org/10.5815/ijmecs.2019.04.04
12. Cheah, C.S., Leong, L.: Investigating the redundancy effect in the learning of C++ computer programming using screencasting. Int. J. Mod. Educ. Comput. Sci. (IJMECS) **11**(6), 19–25 (2019). https://doi.org/10.5815/ijmecs.2019.06.03
13. Alvarez-Dionisi, L.E., Balza, M.M.R.: Teaching artificial intelligence and robotics to undergraduate systems engineering students. Int. J. Mod. Educ. Comput. Sci. (IJMECS) **11** (7), 54–63 (2019). https://doi.org/10.5815/ijmecs.2019.07.06
14. Kuznetsov, A.P., Roman, Yu.P., Stankevich, N.V., Tyuryukina, L.V.: Pulse synchronization and synchronization in coupled systems: new aspects of the classical problem. Izv. Universities "PND", vol. 16, № 3, pp. 88–111 (2008)
15. Kuznetsov, A.P., Seliverstova, E.S., Trubetskov, D.I., Tyuryukin, L.V.: The phenomenon of the van der Pol equation Izv. universities. "PND", vol. 22, no. 4 (2014)
16. Kryukov, V.I.: The role of the hippocampus in long-term memory: a system-dynamic approach. J. High. Nerv. Act. **57**(3), 261–278 (2007)
17. Nekorkin, V.I.: Nonlinear oscillations and waves in neurodynamics. Phys.-Uspekhi **178**(3), 313–323 (2008). UFN
18. Mazurov, M.E.: Nonlinear synchronization and rhythmogenesis in electrically excitable heart systems. Thesis for the degree of Dr.Sc. Pushchino, p. 273 (2007)
19. Mazurov, M.E.: A geometric method for studying the solution of diophantine inequalities in problems of synchronizing relaxation oscillations. IVUZ. sir Maths. №7. C.13–20 (1989)
20. Mazurov, M.E.: Solution of diophantine inequalities in problems of the synchronization of relaxation oscillations. ZhVM MF AS USSR **31**(11), 1619–1636 (1991)

21. Mazurov, M.E.: New periodic solutions of non-autonomous systems of nonlinear differential equations with a small parameter to derivatives and their stability. Dokl. Akad. Nauk SSSR **442**(1), 33 (2012)
22. Mazurov, M.E.: Synchronization of relaxation self-oscillatory systems, synchronization in neural networks. Izv. RAN Phys. Ser. **82**(1), 83–87 (2018)
23. Mishchenko, U.F., Rozov, N.Kh.: Differential equations with a small parameter and relaxation oscillations, p. 247 (1975). M.: Nauka
24. Pikovsky, A., Rosenblyum, M., Kurts, U.: Synchronization, a fundamental non-linear phenomenon, p. 508 (2003). M.: Technosphere
25. Khinchin, A.Ya.: Continued fractions, GIFML (1978). M.: "Science"
26. Chik, D., Borisyuk, R.M., Kazanovich, Ya.B.: Synchronization modes in the Hodgkin - Huxley neuron network with a central element. Math. Biol. Bioinform. **3**(1), 16–35 (2008)
27. Fitzpatrick, J.S., Akopian, G., Walsh, J.P.: Short-term plasticity at inhibitory synapses in rat striatum output. J. Neurophysiol. **85**, 2088–2099 (2001)
28. Fries, P., Schroeder, J.-H., Roelfsema, P.R., Singer, W., Engel, A.K.: Correlation of stimulus selection. J. Neurosci. **22**, 3739–3754 (2002)
29. Hodgkin, A.L., Huxley, A.F.: A quantitative description of membrane current and its application to conduction and excitation in nerve. J. Physiol. **117**, 500–544 (1952)
30. Fries, P., Schroeder, J.-H., Roelfsema, P.R., Singer, W., Engel, A.K.: Correlation of stimulus selection. J. Neurosci. **22**, 3739–3754 (2002)
31. Gray, C.M., Konig, P., Engel, A.K., Singer, W.: Oscillatory responses in cat visual cortex exhibit intercolumnar synchronization. Nature **388**, 334–337 (1989)
32. Klein, R.: Inhibitory tagging system facilitates visual search. Nature **334**, 430–431 (1988)
33. Takeda, Y., Yagi, A.: Inhibitory tagging stimuli remain visible. Percept. Psychophys. **62**, 927–934 (2000)
34. Zucker, R.S., Regehr, W.G.: Short-term synaptic plasticity. Ann. Rev. Physiol. **64**, 355–405 (2002)
35. Wiener, N., Rosenblueth, A.: The Arch. Inst. Cardiologia de Mexico. **205**(16), 3–4 (1946)

The Electrical Model of Multicellular Systems Based on Circuit Simulation Techniques

R. R. Aliev[1,2], M. M. Gourary[3(✉)], and S. G. Rusakov[3]

[1] Institute of Theoretical and Experimental Biophysics of Russian Academy of Sciences, Moscow, Russia
[2] Moscow Institute of Physics and Technology, Dolgoprudny, Russia
[3] Institute for Design Problems in Microelectronics of Russian Academy of Sciences, (IPPM RAS), Moscow, Russia
gourary@yandex.ru

Abstract. The paper presents the model for the evaluation of the electric field in the cellular tissue. The model in the form of an electrical network is obtained by modifying the known transport lattice model. The network parameters are defined by applying the finite volume method to the Maxwell equations for an electric field in a homogeneous medium. It is shown that the main limitation of the known form of transport lattice method is the stepped approximation of the cell surface. To eliminate the error a new approach based on projecting the steps on the membrane surface is developed. Numerical experiments by the circuit simulator confirmed the error decrease due to the proposed approach.

Keywords: Cellular tissue · Cell membrane · Finite Volume Method · Transfer function · Circuit simulation · Electric network

1 Introduction

The study of the electrical activity of tissues is one of the most important problems in biological physics. The most of the theoretical studies of electrically excitable cells repeated the Hodgkin-Huxley formalism after creation of Hodgkin and Huxley (H-H) model for the propagation of the nerve impulse [1]. Extensions and modifications of the model consisted in taking into account additional membrane currents found in electrically excitable cells, such as cardiomyocytes and neurons [2]. However the obvious limitation of the H-H model when considering individual cells is the assumption of equipotentiality of the cytoplasm of the cell, for example cardiomyocytes. This assumption is sufficiently substantiated when considering the low frequency processes in normal physiological conditions. However the equipotentiality of the cytoplasm of a cell is disturbed in the presence of high-frequency oscillations or in the presence of rapidly changing fields, for example in the discharge of a pacemaker. In this case the study of the development of tissue electroporation as the process providing and accompanying successful myocardial defibrillation is of great importance.

Other important applications of the analysis of the tissue electromagnetic fields are MRI (magnetic resonance imaging) investigations [3, 4].

Z. Hu et al. (Eds.): AIMEE 2019, AISC 1126, pp. 253–262, 2020.
https://doi.org/10.1007/978-3-030-39162-1_23

The paper discusses an approach to obtain the distribution of potential within a cell providing a link to traditional models such as the H-H model describing membrane currents. It is proposed to evaluate charge transfer in cells by applying circuit simulation tools [5, 6] to a tissue model approximated by an electrical network. The motivation for this approach is a high degree of automation of the analysis of electrical characteristics and a wide range of modeling modes in the time and frequency domains, which are currently provided by circuit simulators.

The presentation of the cellular transport mechanisms by electrical equivalents was proposed in [7] and the technique was further developed in [8–13]. The model approximates the continuous electric field in the tissue by discrete nodal potentials of the network defined by the Kirchhoff laws. However, the model is heuristic since it is not explicitly derived from the field equations. Another drawback of the proposed model is the poor approximation of the cell surface leading to the simulation inaccuracy.

This paper aims to establish the relationship of the initial tissue field equations with the electrical equivalents of the model and providing the correct representation of the cell surface in the model. The goal of the research is the further development of algorithms that increase the productivity and accuracy of modeling bioelectric processes in multicellular structures of arbitrary shape by methods of circuit simulation.

The rest of the paper is organized as follows. Section 2 describes the transport lattice model and presents its shortcomings. Section 3 contains the model derivation by applying the finite volume method to the Maxwell equations in the electrolyte medium. Section 4 presents a new approach to membrane representation that essentially reduces the simulation error. Section 5 shows results of testing experiments.

2 The “Transport Lattice” Model of Cellular Tissue

The cellular tissue is a set of cells that are in a intercellular environment of an external electrolyte. Each cell consists of a membrane and the cytoplasm (internal electrolyte) contained in it. The membrane thickness is much smaller than the cell size.

To analyze the effect of an electric field on tissue the “transport lattice” model was proposed in [7] and further developed in [6–12]. The model presents cellular tissue by an electrical circuit of two-terminal sections connecting adjacent nodes of a rectangular grid (Fig. 1a). The structure of the sections (Fig. 1b–d) can be described as follows.

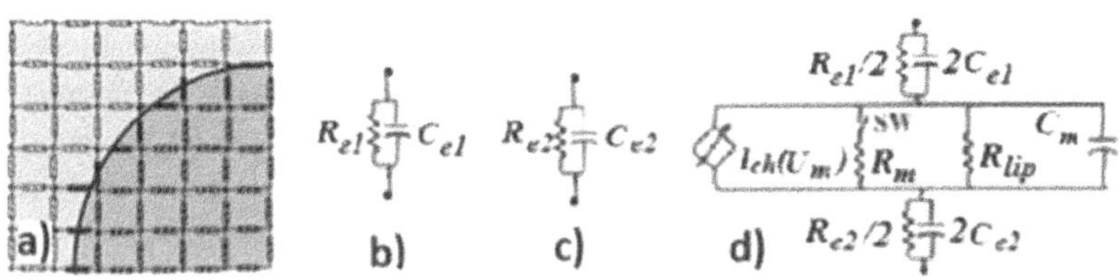

Fig. 1. The electrical model of the field in the cellular tissue [7]: (a) view of the circuit network near the membrane; (b) network between the external nodes of the cell; (c) network between the internal nodes of the cell; (d) network between the nodes on opposite sides of the membrane.

If both nodes are on the same side of the membrane belong to the same electrolyte, then the two-terminal network is represented by the RC section (Fig. 1b, c). The resistance and capacitance values of the section are determined by the corresponding values of the cube with side l equal to the grid step:

$$R_e = \rho_e/l, \;\; C_e = \kappa_e \varepsilon_0 l. \tag{1}$$

Here ρ_e, κ_e- are the electrolyte resistivity and its relative permittivity correspondingly, ε_0- the absolute dielectric constant of the vacuum.

If the terminals are on opposite sides of the membrane of thickness d then the network between them is a series connection (Fig. 1d) of three RC sections. The series connection contains components $R_e/2$, $2C_e$ representing the semicubes $l \times l \times l/2$ of external and internal electrolytes and the central component defining a lipid membrane of thickness d and area l^2 and voltage U_m.

The central component in Fig. 1d includes the parallel connection of subnetworks:

- RC chain to model the membrane body

$$R_{lip} = \rho_{lip} d/l^2, \;\; C_m = \kappa_m \varepsilon_0 l^2/d, \tag{2}$$

- nonlinear model of ion channels with a current-voltage characteristic

$$I_{ch}(U_m) = U_m\left(\frac{g_m^+ + g_m^-}{2}\right) + w\left(\frac{g_m^+ - g_m^-}{2}\right)\ln\cosh\left(\frac{U_m - U_{m0}}{w}\right) + I_c, \tag{3}$$

- electroporation model in the form of a series connection of the resistance and a switch. The switch closure occurs at a voltage $U_m = 0.5$ V on the membrane

$$R_m = 10^3 \rho_{e1} d/l^2 \text{ at } U_m \geq 0.5\text{V}. \tag{4}$$

The parameters κ_m, ρ_{lip} in the expressions (2–4) are the relative permittivity and the membrane resistivity correspondingly, g_m^+, g_m^- are the asymptotic conductivities with forward and backward bias of the ion channel, U_{m0} is the resting potential of the cell, w = 22 mV, I_c = 36 pA - parameters of the approximation.

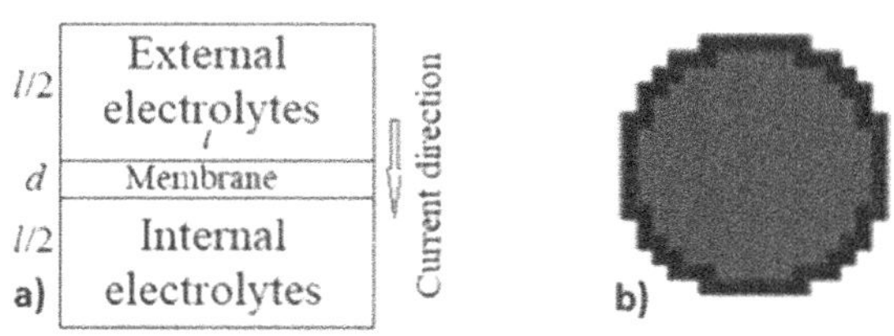

Fig. 2. Elementary cube corresponding to the electrical branch between the nodes located on different sides of the membrane (a), stepped approximation of the curved surface of the cell.

Note that the presented section Fig. 1d corresponds to a cube with side l, which is divided into two equal parts by a membrane of thickness d. The membrane is perpendicular to the current direction (see Fig. 2a). This representation is explained by the stepped approximation of the curvilinear cell surface (Fig. 2b).

Thus, the disadvantages of the heuristic model in Fig. 1 are the lack of its substantiation by the equations of the tissue electric field, as well as the stepped approximation of the membrane, leading to simulation errors.

3 Substantiation of the Representation of the Model

The model based on the Admittance Method is often used for the analysis of biological structures [14–16]. The method represents the electrical network by the division of the tissue into parallelepipeds and connecting each pair of adjacent vertexes by the RC chain (Fig. 3a).

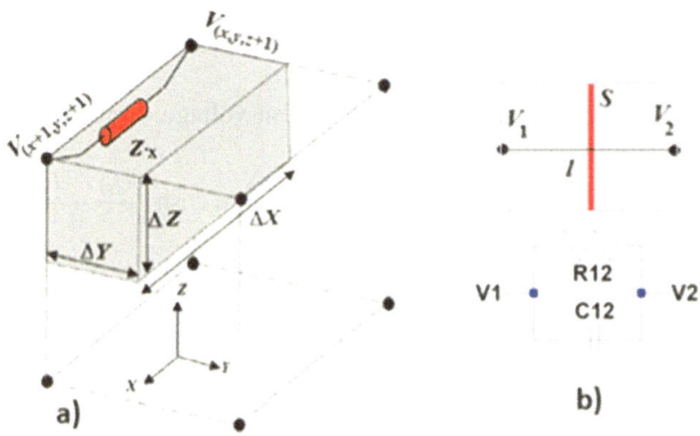

Fig. 3. (a) Equivalent circuit for the representation of a unit cube in a homogeneous tissue site by the Admittance Method; (b) equivalent RC circuit for representing the current through a face of a cube in the finite volume method.

The expressions for the resistance and capacitance of a parallelepiped (with dimensions ΔX, ΔY, ΔZ) along the coordinate axes have the form (e.g. for the x-axis):

$$R_x = \rho_e \frac{\Delta X}{\Delta Y \cdot \Delta Z}, \; C_x = \kappa_e \varepsilon_0 \frac{\Delta Y \cdot \Delta Z}{\Delta X}.$$

These expressions for the unit cube coincide with (1) for the models in Fig. 1b, c. This indicates that the expression (1) corresponds to the Admittance Method. However, the tissue site containing both sides of the membrane (Fig. 1d) generated in accordance with the Admittance Method, should contain RC chains parallel to the membrane. Since there are no such chains in the model in Fig. 1, the model cannot be regarded as obtained by the Admittance Method.

To derive a tissue model in the form of electrical equivalents, we consider the Maxwell equations for the electric field in a homogeneous medium

$$\text{A} : \nabla j = 0, \text{B} : j = \sigma E + \varepsilon \frac{\partial E}{\partial t}, \text{C} : E = \nabla V. \quad (5)$$

Here (5A) represents the charge conservation law (the zero current divergence), (5B) defines the total current density as the sum of conduction and displacement currents, (5C) determines the field strength (E) as the potential (V) gradient.

One can apply the finite volume method (FVM) [17] to the approximate transform of (5) into a system of differential-algebraic equations. FVM splits the spatial domain into a set of elementary volumes (cubicles) and performs the following operations:

- integration of (5A) by the volume of the cubicle to express the result as a surface integral of the current density due to Gauss theorem;
- representation of the current through the cubicle face center as a product of the current density in the center of the face by the face area;
- obtaining the field strength E in the cubicle face center using a finite-difference replacement of (5C) and the potentials in the centers of neighboring cubicles;
- estimation of the current density in the cubicle face center by (5B).

The application of these principles to evaluate the current I_S through the cubicle face area S (Fig. 3b) leads to the expressions

$$E_S = \frac{V_2 - V_1}{l}, \; I_S = S \cdot j_S = \frac{S \cdot \sigma}{l}(V_2 - V_1) + \frac{S \cdot \varepsilon d(V_2 - V_1)}{l \quad dt}. \quad (6)$$

Here V_1, V_2 are the potentials of neighboring nodes, l is the distance between nodes, E_S is the field strength at the center of the face, j_S is the current density through the face. From (6) we can obtain an expression for the current I_S in the form

$$I_S = G_{12}(V_2 - V_1) + C_{12}\frac{d(V_2 - V_1)}{dt}, \quad (7)$$

where

$$G_{12} = 1/R_{12} = S \cdot \sigma/l, \; C_{12} = S \cdot \varepsilon/l. \quad (8)$$

We have $S = l^2$ for the RC chain (7) representing the cubicles between the model nodes inside the electrolyte (Fig. 1b, c). In this case (8) coincides with (1) for the parameters of two-terminal chains

To analyze a cube divided by a membrane (Fig. 2a) it is necessary to consider each half of it as two cubes. One cube with dimensions $l/2 \times l \times l$ occupies the entire volume of the semicube. The second one is adjacent to the membrane and has an infinitesimal thickness with dimensions $0 \times l \times l$. Then the distance between the nodes in the centers of these cubicles is $l/2$ that corresponds to the parameters of RC chains in Fig. 1d in the direction of normal to the plane of the membrane. In the direction parallel to the membrane, RC chains are absent. This is due to the infinite small thickness of membrane cubicles leading to zero values of conductivity and capacitance.

Thus, we showed that the model in the form of Fig. 1b, c, d can be derived by applying the FVM to the Maxwell equations for the electrolyte medium.

4 Correction of the Transport Lattice Model

The main limitation of the method under discussion is the stepped approximation of the cell surface (Fig. 2b). This leads to an incorrect estimate of the surface area.

For example, in the two-dimensional case, the length of a circle stepped approximation is equal to the perimeter of the circumscribed square $-L_{step} = 4D$. The difference between the length of a circle of diameter D ($L_{cir} = \pi D$) and such an approximation is determined by the ratio $L_{step}/L_{cir} = 4/\pi \approx 1.27$. In the three-dimensional case the ratio of the area of a sphere ($S_{sph} = \pi \cdot D^2$) to the area of the surface of the circumscribed cube is equal to $L_{step}/L_{cir} = 6/\pi \approx 1.91$.

Figure 4a demonstrates another example of a significant error due to the stepped approximation of a plane membrane that runs at an angle of 45° to the coordinate axis. Estimating the current through the membrane at constant potentials on each side, based on its stepped approximation (I_{step}), exceeds the true value (I_{45}) by 1.41 times. This ratio also cannot be reduced by decreasing the step size.

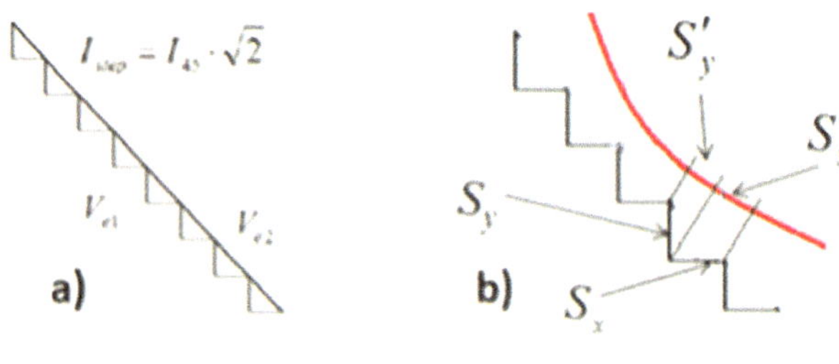

Fig. 4. (a) Error in the area value under a stepped approximation of a flat membrane at an angle of 45° to the axes; (b) the step area projection onto the membrane.

The analysis showed that the error in the area estimation leads to a membrane current error, which distorts the model parameters. In order to exclude the error due to the stepped approximation of the membrane without using complex non-rectangular grids, the following modification of the transport lattice method is proposed.

When estimating the current through the step plane, it is necessary to take into account not its actual area S but the area of its projection onto the original membrane: $S'_{step} = S_{step} \cdot (n_{step} \cdot n)$, where n_{step} is the unit vector of the normal to the step surface, n is the unit vector of the normal to the surface of the membrane. In the two-dimensional case $n = [\cos\alpha, \sin\alpha]$, where α is the angle between the normal n_{step} and n. Thus, it is obtained: for the step plane along the x-axis $S'_x = S_x \cdot \cos\alpha$, and for the step plane along y-axis $S'_y = S_y \cdot \sin\alpha$ (Fig. 4b).

Thus, the expression for the membrane current through the plane of the step $I_{step} = j_m(V_m) \cdot S_{step}$ should be replaced by $I_{step} = j_m(V_m) \cdot S'_{step}$. The changes in the area estimation require changes in the estimation of the model parameters (2–4). One needs to multiply values of R_{lip} (2), R_m (4) by *S/S'* and the value of I_{ch} (U_m) (3) by *S'/S*.

5 Numerical Experiments

5.1 Verification of the Modified Model

To verify the proposed improvement of the method the computational procedure has been developed. The procedure performs three options for simulating a spherical cell between parallel electrodes:

- simulation based on the original model (Fig. 1);
- simulation based on the proposed modification;
- simulation using a theoretical model for a spherical cell in an ideal homogeneous field (formulas are given in [18]).

The linear model of the membrane that included the RC chain (2) only was applied. The result of the simulation is the frequency transfer function from the field strength in the empty space to the field strength in the point of the spherical membrane nearest to the electrode. The results for different models are shown in Fig. 5a. The errors of initial and modified models with respect to the theoretical dependence are shown in Fig. 5b. Thus, the results of computational experiments show that the proposed approach significantly improves the accuracy of the model.

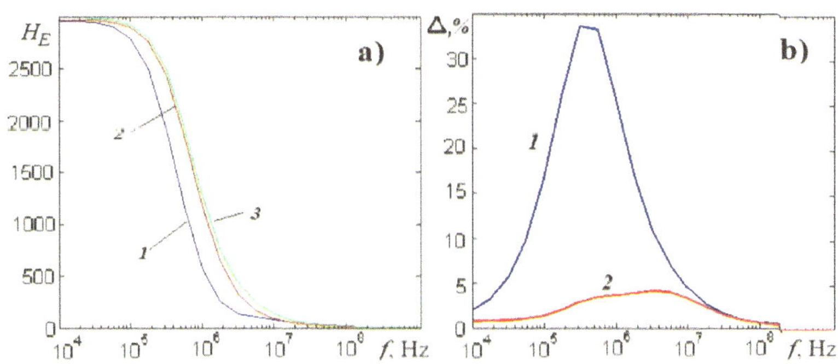

Fig. 5. (a) the transfer function H_E with respect to the field strength calculated on the base of the initial model (1), on the base of the proposed modification (2), using the theoretical model (3); (b) relative error (%) of the initial model (1) and of the of the modified model (2).

5.2 Simulation Examples

The presented model in the form of an electrical network was also evaluated by applying circuit simulator to the analysis of a single spherical cell between two electrodes. The simulation was performed in both frequency and time domains.

The frequency transfer characteristics were simulated using the linear membrane model. To provide the model linearity nonlinear elements representing ion channels (3) and the electroporation process (4) were removed from the network (Fig. 1d). Figure 6a demonstrates the calculated characteristic for a three-dimensional model of a spherical cell with 29 × 29 × 29 of 3D spatial points. For the comparison, the calculated frequency response for the two-dimensional model with 29 × 29 2D spatial points is shown in Fig. 6b.

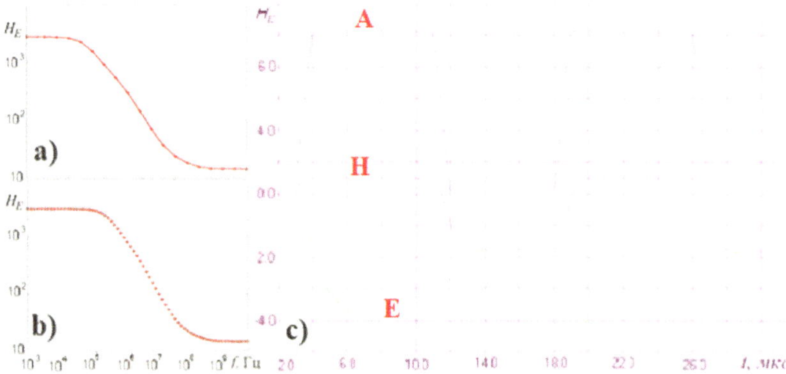

Fig. 6. Calculated frequency transfer characteristics for three-dimensional (a) and two-dimensional (b) models of a spherical cell; results of simulation in the time domain of the effect of pulses on the cellular tissue (c). Input pulses (A), graphs of the relative field strength in the upper (H) and lower (L) points of the spherical membrane.

Figure 6c presents the results of applying the circuit simulator to the analysis of the complete nonlinear two-dimensional cell model in the time domain. The example illustrates the possibility of simulating transient processes in the tissue under pulsed excitation. One can see from in Fig. 6c that the voltage across the membrane at the point H nearest to the electrode is limited by 0.5 v due to a sharp drop of the membrane resistance when the switch presenting the electroporation (4) is shorted.

6 Conclusions

The paper considers the problem of the evaluation of the voltage distribution in biological tissues. The effective approach to provide the numerical evaluation of the distribution consists in obtaining a tissue model in the form of an electrical circuit from discrete components and its subsequent analysis by circuit simulation tools.

The known transport lattice model provides the network representation of the electric field in the tissue but the insufficient accuracy prevent effective application of the model. Determining the source of poor accuracy is complicated by the heuristic nature of the model and its insufficient justification.

Therefore, at the first stage of the research the connection between the structure of transport lattice model and Maxwell equations in the tissue is investigated. The investigation showed that parameters of the network components could be derived by the application of Finite Volume Method to Maxwell equations of the tissue field. Membranes and electrodes define the boundary conditions.

The insufficient accuracy of the transport lattice model is explained in the paper by the stepped approximation of the membrane, which leads to the incorrect estimate of the membrane current due to the irreducible error in evaluating the area of the curvilinear surface.

The modification of the model that eliminates this error is proposed. It is based on evaluating the area of the projection of the step plane onto the membrane surface to provide the correct value of the current density through the membrane. Proposed approach replaces the original step plane area in expressions for model parameters currents by the plane projection area.

Traditional and proposed approaches were applied to the simulation of the spherical cell in the homogeneous field. The comparison of the simulation results with the evaluations by the known accurate analytical expression demonstrated eightfold reducing the error due to the proposed approach.

Additional numerical experiments have demonstrated the effectiveness of the application of circuit simulation tools to the analysis of tissue electric properties.

Acknowledgment. The study was financially supported by the Russian Foundation for Basic Research (project no. 19-29-03012).

References

1. Hodgkin, A.L., Huxley, A.F.: A quantitative description of membrane current and its application to conduction and excitation in nerve. J. Physiol. **117**, 500–544 (1952)
2. Aliev, R.R.: Komp'yuternoe modelirovanie ehlektricheskoj aktivnosti serdca (in Russian: Computer Simulations of Electrical Activity of the Heart). Uspehi fiziologicheskih nauk **41** (3), 44–63 (2010)
3. Kalaiselvi, T., Kalaichelvi, N., Sriramakrishnan, P.: Automatic brain tissues segmentation based on self initializing K-means clustering technique. Int. J. Intell. Syst. Appl. (IJISA) **9** (11), 52–61 (2017)
4. Thejaswini, P., Bhat, B., Prakash, K.: Detection and classification of tumour in brain MRI. Int. J. Eng. Manufact. (IJEM) **9**(1), 11–20 (2019)
5. Ruparelia, V., Chakraverty, M., Desai, S.S., Harisankar, P.S.: Performance comparison of commercially available RF analog and mixed signal simulation tools using Benchmark circuits. In: Anguera, J., Satapathy, S.C., Bhateja, V., Sunitha, K.V.N. (eds.) Microelectronics, Electromagnetics and Telecommunications. LNEE, vol. 471, pp. 443–451. Springer, Singapore (2018). https://doi.org/10.1007/978-981-10-7329-8_45
6. Khari, M., Kumar, R., Le, L-N., Chatterjee, J.M.: Interconnect network on chip topology in multi-core processors: a comparative study. Int. J. Comput. Netw. Inf. Secur. (IJCNIS).**9** (11), 52–62 (2017)
7. Gowrishankar, T.R., Weaver, J.C.: Proc. Natl. Acad. Sci. U.S.A. **100**(6), 3203–3208 (2003)

8. Gowrishankar, T.R., Stewart, C., Weaver, J.C.: Proceedings of the of 26 Annual International Conference of the IEEE Engineering in Medicine and Biology Society, pp. 5444–5446 (2004)
9. Stewart, D.A., Gwrishankar, T.R., Weaver, J.C.: Cylindrical cell membranes in uniform applied electric fields: validation of a transport lattice method. IEEE Trans. Biomed. Eng. **52** (10), 1643–1653 (2005)
10. Stewart, D.A., Gowrishankar, T.R., Weaver, J.C.: Three dimensional transport lattice model for describing action potentials in axons stimulated by external electrodes. Bioelectrochemistry **69**(1), 88–93 (2006)
11. Gowrishankar, T.R., Smith, K.C., Weaver, J.C.: Transport-based biophysical system models of cells for quantitatively describing responses to electric fields. Proc. IEEE **101**(2), 505–517 (2013)
12. Morshed, B., Shams, M., Mussivand, T.: Deriving an electric circuit equivalent model of cell membrane pores in electroporation. Biophys. Rev. Lett. **8**(1&2), 21–32 (2013)
13. Lamberti, P., Citro, N., Egiziano, L., Tucci, V.: A coarse 3D lattice network modeling electroporation phenomenon in an excitable cell. In: 14th International Conference on Synthesis, Modeling, Analysis and Simulation Methods and Applications to Circuit Design (SMACD), pp. 1–4 (2017)
14. Armitage, D.W., LeVeen, H.H., Pethig, R.: Radiofrequency-induced hyperthermia: computer simulation of specific absorption rate distributions using realistic anatomical models. Phys. Med. Biol. **28**(1), 31–42 (1983)
15. Gandhi, O.P., Lazzi, G.: Impedence method for calculation of power deposition patterns in magnetically induced hyperthermia. IEEE Trans. Biomed. Eng. BME **31**(10), 644–651 (1984)
16. Eberdt, M., Brown, P.K., Lazzi, G.: Two-dimensional SPICE-linked multiresolution impedance method for low-frequency electromagnetic interactions. IEEE Trans. Biomed. Eng. **50**(7), 881–889 (2003)
17. Young, J.L., Nelson, R.O., Gaitonde, D.V.: A detailed examination of the finite-volume, time-domain method for Maxwell's equations. Progress Electromagnet. Res. **28**(6), 231–252 (2000)
18. Kotnik, T., Miklavcic, D.: Second-order model of membrane electric field induced by alternating external electric fields. IEEE Trans. Biomed. Eng. **47**(8), 1074–1081 (2000)

Semi-phenomenological Approach to Surface-Bonded Chiral Nanostructures Creation Based on DNA-origami

Veronika S. Beliaeva[1], Olga A. Chichigina[1], Dmitriy S. Klyuev[2], Anatoly M. Neshcheret[2], Oleg V. Osipov[2], and Alexander A. Potapov[3,4](✉)

[1] Lomonosov Moscow State University, 1-63 Leninskie Gory St., 119991 Moscow, Russian Federation
shfordogs@gmail.com, chichuch@yandex.ru
[2] Povolzhskiy State University of Telecommunications and Informatics, 23 L. Tolstoy St., 443010 Samara, Russian Federation
klyuevd@yandex.ru, neshchereta@gmail.com
[3] V.A. Kotelnikov Institute of Radio Engineering and Electronics, 11-7 Mokhovaya St., 125009 Moscow, Russian Federation
potapov@cplire.ru
[4] Joint-Lab of JNU-IREE RAS, JiNan University, Guangzhou, China

Abstract. In this work the statistical properties of quasi-periodic structures disturbed by thermal fluctuations are investigated. A chain of DNA as an example of such structure is under consideration. This chain interacts with some chemical elements and creates great variety of stable nanostructures. The process is called DNA-origami. The problem of interaction of DNA-molecule with periodic flat surface is solved by using a semi-phenomenological model. The places of DNA attachments to the surface correspond to a quasi-periodic sequence of random points. The model allows determinate the optimal frequency of chemical periodical patterning on a substrate.

Keywords: DNA-origami · Statistical distribution · Chirality · Nanomaterial · Renewal process

1 Introduction

The interaction of DNA molecules with mica surface is due to the presence of nickel ions, which appear as "bridge" between DNA molecules and mica surface. Mica is laminate mineral with negatively charged surface. DNA is negatively charged polymer [1, 2].

The DNA origami technique has enabled the realization of almost arbitrarily shaped molecular objects, which can further be decorated with functional molecules or nanoparticles with a spatial resolution of only a few nanometers [10]. For example, cations of metals make an effective bridge between negatively charged DNA of molecules and negatively charged mica surface [1, 2].

Z. Hu et al. (Eds.): AIMEE 2019, AISC 1126, pp. 263–272, 2020.
https://doi.org/10.1007/978-3-030-39162-1_24

An investigation of DNA interaction with surfaces can be used as a direction of research of higher order DNA self-assembly for massive arrays of nanodevices creation [3]. For example, rectangle fragments of DNA structures are fixed by metal cations on a mica surface. Firstly, the rectangles are located on the substrate, they do not move because of divalent magnesium ions. Then the diffusion is developed by the increase in the monovalent sodium cations concentration. The self-assembly starts. The resulting structures can be fixed by adding magnium ions again. This method of lattice structures from elements of rectangle DNA-origami creation is fast, simple and can be carried out isothermally [4].

If we want to use such a geometrical property of DNA as chirality [5], it is necessary to keep in mind that the lattice length of the cover, which DNA is located on, should be much less than the electromagnetic wavelength, which radiates the DNA substrate. However, there are some ways to mitigate this requirement, for example, by passing a constant electric current across the substrate. Due to the interaction of DNA with the substrate, leading to self-assembly, it is possible to obtain special anisotropic planes with an explicitly selected direction of the arrangement of surface-bound chiral nanostructures specially created from DNA. This affects the way the circular dichroism demonstrates itself in such a construction. DNA rods take a vertical position in a special solution and horizontal when the solution is removed. Depending on this, there will be one or another response in the spectrum of the traversed light [6, 7].

DNA-origami is also used in biorobots creation. The so-called molecular spiders, consisting of streptavidin molecules and three-component compounds, provide deoxyribose as a "leg", enabling its structure to move along the surface on which DNA-origami is applied. If DNA-origami has been pre-built appropriately, then this biorobot would move across the substrate, it would execute commands such as "start", "follow", "turn" and "stop". The sequence of these commands depends on the location of the nucleotides on the surface [8].

Thus, the main aim of this work is to develop methods for creating surface-bound chiral nanostructures based on DNA origami, by using a semi-phenomenological approach that allows to predict the places of nucleotide attachment.

This paper consists of an introduction, the main part and conclusions. In the main part we consider the problem statement, the method of its solution and numerical results. Conclusions express basic results.

2 Problem Formulation

Taking into consideration a set of linear slightly deformed polymers or nanostructures. They can be presented as a well-known model of a chain of N identical stiff rods with μ length, witch connected by hinges allowing small deflection angles. This we can measure the average length l_N (distance between ends) and variance σ_N^2 of these linear structures depending on N. Being straightened in a line, this structure is periodic with the period equal to μ. This structure length is $l_N = N\mu$ and variance is zero. Thermal fluctuations deform this structure. The average length decreases $l_N < N\mu$ and variance increases $\sigma_N^2 > 0$. In the stationary case, the dependence of the probability distribution density function on the temperature T is determined by the Boltzmann distribution, where the deformation energy $E_d(l)$ has a minimum when the chain is straight

$$w(l) = C \exp\left(-\frac{E_d(l)}{kT}\right). \tag{1}$$

Some patterning of parallel lines with a fixed period T is plotted on the surface (Fig. 1). The chain is arranged perpendicular to these lines. The chain interacts with the surface only by the hinge joints, if they are close enough to these lines. The relation of the periods of the pattern and the quasi-periodic chain structure determine their strong-weak interaction.

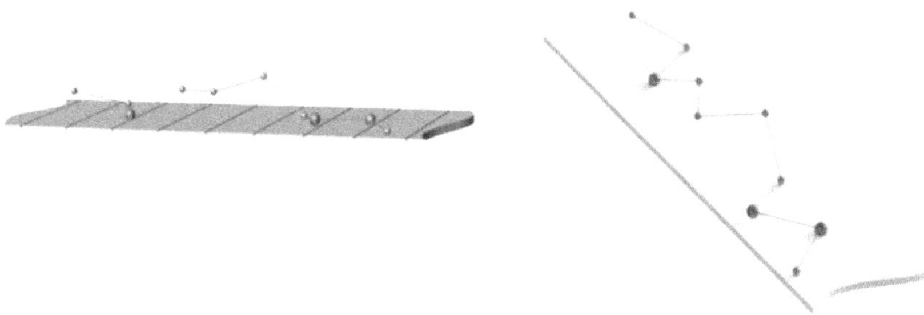

Fig. 1. Model of a quasi-periodic structure on the surface with periodic pattern

The novelty of the proposed approach is connected with the application of statistical radiophysics methods, namely, the theory of pulse processes for a quasi-periodic structure description [9, 10]. The energy and, correspondently, the probability of interaction with the periodic structure of a given frequency will be proportional to the spectral density of the quasi-periodic process corresponding to the chain. For example, DNA-origami needs a strong attachment of a chain to the surface. Meanwhile, weak and regulated connection is better for chiral structures that are switched under the action of polarized light.

At this stage of the research, it is important to create simple basic models that can be investigated analytically and clarify the basic patterns. In the future, it will be possible to create models that are more realistic.

3 Renewal Process

An renewal random pulse process is a process with independent and equally distributed intervals between adjacent pulses. Consider a pulsed renew process $\eta(t)$ with a shift parameter μ. The each pulse coordinate corresponds to the position of the chain node along the axis perpendicular to the notches. The Poisson process is a special case of the renewal process [11]. Such processes are also called random points. Coordinates of pulses are random values x_n. The intervals between neighboring, following one another pulses $v = x_n - x_{n-1}$ are also independently equally distributed random variables.

The average distance between two consecutive pulses $\langle v \rangle = T$ is fixed, i.e. conditional period T is constant value. The main characteristics $m_1 = l_N/N$, $\sigma^2 = \sigma_N^2/N$ set the distribution parameters for the intervals between neighboring pulses $w(v)$.

In this semi-phenomenological approach, we will consider two distributions: a three-parameter gamma distribution and a cosine distribution. These distributions were chosen and based on a universality consideration: the fact is that many continuous distributions, for example, normal or exponential ones, can be approximated by a gamma distribution due to a large number of variable parameters. As for the cosine distribution, it is impossible to approximate this distribution with the gamma one, thus, using a sample of two distributions at this stage of work, a large number of continuous dependencies can be simulated. Moreover, unlike the gamma distribution, the cosine distribution is bounded on both sides along the x axis. This allows to avoid negative projections on the axis (i.e., those cases where the i-th node of the structure is to the right of the i + 1-st one on the substrate).

In the future, calculations will also be carried out for other probability distributions, for example, Weibull, Pareto, etc.

3.1 A Process in a Shift Gamma Distribution (Three-Parameter) Between Pulses

Let us consider three-parameter gamma distribution

$$w(v) = \frac{(-v+\mu)^{n-1}}{\Gamma(n)\alpha^n}\exp\left(-\frac{-v+\mu}{\alpha}\right), \tag{2}$$

where n is a shape parameter, α is a scale parameter. The parameter μ is determined from the experiment as the ratio of the length of the polymer to which this length tends when the temperature decreases, to the number of chain links. A random value v varies within $(-\infty, \mu)$.

The main characteristics will be equal to

$$m_1 = -n\alpha + \mu,\ \sigma^2 = n\alpha^2. \tag{3}$$

Here $m_1(x)$ is the first moment of a random value x. To calculate the spectral density, let us calculate the characteristic function

$$c(\omega) = m_1(\exp(i\omega v)). \tag{4}$$

In this regard, the characteristic function in case of gamma-distribution is determined as:

$$c(\omega) = \int_{-\infty}^{\infty} \exp(i\omega v)\frac{(-v+\mu)^{n-1}}{\Gamma(n)\alpha^n}\exp\left(-\frac{-v+\mu}{\alpha}\right)\theta(v)dv = \frac{\exp(i\omega\mu)}{(1+iw\alpha)^n}. \tag{5}$$

Here the integral by definition of the gamma function was taken into account

$$\int_0^{\infty} \xi^{n-1} \exp(-\xi\{1/\alpha + i\omega\})d\xi = \frac{1}{\left(\frac{1}{\alpha} + i\omega\right)^n} \int_0^{\infty} x^{n-1} \exp(-x)dx = \frac{\Gamma(n)}{\left(\frac{1}{\alpha} + i\omega\right)^n}. \quad (6)$$

The spectral density is according to the formula:

$$S(\omega) = \frac{f_0^2}{2\pi T} \frac{1 - |c(\omega)|^2}{|1 - c(\omega)|^2}. \quad (7)$$

The properties of the characteristic function imply a relationship between the spectral density at the zero point with other characteristics (expected value, variance, etc.), which is described in general terms by the following expression:

$$S(\omega \to 0) = \frac{f_0^2}{2\pi T} \frac{\sigma^2}{m_1^2}. \quad (8)$$

In this case, the limit of this periodic function (8) will be equal to

$$S_p(\omega) = \frac{f_0^2}{2\pi T} \frac{\sigma^2 \omega^2}{2(1 - \cos(m_1 \omega))}. \quad (9)$$

Substituting (5) into (7), we finally get

$$S(\omega) = \frac{f_0^2}{2\pi T} \frac{(1 + \alpha^2\omega^2)^n - 1}{(1 + \alpha^2\omega^2)^n + 1 - 2(1 + \alpha^2\omega^2)^{n/2} \cos(-\omega\mu + n \arccos(1 + \alpha^2\omega^2)^{-1/2}))}. \quad (10)$$

Below in Fig. 2, 3, 4, 5, 6 and 7, illustrations of the impulse process are presented for various parameters of the gamma distribution. The first Fig. (a) of the triple represents the probability arrangement of the subsequent nucleotide location on the substrate. The vertical line on the graph corresponds to the location of the previous nucleotide. The second Fig. (b) is the result of the described process, i.e. illustration of the molecules arrangement with given distribution parameters. The frequency of notches corresponds to the first maximum of the spectrum density in the third Fig. (c).

The spectrum density will have maxima and minima corresponding to the frequencies of maximum or, conversely, minimum adhesion of the polymer to the substrate. Due to the existence of a large number of scientific problems associated with the application of DNA to a substrate, both cases are of interest.

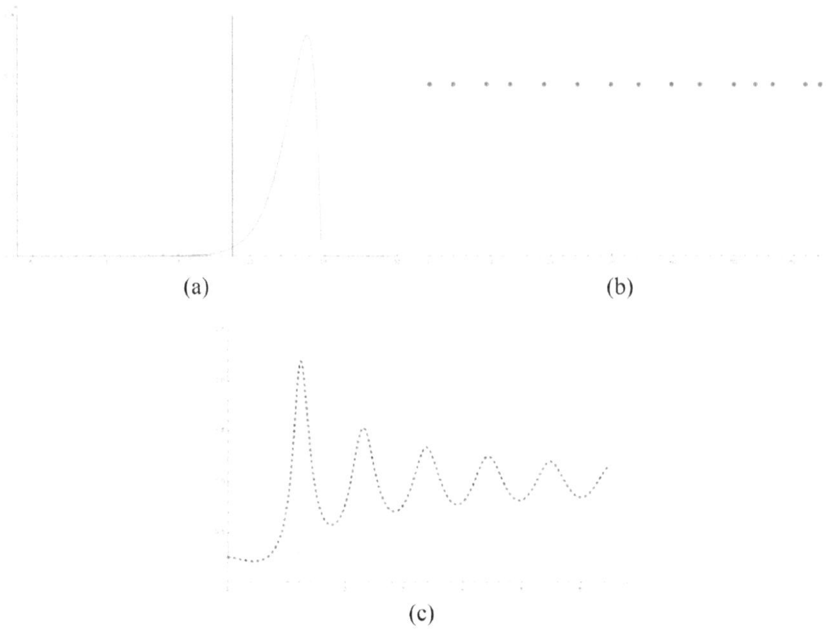

Fig. 2. $\alpha = 2$; $n = 1$. (a) - PDF, (b) - process, (c) - spectrum density

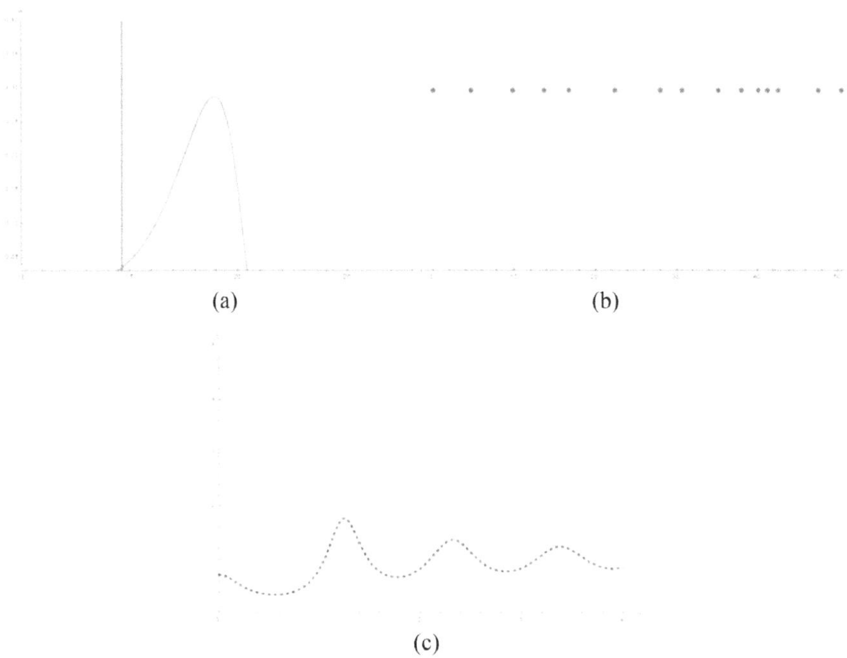

Fig. 3. $\alpha = 2,75$; $n = 1$. (a) - PDF, (b) - process, (c) - spectrum density

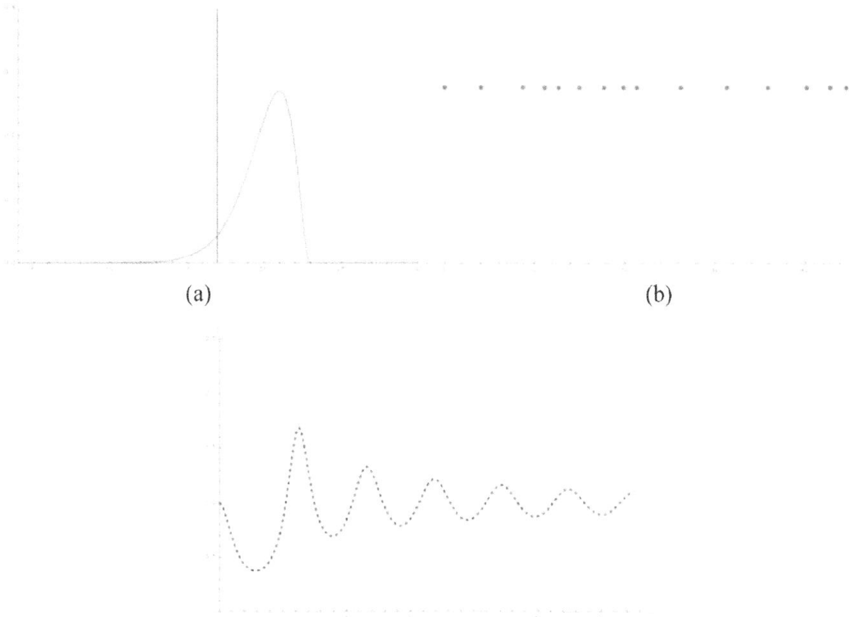

(a) (b)

(c)

Fig. 4. $\alpha = 3$; $n = 1$. (a) - PDF, (b) - process, (c) - spectrum density

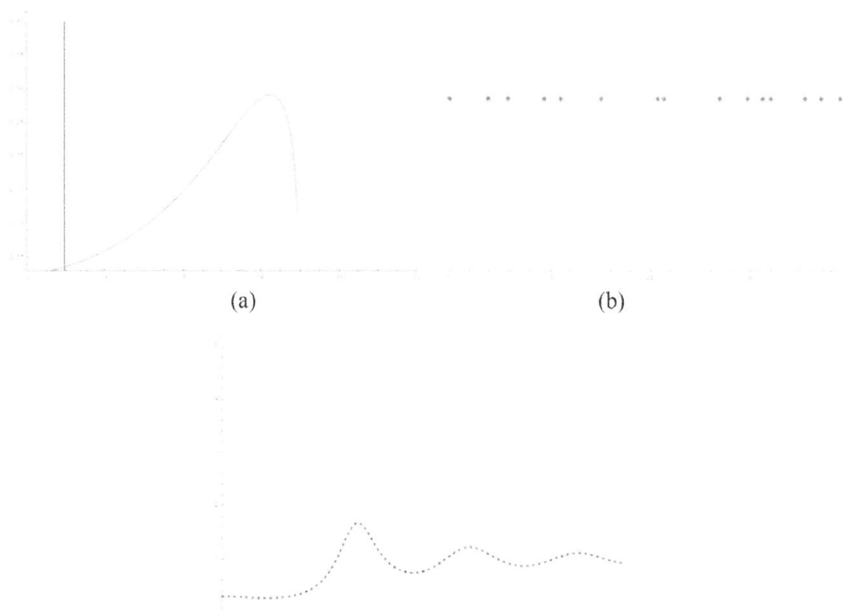

(a) (b)

(c)

Fig. 5. $\alpha = 1.4$; $n = 1.8$. (a) - PDF, (b) - process, (c) - spectrum density

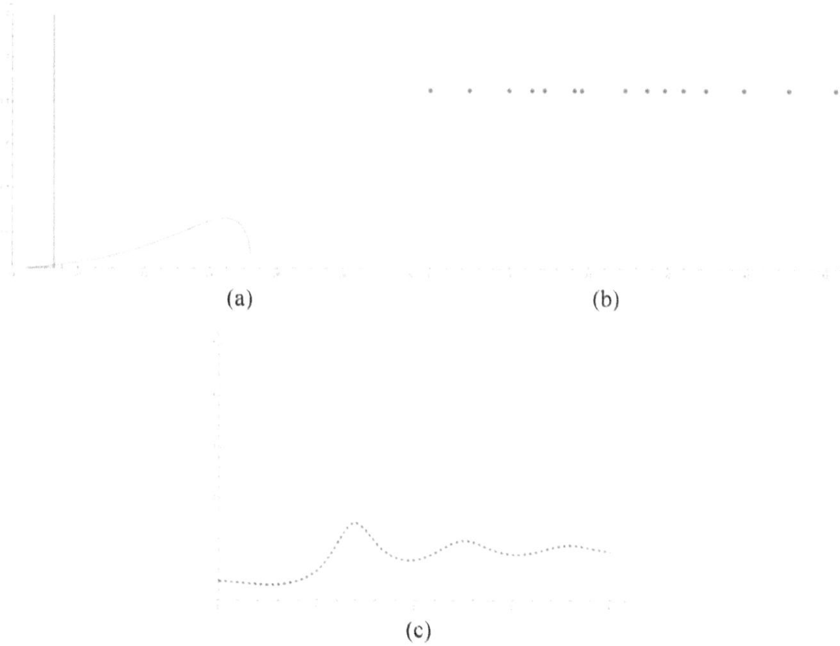

Fig. 6. $\alpha = 1.4$ $n = 2$. (a) - PDF, (b) - process, (c) - spectrum density

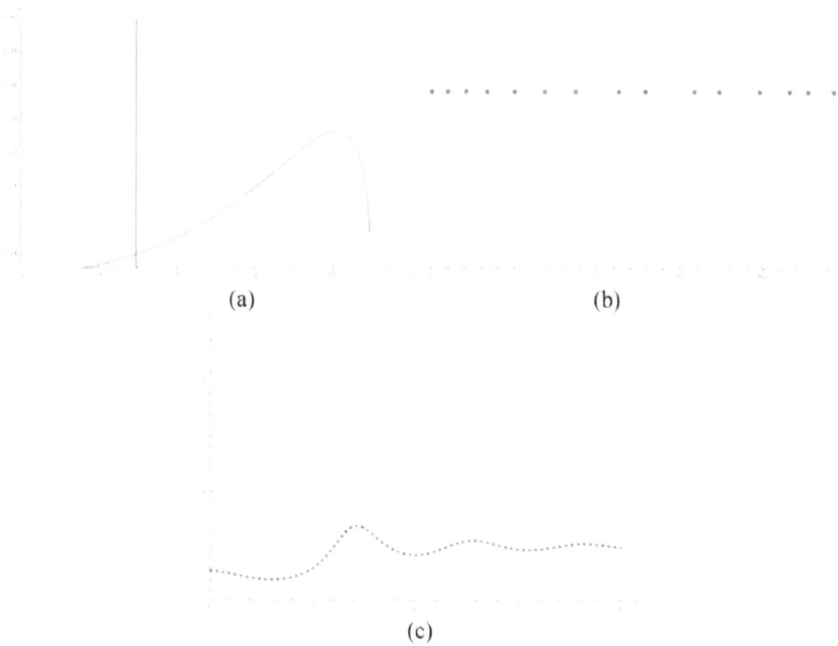

Fig. 7. $\alpha = 1.4$; $n = 2.25$. (a) - PDF, (b) - process, (c) - spectrum density

3.2 Process with a Shift Distribution Between Pulses in Half a Cosine

Now let us consider an renewal process $\eta(t)$ with shift parameter μ, besides $\mu - \alpha \leq v \leq \mu$. The average between two neighboring pulses is $T = m_1 = \mu - \alpha + 2\alpha/\pi$, and variance is $\sigma^2 = (2\alpha/\pi)^2(\pi - 3)$.

The probability density function of the cosine distribution is described by the following expression

$$w(v) = \frac{\pi}{2\alpha}\cos\left(\frac{\pi}{2\alpha}(-v + \mu)\right). \tag{11}$$

The spectrum density is

$$S(\omega) = \frac{f_0^2}{2\pi T}\times \frac{\varsigma(\omega)^4 - 3\varsigma(\omega)^2 + 2\varsigma(\omega)\sin(\alpha\omega)}{2 + \varsigma(\omega)^4 - \varsigma(\omega)^2 - 2\varsigma(\omega)\sin(\alpha\omega) - \xi(\omega)\cos(\mu\omega - \alpha\omega) + \varsigma(\omega)\xi(\omega)\sin(\mu\omega)},$$
$$\varsigma(\omega) = \frac{2\alpha\omega}{\pi},$$
$$\xi(\omega) = 2\left(1 - \varsigma(\omega)^2\right). \tag{12}$$

The spectrum density of the process with cosine distribution for different values of the shift parameter μ is shown in Fig. 8.

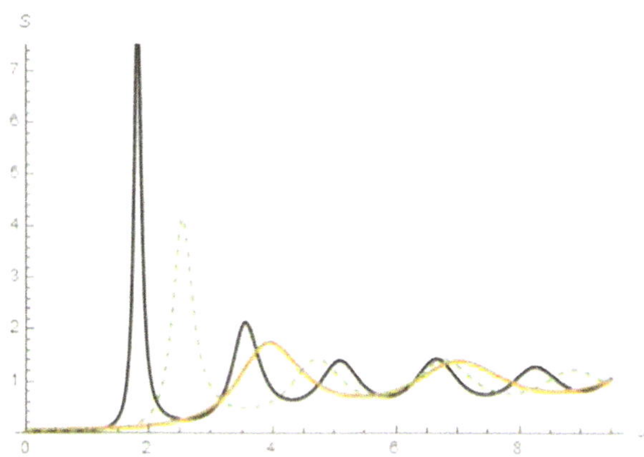

Fig. 8. Cosine spectrum density for values $\mu = 2, 3, 4$.

Let us notice that the spectrum density in the case of the cosine distribution has also explicit maxima and minima.

4 Conclusions

The results obtained using the two distributions, despite the fundamental difference in the qualitative characteristics of the probability density, allow to determine the optimal frequency of notching on the substrate. With this approach, it is also possible to determine the degree of DNA adhesion to the substrate by changing the temperature and thus shifting the maxima and minima of the spectral density. Further development of this approach is associated with finding suitable distributions, including distributions with "heavy tails" and initial parameters depending on the specific task.

References

1. Liu, L., Li, Y., Wang, Y., et al.: Regulating DNA self-assembly by dna-surface interactions. ChemBioChem **18**(24), 2404–2407 (2017). https://doi.org/10.1002/cbic.201700545
2. Aithala S.P., Aithalb, S.: Nanotechnology innovations and commercialization-opportunities, challenges & reasons for delay. Int. J. Eng. Manufact. (IJEM), **6**, 15–25 (2016). Published Online November 2016 in MECS. (http://www.mecs-press.net). https://doi.org/10.5815/ijem.2016.06.02
3. Khalil, M.I.: A new heuristic approach for DNA sequences alignment. Int. J. Image, Graph. Signal Process. (IJIGSP) **12**, 18–23 (2015). Published Online November 2015 in MECS. (http://www.mecs-press.org/). https://doi.org/10.5815/ijigsp.2015.12.03
4. Woo, S., Rothemund, P.W.K.: Self-assembly of two-dimensional DNA origami lattices using cation-controlled surface diffusion. Nature Commun. **5**(1), 1–10 (2014). https://doi.org/10.1038/ncomms5889
5. Tverdislov, V.A., Malyshko, E.V., et al.: Periodic system of chiral structures in molecular biology. Biophysics **63**(3), 421–434 (2017). https://doi.org/10.1134/S0006350917030228
6. Schreiber, R., Luong, N., Fan, Z., et al.: Chiral plasmonic DNA nanostructures with switchable circular dichroism. Nature Commun. **4**(1), 1–6 (2013). https://doi.org/10.1038/ncomms3948
7. Kaur P., Aggarwal S.K., De A.: Design and investigation of circularly polarized RMPA with chiral metamaterial cover. Int. J. Wirel. Microwave Technol. (IJWMT) **3**, 61–70 (2016). Published Online May 2016 in MECS. (http://www.mecs-press.net). https://doi.org/10.5815/ijwmt.2016.03.07
8. Lund, K., Manzo, A.J., Dabby, N., et al.: Molecular robots guided by prescriptive landscapes. Nature **465**(7295), 206–210 (2010). https://doi.org/10.1038/nature09012
9. Daley, D.J., Vere-Jones, D.: An Introduction to the Theory of Point Processes. Volume I: Elementary Theory and Methods. 471 p. Springer, New York (2003). https://doi.org/10.1007/b97277
10. Akhmanov, S.A., Diakov Y.Y., Chirkin A.S.: Vvedenie v statisticheskuyu radiofiziku i optiku [Introduction to statistical radiophysics and optics]. 640 p. Nauka, Moscow (1981). (In Russian)
11. Haight F.A.: Handbook of the Poisson Distribution. 168 p. Wiley, New York (1967)

Engineering in the Scientific Music Therapy and Acoustic Biotechnologies

Sergey V. Shushardzhan[1(✉)] and Sergey V. Petoukhov[2]

[1] FSBI «National Medical Research Center of Rehabilitation and Balneology» of the Ministry of Healthcare of the Russian Federation, Novy Arbat street № 32, Moscow, Russia
medart777@yandex.ru

[2] Mechanical Engineering Research Institute, Russian Academy of Sciences, M. Kharitonievsky pereulok, 4, Moscow, Russia
spetoukhov@gmail.com

Abstract. The article is devoted to fundamentals and achievements of scientific musical therapy and acoustic bitechnologies developed in Russia and used now in many countries. This scientific-technologic direction has received in 2019 year from European Union a special grant for further developing thematic international cooperation: "Comprehensive multiprofesional approach to the treatment the patients using the elements of the scientific music therapy". The article describes some Russian patented solutions in this fields and also theoretical and technological approaches for further increasing their effectiveness.

Keywords: Resonances · Musical therapy · Acoustic biotechnology · Genetics · Matrices

1 Introduction

Thoughts about key significance of harmonious vibrations in organization of the world exist from ancient time. According to notions of Ancient Chinese, music is present at origin of the world and plays a space role: music represents a microcosm reflecting a structure of the Universe. Since ancient times, music has been used as a means of influencing the physiological state of a person. In particular, in ancient China, music therapy was strongly developed.

Nowadays, additional interest in music and the physiological mechanisms of its impact on humans is caused by the use of music in the highly profitable transnational media business. This use relies on the increased sensitivity to music of many people, music lovers, as well as the ability of music to unite huge audiences of listeners. Living organisms can be considered as informational essences whose main task is to transfer genetic information to descendants. Material physico-chemical substances of their bodies are minor factors contributing to the implementation of this basic information task. The affinity of living organisms with music is one of the manifestations of living organisms as informational essences, built on certain harmonic principles of cooperative unification of parts into a single whole.

Science has long been involved in the physiological mechanisms of music perception [1]. People compose and listen the music from deep antiquity. More than

Z. Hu et al. (Eds.): AIMEE 2019, AISC 1126, pp. 273–282, 2020.
https://doi.org/10.1007/978-3-030-39162-1_25

30 thousand years ago, long before the advent of arithmetic, our ancestors had already played stone flutes and bone harps. So, the bone flute found in France is at least 32 thousand years old. Modern researchers are studying the effect of music on humans, animals, plant and protozoa, using the achievements of scientific and technological progress for developing not only a theoretical knowledge but also new biotechnologies (see for example [2–6]). Our article pays a special attention to development of scientific musical therapy and acoustical biotechnologies using, in particular, achievements in the field of genetical coding and bioinformatics.

2 The Theory of Resonances and Physiology

Any living organism is a great chorus of coordinated oscillatory processes (mechanical, electrical, piezoelectrical, biochemical, etc.), which are connected with their genetic inheritance along chains of generations. One can mention here that mechanical and electrical oscillations in living bodies are closely connected because many tissues are piezo-electrical (nucleic acids, bone, actin, dentin, tendons, etc.). Since ancient times, chrono-medicine believes that all diseases are the result of disturbances in the ordered set of oscillatory processes.

The notion "resonance" is one of basic notions of classical and quantum physics. Genetic molecules belong to the world of principles of quantum mechanics where the notion «resonance» has the fundamental meaning as E. Schrodinger emphasised [7, p. 115]: "*The one thing which one has to accept and which is the inalienable consequence of the wave-equation as it is used in every problem, under the most various forms, is this: that the interaction between two microscopic physical systems is controlled by a peculiar law of resonance*".

Nobel laureate Pauling has developed the theory of resonances in structural chemistry and claimed that resonances in molecules of living bodies should be very essential for biological phenomena. His book [8] about this theory is the most quoted among scientific books of the XX century. The theory was developed to explain the formation of hybrid bonds in molecules. His theory uses the fundamental principle of a minimal energy because - in resonant combining of parts into a single unit - each of members of the ensemble requires less energy for performing own work than when working individually. An organism during its life on genetic basis should solve algorithmic problems of two types: (1) informational, providing coordinated energy processes; (2) energetic, providing information processes. Systems of resonances can be used as a common basis of such "two-faced" algorithms since resonances are associated both with oscillatory energy and with informatics of communications.

Russian biophysicist Schnoll [9] noted: "*From possible consequences of interaction of macromolecules of enzymes, which are carrying out conformational (cyclic) fluctuations, we shall consider pulsations of pressure - sound waves. The range of numbers of turns of the majority of enzymes corresponds to acoustic sound frequencies. We shall consider ... a fantastic picture of "musical interactions" among bio chemical systems, cells, bodies, and a possible physiological role of these interactions. ...... It leads to pleasant thoughts about nature of hearing, about an origin of musical perception and about many other things, which belong to area of biochemical aesthetics already*".

Our research is associated with "biochemical aesthetics". An organism can be seen as a musical synthesizer with multiple settings of inherited resonant modes [9–12]. Music is a game with acoustic resonances, to which people are remarkably predisposed. Throughout tens of thousands of years, people create musical instruments, adjusting them to specific systems of resonances. Over the centuries, people have learned to combine individual instruments and singers into orchestras and choirs as coordinated oscillating systems with an increased number of degrees of freedom. G. Leibniz declared that music is arithmetic of soul, which computes without being aware of it.

Every organism is endowed with the inherited ability to tune into resonances and to use resonances as carriers of information, the example of which is communication among human persons by speech and vocal sing. There are evidences in favor that genetic informatics is also based on the principles of the theory of resonances, the study of which led to the emergence of a scientific direction called "multi-resonance genetics" [12–14].

From a formal point of view, a living organism is an oscillatory system with a great number of degrees of freedom. Theory of oscillations uses mathematics of matrices to study resonant characteristics of oscillatory systems with many degrees of freedom (see, for example, [15]). The mentioned scientific field of multi-resonance genetics uses matrices to study genetic phenomena.

Matrices possess a wonderful property to express resonances, which sometimes is called as their main quality [16, 17]. The expression $y = A*S$ models the transmission of a signal S via an acoustic system A, represented by a relevant matrix A. If an input signal is a resonant tone, then the output signal will repeat it with a precision up to a scale factor $y = \lambda*S$ by analogy with a situation when a musical string sounds in unison with the neighboring vibrating string. In the case of a matrix A, its number of resonant tones S_i corresponds to its size. They are called its eigenvectors, and the scale factors λ_i with them are called its eigenvalues or, briefly, spectrum A. Frequencies $\omega_i = \lambda_i^{0.5}$ [15] are defined as natural frequencies of the system, and the corresponding eigenvectors are defined as its own forms of oscillations (or simply, natural oscillations). Eigenvalues and eigenvectors are fundamental notions in the theory of mechanical, electrical and other oscillations at macroscopic or microscopic levels (see for example [15, p. 61].

Applications of such matrix approaches allowed us discovering hidden relations of structures of genetic systems with the theory of resonances of oscillatory systems with many degrees of freedom [12–14]. We apply received results in our study of actual problems of the scientific musical therapy and acoustic biotechnologies considered below.

3 Regarding the Scientific Music Therapy

Therapeutic use of music has a long history. Hippocrat, Aristotel, Confucius and others ancient sages a thousand years ago were trying to treat by music nervous and mental patients. There are a lot of documentary mentions. They refer to different periods and civilizations, and give a clear idea that music in medicine has been used, but empirically, the explanations of therapeutic effects were based on myths, metaphysical theories or religious views.

In the XX century music therapy has been widely practiced in various European countries and USA. National Music Therapy Associations were established in England, Germany, Holland, France, Belgium, Italy, etc. Music therapy was recognized in the U.S.A. after the second world war, when the music was successfully applied in the treatment of emotional disorders among war veterans. Currently, there are around 3500 professional music therapists registered in the U.S.A. and the need for specialists that profile has been steadily increasing. Currently, more than 100 universities and colleges all over the world offer educational programs, after which students receive a Bachelor's degree, Master's or Doctoral degrees in this field.

Scientific Music Therapy (SMT) is the new interdisciplinary direction of music therapy based on the synthesis of Western and Eastern medicine, arts, natural sciences and modern technologies. Fundamentals of SMT was developed in Russia - by efforts of the first author of this article S.V. Shushardzhan - in the early nineties of the last century. In Russia, various researches and applications of SMT are concentrated in the Science and Research Center for Music Therapy and Healthcare Technologies in Moscow (www.doctor-art.ru). This basic Center is the founder of the European Academy for Music Therapy (Bulgaria) and has a close cooperation with teams of specialists in many countries where many followers of the SMT appeared in last years: Germany, Estonia, Slovakia, Israel, Bulgaria, Poland, USA, England, Seychelles, Morocco etc.

In this 2019 year SMT received the new international confirmation of its importancy: the European Union has issued a special grant on the development and use of SMT to an international group developing and practicing Russian achievements in the field of SMT: "Comprehensive multiprofesional approach to the treatment the patients using the elements of the scientific music therapy" in relation to the Programme Erasmus+. The start application for the international grant was submitted by the Comenius University in Bratislava and the First Clinic of Acupuncture and Natural Medicine of G. Solár in Šamorín, Slovakia, which are enthusiasts of the Russian achievements of SMT and have been working in cooperation with the aforementioned Science and Research Center for Music Therapy and Healthcare Technologies in Moscow for a number of years, in particular, on the basis of Russian technologies, 11 Russian patents and some know-how (see for example patents [18–26]. This grant gives additional opportunities to develop methods and devices of SMT with extended education of young specialists in this perspective direction. SMT has a preference to study the features of complex body reactions - psychological, physiological and biophysical - to music by various modern diagnostic technologies.

In the course of these researches, there were discovered 3 main musical-acoustic algorithms (S-, HR-, T-) bioregulators, which cause characteristic changes in the state of the nervous system (S - sedative, T - tonic and HR - harmonizing) and in the level of hormones in the blood. The results of our many years scientific research and the world experience of music therapy were generalized in the form of a Neuro-Hormonal-Resonance Theory of Acoustical Influences, which was proposed in 2005 (S. Shushardzhan), become theoretical basement of SMT and gained international recognition. Scientific Music Therapists have a clear idea of the algorithmic mechanisms of music influence on a person, knows how to carry out the necessary diagnostics and choose the best method or technology needed in each specific case. SMT provides more qualified help and a significantly higher efficacy in treatment compared to intuitive-empirical approaches.

Significant attention in our researches was paid to study acoustic influences to human skin with its reflex points (acupuncture points) by using special devices (Fig. 1). In the course of these researches, the following parameters were studied: capillary circulation; temperature; pain sensitivity; electrical conductivity; physiological reactions; clinical effects. For this study the innovative hardware-software complex "AkuTon" was created, which capable to provide music therapy with simultaneous acoustic-magnetic-vacuum music therapy impact giving effects of analgesia. The patented complex AkuTon relieves pain in the back, joints and muscles; relieves stress; increases sexual activity. In the result of the conducted study, regularities of acoustic influences to the skin and reflex points were additionally established.

Fig. 1. Using the patented device "AkuTon" for musical-acoustic magnetic-vacuum effects on acupuncture points and reflex zones [23].

Fig. 2. The patented hardware-software complex "Bonny-Grand" for «Meso-Forte» music therapy [22, 24].

In the course of our studying acoustic influences to human skin, an exclusive and patented hardware-software complex "Bonny-Grand" was also created for anti-aging, wellness and stabilizing the emotional state with so called Meso-Forte therapy using special music algorithms (Fig. 2). This complex includes mask-converter of acoustic waves and «Meso-Forte» music therapy software (see patents [22, 24]).

Steps of Meso-Forte therapy procedure are the following: a cosmetic mask is applied on human face; mask-converter is superimposed; headphones are put on top; special music programs start to sound and act via mask-converter and headphones simultaneously. The procedure lasts 20–25 min. In the result, effects of full relaxation and skin smoothing are provided. Anti-aging effect in Meso-Forte therapy is not only cosmetic. The main mechanism is synergy of general (internal) and local (skin) anti-aging reactions providing the following: stabilization of brain activity, optimizing the hormones level and adaptation features, adaptation of blood circulation, stimulation of tissue regeneration, deep penetration of cosmetics into the skin, internal rejuvenation and skin rejuvenation.

The Russian achievements in SMT are also used for a creation of fully equipped office for health improvement and creative development for children, as well for correction of neurological disorders: autism, mental delay, speech disorders. Scientific music therapy has proved treating and recovery effects: it is perfectly combined with analgesic therapy, drugs, balneology treatment, physiotherapy, massage, reflexology, physiotherapy exercises, etc.

One of perspective directions of engineering creativity in the field of SMT is a creation of "robots-music therapists" for use in many hospitals, medical and community centers, kindergartens and schools, etc. The first Russian "robot-music therapist" or rehabilitologist with music therapy option was already created (Fig. 3).

Fig. 3. The first "robot-music therapist" created in Russia.

This unique robot possessed the following standard abilities: movement on specified routes; easy and witty communication; identification the client's age; definition of emotional state; selection of the desired track of music therapy program; creation acoustically favorable environment at public places. In addition, the robot can make the following: service for bed patients; delivery of drinks, food, medicine; rendering of any background information. The use of such robots has advantages: effective psychotherapeutic and health care; attracting clients (the robot is equally loved by both adults and children); creating a favorable image of the company; tirelessness, no maintenance costs. The robot can be programmed on an individual order for solving problems for each specific case.

It is obvious that such robot can be improved more and more by achievements of artificial intelligence, which provide its advanced features, for example, the following: independent and targeted movement along complex trajectories, for example, in hospitals from ward to ward to serve a multitude of patients; adequate communication with different people; high-level general assessment of the situation and interlocutors, etc.

4 Regarding Acoustic Biotechnologies

Regeneration of organs and tissues, as well as the intensification of growth bioprocesses are among the most important tasks of modern biotechnologies with a multitude of prospects in the fields of medicine, pharmacological production, agricultural industry, etc. From 1996 year our team studies acoustic influences with elements of music on experimental cell cultures in vitro, including tumor cells, microbes, blood cells and other biological objects. In the result, it has been shown that certain acoustic parameters are able to intensificate the activity of various cell cultures, while others are able to inhibit this activity. Figure 4 shows one of our acoustic experiments with blood cells, which were placed in medical tubes to provide controlled acoustic stimuli to them through headphones.

Fig. 4. The experimental device for studying acoustic influences to blood cells in vitro.

In such experiments, special activating algorithms of acoustic musical impact were detected, the use of which during only one hour allowed to increase the total number of Leukocytes on average 4.7 times and number of immature granulocytes increased in 18.3 times!

Three main types of algorithms of acoustic impacts, which make significant changes in the level of hormones in blood, were found: S-algorithms, HR-algorithms and T-algorithms. In the result, the method of neuro-hormonal correction and rejuvenation by musical-acoustic algorithms has been developed and patented [18–21]. Using musical-acoustic algorithms can provide control for the physiological reactions and emotional state.

Some organisms have a remarkable ability to regenerate lost parts of their bodies. For example, the lizard grows a new tail instead of its lost tail. Many such biological facts give reason to think about the possibility of creating methods for awakening the potencies hidden in the cells of the body to regenerate tissues and organs, including in humans. Such methods are related to the subject of "biological prosthetics" and they are associated with the hope of ridding people of technical prostheses in the future. The search for such methods of regenerative medicine is conducted in laboratories around the world. Our studies of acoustic-musical influences are directly related to this important topic. At the same time, we additionally use in our works the new knowledge on connections of the molecular-genetic system with the relations of musical harmony, which was also used for development of genetic musical systems and the so-called "genetic music" based on them [10–13, 27–29]. The peculiarities and possibilities of the corresponding new direction of musical culture are been studying in detail and used in the "Center for Interdisciplinary Studies of Musical Creativity" of the Moscow State Conservatory. Various scientific approaches [30–35] are used in this work including approaches from theories of noise-immune coding and resonances of oscillatory systems, matrix analysis, genetics, etc.

5 Some Concluding Remarks

Music is perceived through multi-level acoustic-jet systems, including hearing organ, skin, other organs (cells), reflex points. Musical therapy uses all these physicological channels. Mathematical methods of bioinformatics and theory of resonances of oscillatory systems with many degrees of freedom are useful for further developing scientific musical therapy and acoustic biotechnologies.

In the USA, Canada, Australia and Europe, music therapy is a recognized state specialty, with a developed system of university and postgraduate education, where more than 100 universities conduct intensive training of music therapists in music faculties according to programs at all levels: undergraduate, graduate and doctoral studies.

In Russia, there is no specialty "music therapy". However, after the recognition by the Ministry of Health of the Russian Federation of the methods of music therapy, [36], professional training in scientific music therapy began. There are "Higher Courses of Music Therapy - 3 levels of education", which are the cycle of postgraduate education for those who already have a higher basic education, for example: medical, psychological, musical, pedagogical, etc. Training is conducted with the participation of the European Academy of Music Therapy (Bulgaria). Upon completion, diplomas are issued that have official recognition in the European Union. The Kazan State Institute of Culture conducts admission to the undergraduate degree 44.03.01 "Pedagogical Education", the profile "Musical Therapy", as well as admission to the magistracy 55.04.06 "Music Science and Applied Music", the profile "Musical Pedagogy and Music Therapy" ". In 2017, the Department of Music Therapy was opened at the Moscow Social and Pedagogical Institute with a retraining program in the specialty "Clinical psychologist, music therapist".

Scientific music therapy and acoustic biotechnologies, accounting genetically inherited features of biological organisms, have significant prospects for further development in Russia and other countries of the world.

References

1. Weinberger, N.M.: Music and brain. Sci. Amer. **291**(5), 88–95 (2004)
2. Wang, B.C., Chen, X., Wang, Z., Fu, Q.Z., Zhou, H., Ran, L.: Biological effect of sound field stimulation on paddy rice seeds. Colloids Surf. B **32**, 29–34 (2003)
3. Zhao, H.C., Wang, B.C., Cai, S.X., Xi, B.S.: Effect of sound stimulation on the lipid physical states and metabolism of plasma membrane from chrysanthemum callus. Acta Bot. Sin. **44**, 799–803 (2002)
4. Jia, Y., Wang, B.C., Wang, X.J., Wang, D.H., Duan, C.R., Toyama, Y., Sakanishi, A.: Effect of sound wave on the metabolism of chrysanthemum roots. Colloids Surf. B **29**, 115–118 (2003)
5. Zhao, H.C., Wu, J., Zheng, L., Zhu, T., Xi, B.S., Wang, B., Cai, S., Younian, W.: Effect of sound stimulation on Dendranthema morifolium callus growth. Colloids Surf. B **29**, 143–147 (2003)
6. Matsuhashi, M., Otani, S., Kaneko, T.: Process for culturing microorganisms or cells using sound waves. United States Patent 5955334, pp. 15–17 (1999)
7. Schrödinger, E.: Are there quantum jumps? Part I. Br. J. Philos. Sci. **3**(10), 109–123 (1952). Published by: Oxford University Press on behalf of the British Society for the Philosophy of Science Stable. http://www.jstor.org/stable/685552
8. Pauling, L.: The Nature of the Chemical Bond and the Structure of Molecules and Crystals: An Introduction to Modern Structural Chemistry. 2nd edn, 664 p. Oxford University Press, London (91940)
9. Shnoll, S.: Physical-chemical factors of biological evolution [in Russian, «Fiziko-himicheskie factory biologicheskoy evolutsii»]. Moscow, Nauka (1979)
10. Darvas, G., Koblyakov, A., Petoukhov, S., Stepanyan, I.: Symmetries in molecular-genetic systems and musical harmony. Symmetry: Cult. Sci. **23**(3–4), 343–375 (2012)
11. Petoukhov, S.V.: Music and the modeling approach to genetic systems of biological resonances. In: Lecture at the International conference «ISIS Summit Vienna 2015. The information society at the crossroads», Vienna Austria, 3–7 June 2015. http://sciforum.net/conference/70/paper/2812
12. Petoukhov, S.V.: Resonances and genetic biomechanics. Symmetry: Cult. Sci. **26**, (3), 379–397 (2015). http://petoukhov.com/PETOUKHOV_IN_SCS_2015.pdf
13. Petoukhov, S.V.: The system-resonance approach in modeling genetic structures. Biosystems **139**, 1–11 (2016)
14. Petoukhov, S.V., Petukhova, E.S.: The concept of systemic-resonance bioinformatics. Resonances and the quest for transdisciplinarity. In: Burgin, M., Hofkirchner, W. (eds.) Information Studies and the Quest for Transdisciplinarity. World Scientific Series in Information Studies, vol. 9, chap. 17, pp. 467–487, World Scientific (2017)
15. Gladwell, G.M.L.: Inverse Problems in Vibration. Kluwer Academic Publishers, London (2004)
16. Bellman, R.: Introduction to Matrix Analysis. Mcgraw-Hill Book Company, Inc., New-York (1960)
17. Balonin, N.A.: New course on the theory of motion control (Novyi kurs teorii upravleniia dvizheniem). Saint Petersburg State University, Saint Petersburg (2000). (in Russian)
18. Shushardzhan, S.V.: The method of changing the activity of microorganisms in vitro. Patent No. 2195493. Russian Patent and Trademark Agency (2002)
19. Shushardzhan, S.V.: The method for potentiating the therapeutic effect by a combination of musical-acoustic and medicinal effects. Patent No. 22240146. The Federal Service for Intellectual Property, Patents and Trademarks (2004)

20. Shushardzhan, S.V.: The method for correcting the state of coagulation of human blood. Patent No. 2336098. The Federal Service for Intellectual Property, Patents and Trademarks (2008)
21. Shushardzhan, S.V.: The method of healing and rejuvenation of the skin. Patent number 2429026. Registered in the Russian State Register of Inventions (2011)
22. Shushardzhan, S.V.: Software and acoustic complex Profi-Grand. Patent number 126602. Registered in the State Register of Inventions of the Russian Federation (2013)
23. Shushardzhan, S.V., Shushardzhan, R.S.: Device for musical acoustic magnetic-vacuum effects on acupuncture points and reflexogenic zones Akuton. Patent number 129398. Registered in the State Register of Inventions of the Russian Federation (2013)
24. Shushardzhan, S.V.: Device for skin rejuvenation and recovery Bonnie Grand. Patent number 129820. Registered in the Russian State Register of Inventions (2013)
25. Shushardzhan, S.V.: The method of enhancing the growth of leukocyte mass and the complex correction of the blood in Vitro. Patent number 2518534. Registered in the State Register of Inventions of the Russian Federation (2014)
26. Shushardzhan, S.V.: The method of neuro-hormonal correction and rejuvenation with the help of musical-acoustic effects. Patent No. 2518538. Registered in the State Register of Inventions of the Russian Federation (2014)
27. Petoukhov, S.V.: Matrix genetics, algebras of the genetic code, noise immunity. M., RCD, 316 p. (2008). (in Russian)
28. Petoukhov, S.V., He, M.: Symmetrical Analysis Techniques for Genetic Systems and Bioinformatics: Advanced Patterns and Applications. IGI Global, Hershey (2009)
29. Hu, Z.B., Petoukhov, S.V.: Generalized crystallography, the genetic system and biochemi cal esthetics. Struct. Chem. **28**(1), 239–247 (2017). https://doi.org/10.1007/s11224-016-0880-0
30. Angadi, S.A., Hatture, S.M.: Biometric person identification system: a multimodal approach employing spectral graph characteristics of hand geometry and palmprint. Int. J. Intell. Syst. Appl. (IJISA), **3**, 48–58 (2016). http://www.mecs-press.org/ijisa/ijisa-v8-n3/IJISA-V8-N3-6.pdf
31. Sahana, S.K., AL-Fayoumi, M., Mahanti, P.K.: Application of Modified Ant Colony Optimization (MACO) for multicast routing problem. Int. J. Intell. Syst. Appl. (IJISA), **4**, 43–48 (2016). http://www.mecs-press.org/ijisa/ijisa-v8-n4/IJISA-V8-N4-5.pdf
32. Algur, S.P., Bhat, P.: Web video object mining: a novel approach. Int. J. Intell. Syst. Appl. (IJISA), **4**, 67–75 (2016). http://www.mecs-press.org/ijisa/ijisa-v8-n4/IJISA-V8-N4-8.pdf
33. Hata, R., Akhand, M.A.H., Islam, M.M., Murase, K.: Simplified real-, complex-, and quaternion-valued neuro-fuzzy learning algorithms. Int. J. Intell. Syst. Appl. (IJISA), **10**(5), 1–13 (2018). https://doi.org/10.5815/ijisa.2018.05.01
34. Awadalla, M.H.A.: Spiking neural network and bull genetic algorithm for active vibration control. Int. J. Intell. Syst. Appl. (IJISA), **10**(2), 17–26 (2018). https://doi.org/10.5815/ijisa.2018.02.02
35. Abuljadayel, A., Wedyan, F.: An approach for the generation of higher order mutants using genetic algorithms. Int. J. Intell. Syst. Appl. (IJISA) **10**(1), 34–35 (2018)
36. Razumov, A.N., Shushardzhan, S.V.: Methods of music therapy (manual for doctors). RNTSVM and K, Ministry of Health of the Russian Federation, Moscow (2003)

User Keystroke Authentication and Recognition of Emotions Based on Convolutional Neural Network

Ihor Tereikovskyi[1](✉), Liudmyla Tereikovska[2], Oleksandr Korystin[3], Shynar Mussiraliyeva[4], and Aizhan Sambetbayeva[4]

[1] National Technical University of Ukraine "Igor Sikorsky Kyiv Polytechnic Institute", Kiev, Ukraine
terejkowski@ukr.net
[2] Kyiv National University of Construction and Architecture, Kiev, Ukraine
tereikovskal@ukr.net
[3] Scientifically Research Institute of the Ministry of Internal Affairs, Kiev, Ukraine
alex@korystin.pro
[4] Al-Farabi Kazakh National University, Almaty, Kazakhstan
mussiraliyevash@gmail.com,
aizhan.sambetbaeva@kaznu.kz

Abstract. The article is devoted to the problem of improving Biometric identification systems based on Keystroke Dynamics for recognizing emotions and authenticating users of information systems through the implementation of modern neural network solutions based on Convolutional Neural Network (CNN). It is established that the difficulties of such implementation are associated with coding the keystroke parameters to a form suitable for CNN processing. A coding procedure based on the presentation of fixed-size keystroke parameters in the form of a color square image is proposed. Each encoded text symbol corresponds to a separate point of the image and is characterized using the corresponding ASCII code and keystroke parameters such as the key hold time and the time between keystrokes. Experimental studies showed that the proposed coding procedure made it possible to use CNN for analyzing Keystroke Dynamics and achieve recognition error of emotions and personality at the level of the best modern recognition systems.

Keywords: Recognition of emotions · Biometric authentication · Keystroke Dynamics · Convolutional neural network

1 Introduction

Currently, one of the most urgent tasks in the field of information technology is the development of effective emotion recognition tools (ERT) for users of information systems (IS) for various purposes. These tools are necessary, for example, for the operational monitoring of operators of critical infrastructure, where a significant

Z. Hu et al. (Eds.): AIMEE 2019, AISC 1126, pp. 283–292, 2020.
https://doi.org/10.1007/978-3-030-39162-1_26

number of accidents and emergency incidents are associated with a violation of their emotional state [1, 2]. Another example is the use of ERT in distance learning IS for automatic control of the perception of educational materials.

ERT on the basis of Keystroke Dynamics has a quite broad prospects because of using standard peripheral equipment to obtain biometric characteristics, widespread using of text information in information systems, inalienability of the owner's identity, complexity of biometric information falsification, the possibility of covert monitoring during of professional activity [1–4].

2 Analysis of Literature Sources in the Field of Research

The concept of Keystroke Dynamics can be understood as user's individual biometric behavioral characteristic, which determines the features of typing text from the keyboard [4, 5]. It is believed that each person has his own keystroke dynamics, which is described by the speed and dynamics of typing, the transition times between several (two or more) keys and typing errors....

When using universal keyboard input tools, the key code, the character corresponding to the key, the time the key is pressed and the time it is released can be registered for determining the keystroke parameters. In specialized systems of keyboard input for this, parameters that characterize the user's pressure on the key and the speed/acceleration of pressing can additionally be used. In the future, before entering the recognition system, these parameters are subject to filtering and primary processing, during which some general keystroke parameters are calculated.

A typical filtering procedure is given in [3–6] and consists in applying the so-called frequency, temporal, and keyboard filters. Analysis of sources [4–12] suggests that today there is no generally accepted methodology for forming a set of input data of the Keystroke Dynamics recognition module. In this case, a common feature of the most well-known Keystroke Dynamics recognition systems is the use of filtered values of the hold time of individual keys (TIK) and the time between pressing two and three separate keys (TMK) as input data. It is also possible to determine the dependence of the input data nomenclature on the mathematical support of the recognition process, which reduces to a comparison of the input sequence with the Keystroke Dynamics reference of a certain emotion of the IS user. There are two types of models of recognition: by a predetermined text and by the text of arbitrary content. In both cases, to determine Keystroke Dynamics patterns, user must enter one or several fragments of the same text several times.

Most of the known patterns are statistical models of Keystroke Dynamics parameters, for example, based on the normal or bimodal distribution [4–6]. In the case of Keystroke Dynamics recognition on the basis of a certain text fragment, the basis of the patterns, as a rule, is made up of the indicators of TIK and TMK concerning the sequential order of keystrokes. When recognizing an arbitrary content on the basis of a fragment of text, the patterens, for the most part, are formed on the basis of the statistical indicators of the TIK and TMK of individual stable sequences of keyboard events, reflecting the features of the typing dynamics of an individual user. However, in [4, 7], on the basis of experimental studies, the unsatisfactory adaptation of statistical models to changes in the user's emotional state and the complexity of their formation in the case of analyzing a text fragment of arbitrary content was proved. It is also shown that the complexity of the Keystroke Dynamics recognition problem is explained by the need to analyze rather large volumes of multidimensional data. At the same time, just from the point of view of the proven efficiency of analyzing multidimensional data, a promising direction for improving Keystroke Dynamics recognition systems is the use of neural network models [8–10]. This is confirmed by the data of works [6, 11], in which, along with the positive results of the application of neural network models, their limitations associated with the difficulties of forming the nomenclature of input parameters when analyzing the Keystroke Dynamics of an arbitrary text. You can also note the obsolescence of the used neural network models of the multilayer perceptron type, probabilistic neural networks, Kohonen maps, Hopfield networks, RBF networks and fully connected deep neural networks with direct propagation of the signal [4, 11, 12]. It is also worth noting the possibility of Keystroke Dynamics recognizing with the help of neural network models of type LSTM [13, 14], which allows processing texts of arbitrary length. However, the construction of such neural network models is associated with the complexity of the formation of an academic sample.

At the same time, based on the methods of developing neural network information protection tools, we can assume the expediency of using the neural network models type of convolutional neural network (CNN) in ERT. So in [14–17], a method for converting keystroke data into an image was proposed for subsequent use in CNN for user authentication. There is an opportunity on fixed texts to achieve recognition accuracy of 96.8%. Similar results were also obtained in [18], in which CNN was used to analyze Keystroke Dynamics in order to increase the resilience of user password protection. At the same time, in the works [14–16] listed above, the problems of the influence of the user's emotional state on the authentication accuracy, as well as the coding of CNN input parameters, are not fully covered. Therefore, the main objective of this study is to develop a procedure for coding the input parameters of a convolutional neural network to recognize the emotions and personality of the user using keyboard handwriting.

3 Coding Procedure

A feature of the classic version of CNN is the need to present the input information in the form of a square image. The basic version uses a black and white image. More sophisticated options involve the use of three-dimensional gray and color images. This feature imposes a significant limitation on the use of CNN - the ability to analyze Keystroke Dynamics on text fragments with a fixed number of characters. In this case, the general formulation of the problem of recognizing the emotions of the user of the IS implies the need to analyze both a predetermined fragment of the text and a fragment of the text of arbitrary content. The first case can be correlated with monitoring the emotional state of a user when he enters password data. The second case correlates with the current monitoring of the user's personal/emotional state when he enters text information from the keyboard. In accordance with the data of [4, 6, 15], such monitoring is feasible due to the analysis of Keystroke Dynamics with the introduction of text of a fixed length even when using relatively simplified statistical models. Thus, the above restriction on a fixed number of characters practically does not adversely affect the functionality of CNN. In the first case, the number of characters in the password data is already a fixed value. In the second case, the volume of a fragment of text of arbitrary content to be analyzed is limited to a predetermined number of characters.

Let us consider the principle of coding Keystroke Dynamics parameters into a black and white square image on which the proposed coding procedure is based. It is proposed to correlate the ordinate axis with the keyboard layout—ASCII codes of keys or characters corresponding to keys. X-axis proposed to correlate with the entered text. Thus, a single character of the entered text will correspond to one separate image point. On the axis of the co-ordinate of the encoded symbol corresponds to the position (number) of this symbol in the text. The coordinate of the ordinate corresponds to the position of the character on the keyboard/ASCII code of the previous character in the text. It is assumed that the first character on the keyboard corresponds to the space character. In the case when the number of characters of the text will be more than the number of characters on the keyboard/ASCII codes, then to save the square shape, the figure above the axis of ordinates is complemented by strings that correspond to the space character. If the number of characters of the text is less than the number of characters on the keyboard, then to save the form, the picture on the right will be supplemented with columns with space characters.

An illustration of the proposed method is Fig. 1, which shows a two-dimensional black-and-white image of the encoded text "AUTOMOBILE".

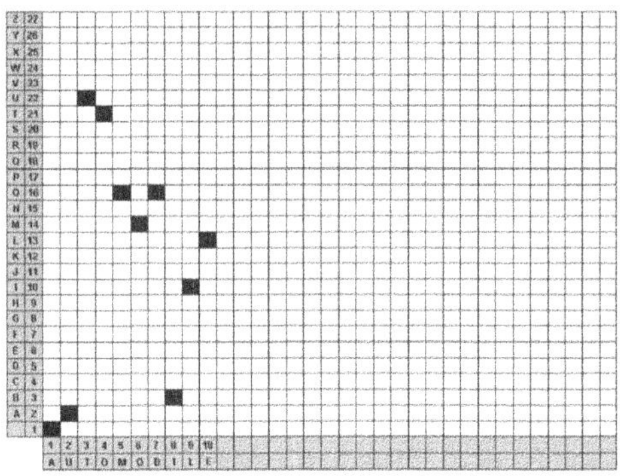

Fig. 1. Black and white display of the coded word "AUTOMOBILE"

To simplify the demonstration, the coding adopted the assumption that it is necessary to analyze the text, which consists exclusively of the capital letters of the English alphabet and the space character. In this figure, auxiliary fragments are highlighted in gray on which the characters/character number on the keyboard and the character number/character of the text to be encoded are displayed along the vertical axis. For clarity, the individual points of the image are separated by straight lines.

Each dot of the image filled with black corresponds to the coded value of the text symbol. For example, the symbol "T" corresponds to the black point of the image located at the intersection of the vertical, drawn from position No. 3 ("T") along the axis of the X-axis, and horizontal, drawn from position No. 22 ("U") to the axis of ordinates. This is explained by the fact that in the word "AUTOMOBILE" the letter "T" follows the letter "U". In numerical form, this figure represents a square matrix. The black spots of the pattern correspond to the elements of the matrix equal to 1, and the bright points to the elements of the matrix equal to 0.

The coding of the entered text in the form of a black and white image does not allow to realize the recognition of Keystroke Dynamics, which involves analysis, as the minimum of one of the main parameters of Keystroke Dynamics. Therefore, the proposed coding procedure involves presenting the entered text as a colored square image with a multichannel raster. In the base case, the procedure assumes two raster channels. Each point of such an image should characterize the entered symbol and one of the Keystroke Dynamics parameters relating to this symbol.

Based on the results [3–6], the expediency of the use of indicators of a TIK or TMK is determined, the calculation of which is implemented using the expressions:

$$y_r(i) = t_d(i) - t_u(i), \tag{1}$$

$$y_b(i, i+1) = t_d(i) - t_d(i+1), \tag{2}$$

where y_r – key hold time; t_d – key pressed time; t_u – key release time; y_b – time between successive pressing of two keys; i – keystroke number when entering text.

We note the complexity of using and, which is explained by the necessity of their registration in universal computer systems with an accuracy of millisecond. Therefore, in accordance with the recommendations of [3, 6], a Windows-oriented program was developed that allows registering values with an error equal to the duration of 50 process cycles. For an example in Figs. 2 and 3, it is shown the histograms of the TIK

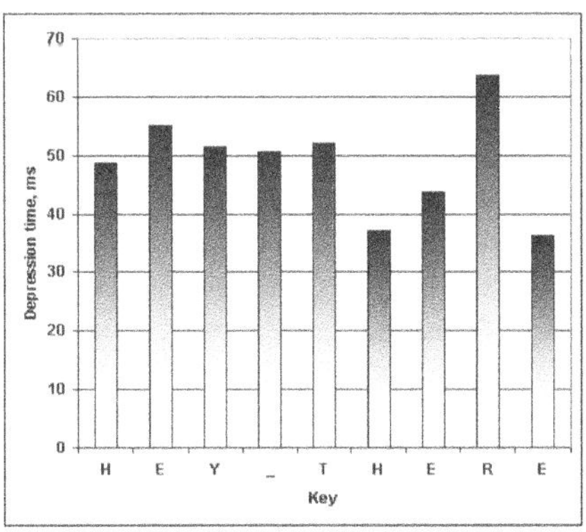

Fig. 2. Histogram of values of TIK for the text "HEY THERE"

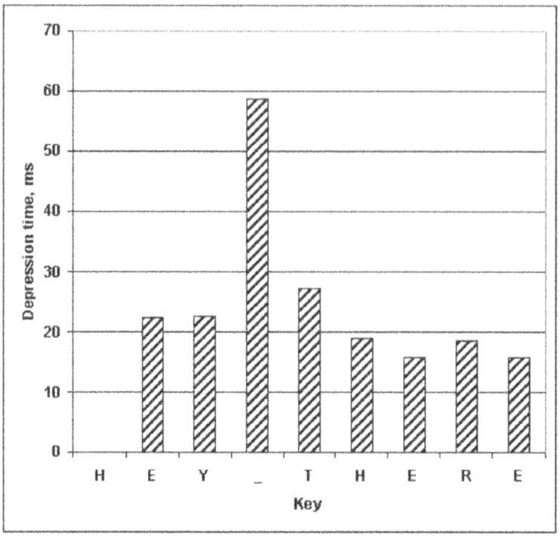

Fig. 3. Histogram of TMK values for the text "HEY THERE"

and TMK values for the text "HEY THERE", the input of which was recorded using the specified program.

Using of this program allowed to establish that the values of the TIK and TMK do not exceed 500 ms. Values of TIK and TMK which exceed this value should be filtered. It should be noted a significant dependence of TIK and TMK on the keyboard type.

So, according to [6, 8], on the keyboard shortcuts (laptop keyboard), on average, $y_r = 100$ ms, $y_d = 150$ ms, and for a keyboard with a long key stroke (standard keyboard), $y_r = 150$ ms, $y_d = 200$ ms. The y_r and y_d values obtained in the course of this study are approximately 1.5–2 times smaller. Therefore, the normalization of TIK and TMK to average values may entail errors associated with the characteristics of the keyboard used. Thus, at the entrance to CNN it is advisable to submit the absolute filtered values of the TIK and TMK. In the first approximation, the introduced symbol is proposed to be represented as a number corresponding to the ASCII code. To illustrate the results of the coding procedure described in Fig. 4 is a fragmentary representation of the display of the text "HEY THERE", encoded using TIK. Unlike Fig. 1 each point of the image corresponding to the encoded value of the character of the text, is characterized by two numbers recorded in the corresponding cell. The first digit is the ASCII – code of the character entered, and the second is the TIK. For example, the "Y" symbol corresponds to a dot in an image located at the intersection of the third column with the top row. The values of 121 (ASCII – code) and 51 (TIK) are presented in the corresponding cell.

A variant of the coding procedure for the entered text in the form of a colored square image with a multichannel raster is the development of the basic case in the direction of adding channels that correspond to the parameters of the CP. The coding result in numerical form is a multidimensional matrix, the depth of which is equal to the number of parameters used in the Keystroke Dynamics recognition. In order to verify the proposed coding procedure, we performed experiment for user keystroke

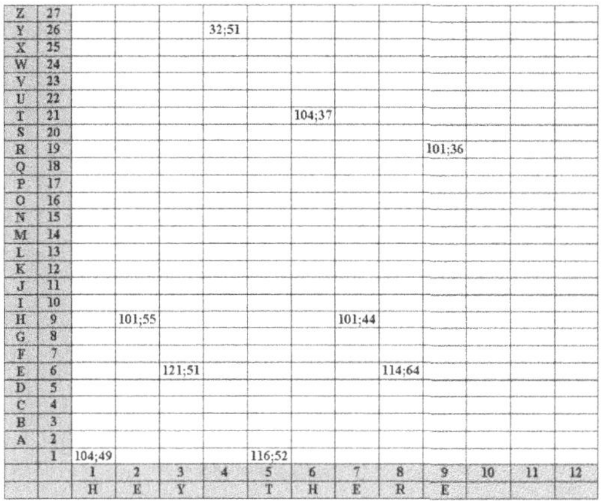

Fig. 4. Display of the text "HEY THERE" encoded using the TIK parameter as an image with a two-channel raster

authentication and recognition of his emotional state. The following terms and conditions apply to CNN: the number of recognizable users is 8; three type of emotions are to be recognized - neutral, joy and fear; text may consist of capital letters of the English alphabet and 6 punctuation marks (a total of 32 characters); the length of analyzed texts is 28 characters; keystroke dynamics is described by the TIK and TMK parameters. These conditions are determined from the standpoint of the estimated nature of experimental studies, simplifying the formation of a training sample and the possibility of a correct representation of the CNN input field. Based on the recommendations [9, 10, 19] the well-known LeNet architecture with the following parameters was used for the experiments: input field size – 32 $\times$ 32, number of output neurons - 11, number of convolution layers - 2, number of subsampling layers - 2, the number of fully connected layers is 2, the size of the convolution kernels is 5 $\times$ 5, the number of convolution maps in the first layer is 6, the number of convolution maps in the second layer is 16, the number of neurons in the first fully connected layer is 120, the number of neurons in the second fully connected layer is 84. The neural network model was implemented using the MATLAB 2018 application software package. For its training, a database of filtered keyboarding samples was used, corresponding to the three indicated emotions for 8 people. 80 records of keyboarding (10 entries per person) for the same text was used to represent one emotion. Watching appropriate multimedia provided a certain emotion of tested users. 90% of the database records were used to form the training sample, the remaining 10% were used for the test sample. The magnitude of errors in recognition of the user's identity and emotional state, calculated from the results of the experiments performed are shown in Fig. 5.

Note that although the experiments in this study used rather outdated neural network solutions, the errors obtained correlate with the errors of the best modern systems [3, 6, 11–22]. At the same time, the recognition accuracy was negatively affected by the relatively small amount of the training sample and the use of the outdated type of CNN architecture. This suggests that a fairly low recognition error was achieved due to the

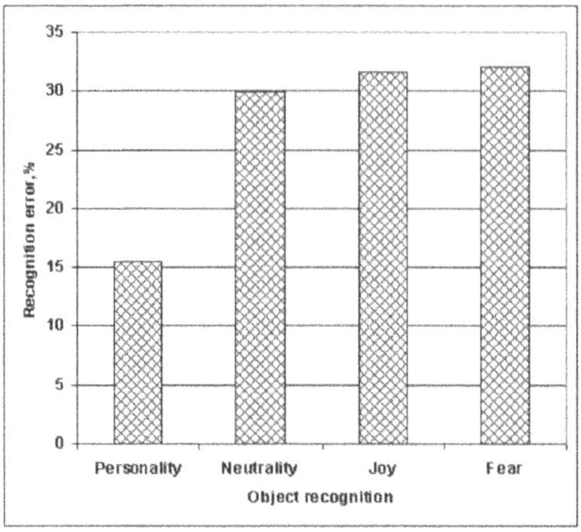

Fig. 5. Recognition error histograms

use of the proposed Keystroke Dynamics coding procedure. In addition, the experiments conducted showed the possibility of integral recognition of the emotions and personality of the operator. Thus, the results of the experiments confirm the promise of using the proposed procedure of coding, both in emotion recognition systems, and in user authentication systems. It can also be argued that the developed Keystroke Dynamics coding procedure can be used as the first stage of the method of adapting the architectural parameters of modern CNN types to the task of integrated recognition of emotions and user authentication by Keystroke Dynamics.

4 Conclusions

The article is devoted to solving the problem of improving the analysis systems on the base of Keystroke Dynamics for recognizing emotions and authenticating users of information systems through the introduction of modern neural network solutions based on convolutional neural networks. It has been established that the difficulties of such an implementation are connected with the need to encode the parameters of keyboard typing to a form suitable for processing a convolutional neural network. The coding procedure based on the representation of the parameters of the keyboard typing of the entered text of a fixed size in the form of a color image is substantiated. Each encoded character of the text is related to a separate point of the image and is characterized using the corresponding ASCII code and Keystroke Dynamics parameters. At the same time, the coordinate of the encoded symbol corresponds to the position of the given symbol in the text along the x-axis. The coordinate on the vertical axis corresponds to the position of the character on the keyboard of the previous character in the text. Experimental studies showed that the proposed coding procedure made it possible to use CNN for analyzing Keystroke Dynamics and achieve recognition error of emotions and personality at the level of the best modern recognition systems. It is proposed to correlate the paths of further research with the development of a method for adapting the architectural parameters of modern types of convolutional neural networks to the problem of integral recognition of emotions and user keystroke authentication, which will improve the accuracy of the corresponding recognition systems.

References

1. Gnatyuk, S.: Critical aviation information systems cybersecurity. In: Meeting Security Challenges Through Data Analytics and Decision Support, NATO Science for Peace and Security Series, D: Information and Communication Security, vol. 47, no. 3, pp. 308–316. IOS Press Ebooks (2016)
2. Gnatyuk, S., Sydorenko, V., Aleksander, M.: Unified data model for defining state critical information infrastructure in civil aviation. In: Proceedings of the 2018 IEEE 9th International Conference on Dependable Systems, Services and Technologies (DESSERT), Kyiv, Ukraine, 24–27 May 2018, pp. 37–42 (2018)
3. Kobojek, P., Saeed, K.: Application of recurrent neural networks for user verification based on keystroke dynamics. J. Telecommun. Inf. Technol. **N3**, 80–90 (2016)
4. Ivanov, A.I.: Nejrosetevye algoritmy biometricheskoj identifikacii lichnosti. Kn. 15: Monografiya/A.I. Ivanov. – M.: Radiotehnika (2004). 144 s
5. Koshevaya, N.A., Maznichenko, N.I.: Podhod k povysheniyu nadezhnosti identifikacii polzovatelej kompyuternyh sistem po dinamike napisaniya parolej Sistemi obrobki informaciyi, vipusk **6**(122 c), 140–146 (2014)
6. Savinov, A.N. Matematicheskaya model mehanizma raspoznavaniya klaviaturnogo pocherka na osnove Gaussovskogo raspredeleniya/A.N. Savinov, I.G. Sidorkina // Izvestiya Kabardino-Balkarskogo nauchnogo centra RAN. Vyp. I. - Nalchik: Kabardino-Balkarskij nauchnyj centr RAN, - S, pp. 26–32 (2013)
7. Aitchanov, B., Korchenko, A., Tereykovskiy, I., Bapiyev, I.: Perspectives for using classical neural network models and methods of counteracting attacks on network resources of information systems. News of the national academy of sciences of the republic of Kazakhstan series of geology and technical sciences 2017, vol. 5, no. 425, pp. 202–212 (2017)
8. Maheshwary, S., Ganguly, S., Pudi, V.: Deep secure: a fast and simple neural network based approach for user authentication and identification via keystroke dynamics. In: Conference: IWAISe, International Joint Conference on Artificial Intelligence (IJCAI), Melbourne, Australia, pp. 34–40 (2017)
9. Dychka, I., Tereikovskyi, I., Tereikovska, L., Pogorelov, V., Mussiraliyeva, S.: Deobfuscation of computer virus malware code with value state dependence graph. In: Hu, Z., Petoukhov, S., Dychka, I., He, M. (eds.) ICCSEEA 2018. AISC, vol. 754, pp. 370–379. Springer, Cham (2019). https://doi.org/10.1007/978-3-319-91008-6_37
10. Tereikovskyi, I., Chernyshev, D., Tereikovska, L.A., Mussiraliyeva, S., Akhmed, G.: The procedure for the determination of structural parameters of a convolutional neural network to fingerprint recognition. J. Theor. Appl. Inf. Technol. **97**(8), 2381–2392 (2019)
11. Akhmetov, B., Tereykovsky, I., Doszhanova, A., Tereykovskaya, L.: Determination of input parameters of the neural network model, intended for phoneme recognition of a voice signal in the systems of distance learning. Int. J. Electron. Telecommun. **64**(4), 425–432 (2018)
12. Alghamdi, S.J., Elrefaei, L.A.: Dynamic user verification using touch keystroke based on medians vector proximity. In: 2015 7th International Conference on Computational Intelligence, Communication Systems and Networks (CICSyN), pp. 121–126. IEEE (2015)

13. Bo, C., Zhang, L., Jung, T., Han, J., Li, X.-Y., Wang, Y.: Continuous user identification via touch and movement behavioral biometrics. In: 2014 IEEE 33rd International Performance Computing and Communications Conference (IPCCC), pp. 1–8. IEEE (2014)
14. Deng, Y., Zhong, Y.: Keystroke dynamics advances for mobile devices using deep neural network. GCSR **2**, 59–70 (2015)
15. Xiaofeng, L., Shengfei, Z., Shengwei, Y.: Continuous authentication by free-text keystroke based on CNN plus RNN. Proc. Comput. Sci. **147**, 314–318 (2019)
16. Liu, M., Guan, J.: User keystroke authentication based on convolutional neural network. Commun. Comput. Inf. Sci. **971**, 157–168 (2019)
17. Lin, C.-H., Liu, J.-C., Lee, K.-Y.: On neural networks for biometric authentication based on keystroke dynamics. Sens. Mater. **30**(3), 385–396 (2018)
18. Çeker, H., Upadhyaya, S.: Sensitivity analysis in keystroke dynamics using convolutional neural networks. In: 2017 IEEE Workshop on Information Forensics and Security (WIFS), 4–7 December 2017, pp. 1–6 (2017)
19. Tereykovska, L., Tereykovskiy, I., Aytkhozhaeva, E., Tynymbayev, S., Imanbayev, A.: Encoding of neural network model exit signal, that is devoted for distinction of graphical images in biometric authenticate systems. News of the national academy of sciences of the republic of Kazakhstan series of geology and technical sciences 2017, vol. 6, no. 426, pp. 217–224 (2017)
20. Malik, J., Girdhar, D., Dahiya, R., Sainarayanan, G.: Reference threshold calculation for biometric authentication. Int. J. Image, Graph. Signal Process. (IJIGSP), **6**(2), 46–53 (2014)
21. Oyedotun, O.K., Dimililer, K.: Pattern recognition: invariance learning in convolutional auto encoder network. Int. J. Image, Graph. Signal Process. (IJIGSP) **8**(3), 19–27 (2016)
22. Sassi, A., Ouarda, W., Amar, C.B., Miguet, S.: Sky-CNN: a CNN-based learning approach for skyline scene understanding. Int. J. Intell. Syst. Appl. (IJISA), **11**(4), 14–25 (2019)

Advances in Technological and Educational Approaches

A Control Strategy for Vehicles in a Traffic Flow Aimed at the Fastest Safe Motion

Andrey M. Valuev(✉)

Mechanical Engineering Research Institute of the Russian Academy of Sciences, 4, Malyi Kharitonievsky Pereulok, 101990 Moscow, Russia
valuev.online@gmail.com

Abstract. The paper discusses the issue of driving a car in a traffic flow that provides the highest speed under conditions of security guarantees. The basic assumption is an accurate knowledge of the speed of the controlled vehicle and its predecessor and the distance to it. The way to define the control law implementing the formulated objective is proposed for a separate vehicle as well as for a chain of connected autonomous vehicles. The results obtained can be used in two ways: (1) as a control program (in the "smart advice-tick" mode for a vehicle controlled by the driver and in the autopilot mode); (2) for estimating the maximum throughput of sections of the urban road network by computer simulation. In the latter case, an adequate means is the use of the obtained control as an element of the general "microscopic" model of traffic flow, which it is expedient to formulate in the form of a hybrid dynamical system—an event-switched process. Computational and information aspects as well as perspectives of the approach development are discussed.

Keywords: Vehicular traffic flow · Safety · Intelligent control system · Control law · Connected autonomous vehicles · Road throughput

1 Introduction

Intellectualization of control and decision making in complicated environments, especially including both elements with human and automatic controls or action of assemblies of intelligent robots is now the realm of intensive research [1–5]. Problems related with transport systems are among research topics [4, 5]. In more narrow sense, nowadays emerges the problem of driving automation that is treated very wide, both in theoretical and practical ways. Concepts of advanced driver assistance systems (ADASs), adaptive cruise control (ACC) and even connected autonomous vehicles (CAVs) are developed either separately or in combination [6, 7]. Some features of the proposed systems are already implemented in some new car models. The related problem of data transmission in Vehicle-to-Vehicle (V2V) and Vehicle-to-Infrastructure (V2I) communication is being solved as well [8, 9]. On the other hand, emerging opportunities to improve vehicle driving must lead both to better safety of their passengers and to the whole traffic situation improvement. The latter mean the improvement of conditions for satisfying social demands for transportation.

Z. Hu et al. (Eds.): AIMEE 2019, AISC 1126, pp. 295–304, 2020.
https://doi.org/10.1007/978-3-030-39162-1_27

The driving control choice description is incorporated in car-following models that were developed for decades and serve as the most important way of "microscopic" traffic simulation [10–12]. There are several approaches to them. The empirical model by Tanaka and similar ones express average choice of drivers in real traffic. On the contrary, the well-known "intelligent driver model" by Treiber tries to express the reasonable driving and may serve as the recommendation to drivers. Recently the practical approach was proposed under the title "model predictive control" (MPS) [6, 7] that is aimed at many objectives including not only collision avoidance but fuel economy and ride comfort as well. All these techniques aimed at development of recommendations for the vehicle control are, however, short-sighed. They do not consider the local traffic situation evolution for longer periods from the viewpoint of the present vehicle resources use to avoid dangerous situations and achieve the goals.

The proposed approach to strategic control in vehicle driving is aimed to combine both aims of traffic processes participants, namely the safety maximization and the trip time minimization that, in turn, must lead to better use of the road network capacity. It lies in the field of "car-following models". In the paper the model of driving control is proposed that maximizes the vehicle speed in a traffic flow with guaranteed absence of collisions. Such model is developed not only for an individual vehicle but for a chain of connected autonomous vehicles as well.

Section 2 introduces the original principal approach for determination of the fastest safe trajectory for a vehicle in a traffic flow. Section 3 presents the problem setup and solution method for the case of independent vehicles on a lane. Section 4 proposes its generalization for a chain of CAVs and regards its computational and informational issues. In Sect. 5 prospects of the approach development are discussed.

2 Aims and Restrictions for the Control Choice

In modern conditions of the urban road network (URN), the main motive for most drivers should be the goal to reach their destination as soon as possible, without exposing their lives and health to danger. It is this striving that leads to a congestion of traffic flows, attempts to choose a lane, on which it is possible to move somewhat faster in the next few minutes. In reality, a typical driver does not consciously violate traffic regulations, but with regard to the assessment of safety conditions, he acts intuitively, based on inaccurate and late-arriving information [10]. As to empirical data, the actual variety of driving styles typical for Moscow highways shows the Fig. 1 [13]. Note that styles with minimal (deficient) headways demand instantaneous reaction and lead to collisions in the case of drivers' mistakes whereas styles with maximal (redundant) headways lead to underemployment of road capacity measured in tens of percent.

The purpose of the present paper is to formulate and solve the problem of the a car driver (or the autopilot replacing him) to choose an acceleration control ensuring the fastest possible movement with a guaranteed absence of a collision with the previous car (leader) and avoiding violation of speed restrictions. In this case, it is assumed that the driver (autopilot) knows the current values of the coordinates and speeds of his own car and the leader; with regard to acceleration, it is assumed knowledge of the range of

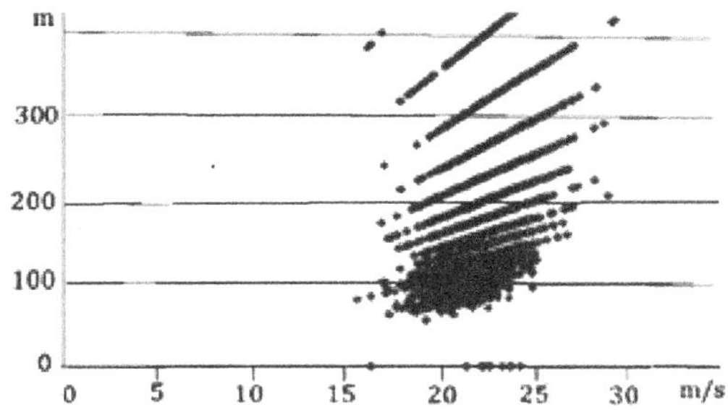

Fig. 1. Dependencies between the speed and the headway for different types of driving

possible and acceptable values of acceleration for your own car, whereas for the leader only the maximum deceleration is known.

3 Control Law for Fastest Safe Motion in the Case of Single-Lane Traffic of Independent Vehicles

First of all, consider the succession of moving vehicles along a straight horizontal road lane during the time interval $[T_0, \infty)$. Their motion is subject to equations

$$\ddot{x}_i = a_i(t) \tag{1}$$

with the acceleration and the speed satisfying restrictions

$$-b_{\max i} \le a_i(t) \le a_{\max i}, \tag{2}$$

$$0 \le \dot{x}_i(t) \le v_{\max}. \tag{3}$$

We begin with the motion of a pair of successive vehicles traditionally named the "leader" and the "follower" and use for them values $i = 0$ and $i = 1$, resp. The problem in question is to find the control law for the follower providing the fastest safe motion, the safety consisting in satisfying the condition of the collision absence for any $t \in [T_0, \infty)$

$$x_{i+1}(t) \le x_i(t) - L_i \tag{4}$$

for an arbitrary possible motion of the leader. This control law must be based on knowledge of the current positions and speeds of both vehicles; for the follower a narrower range for the acceleration $[-b_{\text{norm}\max 1}, a_{\text{norm}\max 1}]$ is considered acceptable for selection. The latter difference results from the lack of exact information that the follower has about the leader.

We call the quadruple $\{x_i(T_0), x_{i+1}(T_0), \dot{x}_i(T_0), \dot{x}_{i+1}(T_0)\}$ *a safe configuration* in the phase space for the pair "leader i – follower $i + 1$" if for any choice of $a_i(t)$, $t \in [T_0, \infty)$ due to (2) and (3) there exists $a_{i+1}(t)$, $t \in [T_0, \infty)$, satisfying

$$-b_{\text{norm max}\, i+1} \leq a_i(t) \leq a_{\text{norm max}\, i+1}, \tag{5}$$

such that for the $(i + 1)$-th vehicle for any $t \in [T_0, T_1]$ conditions (3) and (4) are satisfied. Let's call the dependence $A_{i+1}(x_i(t), x_{i+1}(t), \dot{x}_i(t), \dot{x}_{i+1}(t))$ *a safe positional control* (SPC) for the $(i+1)$-th vehicle if with any safe configuration as an initial state and with any choice of $a_i(t)$, $t \in [T_0, T_1]$, in accordance with (2) and (3), the trajectory of $(i+1)$-th vehicle generated by the control $a_{i+1}(t) = A_{i+1}(x_i(t), x_{i+1}(t), \dot{x}_i(t), \dot{x}_{i+1}(t))$ and this control itself satisfy (3), (4), (5) for all $t \in [T_0, T_1]$. In other words, a safe positional control is a control law for the acceleration of the follower vehicle in the form of a dependence on current positions and velocities of the both leader and follower vehicles that guarantee the absence of collisions of these vehicles independently of any possible motion of the leader. It may be applied if current values of $x_i(t), x_{i+1}(t), \dot{x}_i(t), \dot{x}_{i+1}(t)$ form a safe configuration; this restriction is natural since in the situation when the follower velocity is too great and the distance between vehicles is too small there is no control for the follower that avoids collision.

To establish relationships that determine the safe configuration we first of all note that it depends not on positions of both vehicles as such but on

$$S_i(t) = x_i(t) - L_i - x_{i+1}(t). \tag{6}$$

As the safety condition (4) is $S_i(t) \geq 0$ for all $t \in [T_0, \infty)$ let us look carefully on (6). Coordinates of both vehicles enter the expression independently. So for the minimum value of $S_i(t)$ (for a certain t and a certain value of $x_{i+1}(t)$) $x_i(t)$ must have the minimum possible value and, v.v., for the maximum value of $S_i(t)$ (for a certain t and a certain value of $x_i(t)$) $x_{i+1}(t)$ must have the minimum possible value. From (1)–(3) follows that the below formula yields the minimum possible $x_i(t)$ denoted by $\dot{x}_i^*(t)$ for all $t \in [T_0, \infty)$:

$$a_i^*(t) = -b_{\max i}, \text{if } \dot{x}_i^*(t) > 0, \quad \text{otherwise } a_i^*(t) = 0. \tag{7}$$

In other words, $\{x_i^*(t), t \in [T_0, \infty)\}$ is the trajectory of the fastest braking for the i-th vehicle. In turn, the guaranteed maximum of minimum distance $\inf_{t \geq 0} S_i(t)$. between the leader and the follower is reached when the follower fulfils the fastest braking satisfying (6). The value $\inf_{t \geq 0} S_i(t)$ depends on initial values of $S_i(t)$ and speeds of both vehicles; setting $i = 0$ and $T_0 = 0$, we denote it $F_1(S_0(0), v_0(0), v_1(0))$. From (1), (3), (7) the following formulas for its calculation are derived:

1. The case $b_{\max 0} \geq b_{\max 1}$ (it is typical when both vehicles are of the same category). The earliest time of the i-th vehicle stop is $t_i^* = v_i(0)/b_{\max i}$. If $t_1^* < t_0^*$, then the minimum distance between the leader and the follower is achieved at the moment of the follower stop, so

$$F_1(S_0(0), v_0(0), v_1(0)) = S_0(0) - \frac{v_1(0)t_1^*}{2} + \left(v_0(0) - \frac{b_{\max 0} t_1^*}{2}\right)t_1^*. \tag{8}$$

Otherwise the minimum distance is

$$F_1(S_0(0), v_0(0), v_1(0)) = S_0(0) + \frac{v_0(0)t_0^*}{2} - \frac{v_1(0)t_1^*}{2}. \tag{9}$$

2. The case $b_{\max 0} < b_{\max 1}$ (it is possible when the leader is a heavy truck). If $v_0(0) \geq v_1(0)$, then

$$F_1(S_0(0), v_0(0), v_1(0)) = S_0(0). \tag{10}$$

Otherwise, the earliest moment at which non-zero speeds of the both may become equal is determined: $t_{01}^* = (v_1(0) - v_0(0))/(b_{\max 1} - b_{\max 0})$. If $t_{01}^* \geq t_0^*$, then speeds do not become equal until both vehicles stop. The case is analogous to the above and $F_1(S_0(0), v_0(0), v_1(0))$ is computed by (9), otherwise the minimum distance is

$$F_1(S_0(0), v_0(0), v_1(0)) = S_0(0) + (v_1(0) - v_0(0))t_{01}^*/2. \tag{11}$$

As the condition

$$F_1(S_i(T_0), v_i(T_0), v_{i+1}(T_0)) \geq 0 \tag{12}$$

means that the safety condition (4) may be satisfied for any $t \in [T_0, \infty)$ by the follower control (7) independently of the possible motion of the leader, each combination of $S_i(T_0), v_i(T_0), v_{i+1}(T_0)$ is treated as a safe configuration.

It must be emphasized that if (12) is satisfied for $t = T_0$, then with the adequate control choice the condition

$$F_1(S_i(t), v_i(t), v_{i+1}(t)) \geq 0 \tag{13}$$

would be satisfied as well for any $t \in [T_0, \infty)$ (it follows from the above consideration) and namely (13) guarantee the possibility of safe motion on $[t, \infty)$. It is easy to see that retaining the value $F_1(S_i(t), v_i(t), v_{i+1}(t))$ non-negative but very small provides the *fastest safe motion* of the follower. Ideally, $F_1(S_i(t), v_i(t), v_{i+1}(t))$ should be permanently zero value; however, it is practically impossible without an accurate measurement of the leader vehicle acceleration. If it were possible we could define the follower control for the fastest safe motion in this way:

$$a_{\mathrm{FSM}i+1}(t) = -\left((\dot{x}_i - \dot{x}_{i+1}(t)) \cdot \frac{\partial F_1(S_i(t), v_i(t), v_{i+1}(t))}{\partial S_i(t)} + \frac{\partial F_1(S_i(t), v_i(t), v_{i+1}(t))}{\partial v_i(t)} a_i(t)\right) \Big/ \frac{\partial F_1(S_i(t), v_i(t), v_{i+1}(t))}{\partial v_{i+1}(t)} \tag{14}$$

$$f_i(t) = F_1(S_i(t), v_i(t), v_{i+1}(t)), \tag{15}$$

$$a_{i+1}(t) = \begin{cases} a_{\text{norm max}\, i+1}, & \text{if} f_i(t) > 0 \text{ and } \dot{x}_{i+1}(t) < v_{\max}, \\ 0, & \text{if} f_i(t) > 0 \text{ and } \dot{x}_{i+1}(t) = v_{\max}, \\ a_{\text{FSM}i+1}(t), & \text{if} f_i(t) = 0. \end{cases} \quad (16)$$

The formula (14) may be used, however, for theoretical estimations when a definite leader control is known (for our purposes, it is mainly $a_i^*(t)$). For practical use, the approximation of the ideal follower control may be performed using formulas (14) and

$$a_{i+1}(t) = \begin{cases} a_{\text{norm max}\, i+1}, & \text{if} f_i(t) \geq \varepsilon \text{ and } \dot{x}_{i+1}(t) < v_{\max}, \\ 0, & \text{if} f_i(t) \geq \varepsilon \text{ and } \dot{x}_{i+1}(t) = v_{\max}, \\ -b_{\text{norm max}\, i+1}, & \text{if } f_i(t) < \varepsilon. \end{cases} \quad (17)$$

Thus the control due to formulas (14), (15), (17) with a reasonable small ε gives the practically applied approximation of the fastest safe motion trajectory.

4 Generalization of the Above Results for Chains of CAVs

Informational exchange in a succession of vehicles on a road and the possibility to coordinate driving principally changes the task of providing safe traffic. Vague expectations with respect to the preceding vehicle driving lead to excessive caution which, in turn, causes insufficient use of road networks. To avoid this situation, the informational exchange and control coordination in chains of CAVs may serve. In the succession of CAVs only for the head vehicle the control choice is the same as for an individual vehicle. Next CAVs benefit from the use of the information on programs of their predecessors' driving.

The idea to define conditions on safe configuration stays the same and may be principally applied to CAVs' chains of an arbitrary length. Corresponding functions $F_2(S_0(T), S_1(T), v_0(T), v_1(T), v_2(T)), F_3(S_0(T), S_1(T), S_2(T), v_0(T), v_1(T), v_2(T), v_3(T))$ etc. are defined recursively. To define on-line the fastest safe controls for CAVs in the chain it is sufficient, however, to define not the general formulas for $F_i(S_0(T), \ldots, v_{i+1}(T))$ but only their numerical values for the current chain state in the moment of time T. It is fulfilled in the following way.

With the use of $F_1(S_0(T), v_0(T), v_1(T))$ we can determine for the 1st vehicle in the queue the control and the trajectory of the *fastest safe pursuit* (FSP) of its leader (that being the predecessor of the CAVs' chain) supposing that the leader trajectory is $x_1^*(t)$. For the queue head it is determined by generic formulas (14)–(16) in which it is necessary to substitute the above motion of the leader. For the further consideration it is necessary to view the particular forms of the formula (14). Analyzing formulas (8)–(11) we establish that there are only two cases when $F_1(S_0(T), v_0(T), v_1(T)) = 0$: in the 1st one the leader and the follower stop with the minimal admissible distance between them, the follower not earlier than the leader, in the 2nd one at the moment when the follower reaches the leader their speeds become equal; it is possible only if

$$b_{\max 0} \leq b_{\text{norm}\max 1}. \tag{18}$$

Let us denote T_{01} the moment of the leader stop. So, according to (16), there are four cases for the FSP trajectory of the 1st vehicle in the chain in the case where $F_1(S_0(T), v_0(T), v_1(T)) \geq 0$:

1. for $T \leq t < T_{11}$ $a_1(t) = a_{\text{norm}\max 1}$; $\dot{x}_1(T_{11}) = v_{\max}$; for $T_{11} \leq t < T_{12}$ $a_1(t) = 0$; for $T_{12} \leq t < T_{13}$ $a_1(t) = -b_{\text{norm}\max 1}$; $T_{13} \geq T_{01}$; $S_0(T_{13}) = 0$; $\dot{x}_1(T_{13}) = 0$; $NS(1) = 3$;
2. for $T \leq t < T_{11}$ $a_1(t) = a_{\text{norm}\max 1}$; $\dot{x}_1(T_{11}) < v_{\max}$; for $T_{11} \leq t < T_{12}$ $a_1(t) = -b_{\text{norm}\max 1}$; $T_{12} \geq T_{01}$; $S_0(T_{12}) = 0; \dot{x}_1(T_{12}) = 0$; $NS(1) = 2$;
3. for $T \leq t < T_{11}$ $a_1(t) = a_{\text{norm}\max 1}$; $\dot{x}_1(T_{11}) = v_{\max}$; for $T_{11} \leq t < T_{12}$ $a_1(t) = 0$; for $T_{12} \leq t < T_{13}$ $a_1(t) = -b_{\text{norm}\max 1}$; $\dot{x}_1(T_{13}) = \dot{x}_0(T_{13})$; $T_{13} < T_{01}$; $S_0(T_{13}) = 0$; for $T_{13} \leq t < T_{14} = T_{01} a_1(t) = -b_{\max 0}$; $NS(1) = 4$;
4. for $T \leq t < T_{11}$ $a_1(t) = a_{\text{norm}\max 1}$; $\dot{x}_1(T_{11}) < v_{\max}$; for $T_{11} \leq t < T_{12}$ $a_1(t) = -b_{\text{norm}\max 1}$; $\dot{x}_1(T_{12}) = \dot{x}_0(T_{12})$; $T_{12} < T_{01}$; $S_0(T_{12}) = 0$; for $T_{12} \leq t < T_{13} = T_{01} a_1(t) = -b_{\max 0}$; $NS(1) = 3$.

For all cases $\dot{x}_1\left(T_{1NS(1)}\right) = 0$, $a_1(t) = 0$ for $t \geq T_{1NS(1)}$. Cases 3 and 4 are possible only if (18) is valid. For convenience, we set $T_{i0} = T$ for each i in our consideration. To know what case takes place and establish values of $NS(1)$, $T_{11}, \ldots, T_{1NS(1)}$ the following calculations are fulfilled.

First of all, we note that for $T_{ij-1} \leq t \leq T_{ij}$

$$x_j(t) = x_j\left(T_{ij-1}\right) + \dot{x}_j\left(T_{ij-1}\right) * (t - T) + a_{ij} * \left(t - T_{ij-1}\right)^2/2. \tag{19}$$

If we assume that case 1 or 3 takes place, then the value of T_{11} is calculated via $\dot{x}_1(T_{11}) = v_{\max}$; for case 1 both T_{12} and T_{13} are calculated by the pair of equations $S_0(T_{13}) = 0$ and $\dot{x}_1(T_{13}) = 0$ by substitution the relationship between T_{12} and T_{13} from (19). For case 3 we use $\dot{x}_1(T_{13}) = \dot{x}_0(T_{13})$ as the second equation for calculation T_{12} and T_{13}. For cases 2 and 4 the pair (T_{11}, T_{12}) plays the same role as (T_{12}, T_{13}). Cases 1 or 3 takes place if $F_1(S_0(T_{11}), v_0(T_{11}), v_1(T_{11})) \geq 0$, otherwise cases 2 and probably 4 are being tested. After all the above calculations we obtain all the necessary variables that describe the FSP trajectory for the known initial state $S_0(T), v_0(T), v_1(T)$.

In analogous way the FSP trajectory is obtained for the next i. The simplest problem is to establish the value of $F_i(S_0(T), S_1(T), \ldots, S_{i-1}(T), v_0(T), \ldots, v_i(T))$ having the FSP trajectory for i. We assume that the i-th vehicle trajectory is $x_i^*(t)$. Thus from (19) it follows for both leader and follower trajectories that on each segment with constancy of both accelerations values $x_{i-1}(t)$ and $x_i(t)$ and thus $S_{i-1}(t)$ are polynomials of t of degrees not exceeding 2; so the minimum of $S_{i-1}(t)$ on the time segment may be calculated from a linear equation with respect to t unless it is reached on its end points. Further, to calculate the FSP trajectory for the i-th vehicle in the CAVs' chain, similar computations must be performed as for the 1st vehicle. It must be noted that for each i the following relationship linking the number of switches for subsequent vehicles takes place:

$$NS(i) \leq NS(i-1) + 3. \tag{20}$$

It must be emphasized that the FSP trajectories being calculated in most cases are not used directly for the control determination but only for determination of $F_i(S_0(T), S_1(T), \ldots, S_{i-1}(T), v_0(T), \ldots, v_i(T))$ values. Necessary calculations for a separate FSP trajectory are not laborious. The number of equations to be solved is few times $NS(i)$ that is not greater than $3i+1$ according to (20); the equations themselves are simple. Implementation of the proposed way of the control determination for chains of CAVs requires informational exchange in the form of a data stream from the preceding CAVs to succeeding ones beginning from the chain head. Not only the information on states of previous cars but also parameters of calculated FSP trajectories must be transferred from each leader to its follower. Fortunately, the total count of parameters needed in the FSP trajectory computation for the i-th vehicle from the FSP trajectory for the (i-1)-th vehicle and phase variables of all previous vehicles is only few times greater than i. So the established control law implementation is real in both computational and infocommunicational aspects.

5 Possible Applications and Perspectives of Further Development of the Control Strategy

The results obtained can be used in two ways: (1) as a control program (in the "smart advice-tick" mode for a vehicle controlled by the driver and in the autopilot mode); (2) for estimating the maximum throughput of sections of the urban road network. In the latter case, the most convenient way is computer simulation of the uneven flow of different vehicles implementing the proposed control laws, preferable with the use of relevant "microscopic" model of traffic flows that enables to take into account all the above conditions. It must be emphasized that from the variety of proposed microscopic models the recent model by the author [14] formulated in the form of a hybrid dynamical system—an event-switched process—seems to be the most relevant. We see its preference in two aspects. Firstly, it may process different characteristics of vehicles in the flow. The second aspect in which the model differs from the other ones seems the most important: it admits asynchronous changes of controls by different vehicles that follow from (16). Computational experiments with the model, including simulation of multi-lane traffic with vehicles' shifts between lanes [15] demonstrated its possibilities to catch most aspects that are necessary for the objective formulated here.

Control choice in the case of light cycle switchings may be considered as well in same manner. It must be noted that the knowledge of traffic lights timetable in some aspects simplifies the control problem, but mainly for CAVs' chains. On the contrary, actions of traffic lights increases uncertainty for separate cars since it is difficult to predict which driver makes a decision not to cross the intersection in last seconds of the green phase but to brake before the stop line. Further research is needed to understand this ambiguous situation.

6 Conclusions

In the paper the solution of the problem of establishment of the control law providing the fastest safe motion of the individual vehicle in the traffic flow is proposed and substantiated. It may be implemented with the modern means of the necessary information detection and may serve as the basis of the solution of automated driving problem. Massive implementation of the established control law should simultaneously improve safety and the use of roads' capacities by excluding redundant headways. Valuations of the expected effect require data on the distributions of drivers' behavior parameters, but undoubtedly it has the order of tens of percent.

The emerging technology of CAVs is especially perspective for increasing traffic flows' density without decreasing both average speeds and the safety, although require the development of adequate means of measurement and informational exchange. The established control law for a chain of CAVs must be very useful to enhance the opportunities of the new technology.

References

1. Rashad, L.J., Hassan, F.A.: Artificial neural estimator and controller for field oriented control of three-phase I.M. Int. J. Intell. Syst. Appl. (IJISA), **11**(6), 40–48 (2019)
2. Agrawal, P., Agrawal, H.: Adaptive algorithm design for cooperative hunting in multi-robots. Int. J. Intell. Syst. Appl. (IJISA) **10**(12), 47–55 (2018)
3. Elhoseny, M., Abdulaziz, S., Xiaohui, Y.: Optimizing robot path in dynamic environments using Genetic Algorithm and Bezier Curve. J. Intell. Fuzzy Syst. **33**(4), 2305–2316 (2017)
4. Dennouni, N., Peter, Y., Lancieri, L., Slama, Z.: Towards an incremental recommendation of POIs for mobile tourists without profiles. Int. J. Intell. Syst. Appl. (IJISA), **10**(10), 42–52 (2018)
5. Adebiyi, R.F.O., Abubilal, K.A., Tekanyi, A.M.S., Adebiyi, B.H.: Management of vehicular traffic system using Artificial Bee Colony Algorithm. Int. J. Image, Graph. Signal Process. (IJIGSP) **9**(11), 18–28 (2017)
6. Plessen, M.G., Bernardini, D., Esen, H., Bemporad, A.: Spatial-based predictive control and geometric corridor planning for adaptive cruise control coupled with obstacle avoidance. IEEE Trans. Control Syst. Technol. **26**(1), 38–50 (2017)
7. Wang, C., Gong, S., Zhou, A., Li, T., Peeta, S.: Cooperative adaptive cruise control for connected autonomous vehicles by factoring communication-related constraints. Transportation Research Part C: Emerging Technologies (2019, in press). https://doi.org/10.1016/j.trc.2019.04.010
8. Zhou, C., Weng, Z., Chen, X., Zhizhe, S.: Integrated traffic information service system for public travel based on smart phones applications: a case in China. Int. J. Intell. Syst. Appl. (IJISA) **5**(12), 72–80 (2013)
9. Goyal, K., Kaur, D.: A novel vehicle classification model for urban traffic surveillance using the deep neural network model. Int. J. Educ. Manag. Eng. (IJEME) **6**(1), 18–31 (2016)
10. Treiber, M., Kesting, A.: Traffic Flow Dynamics: Data, Models and Simulation. Springer, Heidelberg (2013)
11. Babicheva, T.S.: The use of queuing theory at research and optimization of traffic on the signal-controlled road intersections. Proc. Comput. Sci. **55**, 469–478 (2015)

12. Glukharev, K.K., Ulyukov, N.M., Valuev, A.M., Kalinin, I.N.: On traffic flow on the arterial network model. In: Kozlov, V.V., et al. (eds.) Traffic and Granular Flow 2011, pp. 399–412. Springer, Heidelberg (2013)
13. Yashina, M.V., Provorov, A.V.: Verification of infocommunication system components for modeling and control of saturated traffic in Megalopolis. In: Zamojski, W., Kacprzyk, J., et al. (eds.) New Results in Dependability and Computer Systems. Advances in Intelligent Systems and Computing, vol. 224, pp. 531–542. Springer, Heidelberg (2013)
14. Valuev, A.M.: Modeling of the transport flow through crossroads with merging and divergence points. In: Proceedings of 2018 Eleventh International Conference Management of Large-Scale System Development (MLSD). Moscow, Russia, 1–3 October 2018. In: Tsvirkun, A. (ed.) IEEE Xplore Digital Library, pp. 1–3 (2018)
15. Solovyev, A.A., Valuev, A.M.: Organization of traffic flows simulation aimed at establishment of integral characteristics of their dynamics. Adv. Syst. Sci. Appl. **18**(2), 1–10 (2018)

Development Approach of Formation of Individual Educational Trajectories Based on Neural Network Prediction of Student Learning Outcomes

Veronika V. Zaporozhko, Denis I. Parfenov(✉), and Vladimir M. Shardakov

Orenburg State University,
13, Prospect Pobedy, 460018 Orenburg, Russian Federation
fdot_it@mail.osu.ru, parfenovdi@gmail.com

Abstract. The study proposed a neural network approach to solving the problem of predicting the results of mastering the educational programs of online learning. The need for prediction is an important component of the decision support system for the intelligent management of the educational process. The proposed approach allows you to make a learning outcomes prediction of the for each student and, if necessary, to dynamically adjust its educational trajectory promptly. The results of the computational experiment to solve the problem of predicting the outcome based on the input data using a multilayer perceptron are presented. The research materials can be used in the design and creation of information systems in which the personalization of the learning process and the automation of the formation process of individual educational trajectories are provided.

Keywords: Neural network technologies · Artificial neural network · Supervised learning · Prediction · Online learning · Automated learning · Individual educational trajectory

1 Introduction

Currently, dynamic processes in the field of e-learning require constant objective assessment, adjustment, and management. Due to the complexity and independence of learning in an online learning platform, intelligent prediction, namely, predicting the development of certain events and evaluating the results, is an important task for a practical solution. One of such methods, allowing to solve the problem, is neural network data analysis.

As part of this study, we applied artificial neural networks to predict which students might experience learning difficulties and deduct from the online learning program.

The purpose of the study is to predict the success of the study of the educational program for each student based on neural networks and, if necessary, to dynamically adjust the individual educational trajectory promptly.

Z. Hu et al. (Eds.): AIMEE 2019, AISC 1126, pp. 305–314, 2020.
https://doi.org/10.1007/978-3-030-39162-1_28

This paper is organized as follows. Section 2 is a background of related work related to the use of artificial neural networks for prediction in education. In Sect. 3, we disclose the problem of intelligent prediction the results of mastering educational programs of online learning and formalize this task. Section 4 deals with the prediction of the investigated data.

2 Related Work

The neural network approach and artificial neural networks are used to predict when solving various applied educational problems.

The main objective of any personalized learning system is the ranking rating of students. As part of the work of Shajith Ikbal and others, students' tendencies toward poor academic performance were investigated and predictive models based on historical data were constructed. The constructed statistical models allow for the adjustment of the educational trajectory in the learning process [1].

A group of researchers led by Okubo used a different approach to student performance prediction. The authors use recurrent neural networks (RNN) to predict student grades. However, the proposed one was studied on a small sample of students, which does not allow one to evaluate this approach about the effect of network retraining [2].

Several researchers propose the construction of a mathematical model to predict student performance to reduce the risk of poor performance at an early stage. For example, Al-Sudani used a combination of various factors to predict student performance using a multilayer neural network to classify students [3].

Prediction of the results of mastering educational programs is especially important for mass open online courses (MOOC). The construction of prediction models is the main tool for the formation of individual educational trajectories within the course. In the study of Xing and Du, it is proposed to use a deep learning algorithm to build a model for predicting the dropout rate of students. However, the study does not address the issue of adjusting educational trajectories for lagging students [4]. To build flexible MOOC (Massive Open Online Course) it is necessary to know in advance the grades of students for various tasks. This is necessary for early detection of performance problems and corrective action. An example of prediction assessments for different types of tasks in one course is considered and target indicators affecting the success of mastering the course are defined [5].

To improve the accuracy of prediction of estimates in several studies it is proposed to pre-divide students into groups according to the way they perceive the content [6]. To improve the accuracy of the learning analytics system through the use of a recurrent neural network, Owen Corrigan, in his study, determined the dependence of academic performance on the intensity of student interaction with various elements of the virtual learning environment [7].

Another effective approach for predicting performance is the use of deep learning methods. In his study, Akhilesh P Patil developed direct-connected neural networks and recurrent neural networks to develop a model to effectively predict a student's average score. Analyzing the data obtained by the authors, we can conclude that recurrent neural networks gave greater accuracy compared to neural networks with a direct connection, as they have a memory and take into account the constancy of student performance [8].

In another study, the recurrent neural network (RNN) is used to predict the outcome of a student based on the time distribution of the data obtained. Researchers predict academic performance in the final semester based on the academic performance obtained in the first and second semester [9].

Thus, the prediction of student performance is becoming more complex due to a large amount of data in educational databases. Currently, there are no effective automated systems that allow analytics of such data and monitoring of performance in real-time.

3 Mathematical Foundations of Artificial Neural Networks

In general, an artificial neural network is a system of artificial neurons connected and interacting with each other. The mathematical model of the neuron is shown in Fig. 1. Let the input signals *xi* be quantitative values characterizing the learning achievements of students and coming from the data store of the online learning platform or other active neurons. Input values can be discrete and take values from the set [0, 1].

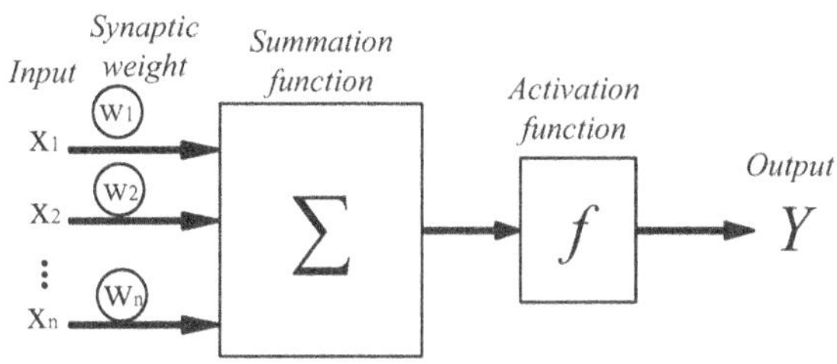

Fig. 1. Mathematical model of the neuron.

The signal strengths x_1, x_2,..., x_n, go to synapses (multipliers). Synapses communicate between neurons, multiply the input signal by a number characterizing the strength of the connection (the weight of the synapse), i.e. w_1x_1, w_2x_2,..., w_nx_n where w_i are the weights (weighting factors) of the corresponding synapses. Then the summation of signals occurs in the body of the neuron and the activation level (potential) of the neuron is calculated by the formula (1):

$$S = \sum_{i=1}^{n} w_i x_i \tag{1}$$

where n is the number of inputs of the neuron;

x_i is the value of the i-th input of the neuron;
w_i is the weight of the i-th synapse.

Then it applies some fixed function f to the sum and outputs the signal of force $Y = f(S)$. Activation function f is designed to calculate the output value of the signal transmitted to other neurons. As an activation function, a logistic function (sigmoid) is used, which can be described by the formula (2):

$$f(S) = \frac{1}{1 + e^{-aS}} \tag{2}$$

where S is the function argument;

α is the slope parameter of the sigmoid activation function.

By changing the parameter α, we can construct functions with different steepness. The range of values of this function belongs to the interval (0, 1).

The sigmoidal function can amplify weak signals better than large ones and prevents saturation from large signals since they correspond to areas of arguments where the sigmoid has a gentle slope.

4 Prediction of the Investigated Data

4.1 Architecture of a Software System for Intelligent Prediction the Results of Learning Online Education Programs

Let us consider the solution of the problem of intelligent prediction the results of mastering educational programs of online learning [10, 11]. The experiment used the information on students studying using distance learning technologies and e-learning at Orenburg State University. The initial data for the analysis is information obtained from the online learning platform data warehouses (see Fig. 2).

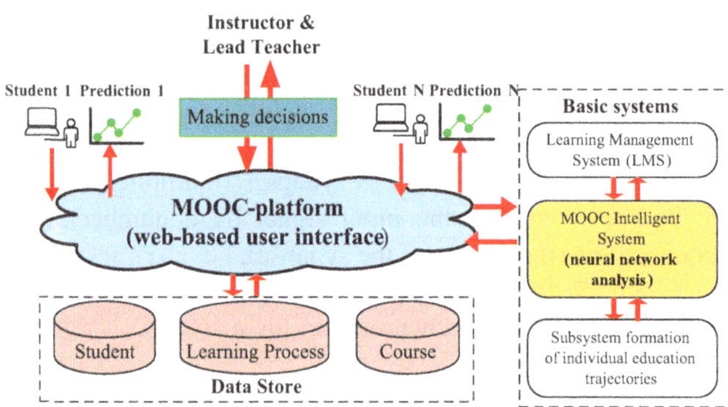

Fig. 2. Architecture of a software system for intelligent prediction the results of learning online education programs.

Student Data Cloud Store is a data sets of students. Course Data Cloud Store is a data sets of courses. Data Cloud Store Learning Process is a data sets of different characteristics describing the learning process (progress of students) based on a specific massive open online course (MOOC). MOOC Intelligent System is a hybrid intelligent system of cloud platform, responsible for the adaptability of the MOOC to the needs, preferences and capabilities of a particular student. In this system more than one method of human intellectual activity simulation is used to solve the problem of personalization in MOOC [6].

However, within the framework of this study, we will limit ourselves to considering only neural network analysis of data analysis, which allows the prediction of the results of mastering the educational programs of online learning.

We have created a software module that is part of the MOOC Intelligent System and is responsible for intelligently predicting the success of the educational program for each student and, if necessary, adjusting the individual educational trajectory. Before testing the functional performance of this module, we conducted a computational experiment to test the proposed methodology for solving the problem using the neural network designer «NeuroNet».

4.2 The Content Statement and the Formalized Description of the Task of Prediction the Results of Mastering the Educational Programs of Online Learning

Each student of an online platform is determined by a number of characteristics that determine his academic achievements at various stages of the educational process. It is required for each student on the basis of neural networks to make a prediction of the success of the educational program and, if necessary, to dynamically adjust the individual educational trajectory in a timely manner [12].

Consider the main stages of solving the problem.

Stage 1. Expert Selection of Significant Characteristics

Initially, we introduce the notation of the most significant characteristics determined by the method of expert assessments.

At the learning entrance, these include the following signs:

- x_1 is the average score of the school certificate;
- x_2 is the average score of the Unified State Exam or entrance examinations;
- x_3 is the average score of the input diagnostics results (diagnostic test, input test work, etc.).

In the process (in progress) of learning, the following features are identified:

- x_4 is the average score of the results of formative evaluation (intermediate tests, current written work);
- x_5 is the average score of the results of summative assessment (final tests, final written works);
- x_6 is the average score of intermediate certification results for the course.

Upon completion of learning (at the exit from learning) the following features are proposed:

- x_7 is the average score of the results of the passage and protection of all types of practices;

- x_8 is the average score of the results of state examinations;
- x_9 is the average score of the result of the defense of final qualifying work (according to the statement of the examination committee).

As outputs of the neural network, the possible results of mastering the educational programs of online learning were taken:

- y_1 if learning is successful, the adjustment of the individual educational trajectory is not required;
- y_2 iflearning fails, requires adjustment of the individual educational trajectory;
- y_3 if the student has significant difficulties in learning and/or has academic debts, requires adjustment of the individual educational trajectory and the construction of an adaptive learning program to prevent premature deduction.

When forming the matrix of initial data for neural network analysis, we translated all the primary data extracted for each of the students into points according to established scales.

Stage 2. The Choice of Neural Network Architecture

In our work, an artificial neural network consists of several layers into which neurons are grouped. On the neurons of the input, the layer is fed measured and transformed values of characteristics, reflecting the student's learning achievements. Signals arriving at the neurons of the input layer are transmitted to the next layer (hidden or output) without conversion (i.e. the activation function is not applied to them). When solving the task, a multilayered network with cross-links is used, which has one hidden layer (see Fig. 3).

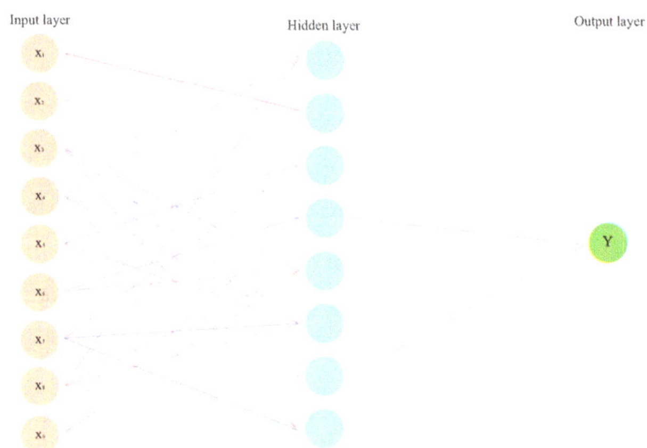

Fig. 3. Schematic image of a multilayer network.

When choosing a neural network architecture, there are 3 layers (input, output and one hidden), which are a multilayer perceptron. The number of neurons in each layer was determined: there are 9 neurons at the input, 1 neuron at the output. The hidden layer contains 8 neurons with a linear activation function. The type of connection between neurons is direct. The function of activating all neurons is the logistic function (sigmoid).

Stage 3. Dividing the Initial Sample into Learning and Test Data Sets. Formation of the Initial Sample for Learning the Neural Network

A total of 419 students were randomly assigned to learning and testing datasets.

The quality of the neural network learning process depends on the quality of the generated learning sample, its ability to correctly solve the task set during operation of the developed software system. Therefore, it is important to carry out the selection of the initial data, since the sample must satisfy several requirements:

(1) correspond to the used structure of the neural network (the number of input and output variables);
(2) contain unique (non-repeating) examples; be consistent and representative.

A fragment of the matrix of the initial data for learning the neural network is presented in Table 1.

Table 1. The values of the main characteristics of the students (data for learning the neural network).

Online learning platform students	Input signals									Output signal
	x_1	x_2	x_3	x_4	x_5	x_6	x_7	x_8	x_9	Y
Student 1	4, 2	75	69	82	78	76	5	4	5	1
Student 2	3, 8	65	58	73	75	72	4	3	4	1
Student 3	4, 8	82	71	72	73	81	5	5	5	1
Student 4	3, 2	58	51	53	52	57	3	3	3	2
…	…	…	…	…	…	…	…	…	…	…
Student n	3, 4	52	48	51	52	63	3	3	0	3

Stage 4. Carrying out the Procedure of Normalization of Variables

To use the neural network model, variables were normalized: ranges of input and output vector values are within [0, 1].

Stage 5. The Learning Process of the Neural Network

Before solving the problem of predicting the results of mastering educational programs of online learning, it is necessary to train (configure) a neural network. The process of network learning consists of determining the set of connections and weights of connections between neurons based on the repeated running of the learning sample through the neural network. The run of the entire learning set is called the era. The learning of the constructed network takes place with the teacher (Supervised learning) since all the examples of the learning sample contain the correct answers (outputs) corresponding to the initial data (inputs). In the process of learning, the weights (coefficients) are adjusted so that the neural network generates the answers that are closest to the correct ones.

The neural network modeling is shown in Figs. 4 and 5, when predicting the results of mastering the educational programs of online learning, was based on data from previous years (for 2016, 2017 and 2018). The learning sample size is 369.

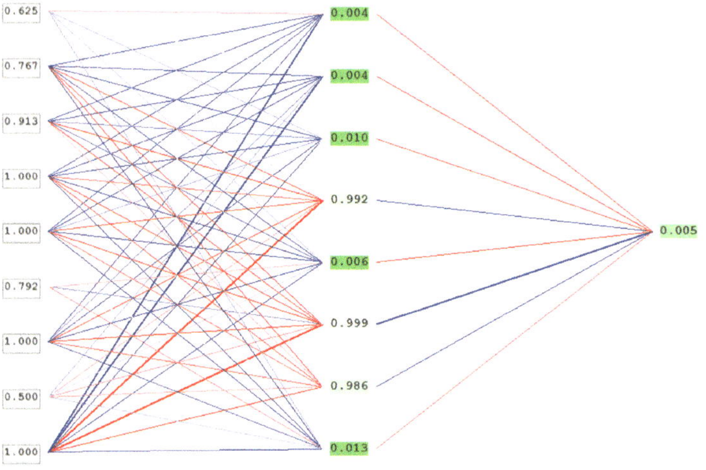

Fig. 4. The process of learning the direct distribution neural network (Feedforward neural network) (step is 700, accuracy is 99,538%).

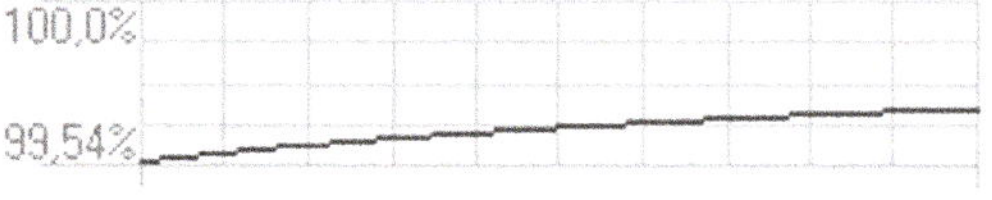

Fig. 5. The activation function of artificial neurons used in solving the problem.

Stage 6. Testing the Neural Network

Testing set as a final set of input signals along with the correct output signals is used in assessing the quality of the neural network. Testing the quality of learning of the neural network must be conducted on an independent sample of examples that did not participate in its learning. The test sample size is 50. According to the test results, the error of the neural network model is estimated. The neural network was configured in such a way that with the help of it you can solve the problem with the required level of error is no more than 2%.

Stage 7. The Use of a Neural Network as a Means of Prediction in Solving the Problem

The trained and tested neural network is used for practical solution of the problem as new source data (values of input variables) arrive. Results from a trained network can be obtained almost instantly and interpreted by an instructor and a leading teacher for later decision making.

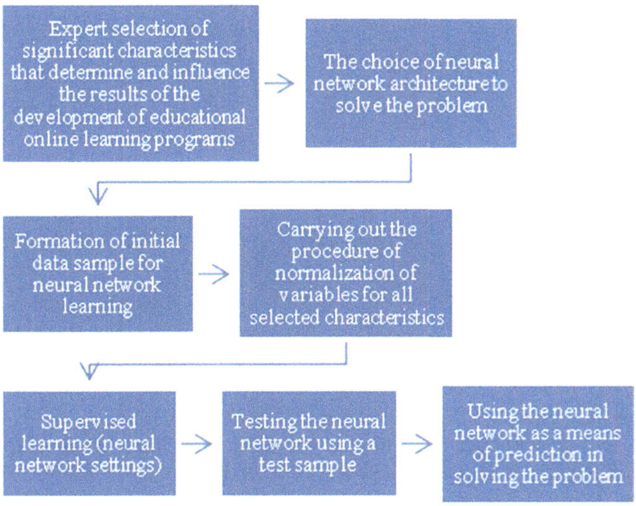

Fig. 6. A generalized scheme of the algorithm for solving the problem of intelligent prediction of the results of mastering the educational programs of online learning.

A generalized diagram of the algorithm for solving the problem of intelligent prediction of the success of learning online learning programs is presented in Fig. 6.

4.3 Evaluation of Neural Network

After learning and cross-validation, the neural network was tested with a set of test data. Thus, the quality assessment of the constructed model took place. The input variable data was transferred to the neural network with no output variable results. The output from the network was then compared with the actual variable data.

When assessing the effectiveness of the neural network, it was found that in 98% of cases the correct answers are given, and in 2% of cases are the wrong answers.

5 Conclusion

As part of the study, a neural network approach was proposed for solving the problem of predicting the results of mastering educational programs of online learning. Testing this approach on the example of real data showed the practical significance of the author's methodology and its suitability for use in the design and creation of information systems in which personalization of the learning process and automation of the process of forming individual educational trajectories are provided. A further research direction is the development of models and algorithms for predicting the results of mastering a specific online course based on artificial neural networks to prevent a premature deduction from it. Thus, it is intended to solve the problem of intelligent management of the learning process in an online environment and promptly to ensure the dynamic adjustment of the individual educational trajectory.

Acknowledgment. The research was conducted with the support of the Russian Foundation for Basic Research (project no. 18-37-00400, 19-47-560011).

References

1. Ikbal, S., Tamhane, A., Sengupta, B., Chetlur, M., Ghosh, S., Appleton, J.: On early prediction of risks in academic performance for students. IBM J. Res. Develop. **59**(6), 5:1–5:14 (2015)
2. Okubo, F., Yamashita, T., Shimada, A., Ogata, H.: A neural network approach for students' performance prediction. In: Proceedings of the Seventh International Learning Analytics and Knowledge Conference, pp. 598–599. ACM, New York (2017)
3. Al-Sudani, S., Palaniappan, R.: Predicting students' final degree classification using an extended profile. Educ. Inf. Technol. **24**(4), 2357–2369 (2019)
4. Xing, W., Du, D.: Dropout prediction in MOOCs: using deep learning for personalized intervention. J. Educ. Comput. Res. **57**(3), 547–570 (2018)
5. Moreno-Marcos, P.M., Muñoz-Merino, P.J., Alario-Hoyos, C., Estévez-Ayres, I., Delgado Kloos, C.: Analysing the predictive power for anticipating assignment grades in a massive open online course. Behav. Inf. Technol. **37**(10–11), 1021–1036 (2018). Advanced Decision Making in Higher Education: Learning Analytics Research and Key Performance Indicators
6. Parfenov, D., Zaporozhko, V.: Automation of controlling of personalization of learning in cloud educational environment based on cluster approach. In: 2018 International Russian Automation Conference, pp. 683–690. IEEE (2018)
7. Corrigan, O., Smeaton, A.F.: A course agnostic approach to predicting student success from VLE log data using recurrent neural networks. In: 12th European Conference on Technology Enhanced Learning Data Driven Approaches in Digital Education, pp. 545–548. Springer (2017)
8. Patil, A., Ganesan, K., Kanavalli, A.: Effective deep learning model to predict student grade point averages. In: Proceedings of the IEEE International Conference on Computational Intelligence and Computing Research, pp. 1–6, Coimbatore. IEEE (2017)
9. Mondal, A., Mukherjee, J.: An Approach to predict a student's academic performance using Recurrent Neural Network (RNN). Int. J. Comput. Appl. **181**(6), 1–5 (2018)
10. Kumar, M., Singh, A.J., Handa, D.: Literature survey on student's performance prediction in education using data mining techniques. Int. J. Educ. Manag. Eng. (IJEME) **6**, 40–49 (2017). MECS Press
11. Kotsiantis, S., Pintelas, P.E.: Predicting students marks in Hellenic Open University. In: Proceedings of the 5th IEEE International Conference on Advanced Learning Technologies, pp. 664–668. IEEE (2005)
12. Suhaimi, N.M., Abdul-Rahman, S., Mutalib, S., Abdul Hamid, N., Ab Malik, A.: Review on predicting students' graduation time using machine learning algorithms. Int. J. Mod. Educ. Comput. Sci. (IJMECS) **11**(7), 1–13 (2019). MECS Press

Study of the Effectiveness of State Support in the Development and Implementation of Neuro-Educational Technologies

T. Bergaliev[1(✉)] and M. Mazurov[2]

[1] Moscow Institute of Physics and Technologies, Moscow, Russia
timurbergaliyev@gmail.com
[2] Russian University of Economics, Moscow, Russia
mazurov37@mail.ru

Abstract. A successful example of a modern school digital laboratory in the field of biophysics and neurotechnologies is the domestic joint development of the DIY kit by BiTronics Lab Company and Laboratory of applied cybernetic systems MIPT. The target audience of consumers of educational neurotechnologies is indicated: schoolchildren, students, specialists of related professions. The results of the implementation of neurotechnologies in the social environment - primary school children - have been studied. A linear regression equation was constructed, which characterizes the dependence of the involved schoolchildren in the work on the study of neurotechnologies from the amount budgetary funds and number of events taken to familiarize with neurotechnologies.

Keywords: Neuroeducation · Neuro DIY kit · Neurotechnology · Elementary school students · Regression dependence

1 Introduction

Currently, one of the fastest growing branches of the digital world in the Russian Federation and in the world is the field of neurotechnologies, including the creation of human-machine interfaces, modern rehabilitation tools, new approaches to the design of robotic and medical equipment, modern the expansion of human capabilities and prosthetics, wearable electronics, remote monitoring tools for human condition, neuro-entertainment and eSports, data analysis based on neural network technologies many others.

The neurotechnology industry is based on interdisciplinary approach in the field of human physiology, biology, physics, human-machine interface design, data analysis technology, engineering and technical skills for debugging and testing complex systems.

In order to meet and study (training pupils and students) in these areas of knowledge, optimal educational tools and approaches should be developed that are organically built into the existing structure of education (integrated into the main and additional educational programs tested in various schools).

Z. Hu et al. (Eds.): AIMEE 2019, AISC 1126, pp. 315–321, 2020.
https://doi.org/10.1007/978-3-030-39162-1_29

The most successful example of modern school digital laboratories in the field of biophysics and neurotechnologies is the domestic joint development of DIY kit by BiTronics Lab company and Laboratory of applied cybernetic systems MIPT. The BiTronicsLab kit is illustrated in Fig. 1.

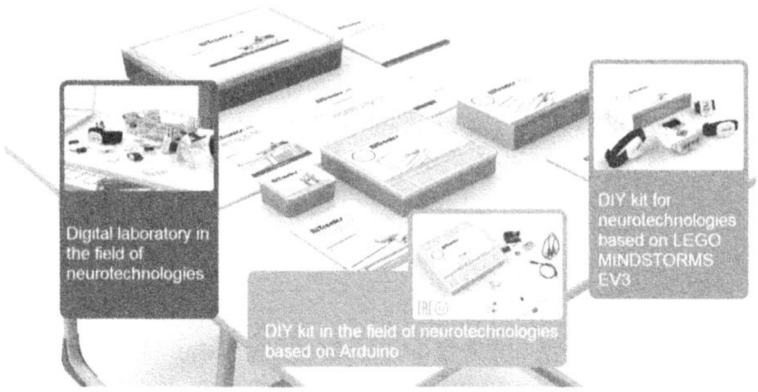

Fig. 1. DIY kit in the field of neurotechnologies BiTronicsLab as an introduction level education tool for younger students

As of the current moment, through the system of state procurement, our laboratories have already been delivered and are successfully used at schools and universities in 20 regions of Russia.

The "Neurotechnology" track of the Olympiad of the National Technological Initiative, training in the field of neurotechnology at the University of the National Technological Initiative has been held on the basis of a digital laboratory in the region for 4 years. The company is an active participant of the Neuronet market, a member of the industry union Neuronet, the company is included in the list of leadership projects supported by the Agency for Strategic Initiatives (ASI) created by the President of the Russian Federation.

This work is carried out as part of the implementation of innovative projects for the development and mastering of new types of high-tech products in order to implement the NeuroNet STI road maps in the direction of the development of neurotechnologies approved by the Presidium of the Presidential Council for Modernization of the Economy and Innovative Development of Russia.

The chosen field of study includes: (1) theoretical information in the field of neuroinformatics; (2) software illustrating basic neurotechnologies; (3) the practical implementation of the neuro-educational system in the form of "iron" or in the form of the material modules of the DIY kit, allowing you to visually explore neurotechnologies live and work with these neurotechnologies. The main significant control result of the implementation of the direction of neuroeducation DK Neuronet is the entering educational market with developed DIY kit. The first generation of DIY kits developed by BiTronics Lab company and laboratory of applied cybernetics systems MIPT includes registration of electrical signals of the human body: EEG (electroencephography), ECG (electrocardiography), EMG (electromyography), and others. It is assumed that this neuro-educational system will be available for schoolchildren over 14 years old.

The second generation of neuro-educational systems should contain basic information about the neurotechnologies of standard neural networks and recently created selective neurotechnologies, give an idea of the in-depth training of neural networks. It is assumed that this neuro-educational system will be available to high school students, students of higher educational institutions, and related specialists professions.

The area of research under consideration includes neuro-educational technologies designed to train a wide range of schoolchildren students and specialists in related professions. An illustration of this circle of people is shown in the photographs of Fig. 2.

Fig. 2. The circle of consumers of neurotechnologies: schoolchildren, students, specialists in related professions

In this article, modeling the effectiveness of government support in the development and implementation of neuro-educational technologies considered as one of the ways to solve problems in the real world.

2 Methods of Modeling in Social Systems

In this paper, we study the results of the introduction of neurotechnologies in the social environment - primary schoolchildren. The general scheme of modeling is illustrated in Fig. 3.

In Fig. 3 below shows the actual development of processes in a complex system, in Fig. 3 above shows the development of processes in the mathematical modeling of the

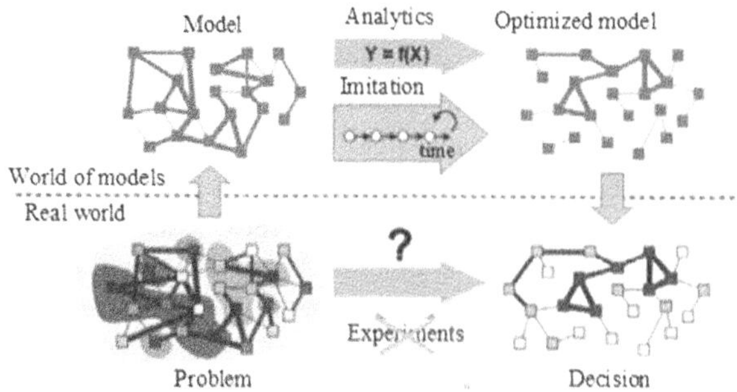

Fig. 3. General simulation scheme

system. The study of processes in a mathematical model allows us to establish processes in a real system without its actual implementation.

Currently, one of the methods for modeling processes in social systems is the agent modeling method [1, 2]. The structural scheme of agent-based modeling is shown in Fig. 4.

Other methods of mathematical modeling of social and economic processes are also possible; these methods are described in more detail in [3–15]. Simulation is used if

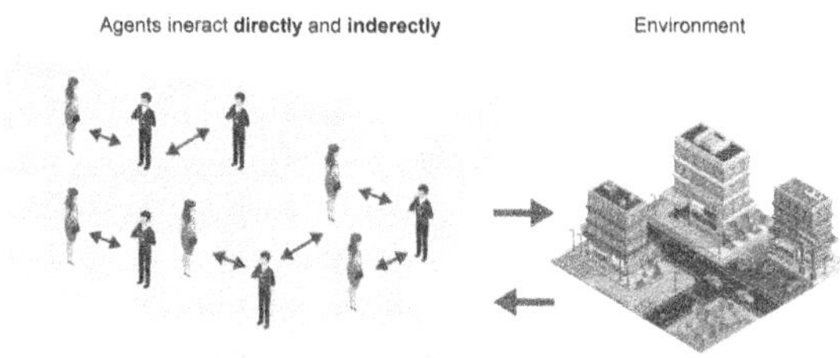

Fig. 4. Agent-based modeling approach

experiments with real objects/systems or their prototyping is impossible or too expensive. Simulation allows you to optimize the system before its implementation. Modeling involves mapping problems from the real world into the world of models (the process of abstraction), analyzing and optimizing the model, finding a solution, and mapping the solution back to the real world. We distinguish between analytical and simulation modeling. The analytical model allows an analytical solution, the dependence of the output on the input can be implemented statistically.

As a result of implementation and entering educational market with the DIY kit by BiTronics Lab for the last 4 years, focus has been done on elementary school students on introducing knowledge of basics of neuroinformatics, applied neurotechnologies and physiology. The results of this work are shown in Table 1.

Table 1. The results of implementation and entering educational market with the DIY kit by BiTronics Lab in schools

Rate\year	2016	2017	2018	2019
Number of events, pieces	25	50	100	120
Number of allocated budget Million rubles	10	20	30	40
Number of schoolchildren involved, person	2000	4500	8000	10000
Number of patents	0	2	3	
The number of teachers involved, person	5	15	40	50

The values of the number of activities, the number of schoolchildren involved, the number of teachers involved for 2019 are shown from the analysis of previous time series for these indicators. A linear regression equation is constructed that characterizes the dependence of the students involved in the study of neurotechnologies y on indicators: the amount of allocated budget funds x_1, the number of measures taken to familiarize themselves with neurotechnologies x_2. Linear regression dependence was obtained as $y = a_0 + a_1x_1 + a_2x_2$, where a_0, a_1, a_2 are the parameters, $a_0 = -273.8$, $a_1 = 115.5$, $a_2 = 47.6$. The resulting regression dependence is shown in Fig. 5.

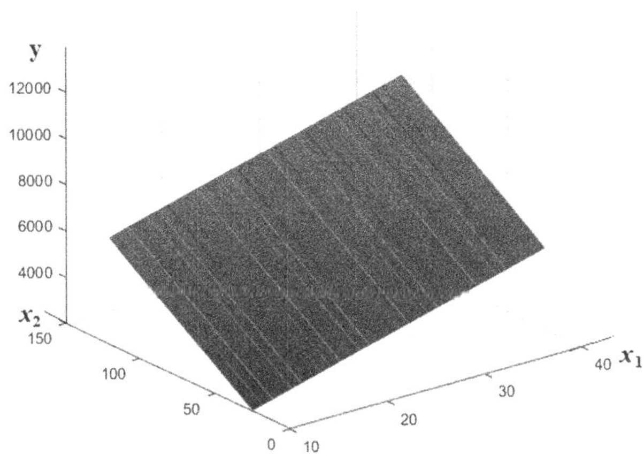

Fig. 5. Regression characterizes the dependence of the involved schoolchildren in the work on the study of neurotechnologies from the amount budgetary funds and number of events taken to familiarize with neurotechnologies

The regression equation found shows that the number of students involved in the study of neurotechnology is proportional to the state funds used and the number of activities carried out to study neurotechnology.

3 Conclusions

A successful example of a modern school digital laboratory in the field of biophysics and neurotechnologies is the domestic joint development of the DIY kit by BiTronics Lab Company and Laboratory of applied cybernetic systems MIPT. The target audience of consumers of educational neurotechnologies is indicated: schoolchildren, students, specialists of related professions. The results of the implementation of neurotechnologies in the social environment - primary school children - have been studied. A linear regression equation was constructed, which characterizes the dependence of the involved schoolchildren in the work on the study of neurotechnologies from the amount budgetary funds and number of events taken to familiarize with neurotechnologies.

References

1. Makarov, V.L., Bakhtizin, A.R.: Social Modeling - New Computer Breakthrough (Agent-Based Models) Economics, 295 p. (2013)
2. Mazurov, M.E.: Identification of Mathematical Models of Nonlinear Dynamic Systems, p. 383. URSS, Moscow (2018)
3. Akkar, H.A., Mahdi, F.R.: Adaptive path tracking mobile robot controller based on neural networks and novel grass root optimization algorithm. Int. J. Intell. Syst. Appl. (IJISA) **5**, 1–9 (2017). https://doi.org/10.5815/ijisa.2017.05.01
4. Mazurov, M.E.: Intelligent recognition of electrocardiograms using neuron networks and deep learning. In: International Conference of Artificial Intelligence, Medical Engineering, Education, Moscow, Russia, pp. 182–198 (2017)
5. Kaur, N., Singh, A.: Analysis of the vascular pattern recognition using the neural network. Int. J. Math. Sci. Comput. (IJMSC) **1**(3), 9–19 (2015). https://doi.org/10.5815/ijmsc.2015.03.02
6. Bhanuprakash, C., Nijagunarya, Y.S., Jayaram, M.A.: Clustering the neural net-work approach. Int. J. Intell. Syst. Appl. (IJISA) **3**, 34–40 (2017). https://doi.org/10.5815/ijisa.2017.03.05
7. Mansor, M.A., Kasihmuddin, M.S.M., Sathasivam, S.: Enhanced hopfield network for pattern satisfiability optimization. Int. J. Intell. Syst. Appl. (IJISA) **11**, 27–33 (2016). https://doi.org/10.5815/ijisa.2016.11.04
8. Lawrence, S., Giles, C.L., Tsoi, A.C., Back, A.D.: Face recognition: a convolutional neural network approach. IEEE Trans. Neural Netw. Pattern Recogn. **8**(1), 1–24 (1997). http://www.neci.nec.com/lawrence
9. Iyanda, A.R., Ninan, O.D., Ajayi, A.O., Anyabolu, O.G.: Predicting student academic performance in computer science courses: a comparison of neural network models. Int. J. Mod. Educ. Comput. Sci. (IJMECS) **10**(6), 1–9 (2018). https://doi.org/10.5815/ijmecs.2018.06.01
10. Alkhathlan, A.A., Al-Daraiseh, A.A.: An analytical study of the use of social networks for collaborative learning in higher education. Int. J. Mod. Educ. Comput. Sci. (IJMECS) **9**(2), 1–13 (2017). https://doi.org/10.5815/ijmecs.2017.02.01
11. Moshref, M., Al-Sayyad, R.: Developing ontology approach using software tool to improve data visualization (case study: computer network). Int. J. Mod. Educ. Comput. Sci. (IJMECS) **11**(4), 32–39 (2019). https://doi.org/10.5815/ijmecs.2019.04.04

12. Cheah, C.S., Leong, L.: Investigating the redundancy effect in the learning of C + + computer programming using screencasting. Int. J. Mod. Educ. Comput. Sci. (IJMECS) **11** (6), 19–25 (2019). https://doi.org/10.5815/ijmecs.2019.06.03
13. Nurafifah, M.S., Abdul-Rahman, S., Mutalib, S., Hamid, N.H.A., Ab Malik, A.M.: Review on predicting students' graduation time using machine learning algorithms. Int. J. Mod. Educ. Comput. Sci. (IJMECS) **11**(7), 1–13 (2019). https://doi.org/10.5815/ijmecs.2019.07.01
14. Alvarez-Dionisi, L.E., Mittra, M., Balza, R.: Teaching artificial intelligence and robotics to undergraduate systems engineering students. Int. J. Mod. Educ. Comput. Sci. (IJMECS) **11** (7), 54–63 (2019)
15. Adekunle, S.E., Adewale, O.S., Boyinbode, O.K.: Appraisal on perceived multimedia technologies as modern pedagogical tools for strategic improvement on teaching and learning. Int. J. Mod. Educ. Comput. Sci. (IJMECS) **11**(8), 15–26 (2019). https://doi.org/10.5815/ijmecs.2019.08.02

An Improvement of Remotely Piloted Aircraft Systems by Identifying Potential Radio-Controlled Areas

Olena Kozhokhina(✉), Roman Odarchenko, and Liudmyla Blahaia

National Aviation University, Kosmonavta Komarova Avenue, 1, Kiev 03058, Ukraine
kozhokhina@gmail.com,
{odarchenko.r.s, b_ludmila}@ukr.net

Abstract. Remotely piloted aircraft systems are a new component of the aviation system and based on cutting-edge developments in aerospace technologies. Research has shown that the responsibilities of the operator of remotely piloted aircraft systems could be over-specified and often deal with extensive information. This paper aims to determine how to keep a sufficient level of information component of the reliability of the operator. In this context, methods of human operator reliability improving were considered. Based on a review of the literature on human factors, it was determined that redundancy could increase the reliability of human-operator. The results indicate that there are hardware, information, algorithmic and time redundancy. On this basis, it is recommended to use the identification of potential radio-controlled areas as algorithmic redundancy for the operator. Further research is needed to identify other factors that could strengthen the effectiveness and reliability of the operator.

Keywords: Remotely piloted aircraft systems · Reliability · Operator · Aviation safety · Information unload and overload · Remote pilot · 5G

1 Introduction

The principal purpose of civil aviation is to ensure the safety and regularity of aircraft operations. This purpose is relevant at all stages of the life cycle of the aircraft and its equipment.

The safety and regularity of flights depend on the reliability and efficiency of the functioning of the technical means of radio engineering support, the aircraft power plant of an aircraft and others.

Remotely piloted aircraft systems (RPAS) are a new component of the aviation system, one that the International Civil Aviation Organization (ICAO), States and the industry are working to understand, define and ultimately integrate. These systems are based on cutting-edge developments in aerospace technologies, offering advancements, which may open new and improved civil/commercial applications as well as improvements to the safety and efficiency of all civil aviation. The safe integration of RPAS into non-segregated airspace is a long-term activity with many stakeholders adding their expertise on such diverse topics as licensing and medical qualification of

Z. Hu et al. (Eds.): AIMEE 2019, AISC 1126, pp. 322–333, 2020.
https://doi.org/10.1007/978-3-030-39162-1_30

remote pilots, technologies for detect and avoid systems, frequency spectrum (including its protection from unintentional or unlawful interference), separation standards from other aircraft and development of a robust regulatory framework [19].

Responsibilities of the operator of RPAS or remote pilot could be over-specified and under specific circumstances leads to mistakes. The operator often operates pressed on time and processing extensive information.

Unfortunately, most of the current practices of reliability prediction of a human-operator is relating to such areas as the atomic energy industry or the chemical industry [2–4]. These practices do not take into account not only the unique aspects of the aviation industry but also poorly agree with the critical factors that have an impact on the operator of RPAS [1, 6].

Nevertheless, it should be noted that some of the practices were updated taking into account needs and particular aspects of the aviation industry, for example, Connectionism Assessment of Human Reliability (CAHR) [7] was adapted for air traffic control tasks in Eurocontrol. In some cases like in [15], accidents in the nuclear industry were compared with aviation incidents. Alternatively, some practice is aimed at studying particular components of the human operator in aviation [18].

The most flexible methods, able to apply to specific maintenance tasks is a decomposition or structuring, for example, SLIM-MAUD Success likelihood index method using multi-attribute utility decomposition [8], workflow scheduling [28], reporting [5], adaptive neuro-fuzzy inference system [29] or structural decomposition of operator activity [9, 10]. Due to these researches, a range of factors allowing evaluating reliability human operator as well as determining the most criticality quality-relevant factors was detected. Such as, for example, large amounts of information that the operator of RPAS deal with, and which can lead to information stress and task skipping, which reduces the safety of RPAS operation.

The safe operation of RPAS necessitates compliance with many requirements. These requirements apply equally to RPAS operations and are intended to mitigate risk to persons and property on the ground and other airspace users.

2 Informational Reliability of Operator of RPAS

In general reliability of remote pilot is a probability of the successful performance of a task, at a given system operating step, at a given time interval [11].

Currently, it is also assured necessity of usage at determining the reliability of human-operator activities not only the results of his performed tasks but also even indicators of the psychological and physiological characteristics [12].

An error of an operator of RPAS can occur in the following cases: the operator aims to achieve some incorrect goals, incorrect operator actions that do not allow achieving the goal, the operator is idle when his participation is necessary [13].

Errors caused by the operator of RPAS can be considered independent and classified by the following four types: functional, operational, information, professional [10].

The information reliability of remote pilot is one of the most critical reliability components, especially in the modern world when the operator should operate with extensive information. Moreover, it also should be noticed that informational reliability is the basis of the operator decision-making process. Wrong decisions, originating from incorrect or missed information, leads to emergency for RPAS.

In addition to this, the RPAS operator also processes a large amount of auxiliary information, the volumes of which may vary depending on the complexity of the task and the mode of operation. The remote pilot also should have a wide range of knowledge, following the functions that in manned aviation performed by pilots, ATC, maintenance engineers, fuelers and others.

Additional unwanted criteria for decreasing of informational reliability of operator is time limits and rush. The operator of RPAS often needs to process information very quickly and make decisions, but at the same time, under external pressure, an erroneous decision can be made leading to aviation accidents.

There are hardware, information, algorithmic and time redundancy.

The redundancy of the operator of RPAS is the additional means and opportunities beyond the minimum necessary for the operator to perform tasks. The purpose of introducing redundancy is to increase the probability level of the operator error-free operation in a certain period.

The overload coefficient can determine information reliability of the remote pilot during a given interval of time. This coefficient is the ratio of the mathematical expectation of the period of continuous operation T_{co} to the expectation of time T_{te}:

$$\psi = \frac{E[T_{co}]}{E[T_{te}]}$$

The volume of information signals measures the duration and quality of operation of the operators. Information overload is the difficulty of understanding the critical situation and making decisions that are the cause by information excess.

The essence of information overload is the amount of incoming useful information exceeds the real possibilities of human-operator. Useful information is necessary for solving problems that ensure a sufficient level of aviation safety. Current researches show that with extensive information, the operator can lose the ability to adequately and reliably assess the situation, and this can prevent to make a correct decision.

The speed of transmission and processing of information by a human is approximately 15–25 bits per second. The same speed cannot be used for describing different human operators; in some conditions, they can transmit information better than in others [26]. However, the operator is not able to maintain such a flow of data continuously. For a twelve-hour shift, the remote pilot can process up to 5000 informational signals, which is a considerable amount of information for perception.

That is why some methods for informational unload of the operator should be used, such as redundancy.

For the operator of RPAS partial automation of tasks, as well as the introduction of additional hint algorithms can be used as algorithmic redundancy. For example, such as taking into account the impact on radio frequencies and estimate radio coverage areas.

3 Impacts on Radio Frequencies (RFs)

Following the ICAO manual for remotely piloted aircraft systems Doc 10019 AN/507 of 2015, the operator (remote pilot) should take into account the following effect on radio frequencies during RPAS operation:

1. Electromagnetic interference (EMI) (e.g. solar flares, volcanic ash, ionospheric activity) may affect the performance of C2 links and GPS reception and should be considered by the remote pilot or operator of RPAS before, and during, flight.
2. The operator of RPAS should consider the available information regarding potential EMI and its impact on the RPAS and flight completion. Additional considerations should be given to possible intentional or inadvertent electronic interference.
3. Operations in areas of high RF transmission/interference (e.g. radar sites, high tension wires) should be avoided unless engineering testing has confirmed that operations in these areas do not impact the safe operation of the RPAS [19].

As can be seen in the modern world, there are many radio technologies, networks that can influence RPAS. That is why in these operational conditions, it is necessary to have a whole vision of the occurred situation. The mathematical model for determining radio-controlled coverage areas for different possible modern wireless technologies was proposed.

4 Mathematical Model of Radio Coverage Areas Estimation for RPAS

As usually, at least two communication systems are placed on RPAS board: duplex or semi-duplex equipment for transmitting command-telemetric information and a simplex data transfer system for payloads. The equipment for transmission of command and telemetry information is intended for the low-speed transmission of command information from the RPAS aboard equipment and the low-speed transmission of telemetric information from the RPAS board to the control centre. The for payload transfer equipment is intended for one-way high-speed transmission from the board of RPAS to control centre.

Communication systems for RPAS meet the following requirements (Table 1).

The conducted analysis showed that the most common frequencies for the transmission of video from the RPAS: 900 MHz; 1.2–1.3 GHz; 2.4 GHz; 5.8 GHz.

Cellular networks can act like a control channel or data transmission between RPAS and the operator. In this case, 'range' of the control channel is limited to the coverage area of the cellular network in the city or countryside. Having lost the connection to the network, RPAS can independently return to the coverage area and select another route.

Table 1. Requirements for RPAS communication systems

Type of communication channel	Transmission of command and telemetry data	Transfer of payload data
Bit error rate (BER)	Not more than 10^{-6}	Not more than 10^{-3}
Data transfer rate	Not more than 56 kbit/s	1–20 Mbit/s (depends on the purpose of the machine and the type of payload)
Data transmission direction	Board-Land; Land-Board	Board-Land

Using the ATM system of GSM/GPRS communications and satellite monitoring technologies, if necessary combined with the ADS-B equipment, allow the display controllers to display all the aircraft located in the VP at low altitudes.

GSM technology is commonly used in monitoring and monitoring systems, and there are excellent reasons for this. However, to monitor drones, promising 5G technology is more suitable because it has low power consumption, which is especially crucial for long RPAS flights. The advantages of using this technology include the high radio signal range, up to 30 km in the open air and up to 8 km in the city. It should be noted the unique bandwidth of the radio signal, which provides a stable connection in hard-to-reach places. The use of this technology is most appropriate in this case federated losses

However, as with the use of GSM, 3G or LTE, 5G or other radio technologies for controlling drones, it remains possible to intercept the control channel, the payload using specialised hardware and software complexes. In these cases, it is possible not only the loss of relevant information but also even the abduction of the RPAS.

The interception of signals from the RPAS is only possible in the so-called controlled area, which is different for different radio technologies and characteristics of transmission systems (transmission speed, transmitter power, bit error rate, type of terrain and others). Usage of the following universal model to determine this controlled area is proposed.

Maximum allowable losses when propagated in the channel level [20]:

$$L = P_{TX} + G_{TX} - P_{RX} - B_{body} + G_{RX} - B_{fid} - IM - L_{slow} - L_{met} - L_{add}, \qquad (1)$$

where P_{TX} – transmitter power; G_{TX} – transmitting antenna gain; P_{RX} – receiver sensitivity; B_{body} – loss in the subscriber's body; G_{RX} – gain of the receiving antenna; B_{fid} – feeder line losses, IM – interference reserve; L_{slow} – stock on slow fading is taken equal to 10.3 dB; L_{add} – additional losses, which consist of losses in antennas and the difference between the receiving and transmit antenna heights can be taken as 1.5 dB; L_{met} – losses due to absorption in atmospheric gases, hydrometeors, fog, etc., dB. L_{met} is determined by the following formula [21, 22]:

$$L_{met} = L_{fog} + L_{hidro} + L_{ag},$$

where L_{fog} – power loss of the radio signal in the fog, dB; L_{hidro} – radio signal losses under the rain (snow) influence, dB; L_{ag} – the magnitude of the radio signal attenuation in atmospheric gases.

Mathematical apparatus for estimation of L_{fog}, L_{hidro} та L_{ag} is shown in [22].

Interference reserve *IM* (dB) is determined as follows:

$$IM = P_{RX} - P_{tr}(10^{-3}).$$

As shown in [15], receiver sensitivity can be represented as follows:

$$P_{RX} = 10 \cdot \lg((E_b/N_0) \cdot k \cdot T \cdot R)$$

where (E_b/N_0) - the signal/noise ratio in the digital systems of communication is the ratio of the signal energy by 1 bit to the noise power density at 1 HZ; *R* – data transfer rate; $k = 1{,}23 \times 10^{-23}$ J/K, *T* – the temperature in Kelvin (absolute temperature).

Thus, the expression (1) can be represented as:

$$L = P_{TX} + G_{TX} - 10 \cdot \lg((E_b/N_0) \cdot k \cdot T \cdot R) - B_{BODY} + G_{RX} - B_{fid} - IM - L_{slow} - L_{met} - L_{add}$$

To determine the signal-to-noise ratio, which does not exceed the probability of occurrence of bit errors, the probability of a bit error in both subchannels for different types of modulation used in the equipment of radio transmitting systems was simulated.

The probability of a bit error BER_0 in the Gaussian channel equals [24]

$$BER_0(\eta) = 0.5 \cdot [1 - \Phi(\sqrt{\alpha \cdot \eta}],$$

$$\Phi(x) = \frac{2}{\sqrt{\pi}} \cdot \int_0^x \exp(-t^2)dt$$

where $\alpha = 2$ and $\alpha = 1$ for binary and quadrature-phase modulations, respectively.

Signal/Noise Ratio (SNR) in the *i*-th own subchannel $\eta_i = \beta_i \cdot \rho_0 \cdot \lambda_i$. Given the normalization of the probability density and by entering the parameter $\rho_i = \beta_i \cdot \alpha \cdot \rho_0$, is the following:

$$BER_i = \frac{1}{2} - \frac{1}{2} \cdot \int_0^\infty f_i(\lambda) \cdot \Phi(\sqrt{\rho_i \cdot \lambda})d\lambda,$$

Consider that the probability of a bit error P_b for BPSK and QPSK is defined by the expression [16]

$$P_b = Q\left(\sqrt{2 \cdot \gamma_b}\right),$$

where $Q(x) = \frac{1}{\sqrt{2\pi}} \cdot \int_0^x \exp\left(\frac{-u^2}{2}\right) du$ – a tabular function whose value is given in [24]; γ_b – bit energy ratio E_b to the spectral density of noise N_O.

For a Gaussian channel and the acceptance of coordinated filters, the probability of a bit error in modulation M-QAM, where $M = 2^k$ and k – the pair is determined as follows [24]

$$BER = \frac{2 \cdot (1 - L^{-1})}{\log_2(L)} \cdot Q\left[\sqrt{\frac{3 \cdot \log_2(L)}{L^2 - 1} \cdot \frac{2 \cdot E_b}{N_0}}\right]$$

where $L = \sqrt{K}$ represents the number of levels of amplitude in one dimension.

Conducting the necessary computations as a result of the probability of a bit error in the strong (first) and weak (second) own channels MIMO - systems with an arbitrary number of transmitting antennas, the following graphic dependencies obtained during the simulation was obtained (Figs. 1 and 2) [23].

To estimate the value of L on the radio propagation route, following the results of the studies, the following guidelines for the use of radio wave propagation models for different frequency ranges can be elaborated (Table 2).

Table 2. Recommendations for using empirical radio wave propagation models for different frequency ranges

Frequency range	Technology	Recommended propagation model
Up to 2 GHz	5G, LTE, WiMAX, GSM, UMTS and others.	Hata model
2,3 GHz; 2,5 GHz; 2,6 GHz	5G, LTE, WiMAX	Hata Cost 231
3,5 GHz	5G, LTE, WiMAX	SUI (Stanford University Interim) (for LOS conditions) and Ericsson (for NLOS conditions)
5 GHz; 5,8 GHz	5G, WiMAX	

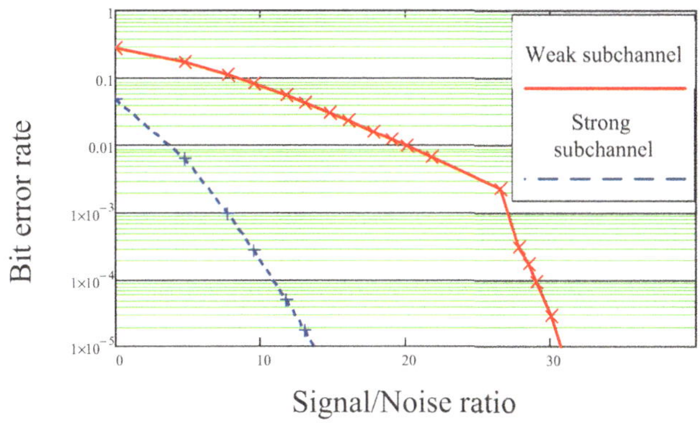

Fig. 1. The logarithmic dependence of the probability of a bit error from the SNR in weak and strong own subchannels for QPSK modulation

Thus, you can express the maximum range for different frequency ranges used in LTE or 5G. For example, for the SUI model [25] for a range above 3.5 GHz:

$$\lg(d/d_0) = (L - (A + X_f + X_h + s))/\, 10\gamma, \text{ для } d > d_0,$$

where d – distance from the base station to the receiving antenna; d_0 = 100 мм; X_f – correction for frequency greater than 2 GHz; X_h – correction for the height of the receiving antenna; s – correction for shading; γ – exponential loss.

The parameter s is between 8.2 and 10.6 dB. Parameter A is defined by:

$$A = 20\lg(4\pi d_0/\lambda),$$

where λ – wavelength; $\gamma = a - bh_b + (c/h_b)$, where h_b –lifting height of the antenna of the base station (m), is in the range from 10 to 80 m. Constants a, b and c depend on the type of terrain, their values are given in [27].

Parameters Xf, Xh can be estimated as:

$$X_f = 6\lg(f/2000),$$

$$X_h = -10,8\lg(h_r/2000),$$

where f – frequency in MHz; h_r – the height of the receiving antenna in meters.

The simulation results show that for city conditions:

1. The coverage radius for LTE technology in the range of 2300–2400 MHz in the operating band of 10, 15 and 20 MHz decreases and makes, respectively:

- For QPSK 1/3 – 3.3, 2.8 and 2.4 km.
- For 16QAM ½ – 1.8, 1.6 and 1.45 km.
- For 64QAM ¾ – 1, 0.8 and 0.7 km.

The corresponding modulation-coding schemes are chosen based on the required quality of service, in particular, the data transfer rate and the bit error rate.

The estimation of the dependence of the communication distance on the frequency at different types of buildings is presented in Fig. 2.

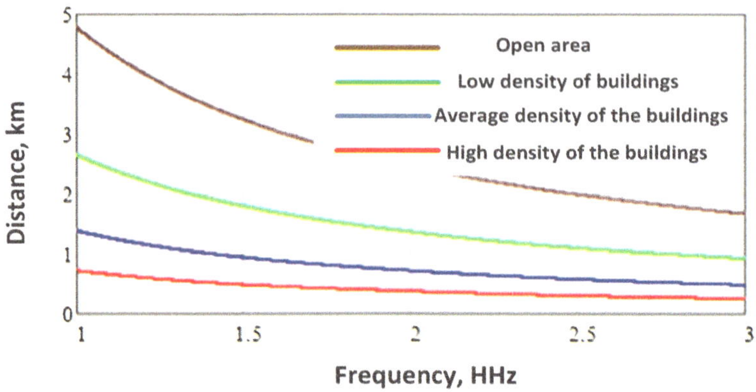

Fig. 2. Dependence of frequency range on different types of building

These modelling results can be used to display all the data in the app and alert operator about potential dangers (Fig. 3). Nevertheless, it can also be built into RPAS intelligence, so that it would autonomously make smarter decisions (Fig. 4).

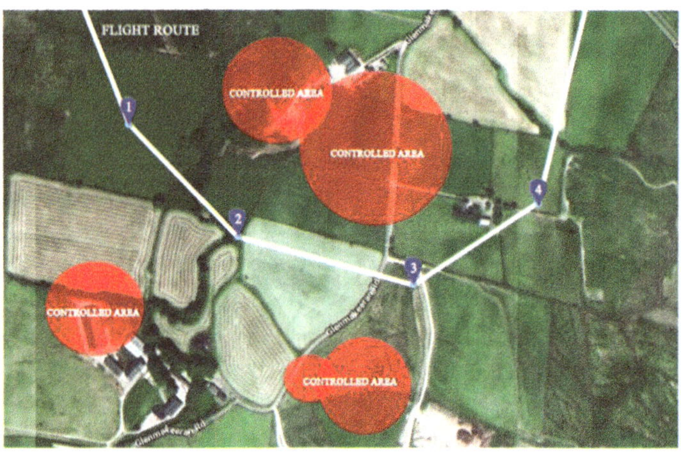

Fig. 3. Display of controlled area in the app for remote pilots

Efficiency of introduction such type of redundancy can be represented by comparison probability of error-free operation of the remote pilots with different redundancy order [13]. It should be noted that first-order redundancy is a single step redundancy.

The probability of error-free operation of the operator is determined by:

$$R(t) = \Phi\left(\frac{\alpha - t}{\beta \cdot \sqrt{\alpha \cdot t}}\right) - \exp\left(\frac{2}{\beta^2}\right) \cdot \Phi\left(\frac{\alpha + t}{\beta \cdot \sqrt{\alpha \cdot t}}\right), \tag{2}$$

where α is the average time of the operator to the first error, β is the coefficient of variation of the operator's time to error, Φ(•) is an integral Laplace function.

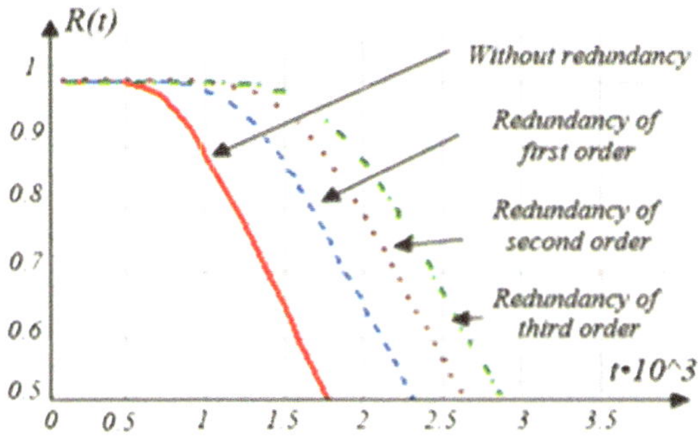

Fig. 4. Probability of error-free operation of the operator

Table 3. Comparison of the probability of error-free operation of the operator equal fifty percent for different redundancy orders in minutes

Redundancy order	R(t) = 0.5 for different redundancy orders in minutes
Without redundancy	1750
First-order redundancy	2350
Second-order redundancy	2590
Third-order redundancy	2830

As can be seen from the results of mathematical modelling even first-order redundancy can improve probability of error-free operation of the operator approximately in 0.7 times (Table 3).

5 Conclusions

This research aimed to identify effective methods for improving informational reliability of the operator of RPAS. Based on the implementation of algorithmic redundancy for the operator, it can be concluded that taking into account the impact on radio frequencies and estimate radio coverage areas can informatically unload the operator of RPAS.

The proposed mathematical model gives the possibility of controlled radio coverage area estimation in the complex conditions of multipath propagation and under the influence of climatic conditions. Results can reduce the informational overload of the remote pilot that leads to informational reliability improvement, increasing the level of aviation safety and flight regularity.

References

1. Reliability theory of ATC systems. Methodical instructions on the study subjects and control tasks, p. 38 (2011). (in Russian), St. Petersburg
2. Swain, A.D.: Accident Sequence Evaluation Program Human Reliability Analysis Procedure. US Nuclear Regulatory Commission, Washington DC (1987)
3. Technical Basis and Implementation Guidelines for a Technique for Human Event Analysis (ATHEANA), Washington DC (2000)
4. Embrey, D.E., Kirwan, B.: A comparative evaluation of three subjective human reliability quantification techniques. In: Coombes, K. (ed.) The Annual Ergonomics Society Conference Proceedings, pp 137–142. Taylor and Francis, London (1983)
5. Taniguchi, A., Onosato, M.: Effect of continuous improvement on the reporting quality of project management information system for project management success. IJITCS **10**(1), 1–15 (2018)
6. Bell, J., Holroyd, J.: Review of human reliability assessment methods. In: Health and Safety Executive, Buxton, pp. 1–90 (2009)
7. Reer, B.: Conclusions from occurrences by descriptions of actions (CODA). In: Drottz Sjöberg, B.M. (ed.) New Risk Frontiers, Proceedings of the 1997 Annual Meeting of the Society for Risk Analysis-Europe, Stockholm (1997)
8. Sträter, O.: The use of incidents for human reliability management. Safety Reliab. **26**(2), 26–47 (2006)
9. Embrey, D.E., Humphreys, P., Rosa, E.A., Kirwan, B. Rea, K.: SLIM-MAUD: an approach to assessing human error probabilities using structured expert judgment, volume I: Overview of SLIM-MAUD. Prepared for the US NRC (1984)
10. Kozhokhina, O., Gribov, V., Rudas, S.: Analytical model of air navigation system operator reliability. In: IEEE 3rd International Conference on Methods and System of Navigation and Motion Control (MSNMC), Proceedings, Kyiv, pp. 170–174 (2014)
11. Kozhokhina, O., Gribov, V. Rudas, S.: Structural reliability of air traffic controllers. In: Proceedings of the National Aviation University, Kyiv, vol. 4, no. 61, pp. 50–56 (2014)
12. Lomov, V.: Dovidnyk z inzhenernoyi psykholohiyi, «Mashynobu-duvannya» , Moscow, p. 368 (1982). (In Russian)
13. Kozhokhina, O., Blahaia, L., Rudas, S., Alexeiev, O.: Informational reliability of radar system operator. In: 17th International Radar Symposium (IRS), Krakow (2016)
14. Bell, J., Holroyd, J.: Review of human reliability assessment methods. In: Health and Safety Executive, Buxton, pp. 1–90 (2009)
15. Alvarenga, M.A.B., Frutuoso, E.M.P.F., Fonseca, R.A.: A critical review of methods and models for evaluating organizational factors in human reliability analysis. Prog. Nucl. Energy **75**, 25–41 (2014)
16. Mitomo, N., Hashimoto, A., Homma, K.: An example of an accident analysis of aircraft based on a human reliability analysis method. In: 2015 4-th International Conference on Informatics, Electronics and Vision, Fukuoka, 15–18 June, p. 1–5 (2015)
17. Lin, Y., Pan, X., He, C.: Human reliability analysis in carrier-based aircraft recovery procedure based on CREAM. In: Proceedings of 2015 the 1-st International Conference on Reliability Systems Engineering, Beijing, 21–23 October, p. 1–6 (2015)

18. Chen, W., Huang, S.: Human reliability analysis for visual inspection in aviation maintenance by a Bayesian network approach. Transp. Res. Rec. **2449**, 105–113 (2014)
19. Manual on RPAS/Doc 10019-AN/507 ICAO, P. 190 (2015)
20. Odarchenko, R.S., Mirutenko, L.V., Dakov, S.: Yu: An improved method for building a LTE support segment. Knowl. Based Technol. **37**(1), 18–26 (2018)
21. Hadidi, M.A., Al-Azzeh, J.S., Tkalich, O.P., Gnatyuk, S.O., Khokhlachova, Y.Y.: Zigbee, bluetooth and Wi-Fi complex wireless networks performance increasing. Int. J. Commun. Antenna Propag. **7**(1), 48–56 (2017)
22. Odarchenko, R., Al Hadidi, M., Al-Azzeh, J.S., Gnatyuk, S., Abakumova, A.: Adaptive regulation of radiated power radio transmitting devices in a modern cellular network, depending on climatic conditions. Contemp. Eng. Sci. **9**(10), 473–485 (2016)
23. Odarchenko, R.S.: Estimation of the probability of bit error and bandwidth of channels of antenna MIMO-systems. Problems of creation, testing, commissioning and operation of complex information systems: Sb. Sciences work Zhytomyr: ZhVI NAU, Vip. 2. pp. 57–67 (2012)
24. Prokys, J.: Digital Communication/Proc. J/Per. from English D. D. Klovsky Moscow, p. 800 (2000)
25. Gradshtein, I., Ryzhik, I.: Tables of integrals, sums, series, and works. Moscow (1971)
26. Pierce D.: Symbols, signals, noises. Patterns and processes of information transfer, Moscow, p. 338 (1967). (in Russian)
27. Maguire, R.: Validating a process for understanding human error probabilities in complex human-computer interfaces. In: Proceedings of the Second Workshop on Complexity in Design, p. 81–89, Glasgow (2005)
28. Nagar, R., Gupta, D.K., Singh, R.M.: Time effective workflow scheduling using genetic algorithm in cloud computing. Int. J. Inf. Technol. Comput. Sci. (IJITCS) **10**(1), 68–75 (2018)
29. Egejuru, N.C., Ogunlade, O., Idowu, P.A.: Development of a mobile-based hypertension risk monitoring system. Int. J. Inf. Eng. Electron. Bus. (IJIEEB) **11**(4), 11–23 (2019)

3-DOF Spherical Parallel Mechanism

Gleb S. Filippov, Victor A. Glazunov, Anna N. Terekhova(✉), Aleksey B. Lastochkin, Robert A. Chernetsov, and Lyubov V. Gavrilina

Mechanical Engineering Research Institute of the Russian Academy of Sciences, 4, Malyi Kharitonievsky pereulok, 101990 Moscow, Russian Federation
filippov.gleb@gmail.com, vaglznv@mail.ru, anya-terehova@bk.ru, lastochkin_aleks@mail.ru, griboedova04@mail.ru

Abstract. The paper presents a spherical parallel mechanism with three dimensions of freedom, the scope of this robot we see in minimally invasive surgery. The mechanism contains three kinematic chains with three rotational kinematic pairs; all rotational actuators have the same basis vector. The paper contains the positioning problem solution, the kinematic analysis based on the screw theory and Gosselin-Angeles approach, the dynamic analysis based on Euler dynamic equations. Based on this data the control algorithm is proposed and numerical simulation performed.

Keywords: Spherical parallel mechanism · Inverse position problem · Kinematic analysis · Dynamics · Control algorithm · Matrix Angeles-Gosseline · Theory of screws

1 Introduction

In spherical parallel robots, the output link rotates round the selected point, which position in space remains unchanged despite of any movement of the output link. This feature is a key operation condition for different positioning and measuring mechanisms, surgical robots and various technological tools.

The first parallel mechanisms appeared early in XX-th century one of the first were the Gough platform or hexapod [1] and the Stewart platform that had three kinematic chains and six degrees of freedom [2], but they still have their place in modern world [3]. Soon the first classifications appeared - one of the first was made by Hunt [4]. Since then many works were published, we want to mention a few that had an influence on this paper. Gosselin and Angeles presented the scheme and parametric synthesis of a spherical mechanism with three degrees of freedom [5], a methodic for kinematic analysis and singularity analysis based on differentiation of the constraint equations [6]. Angeles also proposed parallel manipulators with variable dimensions of freedom [7], in the monograph [8] he considered questions of structure, synthesis, kinematics, and dynamics. Glazunov, Heilo, Tsarkova investigated the control methods and dynamics for planar, serial robots and spherical parallel structure mechanisms [9]. Laryshkin, Glazunov, Erastova investigated the speed and actuator efforts of parallel mechanisms

Z. Hu et al. (Eds.): AIMEE 2019, AISC 1126, pp. 334–344, 2020.
https://doi.org/10.1007/978-3-030-39162-1_31

in singularity proximity [10]. Ganiev, Glazunov, Filippov considered usage of different parallel mechanisms for different technical applications [11]. Nhan, Tang, Heilo, Glazunov considered the oscillations and control algorithm for a spherical parallel manipulator based on the concept of minimization of coordinates, velocities and accelerations oscillations using the inverse dynamic approach [12]. Sadeki, Bourgeois and co-authors presented a spherical parallel manipulator as a robotic component of the hip exoskeleton, providing a complete kinematic correspondence to the 3DOF human hip joint [13]. The Vu paper [14] presented a multi-criteria optimization for spherical manipulator with circular guide design to achieve the best kinematic and dynamic characteristics simultaneously. Cammarata, Calio and coauthors designed a methodic to define dynamic stiffness matrix for the robot in defined position reliable for a high frequency range [15]. Duan, Yang, Cheng presented modeling and analysis of a spherical parallel manipulator with two degrees of freedom [16]. Bai, Hansen and Angeles proposed a methodic to solve the direct positioning problem for spherical parallel robots based on the input-output equations for spherical four-link chains [17]. Algoli, Gaudi and coauthors examined a new method for dynamic simulation of a spherical parallel manipulator using Kane's method and D'Alembert principle presenting the mechanism as a set of elementary bodies and defining actuator moments from the external load [18–22].

In this paper, we consider a spherical parallel mechanism with a circular guide and rotational actuators located with common basis vector. We solve the positioning problem. Using the Gosselin-Angeles approach and kinematic screw theory we provide the kinematic analysis for the robot. The dynamic analysis is carried out on the basis of the Euler equations. We find the control algorithm for a defined output link trajectory, taking into account the elasticity in the actuators and minimizing deviations of position, speed, and acceleration; based on these results we provide the numerical simulation of the output link motion.

2 The Kinematic Analysis

On Fig. 1 the spherical parallel mechanism with three degrees of freedom is presented. The mechanism has a circular guide 1 located on the base, the output link 2, three kinematic chains that have the same structure and contain three rotational kinematic pairs each. The axes of all kinematic pairs intersect at point *O*, any movement of the output link is a rotation around this point. The first kinematic pair of each chain (Fig. 1 marked 3, 4, 5 respectively) is an actuator - they specify movement along the circular guide. The mechanism key design feature is: for each kinematic chain the second joint axis (Fig. 1 marked 6, 7, 8) and the third joint axis (Fig. 1 marked 9, 10, 11) are orthogonal. The center of mass of the mechanism coincides with the center of the output link rotation.

The center of the base coordinate system *OXYZ* we locate in the kinematic pairs' basis vectors intersection point and the axis *OZ* coincides with the axes of the actuators. The actuators movement we define as angles φ_{11}, φ_{21}, φ_{31} around the *OZ* axis - it's the generalized input coordinates of the mechanism. The output link uses the coordinate system $O\xi\eta\zeta$, the positioning angles of the output link in this coordinate system lets designate as α, β, γ respectively.

The kinematic pair's basis vectors are indexed as $\mathbf{e_{kn}}$, where $k = 1, 2, 3$ is the number of the kinematic chain and $n = 1, 2, 3$ is the number of the kinematic pair in the k-th kinematic chain counting from the base:

$$\angle(\mathbf{e}_{12}, \mathbf{e}_{13}) = \angle(\mathbf{e}_{22}, \mathbf{e}_{23}) = \angle(\mathbf{e}_{32}, \mathbf{e}_{33}) = \pi/2 \text{ rad}.$$

Let's define the basis vectors initial position coordinates:

$$\mathbf{e}_{11} = \begin{pmatrix} x_{11} \\ y_{11} \\ z_{11} \end{pmatrix}; \mathbf{e}_{12} = \begin{pmatrix} x_{12} \\ y_{12} \\ z_{13} \end{pmatrix}; \ldots \mathbf{e}_{33} = \begin{pmatrix} x_{33} \\ y_{33} \\ z_{33} \end{pmatrix}.$$

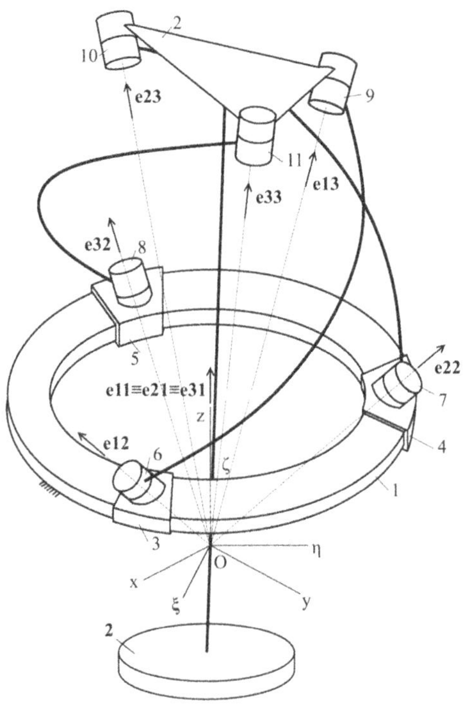

Fig. 1. Spherical parallel mechanism

The output link position is defined using matrix AA. This matrix defines the rotations on angles α, β, γ:

$$\mathbf{AA} = \mathbf{A}_{\gamma} \cdot \mathbf{A}_{\beta} \cdot \mathbf{A}_{\alpha}$$

Therefore, the kinematic pair's basis vectors after an output link movement $\mathbf{et_{kn}}$. We can describe as:

$$\mathbf{et}_{13} = \mathbf{AA} \cdot \mathbf{e}_{13};\ \mathbf{et}_{23} = \mathbf{AA} \cdot \mathbf{e}_{23};\ \mathbf{et}_{33} = \mathbf{AA} \cdot \mathbf{e}_{33},$$

where $k = 1, 2, 3$ is the number of the kinematic chain and $n = 1, 2, 3$ is the number of the kinematic pair in the k-th kinematic chain counting from the base.

As scalar product of orthogonal vectors equals zero, therefore from the scalar product of 2nd and 3rd joints basis vectors we can get implicit functions connecting absolute coordinates α, β, γ and generalized coordinates of kinematic chains φ_{11}, φ_{21}, φ_{31}. Let's define FF_1, FF_2, FF_3 respectively for the first, second and third kinematic chains:

$$FF_1 = \mathbf{et}_{12} \cdot \mathbf{et}_{13};,\ FF_2 = \mathbf{et}_{22} \cdot \mathbf{et}_{23} = 0;,\ FF_3 = \mathbf{et}_{32} \cdot \mathbf{et}_{33} = 0.$$

As the implicit function equals zero its full differential also equals zero, then we can rewrite the partial derivatives and get the matrix velocity equation. The resulting linear equations system can be solved on the approach of known absolute velocities or known generalized velocities.

Matrix Angeles-Gosseline to absolute coordinates and to generalized coordinates:

$$\mathbf{A_{ac}} = \begin{pmatrix} \frac{\partial(FF_1)}{\partial\alpha} & \frac{\partial(FF_1)}{\partial\beta} & \frac{\partial(FF_1)}{\partial\gamma} \\ \frac{\partial(FF_2)}{\partial\alpha} & \frac{\partial(FF_2)}{\partial\beta} & \frac{\partial(FF_2)}{\partial\gamma} \\ \frac{\partial(FF_3)}{\partial\alpha} & \frac{\partial(FF_3)}{\partial\beta} & \frac{\partial(FF_3)}{\partial\gamma} \end{pmatrix};\ \mathbf{A_{gc}} = \begin{pmatrix} \frac{\partial(FF_1)}{\partial\varphi_{11}} & 0 & 0 \\ 0 & \frac{\partial(FF_2)}{\partial\varphi_{21}} & 0 \\ 0 & 0 & \frac{\partial(FF_3)}{\partial\varphi_{31}} \end{pmatrix}.$$

Inverse Jacobian matrix:

$$\mathbf{J_{inv}} = \mathbf{A_{ac}} \cdot \left(\mathbf{A_{gc}}\right)^{-1}$$

Direct Jacobian matrix:

$$\mathbf{J_{dir}} = \left(\mathbf{J_{inv}}\right)^{-1}$$

Let's consider the mechanism position, when the basis vectors of the second kinematic pairs of all kinematic chains are parallel to the base. Now we can define the initial position for calculations. The relative position of the second kinematic pairs are determined by the angles $\angle(\mathbf{e}_{12}, \mathbf{e}_{22}) = {}^{2\pi}/_{3}$ rad; $\angle(\mathbf{e}_{12}, \mathbf{e}_{32}) = {}^{4\pi}/_{3}$ rad; the basis victors of the second and third joints in each kinematic chain are orthogonal $\angle(\mathbf{e}_{12}, \mathbf{e}_{13}) = \angle(\mathbf{e}_{22}, \mathbf{e}_{23}) = \angle(\mathbf{e}_{32}, \mathbf{e}_{33}) = {}^{\pi}/_{2}$ rad; basis vectors of the third joints in kinematic chain $\mathbf{e}_{13}$, $\mathbf{e}_{23}$, $\mathbf{e}_{33}$ have angles $\pi/4$ rad to OXY plane.

So for the first kinematic chain we have kinematic pairs basis vectors coordinates:

$$\mathbf{e}_{11} = \begin{pmatrix} 0 \\ 0 \\ 1 \end{pmatrix};\ \mathbf{e}_{12} = \begin{pmatrix} 1 \\ 0 \\ 0 \end{pmatrix};\ \mathbf{e}_{13} = \begin{pmatrix} 0 \\ \cos(\pi/4) \\ \sin(\pi/4) \end{pmatrix} = \begin{pmatrix} 0 \\ \sqrt{2}/2 \\ \sqrt{2}/2 \end{pmatrix}.$$

Let's write the coordinates of the basis vectors in other position after a t time period.

The second joint:

$$\mathbf{e}_{11} = \begin{pmatrix} \cos(\varphi_{11}) \\ \sin(\varphi_{11}) \\ 0 \end{pmatrix}.$$

The third joint:

$$\mathbf{et}_{13} = \mathbf{AA} \cdot \mathbf{e}_{13} = \begin{pmatrix} (\sqrt{2}/2) \cdot [\sin\gamma \cdot (\sin\alpha - \cos\alpha) + \cos\gamma \cdot \sin\beta \cdot (\sin\alpha + \cos\alpha)] \\ (\sqrt{2}/2) \cdot [\cos\gamma \cdot (\cos\alpha - \sin\alpha) + \sin\gamma \cdot \sin\beta \cdot (\sin\alpha + \cos\alpha)] \\ (\sqrt{2}/2) \cdot [\cos\beta \cdot (\sin\alpha + \cos\alpha)] \end{pmatrix}.$$

Scalar product of vectors $\mathbf{et}_{12}$ and $\mathbf{et}_{13}$:

$$\begin{aligned} FF_1 = \mathbf{et}_{12} \cdot \mathbf{et}_{13} = & \left(\sqrt{2/2}\right) \cdot \cos(\varphi_{11}) \cdot [\sin\gamma \cdot (\sin\alpha - \cos\alpha) + \cos\gamma \cdot \sin\beta \cdot (\sin\alpha + \cos\alpha)] + \\ & + \left(\sqrt{2/2}\right) \cdot \sin(\varphi_{11}) \cdot [\cos\gamma \cdot (\cos\alpha - \sin\alpha) + \sin\gamma \cdot \sin\beta \cdot (\sin\alpha + \cos\alpha)] = 0. \end{aligned}$$

Now we can find generalized coordinates of the first kinematic chain:

$$\varphi_{11} = \operatorname{arctg}\left[\frac{-[\sin\gamma \cdot (\sin\alpha - \cos\alpha) + \cos\gamma \cdot \sin\beta \cdot (\cos\alpha + \sin\alpha)]}{[\cos\gamma \cdot (\cos\alpha - \sin\alpha) + \sin\gamma \cdot \sin\beta \cdot (\cos\alpha + \sin\alpha)]}\right].$$

We get generalized coordinates φ_{21}, φ_{31} for the other chains in the same way:

$$\varphi_{21} = \operatorname{arctg}\left[\frac{\sqrt{3} \cdot \cos\gamma \cdot \cos\beta - \sin\gamma \cdot (\sin 2\alpha + \cos\alpha) - \cos\gamma \cdot \sin\beta \cdot (2 \cdot \cos\alpha - \sin\alpha)}{-\cos\gamma \cdot (\cos\alpha + 2 \cdot \sin\alpha) - \sin\gamma \cdot \sin\beta \cdot (\sin\alpha - 2 \cdot \cos\alpha) - \sqrt{3} \cdot \sin\gamma \cdot \cos\beta}\right] + \pi,$$

$$\varphi_{31} = \operatorname{arctg}\left[\frac{-[\sin\gamma \cdot (\sin 2\alpha + \cos\alpha) + \cos\gamma \cdot \sin\beta \cdot (2 \cdot \cos\alpha - \sin\alpha) + \sqrt{3} \cdot \cos\gamma \cdot \cos\beta]}{[\sqrt{3} \cdot \sin\gamma \cdot \cos\beta - \cos\gamma \cdot (\cos\alpha + 2 \cdot \sin\alpha) - \sin\gamma \cdot \sin\beta \cdot (\sin\alpha + 2 \cdot \cos\alpha)]}\right] + \pi.$$

As an example, lets define the absolute coordinates as $\alpha = \pi/3$ rad., $\beta = \pi/6$ rad., $\gamma = \pi/4$ rad. Then we have the generalized coordinates:

$$\varphi_{11} = -1,277\,\text{rad.};\ \varphi_{21} = 3,42\,\text{rad.};\ \varphi_{31} = 4,539\,\text{rad.}$$

Then we can calculate the partial derivatives and get the Angeles-Gosseline matrix:

$$\mathbf{A_{ac}} = \begin{pmatrix} 0,913 & -0,395 & -0,548 \\ 0,322 & -0,303 & -0,936 \\ 0,35 & 0,217 & -0,964 \end{pmatrix}.$$

If the absolute velocity is defined equal to 1, then we can calculate the generalized velocities:

$$\omega_{\alpha\beta\gamma} = \begin{pmatrix} 1 \\ 1 \\ 1 \end{pmatrix};\ \omega_{123} = \mathbf{J_{gc}} \cdot \alpha\beta\gamma;\ \omega_{123} = \begin{pmatrix} -0,675 \\ 0,708 \\ 0,129 \end{pmatrix}.$$

3 The Dynamic Analysis

The dynamic analysis is based Euler dynamic equations for the spherical mechanism and the results we got in the previous part of the paper. The mechanism mass center coincides with the rotation center of the output link. The mass of kinematic chains is neglected.

Dynamic equations:

$$\begin{cases} J_\xi \cdot \ddot{\varphi}_\xi = M_1 \cdot \dfrac{\partial\varphi_{11}}{\partial\varphi_\xi} + M_2 \cdot \dfrac{\partial\varphi_{21}}{\partial\varphi_\xi} + M_3 \cdot \dfrac{\partial\varphi_{31}}{\partial\varphi_\xi} + \dot{\varphi}_\eta \cdot \dot{\varphi}_\zeta \cdot (J_\zeta - J_\eta); \\ J_\eta \cdot \ddot{\varphi}_\eta = M_1 \cdot \dfrac{\partial\varphi_{11}}{\partial\varphi_\eta} + M_2 \cdot \dfrac{\partial\varphi_{21}}{\partial\varphi_\eta} + M_3 \cdot \dfrac{\partial\varphi_{31}}{\partial\varphi_\eta} + \dot{\varphi}_\xi \cdot \dot{\varphi}_\zeta \cdot (J_\xi - J_\zeta); \\ J_\zeta \cdot \ddot{\varphi}_\zeta = M_1 \cdot \dfrac{\partial\varphi_{11}}{\partial\varphi_\zeta} + M_2 \cdot \dfrac{\partial\varphi_{21}}{\partial\varphi_\zeta} + M_3 \cdot \dfrac{\partial\varphi_{31}}{\partial\varphi_\zeta} + \dot{\varphi}_\xi \cdot \dot{\varphi}_\eta \cdot (J_\eta - J_\xi), \end{cases}$$

J_ξ, J_η, J_ζ – the main Central moments of inertia of the output link relative to the axes ξ, η, ζ;

$\ddot{\varphi}_\xi = \dot{\omega}_\xi$, $\ddot{\varphi}_\eta = \dot{\omega}_\eta$, $\ddot{\varphi}_\zeta = \dot{\omega}_\zeta$ – projections of angular acceleration on the axis of the moving coordinate system associated with the output link and along the main Central axes of inertia;

M_1, M_2, M_3 – the driving torques in the actuators;

$\partial\varphi_{11}/\partial\varphi_\xi$, $\partial\varphi_{21}/\partial\varphi_\xi, \ldots$, $\partial\varphi_{31}/\partial\varphi_\zeta$ – these are partial derivatives relating generalized velocities and projections of the angular velocity of the output link on the axis of the moving coordinate system;

$\dot{\varphi}_{\xi} = \omega_{\xi}$, $\dot{\varphi}_{\eta} = \omega_{\eta}$, $\dot{\varphi}_{\zeta} = \omega_{\varsigma}$ – projections of angular velocity on the axis of the rotating coordinate system associated with the output link and located along the main Central axes of inertia.

The relative power screw moment transmitted from each kinematic chain to the output link, and the kinematic screw of the output link can be presented in both rotating and base coordinate systems. We must note that in this case the power screw has only the moment part. In the base coordinate system, the angular velocity is equal to the vector sum of the angular velocities of all three kinematic pairs of the chain; the axes of the two non-drive joints are orthogonal to the axis of the transmitted moment.

The cross product of the basis vectors of the second and third joints of the first kinematic chain determines the axis of the transmitted moment in the base coordinate system **rt1** ($rt1_x$, $rt1_y$, $rt1_z$):

$$\mathbf{rt1} = \mathbf{et_{12}} \times \mathbf{et_{13}}.$$

The cross product in the rotating coordinate system will give the projection of the coordinates of the transmitted moment on the axis of the rotating coordinate system $\mathbf{rt1_m}$ ($rt1_{\xi}$, $rt1_{\eta}$, $rt1_{\zeta}$):

$$\mathbf{rt1_m} = \mathbf{et_{12m}} \times \mathbf{e_{12}}.$$

Hence, we obtain an equation that connects the angular velocity in the input kinematic pair and the projection of the angular velocities on the axis of the rotating coordinate system. Let's find partial derivatives. For the first chain we have:

$$\omega_{\xi} r1_{\xi} + \omega_{\eta} r1_{\eta} + \omega_{\zeta} r1_{\zeta} = \omega_{11} r1_{z},$$

ω_{ξ}, ω_{η}, ω_{ζ} - angular velocity projections on the axis of the moving system; $r1_{\xi}$, $r1_{\eta}$, $r1_{\zeta}$ - projections of the transmitted moment on the axis of the moving coordinate system; $r1_z$ - projection of the transmitted moment on the axis Z.

From the last equation follows:

$$\frac{\omega_{11}}{\omega_{\xi}} = \frac{r1_{\xi}}{r1_z}; \frac{\omega_{11}}{\omega_{\eta}} = \frac{r1_{\eta}}{r1_z}; \frac{\omega_{11}}{\omega_{\zeta}} = \frac{r1_{\zeta}}{r1_z}.$$

Next, we find the dependency between the coordinates and the transmitted moment, expressed in different coordinate systems. Let's designate some variables to simplify the calculation records:

$$A_1 = \frac{rt1_{\xi}}{rt1_z}; B_1 = \frac{r1_{\eta}}{r1_z}; C_1 = \frac{r1_{\varsigma}}{r1_z}.$$

The dependency for the corresponding partial derivatives for the second and the third chains are determined. Again, we designate the partial derivatives showing the dependency of the angular velocity in the input pair and the angular velocity vector projections on the axis of the coordinate system associated with the output link:

$$\frac{rt2_{\xi}}{rt2_{z}} = A_2; \frac{rt2_{\eta}}{rt2_{z}} = B_2, \frac{rt2_{\varsigma}}{rt2_{z}} = C_2; \frac{rt3_{\xi}}{rt3_{z}} = A_3; \frac{rt3_{\eta}}{rt3_{z}} = B_3; \frac{rt3_{\varsigma}}{rt3_{z}} = C_3.$$

In dynamic analysis we assume that the mass of the output link is 0.5 kg; its moments of inertia are: J_{ξ} = 0,12 kg/m^2, J_{η} = 0,12 kg/m^2, J_{ζ} = 0,24 kg/m^2.

We use the control algorithm that minimizes the deviations of position, speed and acceleration. In this case, the deviation behavior corresponds to the second order of oscillatory link, with "mass", "stiffness" "resistance" proportional to the speed.

Let's define the coefficient specifying oscillation link: Hg = 120, Hg_1 = 7700.

And the absolute coordinates change trajectories are sine:

$$\alpha_d = 0{,}1 \cdot \sin(10 \cdot t),\ \beta_d = 0{,}1 \cdot \sin(10 \cdot t),\ \gamma_d = 0{,}1 \cdot \sin(10 \cdot t);$$

then we have the cosine velocities:

$$\dot{\alpha}_d = \cos(10 \cdot t),\ \dot{\beta}_d = \cos(10 \cdot t),\ \dot{\gamma}_d = \cos(10 \cdot t);$$

and sine accelerations

$$\ddot{\alpha}_d = -10 \cdot \sin(10 \cdot t),\ \ddot{\beta}_d = -10 \cdot \sin(10 \cdot t),\ \ddot{\gamma}_d = -10 \cdot \sin(10 \cdot t).$$

Based on the kinematic Euler equations and the assumed deviation change model, we have an expression for the accelerations provided by the controlled actuators:

$$\ddot{\alpha}_r = \ddot{\alpha}_d - Hg \cdot \left(\omega_{\xi} + \frac{(\omega_{\eta} \cdot \sin\alpha + \omega_{\zeta} \cdot \cos\alpha) \cdot \sin\beta}{\cos\beta} - \dot{\alpha}_d\right) - Hg_1 \cdot (\alpha_d - \alpha_r)$$

Then it is possible to determine the required moments in the actuators M_1:

$$M_1 = \frac{c \cdot (A_2 \cdot B_3 - A_3 \cdot B_2) - b \cdot (A_2 \cdot C_3 - A_3 \cdot C_2) + a \cdot (B_2 \cdot C_3 - B_3 \cdot C_2)}{Z_n};$$

The calculation program uses the calculated moments to determine the real movement.

4 Numerical Simulation

Sine oscillations of the output link in all three coordinates were used as a test motion, the initial conditions are zeros. We calculated data for graphs of α, β, γ coordinates changes, ω_{ξ}, ω_{η}, ω_{ζ} velocities changes and phase trajectories. The Figs. 2, 3 and 4 present the results corresponding to the α angle.

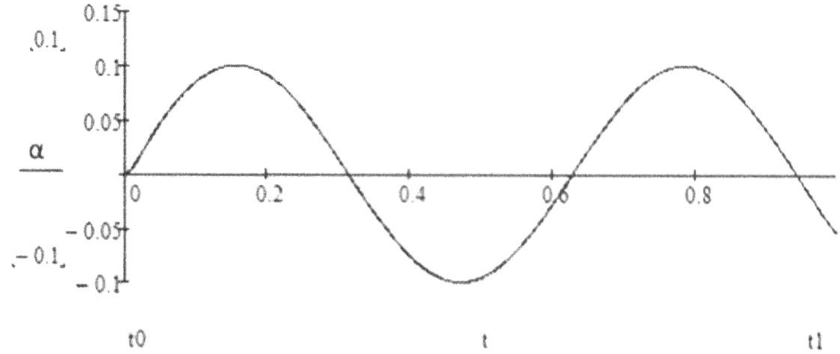

Fig. 2. The graph of change of α coordinate

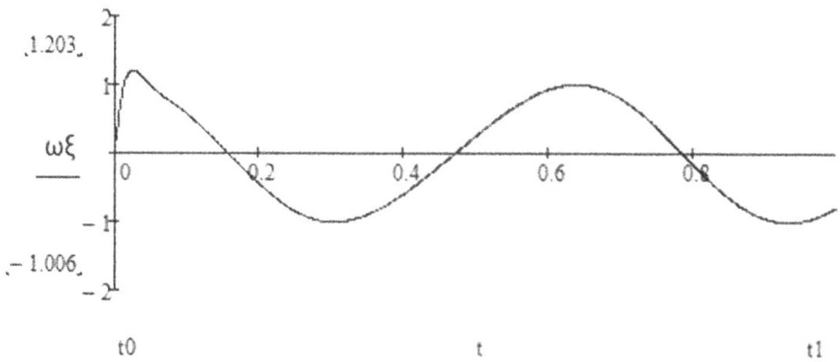

Fig. 3. The graph of change of ω_ξ velocity

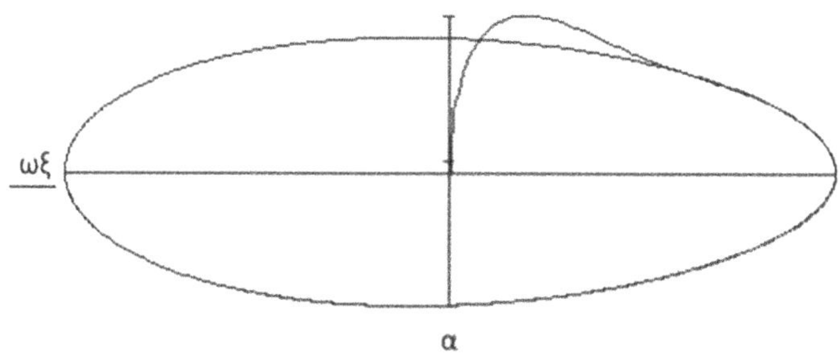

Fig. 4. Phase trajectory

5 Conclusions

The paper presents the 3DOF RRR spherical parallel mechanism; all rotational motors of the mechanism are located along one axis. The positioning problem solution is achieved. In the kinematic analysis the velocities are calculated based on Gosselin-Angeles approach with screw theory involvement. The presented dynamic analysis is based on the Euler dynamic equations. Using the achieved results and control algorithm based on deviation minimization the numerical simulation is calculated.

In the future, it is planned to continue investigation of the mechanism accuracy, the definition of the singularities in the working area and the development of control algorithms for overcoming these singularity zones.

Acknowledgments. The work was supported by the Russian Foundation for basic research (project 16-29-04273).

References

1. Gough, V.E.: Contribution to discussion to papers on research in automobile stability and control and in tyre performance. Autom. Div. Inst. Mech. Eng. **171**, 392–396 (1956)
2. Stewart, D.A.: Platform with six degrees of freedom. Proc. Inst. Mech. Eng. **180**(15), 371–386 (1965)
3. Joumah, A.A., Albitar, C.: Design optimization of 6-RUS parallel manipulator using hybrid algorithm. Int. J. Inf. Technol. Comput. Sci. (IJITCS) **10**(2), 83–95 (2018)
4. Hunt, K.: Structural kinematics of in-parallel-actuated robot arms. ASME J. Mech. Transm. Autom. Des. **105**(4), 705–712 (1983)
5. Gosselin, C., Angeles, J.: The optimum kinematic design of a spherical three-degree-of-freedom parallel manipulator. Trans. ASME J. Mech. Trans. Autom. Des. 202–207 (1989)
6. Gosselin, C.M., Angeles, J.: Singularity analysis of closed-loop kinematic chains. IEEE Trans. Robot. Autom. **6**(3), 281–290 (1990)
7. Angeles, J.: The qualitative synthesis of parallel manipulators. J. Mech. Des. **126**, 617–624 (2004)
8. Angeles, J.: Fundamentals of Robotic Mechanical Systems: Theory, Methods, and Algorithms. Springer, Dordrecht (2006)
9. Glazunov, V.A., Kheylo, S.V., Tsarkov, A.V.: The control complex robotic system on parallel mechanism. In: Smart Electromechanical Systems, pp. 137–146. Springer (2018)
10. Laryushkin, P., Glazunov, V., Erastova, K.: On the maximization of joint velocities and generalized reactions in the workspace and singularity analysis of parallel mechanisms. Robotica **37**, 675–690 (2019)
11. Ganiev, R.F., Glazunov, V.A., Filippov, G.S.: Actual machine science problems and their solutions. Wave technologies, additive technologies, machine-tool construction, robotic surgery. J. Mach. Reliab. **5**, 16–25 (2018)
12. Huu, K.N., Vo, D.T., Kheylo, S., Glazunov, V.: Oscillations and control of spherical parallel manipulator. Nguyen Int. J. Adv. Robot. Syst. (2019)
13. Sadeqi, S., Bourgeois, S., Park, E., Arzanpour, S.: Design and performance analysis of a 3-RRR spherical parallel manipulator for hip exoskeleton applications. J. Rehabil. Assistive Technol. Eng. **4**, 1–11 (2017)
14. Wu, G.: Multiobjective optimum design of a 3-RRR spherical parallel manipulator with kinematic and dynamic dexterities. Model. Identif. Control. **33**(3), 111–122 (2012)
15. Cammarata, A., Calio, I., Urso, D., Greco, A., Lacagnina, M., Fichera, G.: Dynamic stiffness model of spherical parallel robots. J. Sound Vib. **384**, 312–324 (2016)
16. Xuechao, D., Yongzhi, Y., Bi, C.: Modeling and analysis of a 2-DOF spherical parallel manipulator. Sensors **16**, 1485 (2016)
17. Bai, S., Hansen, M., Angeles, J.: A robust forward-displacement analysis of spherical parallel robots. Mech. Mach. Theory **44**(12), 2204–2216 (2009)
18. Elgolli, H., Houidi, A., Mlika, A., Romdhane, L.: Analytical analysis of the dynamic of a spherical parallel manipulator. Int. J. Adv. Manuf. Technol. **101**(1–4), 859–871 (2019)

19. Kumara, V., Sena, S., Roy, S.S., Dasa, S.K., Shomea, S.N.: Inverse kinematics of redundant manipulator using interval newton method. Int. J. Eng. Manuf. (IJEM) **2**, 19–29 (2015)
20. Mohammed, R.H., Elnaghi, B.E., Bendary, F.A., Elserfi, K.: Trajectory tracking control and robustness analysis of a robotic manipulator using advanced control techniques. Int. J. Eng. Manuf. (IJEM) **6**, 42–54 (2018)
21. Krishan, G., Singh, V.R.: Motion control of five bar linkage manipulator using conventional controllers under uncertain conditions. Int. J. Intell. Syst. Appl. (IJISA) **5**, 34–40 (2016)
22. Piltan, F., TayebiHaghighi, S., Sahamijoo, A., Bod, H.R., Jowkar, S., Kim, J.: Adaptive finite-time convergence fuzzy ARX-laguerre system estimation. Int. J. Intell. Syst. Appl. (IJISA) **5**, 27–35 (2019)

Quality Evaluation of Mechanical Experiment Teaching Under the Background of Emerging Engineering Education

Mengya Zhang, Zhiping Liu(✉), Kun Chen, Qingying Zhang, and Jinshan Dai

School of Logistics Engineering, Wuhan University of Technology, Wuhan 430063, China
kmno40311@163.com, {lzp, jinshan.dai}@whut.edu.cn, 690005057@qq.com, kathy8899@126.com

Abstract. Under the background of Emerging Engineering Education, in order to further deepen the education reform, mechanical talents cultivation target and the cultivation system are formulated, the evaluation system of the mechanical experiment teaching quality is constructed. The application of heuristic teaching and the cultivation of students' innovative research ability are more emphasized. The quality of experimental teaching is evaluated and graded by using the method of fuzzy comprehensive evaluation. Finally, an example is analyzed. The results show that the evaluation model is more objective, scientific and comprehensive, and is suitable for the objective evaluation of mechanical experiment teaching quality.

Keywords: Emerging Engineering Education · Mechanical experiment teaching · Quality evaluation

1 Introduction

Since February 2017, the Ministry of Education has actively promoted the construction of Emerging Engineering Education (abbreviated as EEE) to cultivate diversified and innovative outstanding engineering talents, its connotation is to take high moral values establishment and people cultivation as the fundamental task of education in respond to change and shape the future for the construction concept using inheritance and innovation, crossover and integration, coordination and sharing as the main way [1]. Therefore, colleges and universities pay more attention to the practicality of students' knowledge and skills [2].

As a bridge between theory and practice, experimental teaching is particularly important for mechanical students. The professional knowledge of mechanical discipline is complex. Experimental teaching can greatly improve students' understanding and memory of professional knowledge to a large extent, at the same time, students' operational ability is exercised, and students' creative thinking is cultivated [3, 4]. However, at present, colleges and universities do not attach enough importance to experimental teaching. The evaluation of experimental teaching quality often stays in

Z. Hu et al. (Eds.): AIMEE 2019, AISC 1126, pp. 345–353, 2020.
https://doi.org/10.1007/978-3-030-39162-1_32

the intact rate of experimental equipment, the completeness of teachers' teaching plans, classroom discipline, whether the experimental log is filled in or not, and seldom evaluates teachers' teaching process and students' effects, which makes many experimental teaching mere formality and do not pay attention to effects [5]. Therefore, establishing a comprehensive and systematic evaluation system of mechanical experimental teaching is very important to improve the quality of experimental teaching.

2 Establishment of Training System for Mechanical Talents Under the Background of EEE

2.1 Training Objectives of Mechanical Talents

Under the background of the EEE, the students majoring in mechanical engineering should know the basic knowledge of natural science and humanities and social sciences, have good moral quality and social responsibility, have a certain international vision, have solid basic theoretical knowledge and application ability of mechanical specialty, have good engineering practice ability and modern engineering tools using ability, and have team work spirit and leadership potential of engineering and technical personnel [6, 7].

2.2 Training System of Mechanical Talents

Specifically, the social demand for mechanical talents requires solid professional knowledge, strong engineering practice ability and high comprehensive quality [8, 9]. Therefore, a three-legged talent training system consisting of knowledge dimension, ability dimension and quality dimension is proposed (Fig. 1).

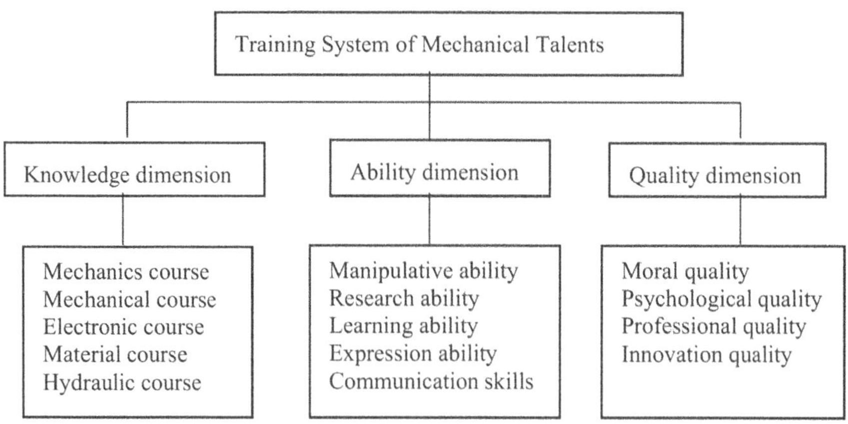

Fig. 1. Training system of mechanical talents

2.2.1 Knowledge Dimension

With the theoretical knowledge of mechanical design as the main line, master theoretical mechanics, material mechanics, mechanical principle, mechanical design, interchangeability and measurement technology, metal technology, electrical and electronic technology foundation, engineering testing technology and application, electromechanical transmission and control, etc.

2.2.2 Ability Dimension

Ability improvement needs experience [10]. Based on the cultivation of experimental practical ability, with the support of various laboratories and practice centers, students' basic ability of engineering design is cultivated, and students' working ideas and innovative ability of engineering design to solve complex engineering problems independently are cultivated [11, 12].

2.2.3 Quality Dimension

Modern mechanical professionals need higher comprehensive quality [13]. Moral quality, psychological quality, professional quality and innovative consciousness are all important comprehensive qualities. Among them, the cultivation of innovative consciousness runs through the whole process of university cultivation, while the innovation of engineering technology is based on extensive knowledge configuration and independent and logical thinking.

3 Establishment of Quality Evaluation Index System for Mechanical Experiment Teaching

Experimental teaching is an important means to cultivate students' abilities and qualities [14]. Mechanical experimental teaching should be guided by the training system of mechanical talents. Combining knowledge dimension, ability dimension and quality dimension, experimental objectives, experimental contents and experimental methods should be established to train students to meet the requirements of training objectives [15]. Therefore, the establishment of the evaluation system should be fully considered in combination with the training system.

3.1 Analysis of Influencing Factors

In order to evaluate the quality of mechanical experiment teaching, it is necessary to analyze the various factors and their internal relations that affect the teaching effect of mechanical experiment [16–18]. The factors affecting the effect of experimental teaching involve many aspects, including the teaching quality of teachers and the learning effect of students, as well as the preparation of experimental teaching conditions and experimental management.

(1) The preparation of teaching conditions mainly includes whether the experimental instruction books and teaching plans are ready, whether the quantity and intact rate of the experimental equipment can meet the experimental requirements,

whether the overall environment of the laboratory is clean and tidy, whether there are potential safety hazards, etc.;

(2) The teaching quality refers to whether the teaching content of teachers is full and novel in the course of experiment; whether teachers use heuristic teaching methods to guide students to think and ask questions [19]; how about the interaction and discussion between teachers and students during course of experiment; and how about the discipline of experiment;

(3) The learning effect refers to whether students have the ability to discover and solve problems, whether they have the ability of comprehensive analysis [20, 21]. In addition, innovation and research ability is particularly important under the background of EEE;

(4) Laboratory management refers to whether there is experimental attendance; the use records of experimental equipment; whether the filling of experimental logs is standardized; whether the experimental report is detailed, etc.

3.2 Establishment of Evaluation Index System

The influencing factors of the quality of mechanical experiment teaching are synthesized, the evaluation indexes are established, and the evaluation index system is established [22], as shown in the Table 1:

Table 1. Evaluation index system of mechanical experiment teaching

Evaluation index system of mechanical experiment teaching	Experimental preparation	Teaching materials
		Equipment intact rate
		Experimental safety
	Experimental process	Experimental content
		Teaching method
		Teacher-student interaction
		Experimental discipline
	Experimental effect	Problem solving ability
		Comprehensive analysis ability
		Innovation capacity
	Experimental management	Attendance management
		Equipment use record
		Report correction situation

4 Fuzzy Evaluation

In order to ensure the quality of experimental teaching, it is necessary to evaluate experimental teaching objectively and scientifically. However, the evaluation of experimental teaching quality is a systematic project, and there is no special method applicable to the evaluation of mechanical experimental teaching quality at present.

The fuzzy comprehensive evaluation method can be applied to the teaching evaluation of mechanical experiment quality, which can evaluate the teaching quality more objectively and accurately.

4.1 Establishment of Fuzzy Relation Matrix

The evaluation index set $U = \{U_1, U_2, \cdots, U_n\}$ refers to the factors in the evaluation index system; the evaluation result set $V = \{V_1, V_2, \cdots, V_n\}$ is the collection of evaluation results; R is the link between U and V, reflecting the fuzzy relationship between evaluation indexes and evaluation results [23, 24]. The fuzzy relation matrix with n evaluation indexes and m evaluation results can be expressed as follows:

$$R = \begin{bmatrix} R_1 \\ R_2 \\ \vdots \\ R_n \end{bmatrix} = \begin{bmatrix} r_{11} & r_{12} & \cdots & r_{1m} \\ r_{21} & r_{22} & \cdots & r_{2m} \\ \vdots & \vdots & \ddots & \vdots \\ r_{n1} & r_{n2} & \cdots & r_{nm} \end{bmatrix} \tag{1}$$

$R_{ij} = u_r(u_i, v_j)$, $0 < R_{ij} < 1$, R_{ij} means: for the evaluation index u_i, the membership degree of the evaluation value is the evaluation result v_j, and the characteristic of the matrix is that the determination of the sum of the elements of each row is 1. R_{ij} is often completed by Delphi method.

4.2 Fuzzy Comprehensive Evaluation

Due to the importance of each evaluation index is different, the weights are different. The weights set $W = \{W_1, W_2, \cdots, W_n\}$ is constructed, and the sum of the weights is 1.

According to the multiplication operation of the fuzzy matrix, the fuzzy evaluation set can be expressed as $B = \{B_1, B_2, \ldots B_m\}$, the fuzzy evaluation set B is calculated [25] according to the following formula:

$$(b_1, b_2, \cdots, b_m) = (\omega_1, \omega_2, \cdots, \omega_n) \times \begin{bmatrix} r_{11} & r_{12} & \cdots & r_{1m} \\ r_{21} & r_{22} & \cdots & r_{2m} \\ \vdots & \vdots & \ddots & \vdots \\ r_{n1} & r_{n2} & \cdots & r_{nm} \end{bmatrix} \tag{2}$$

4.3 Confirmation of Evaluation Results

In this study, the weighted average method is used to solve the final evaluation result V as follows:

$$v = \frac{\sum_{j=1}^{m} b_j v_j}{\sum_{j=1}^{m} b_j} \times 100 \tag{3}$$

The multi-level evaluation is carried out by using the method of fuzzy comprehensive evaluation, and the evaluation result V is obtained. The evaluation result is divided into five grades:

V_1: When $90 < V < 100$, the evaluation grade is excellent;
V_2: When $80 < V < 90$, the evaluation grade is good;
V_3: When $70 < V < 80$, the evaluation grade is medium;
V_4: When $60 < V < 70$, the evaluation grade is poor;
V_5: When $0 < V < 60$, the evaluation grade is extremely poor.

5 Example Analysis

5.1 Background Introduction

After the reform of experimental teaching, the experimental teaching situation of the course "Exchangeability and Measurement Technology" in the second semester of the academic year 2018–2019 is investigated. "Exchangeability and Measurement Technology" is a basic course offered by mechanical engineering major. Its contents are closely related to the design, production and selection of mechanical products. Through experiments, students can combine theory with practice, not only have a better understanding of the interchangeability and measurement technology of mechanical parts and components, but also have a better understanding of the length benchmarks and related standards. At the same time, they can also have a deeper understanding of the principles and methods of measuring instruments and measuring instruments. It is very important for them to master measurement technology and improve their ability to analyze and solve problems.

5.2 Evaluation of Experimental Teaching

For the quality evaluation of the experimental teaching, a questionnaire is designed according to the evaluation index system. There are 135 respondents participated in the survey, including professional teachers, director of the experimental center, senior experimenters, supervisors and students. There are 127 valid questionnaires. The collected questionnaires are counted by the number of person-times according to the project and evaluation level, and then the person-times are converted into percentages.

The experimental teaching evaluation is shown in the Table 2, which is a comprehensive evaluation model. The weight of each evaluation index is determined by the expert group through pairwise comparison, and finally the evaluation value is determined by fuzzy calculation, and the evaluation result and evaluation grade are obtained.

Table 2. Experimental teaching evaluation value

First-level index	Second-level index	Evaluation level				
		Excellent	Good	Medium	Poor	Extremely poor
Experimental preparation	Teaching materials	0.872	0.128	0	0	0
	Equipment intact rate	0.893	0.107	0	0	0
	Experimental safety	0.925	0.075	0	0	0
Experimental process	Experimental content	0.227	0.583	0.108	0.082	0
	Teaching method	0.621	0.214	0.126	0.039	0
	Teacher-student interaction	0.483	0.381	0.122	0.01	0.004
	Experimental discipline	0.312	0.335	0.216	0.118	0.019
Experimental effect	Problem solving ability	0.412	0.387	0.149	0.052	0
	Comprehensive analysis ability	0.274	0.416	0.174	0.084	0.052
	Innovation capacity	0.245	0.386	0.258	0.073	0.038
Experimental management	Attendance management	0.884	0.116	0	0	0
	Equipment use record	0.458	0.379	0.101	0.045	0.017
	Report correction situation	0.783	0.158	0.059	0	0

By the decision of the leading group, the weight set of each evaluation factor is $W = \{0.045, 0.06, 0.045, 0.105, 0.105, 0.07, 0.07, 0.105, 0.105, 0.14, 0.03, 0.045, 0.075\}$.

It is concluded that the comprehensive evaluation of interchangeability and measurement technology is as follows: $B = \{0.49, 0.32, 0.13, 0.05, 0.01\}$.

The evaluation result is: $V = 87.05$.

5.3 Conclusion of Teaching Quality Evaluation

After sorting out the evaluation data and using the method of fuzzy comprehensive evaluation, the evaluation result of "interchangeability and measurement technology" experiment teaching quality is 87.05, which belongs to "good" grade. The evaluation of experiment preparation and experiment management is good. But the comprehensive analysis ability and innovation research ability need to be strengthened.

6 Conclusions

Under the background of EEE, mechanical professional talent cultivation put forward higher request. Combining with the training plan of the school, the new training objectives and training system of mechanical professionals are put forward. Based on this, the evaluation system of mechanical experimental teaching under the background of EEE is established. The quality of experimental teaching is evaluated by using the method of fuzzy comprehensive evaluation, and the evaluation system is scientific and comprehensive. The evaluation process is simple and has certain universality. This method is used as the basis of experimental teaching evaluation to provide support for improving experimental teaching quality.

Acknowledgements. This paper is supported by—(1) Wuhan University of Technology Teaching Research Project "The reform of mechanical experiment teaching method based on BOPPPS model under the background of Emerging Engineering Education" (NO. w2018102); (2) 2019 China Logistics Association Logistics Teaching Reform Project "The construction of port practice platform based on the combination of 3D scene and VR" (NO. JZW2019128).

References

1. Wen, H.Y.: Exploration of experimental teaching in applied undergraduate universities under the background of emerging engineering education. Educ. Res. **10**, 163 (2017). (in Chinese)
2. Kaviyarasi, R., Balasubramanian, T.: Exploring the high potential factors that affects students' academic performance. Int. J. Educ. Manag. Eng. (IJEME) **8**(6), 15–23 (2018)
3. Castillo, R.C.: A paradigm shift to outcomes-based higher education: policies, principles and preparations. Int. J. Sci. Basic Appl. Res. **14**(1), 174–186 (2014)
4. Feng, D.Y., Mo, Y.M.: Discussion on the training mode of mechanical design and manufacturing and automation professionals under the concept of emerging engineering education. Shandong Ind. Technol. **2017**(22), 236–238 (2017). (in Chinese)
5. Zhang, M.Y., Zhang, Q.Y., Wang, Z.Y., et al.: Mechanical experimental platform construction based on BOPPPS model. In: International Conference on Computer Science, Engineering and Education Applications, pp. 194–204. Springer, Cham (2019)
6. Rajak, A., Shrivastava, A.K., Bhardwaj, S., et al.: Assessment and attainment of program educational objectives for post graduate courses. Int. J. Mod. Educ. Comput. Sci. (IJMECS) **11**(2), 26–32 (2019)
7. He, W., Fan, X.H., He, Y.M., et al.: Research on the training mode of electrical information professionals under the background of emerging engineering education. Educ. Mod. **5**(13), 10–12 (2018). (in Chinese)
8. Hong, G., Xu, D.Q.: On the training of high-qualified and application-oriented talents of engineering. Res. High. Educ. Eng. **6**, 44–48 (2010)
9. Sun, L., Liu, Y., Chang, L.: Building of applied innovative talents training system in local colleges. Res. Explor. Lab. **1**, 4555–4559 (2011)
10. Zheng, S., Guan, W., Li, B., et al.: Analysis of Internet of Things talent training and curriculum system innovation. In: International Conference on Education, Management and Computing Technology (ICEMCT-2016). Atlantis Press (2016)

11. Khan, A.A., Madden, J.: Speed learning: maximizing student learning and engagement in a limited amount of time. Int. J. Mod. Educ. Comput. Sci. (IJMECS) **8**(7), 22–30 (2016)
12. Robles, A.C.M.O.: Blended learning for lifelong learning: an innovation for college education students. Int. J. Mod. Educ. Comput. Sci. (IJMECS) **4**(6), 1–8 (2012)
13. Ying, L., Yanfei, X., Ran, H.: Application of QFD in the process of cultivating high-quality practical and innovative talents of mechanical specialty. Res. High. Educ. Eng. **2**, 124–128 (2011). (in Chinese)
14. Xu, G.N.: Exploration and practice of constructing the large platform of experiment teaching for local universities. Res. Explor. Lab. **29**(5), 1–3+11 (2010). (in Chinese)
15. Wang, W., Meng, X., An, Y.: Exploration of the experimental teaching reform under the cultivation model of innovative talents. Exp. Sci. Technol. **2**, 53 (2013)
16. Fetaji, B., Fetaji, M., Ebibi, M., et al.: Analyses of impacting factors of ICT in education management: case study. Int. J. Mod. Educ. Comput. Sci. (IJMECS) **10**(2), 26–34 (2018)
17. Jiang, G., Li, X., Li, G., et al.: Reform of the examination and evaluation system for the mechanical specialty in universities. World Transp. Eng. Technol. Educ. **13**(4), 620–626 (2015)
18. Tam, M.: Outcomes-based approach to quality assessment and curriculum improvement in higher education. Qual. Assur. Educ. **22**(2), 158–168 (2014)
19. Boring, A., Ottoboni, K., Stark, P.: Student evaluations of teaching (mostly) do not measure teaching effectiveness. Sci. Open Res. **1**, 1–11 (2016)
20. Haji, E., Azmani, A., Harzli, M.E.: Using AHP method for educational and vocational guidance. Int. J. Inf. Technol. Comput. Sci. (IJITCS) **9**(1), 9–17 (2017)
21. Tshai, K.Y., Ho, J.H., Yap, E.H., et al.: Outcome-based education–the assessment of programme educational objectives for an engineering undergraduate degree. Eng. Educ. **9** (1), 74–85 (2014)
22. Kotevski, Z., Tasevska, I.: Evaluating the potentials of educational systems to advance implementing multimedia technologies. Int. J. Mod. Educ. Comput. Sci. (IJMECS) **9**(1), 26–35 (2017)
23. Wang, X.Y.: Demonstration effect evaluation of experimental teaching demonstration center based on fuzzy comprehensive evaluation. Lab. Res. Explor. **04**, 132–135 (2015). (in Chinese)
24. Liu, S., Chen, P.: Research on fuzzy comprehensive evaluation in practice teaching assessment of computer majors. Int. J. Mod. Educ. Comput. Sci. (IJMECS) 7(11), 12–19 (2015)
25. Mu, Y.: Evaluation and analysis of experimental teaching quality in universities based on fuzzy comprehensive evaluation. BBS **12**(5), 115–116 (2013). (in Chinese)

The Influencing Factors on the Effective Use of Education APP Under the Background of Education Informatization

Xiaofen Zhou[1(✉)] and Yi Zhang[2]

[1] College of Logistics, Wuhan Technology and Business University, Wuhan 430065, China
103343280@qq.com
[2] School of Business Administration and Tourism Management, Yunnan University, Kunming 650504, China
18208798522@163.com

Abstract. Under the background of educational informationization, educational information system and various informationized teaching tools are changing people's learning methods and lifestyles. However, as one of the important carriers or tools of educational informationization, the effect of educational APP in practical application is not as good as the designer expected. Therefore, how to exert the value of educational APP has become a problem urgently to be solved. This problem, based on the theory of social influence and 241 data collected from the survey as samples, constructs a research model for empirical research. The results show that the individual's exploratory behavior towards educational APP can significantly affect their willingness to continue exploring. Subjective norms and social identity in social influence theory can affect the individual's willingness to explore APP, while group norms have no significant impact. In addition, the results also show that perceived usefulness partially mediates the relationship between exploratory behavior and sustained exploratory willingness. The research in this paper is used for reference in the development and design of educational APP.

Keywords: Education informatization · Education APP · Value realization · Education information system · Social influence theory

1 Introduction

With the rapid development of mobile Internet technology, information technologies have gradually entered the field of education, promoting a major change in learning methods and modes [1]. Personal mobile terminals, such as smart phones and tablets, step into people's learning and work [2], providing learners with favorable resources and learning platforms, and also promoting the development of educational APP. Currently, various kinds of digital education resources, such as MOOC, excellent courses, micro-courses, flipped classes, have become the direct product of the combination of information technology [3], making it possible to call ubiquitous learning and mobile learning [4, 5]. In this context, the "Education Informatization 2.0 Action

Z. Hu et al. (Eds.): AIMEE 2019, AISC 1126, pp. 354–367, 2020.
https://doi.org/10.1007/978-3-030-39162-1_33

Plan" issued by the Chinese Ministry of Education clearly takes the "Platform + Education" service model and the education resources sharing plan as the strategic goal of the national education informatization construction. As a special manifestation of educational informationization, how to exert the use value of educational APP is particularly important.

From the perspective of theoretical research, as an important carrier of digitalization of educational resources, the role of educational APP has been widely recognized. However, in the actual application, the effect is not satisfactory [6]. "Jobs' queries" proposes that "why has IT technology changed almost all fields, but has not had a significant impact on education?" [1]. For this problem, numerous scholars have studied it from different perspectives and achieved corresponding results. One view is that the full realization of educational APP value depends to a large extent on the users themselves. Based on the technology acceptance model, Park *et al.* discussed the influence of system accessibility, self-efficacy and other factors on mobile learning effect [7]. The results of Husiao show that besides the service quality of APP, the entertainment value perceived by users contributes to the sustainable use of educational APP [8]. Chrompton and others argue that students' proficiency in mobile devices, time spent using them, and teachers' guidance are closely related [9].

From the literature retrieved, users' behavior is able to be affected by their internal characteristics and external environment. On the one hand, the particularity of educational APP puts forward corresponding requirements for users. Some studies suggest that when users realize that the application of APP could bring positive benefits for themselves, individuals will show a tendency to use APP continuously [7, 8]. On the other hand, people are embedded in the social environment, and their behavior will be affected by the social environment [10]. The theory of social influence points out that the factors that lead to the change of individual behavior come largely from the external environment, such as the suggestions of other people or group members, etc. [11]. However, in the field of educational informationization, few papers consider the ways of external environmental factors and personal characteristics affect the application of educational APP. For the reasons mentioned above, based on the social influence theory, this study will construct a theoretical model for the effective use of educational APP from the viewpoint of environmental and emotional factors, and explore the elements affecting the effective use of educational APP.

2 Theoretical Basis

2.1 Social Influence Theory

Social Influence Theory is a theoretics used to explain the relationship between individuals, groups and the social environment. It was formally put forward in a report submitted by Kelman to the American Association for the Advancement of Science [11]. At present, academia generally believes that social influence consists of three dimensions: Subjective Norm (SN), Social Identity (SI) and Group Norm (GN) [11, 12]. Subjective norm (SN) means accepting the influence of others in order to get a valuable response from others. Such behavior does not necessarily reflect the real idea of an individual [12]. Social identity (SI) is an attitude or behavior adopted to maintain a stable relationship with others or groups. Identity can be expressed as identification with oneself or group behavior [13]. Group norm (GN) accepts the influence of others because of the consistency of group values and individual values, and this value helps to optimize their own values [12]. Subsequently, Malhotra (2003) and others defined social identity and group norms as emotional factors to change individual behavioral attitudes when they studied the implementation of knowledge management system. Venkatesh and Davis clearly pointed out that subjective norms are a mandatory situational factor, the impact of specific situations on people, the behavior of individuals under external pressure [14]; social identity and group norms are a voluntary emotional factor, which can directly affect the behavior of the individuals [14]. Therefore, the theory of social influence can be regarded as a combination of internal and external factors leading to individual behavior.

2.2 Effective Use and Systematic Exploration

The concept of effective use belongs to the post-adaption phase of information system (IS). Burton-Jones and others think that the essence of IS use is actually effective use, and define the concept of effective use from the perspective of users, information systems and tasks. They think that effective use is task-oriented, and users utilize IS in some way to achieve their goals [15]. Liang based on the theory of effective use [15], think that system exploration can promote the effective use of IS, since system exploration includes learning and adoption [16]. At present, IS presents the characteristics of complexity and functional diversity. Before users reaching their goals by using IS, they must learn and understand the functions and methods of the system. Boudreau found that employees' functional reconfiguration of ERP system can improve their work efficiency [17]. Through in-depth exploration of system functions, users can discover and construct meaningful application methods to improve their efficiency [18].

Although scholars have done a lot of research on system exploration, and got some results, e.g. work autonomy, system complexity, team environment can have a direct or indirect impact on system exploration [16]. Both perceived usefulness and personal traits play a positive role in system exploration [18]. However, to a large extent, this kind of research focuses on enterprise information systems. As a special information system, the research results of educational APP are very limited. Whether users' behavior will be affected by these factors is seldom answered by literature. To make up for this deficiency, our research adds perceived usefulness and systematic exploration into the research model to explore the factors that affect the value of educational APP.

3 Research Model and Hypothesis

Based on the theory of social influence, our study explores the impact of situational factors and emotional elements on individual behavior. Among them, situational factors contain subjective norms; emotional factors include group norms and social identity, shown in Fig. 1.

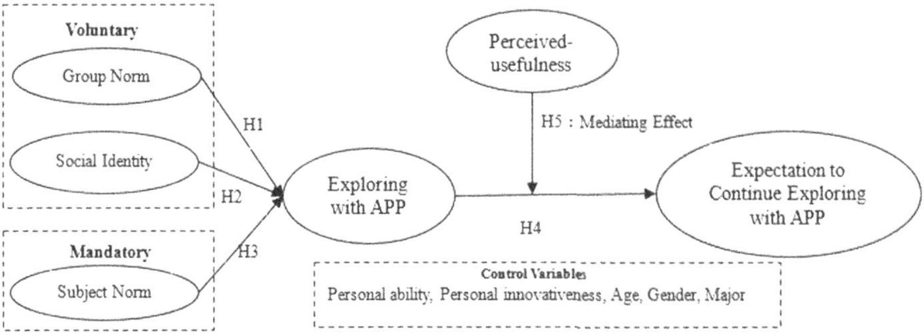

Fig. 1. Social influence and systematic exploring

3.1 Social Influence and Systematic Exploration

The researchers of us believe that group norms and social identity are emotional factors which explain the relationship between individuals, teams and social environment. Group norms emphasize that individuals accept the influence of others because the behavior or values of group members are consistent with their own values [11, 12], which is the result of individuals' recognition of the behavior or values of group members, and this behavior or values contribute to the optimization of their own values [11]. Since the individual is persistent in social activities, the individual's behavior is bound to be affected by others. Time stress theory shows that when time is in a state of urgency, or when individuals realize that time is an externally imposed control, they will actively seek the help of others to achieve their goals [19]. The purpose of using educational APP is to master the knowledge or skills needed more efficiently. If the exploration of the function of APP is regarded as a prior condition for acquiring knowledge, when the group goals are consistent with the individual goals, suggestions from others on the use of educational APP can stimulate individuals to explore new functions. Therefore, the hypothesis is put forward:

H1: Group norms have a significant positive impact on system exploratory behavior.

Social identity refers to the attitude and behavior that individuals adopt in order to maintain the relationship between themselves and others or groups [11, 12]. It is a subjective tendency that individuals are willing to change their inherent behavior by recognizing others' behavior. In psychological research, Levine pointed out that individuals usually achieve their goals in the context of interaction and interdependence with others, and this process of socialization can also affect individual behavior [20]. If an individual has formed a high degree of identity with his group, he will be more willing to listen to the group members' suggestions and take corresponding actions in the process of using educational APP. When individuals realize that new APP functions or methods can bring positive results, in this sense of identity, individuals will make greater efforts to explore or support this behavior. Therefore, the following assumption is proposed:

H2: Social identity has a significant positive impact on system exploratory behavior.

Subject norm is a kind of social behavior that an individual accepts the influence of others in order to obtain praise [11, 12]. It is also a kind of social behavior that an individual takes under the pressure of situation [14]. According to the theory of interpretative level, the current behavior of an individual is more affected by feasibility factors. It shows the problem of how an individual needs to "do" in order to accomplish a task, and it is more affected by situational factors [21, 22]. In the process of using educational APP, individuals often seek advice from people around them because of time pressure. When people around the users think that they have to adopt a specific function of APP or use a specific APP to achieve results for users, even if users themselves are unwilling to change their choice or have a negative attitude towards using a function, they will be willing to accept or probe new APP functions. So, this paper holds that:

H3: Subjective norms have a significant positive impact on system exploratory behavior.

3.2 Exploratory Behavior and Continuous Exploratory Willingness

Ajzen thinks that understanding the current behavior of individuals is of great significance to getting to know future behavioral intentions [23]. According to the theory of interpretative level, time plays an important role in people's daily life, and there are differences between individuals' current and future behaviors [24]. Studies have presented that current behavior is an important variable leading to individual future behavior. Fujita found that when people believe that something is happening in the distant future [25], people interpret behavior based on the results of current behavior rather than means, which indicates that if the individual's exploration of educational APP function can bring positive results, the individual is more likely to focus on future work. Repeat this behavior. In addition, continuous exploratory behavior has the meaning of persistence [26, 27], which indicates that the results of initial behavior will lead to changes in individual future behavior. Shaping behavior theory holds that when a certain behavior can lead to positive results, individuals tend to repeat this behavior [28]. For this reason, this research argues that the individual's exploratory behavior will have a significant impact on the individual's future exploratory willingness. The following assumptions are made:

H4: There is a significant positive effect between systematic exploratory behavior and future exploratory intention.

3.3 The Mediating Role of Perceived Usefulness

Liang et al. have shown that the complexity of information system itself will affect the willingness of individuals to use information system [16]. Bandura's social cognitive theory holds that people's subjective feeling of doing something is the premise of influencing individual behavior choice [29, 30]. When an individual uses APP to achieve a certain goal, the individual will form a subjective evaluation of the system itself and the use process, which will inevitably affect the individual's future behavior methods. According to Bhattacherjee's Expectation Recognition Theory (ECT), it is easier for individuals to motivate their willingness to use information systems continuously when they realize that system functions and usage methods can help them reach their targets better [27, 31]. The theory of planned behavior also shows that individual's subjective attitude can influence individual's future behavior [32]. This means that when an individual realizes that using an APP can help him improve his efficiency, he will be more willing to use the APP continuously. Hence, this paper proposes the following assumptions:

H5: Perceived usefulness mediates exploratory behavior and persistent exploratory intention.

4 Data Collection

4.1 The Collection of Data

Data collection was carried out by questionnaire survey method, and 320 valid questionnaires were distributed to college students. In the process of answers recovery, 79 questionnaires were not adopted because of the lack of data or the difficulty of confirming information. Thus, 241 individual questionnaires were adopted in this experiment, and the effective recovery rate of the questionnaires was 80.71%.

Among the valid samples, 87 were males (36.1%), 154 were females (63.9%) and 153 were under 20 years old, accounting for 63.48% of the total valid samples. Among 241 valid samples, 131 students (54.36%) had business background (e-commerce, human resources management, financial management, business administration, etc.) and 110 students (45.64%) had other professional background. They all have undergraduate degrees, with an age range 18–22.

4.2 Scale Design and Sources

The questionnaire used in this paper consists of two parts: basic information and latent variables designed by the model. Among them, basic information includes: gender, age, major and so on; latent variables include: social identity, group norms, subjective norms, totally eight ones. In order to reflect the rationality of the questionnaire, the design of the scale is based on the classical literature at home and abroad, and corrected properly according to the background of this study. Each item was measured with the 7-point Likert scale.

Table 1. Questionnaire design and sources

Variable	Code	Operational items	Source
Perceived Usefulness (PU)	PU1	I think this APP is very useful	Maruping [33]
	PU2	This APP has powerful functions to solve many problems	
	PU3	This APP can help me improve the efficiency of problem solving	
Group Norms (GN)	GN1	I like to do that	Shen, Cheung [34]
	GN2	In line with my values	
	GN3	This behavior is important	
Subjective Norms (SN)	SN1	Those who matter to me think I should do it	
	SN2	Those who have a major impact on me think I should do this	
	SN3	Those who are important to me agree with me	
Social Identity (SI)	SI1	I will take the initiative to help other students achieve this behavior	
	SI2	I agree with the behavior of other students	
	SI4	I think this kind of behavior is commendable	
Sustainable Willingness to Explore (CETE)	CETE1	In the future, I hope to continue to use more functions	Maruping [33]
	CETE2	In my future study, I expect to continue to spend time and effort exploring new functions	
	CETE3	In my future study, I look forward to continuing to explore new methods	
Exploratory Behavior (IINNO)	INNO1	I usually explore new features of APP applications	Liang, Peng, Xue, Guo and Wang [16]
	INNO2	I'm trying to explore new app applications	
	INNO3	I'm trying to use a new way to use APP applications	

5 Data Analysis and Research Results

5.1 Reliability and Validity Test of Scale

The test of reliability needs to test both the reliability of latent variables and the reliability of scale.

Usually the Cronbach's α coefficient is greater than 0.7, which indicates that the reliability of this latent variable has reached an acceptable level. But Bollen points out that the size of Cronbach's α is influenced by the number of items measured. Therefore, in order to make the results more credible, this study uses Cronbach's α as the basis, and then uses combination reliability (CR) and AVE as three indicators to explain the reliability of reliability (Table 2).

Table 2. Reliability and validity test

Factor	Code	Factor load	Cronbach's α	CR	AVE
Perceived Usefulness (PU)	PU1	0.930	0.931	0.956	0.878
	PU2	0.949			
	PU3	0.933			
Group Norms (GN)	GN1	0.890	0.878	0.924	0.803
	GN2	0.885			
	GN3	0.913			
Subjective Norms (SN)	SN1	0.900	0.865	0.917	0.786
	SN2	0.915			
	SN3	0.842			
Social Identity (SI)	SI1	0.853	0.809	0.887	0.724
	SI2	0.918			
	SI3	0.744			
Sustainable Willingness to Explore (CETE)	CETE1	0.926	0.923	0.951	0.867
	CETE2	0.923			
	CETE3	0.945			
Exploratory Behavior (IINNO)	INNO1	0.922	0.938	0.960	0.889
	INNO2	0.959			
	INNO3	0.947			
Creative Spirit (PII)	PII1	0.856	0.807	0.885	0.720
	PII2	0.858			
	PII3	0.831			
Innovation Ability (ABIT)	ABIT1	0.880	0.859	0.913	0.779
	ABIT2	0.895			
	ABIT3	0.872			

Note: AVE - mean variance extraction; CR - combination reliability;

The factor loads, CR, AVE and α coefficients of each item are shown in Table 3. It can be seen from the table that the factor loads of all items are between 0.744 and 0.959, which are greater than the standard value of 0.7, indicating that the α coefficient has passed the test. According to Fornell and Larcker's point of view [35], combined reliability can be used to measure the internal consistency between latent variables and observed variables, and to avoid the risk of using α coefficient. In the table, all combination reliability values are greater than 0.7, so the combination reliability passes the test. Finally, AVE can be used to test the internal consistency of the model. If the AVE value is greater than 0.5, it means that AVE passes the test. In the table, the minimum AVE is 0.720, so the AVE value is considered to pass the test. To sum up, the reliability of the scale has passed the test.

The validity test consists of Convergent Validity and Discriminate Validity. Fronell pointed out that the convergence validity requirement: a. Factor load of each measurement item > 0.7; b. AVE value > 0.5.

These two conditions meet at the same time, indicating that the convergence validity of the model passed the test [35]. Combining with Table 1, these two conditions are satisfied at the same time, so the convergence validity has passed the test.

Table 3. Coefficient of correlation and square root of AVE

	1	2	3	4	5	6	7	8	9	10	11
1. Age	–										
2. Gender	0.002	–									
3. ABIT	−0.056	0.274	**0.882**								
4. Major	0.073	−0.042	−0.078	–							
5. PII	−0.094	0.254	0.29	−0.098	**0.849**						
6. SN	0.022	−0.041	0.129	−0.114	0.099	**0.886**					
7. IINNO	0.002	0.107	0.293	0.025	0.38	0.248	**0.943**				
8. PU	0.06	−0.166	0.141	−0.016	0.14	0.298	0.165	**0.937**			
9. CETE	0.137	−0.086	0.163	0.091	0.138	0.461	0.492	0.379	**0.931**		
10. SI	0.011	0.111	0.203	−0.029	0.156	0.327	0.442	0.143	0.425	**0.851**	
11. GN	0.074	−0.054	0.231	0.075	0.191	0.4	0.492	0.331	0.608	0.485	**0.896**

Note: The bold characters on the diagonal line represent the square root of AVE.

Discriminate Validity can be determined by examining whether the minimum AVE value is greater than the square of the correlation coefficient between variables. Table 3 shows the correlation coefficient matrix generated by SMART PLS 3.0, in which the black font on diagonal lines is represented as the square root of AVE. Therefore, it is only necessary to determine whether the minimum AVE square root is greater than the maximum correlation coefficient. As can be seen from Table 3, the minimum AVE square root is 0.849, which is larger than the maximum correlation coefficient of 0.608, except for the control variables. Then, the discriminate validity of this paper has passed the test.

In summary, both the reliability and validity of the scale passed the test.

5.2 Testing of Structural Models

In this study, smart pls 3.0 was used to calculate the structural model. Each index explained 38.2% of the exploratory behavior and 48.5% of the persistent exploratory willingness. Path coefficients and saliency results are shown in Fig. 2.

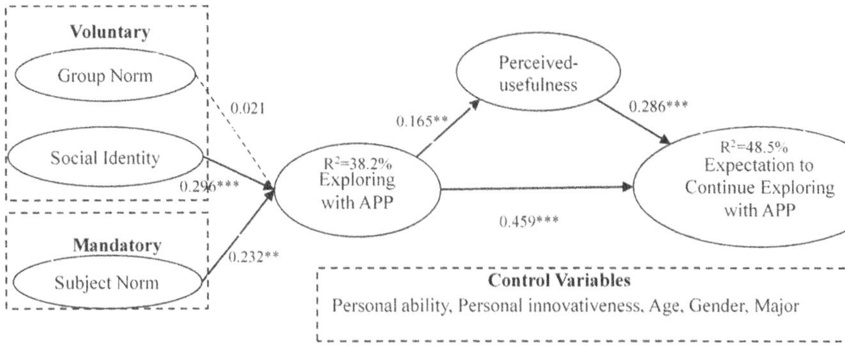

Note: ***: <0.001; **: <0.01; *: <0.05

Fig. 2. Model path system and significance results

Firstly, group norms (β = 0.021, P > 0.05) do not affect individual exploratory behavior in the impact of educational APP, which means that there is no sufficient evidence to show that H1 is supported. Social identity (β = 0.296, P < 0.001) and subjective norms (β = 0.232, P < 0.01) can influence individual exploratory behavior, so H2 and H3 are supported. In addition, exploratory behavior (β = 0.459, P < 0.001) can significantly affect the individual's willingness to continue exploring, hence, the H4 hypothesis is supported.

In addition, Fig. 2 shows that the individual's exploratory behavior can not only directly affect the willingness to continue exploring (β = 0.495, P < 0.001), but also indirectly affect the willingness to continue exploring through perceived usefulness (β = 0.165, P < 0.001; β = 0.286, P < 0.001). Therefore, it can be speculated that perceived usefulness has a partial mediating effect between exploratory behavior and persistent exploratory willingness. Through the Bootstrap function in Smart pls 3.0, under 5000 repeated calculations, the median effect range of "exploratory behavior perceived usefulness persistent exploratory will" is [0.018, 0.100], which does not include 0, indicating that part of the mediating effect is valid, and perceived usefulness shows a positive mediating effect. Then, the H5 hypothesis is further supported.

In summary, the direct and mediating effects of this research are supported.

6 Research Results and Significance

6.1 Research Results

Based on the theory of social influence, this paper constructs a research model, and uses Smart pls 3.0 to demonstrate the hypothesis of the model. The results show that:

(1) Exploratory behavior (β = 0.450, P < 0.001) can have a significant positive impact on the willingness to continue exploring. This shows that the individual's current behavioral outcomes will affect the behavioral intentions, which is consistent with the views of Agrifoglio et al. [26, 27, 36]. In addition, 48.5% of the explanatory power of the structural model shows that when the individual's exploratory spirit and

exploratory ability are set as control variables, the individual's current exploratory behavior can still affect the individual's exploratory intention, which indicates that personal trait [37] is not the factor that affects the use value of educational APP. This is a new discovery of this study and one of the contributions of this paper.

(2) In the dimension of social influence theory, group norms cannot directly have a significant positive impact on current behavior ($\beta = 0.021$, $P > 0.05$), while subjective norms can significantly affect exploratory behavior ($\beta = 0.232$, $P < 0.001$). This shows that, under the influence of time pressure, the acceptance of educational APP will be affected by situational factors, i.e., the help or suggestion of others will affect the individual's willingness to adopt. Social identity ($\beta = 0.296$, $P < 0.001$) can also affect individual exploratory behavior, which proves that both situational and emotional factors have an impact on individual behavior. From the sight of path coefficient, social identity ($\beta = 0.296$) is larger than subjective norm ($\beta = 0.232$), which means that the individual's subjective needs are the fundamental reason to determine the value of educational APP. When an individual has a need for certain knowledge, he will use specific APP only when others suggest. From the above results, therefore, it can be recognized that the value of educational APP depends largely on the needs of users.

(3) The following is a description of the mediating role of perceived usefulness [0.018, 0.100] in exploratory behavior and exploratory intention: User's subjective identity with a particular APP is an important factor in the sustainability of APP use [38], which means, when individuals realize that APP can meet their subjective needs, they are more likely to use that APP repeatedly in the future work.

6.2 Significance of the Research

Based on the theory of social influence, this study studies the use value of educational APP, and explores the factors that affect the value of educational APP, which is concretely to be seen in the following aspects: Firstly, this study introduces the theory of social psychology - social influence theory into the field of educational information system, expanding the scope of social influence theory. Secondly, on the basis of previous studies on educational informationization, this study regards educational APP as a manifestation of educational informationization, and analyses its influencing factors. Thirdly, through empirical research, it is found that the value of educational APP is not only affected by the needs of users themselves, but also by the suggestions of other users.

7 Conclusion

This study is based on the theory of social influence and constructs a research model through empirical research to explore the factors affecting the realization of educational APP value. The conclusion of this study has certain reference value for the research and development of education APP and the promotion and use of students. For R&D institutions, when developing educational APP, they should not only pay attention to the detailed distinction of user group needs, but also pay attention to the continuous updating of application modules and APP usage methods. It is conducive to users' continuous attention to APP and improve the frequency of customers' use. For teachers to guide students to use educational APP, in promoting the use of APP, we should pay attention to learners' acceptance of educational APP is affected by the surrounding environmental factors, and focus on the construction of the scope of group use. However, it should be noted that the object of the questionnaire survey is college students, but the diversity of education APP, the groups faced are also different. Thus, the universality of the survey results of college students remains to be further studied.

Acknowledgement. This research was financially supported by the Teaching Team Project of Hubei Province (Teaching team of the construction of smart logistics curriculum system), the Wuhan Technology and Business University (Teaching team of smart logistics development and management, TDXZ1802) and the Subject of Educational Science Planning in Hubei Province in 2018 (Research on Online Course Teaching Design of Applied Universities, 2018GB121).

References

1. Li, M., Kong, W., Li, Z.: Ask "jobs' question": how to measure the role of education informatization. Mod. Distance Educ. Res. **3**, 3–10 (2017). (in Chinese)
2. Cai, H., Yin, H., Chen, M.: What factors affect teachers' use of digital education resources? - Perspective on the construction and development of educational informatization in China in the intelligent age. Audiov. Educ. Res. **7**, 60–69 (2019). (in Chinese)
3. Ren, Y., Gu, X.: Educational technology: questions and answers to discipline development. Educ. Res. **40**(1), 141–152 (2019). (in Chinese)
4. Sharples, M., Corlett, D., Oliver, W.: The design and implementation of a mobile learning resource. Pers. Ubiquit. Comput. **6**(3), 220–234 (2002)
5. Hwang, G.-J., Tsai, C.-C., Yang, S.J.: Criteria, strategies and research issues of context-aware ubiquitous learning. J. Educ. Technol. Soc. **11**(2), 81–91 (2008)
6. Ke, Q., Wang, P., Zhang, J.: The supply mode of digital education resources, Yao classification framework and development countermeasures. Res. Audio-Vis. Educ. **39**(3), 68–74 (2018). (in Chinese)
7. Park, S.Y., Nam, M.W., Cha, S.B.: University students' behavioral intention to use mobile learning: evaluating the technology acceptance model. Br. J. Educ. Technol. **43**(4), 592–605 (2012)
8. Hsiao, K.L., Huang, T.C., Chen, M.Y., et al.: Understanding the behavioral intention to play austronesian learning games: from the perspectives of learning outcome, service quality, and hedonic value. Interact. Learn. Environ. **1**, 1–14 (2017)
9. Crompton, H., Burke, D., Gregory, K.H.: The use of mobile learning in PK-12 education: a systematic review. Comput. Educ. **110**, 51–63 (2017)
10. Granovetter, M.: Economic action and social structure: the problem of embeddedness. Am. J. Sociol. **91**(3), 485–510 (1985)
11. Kelman, H.C.: Interests, relationships, identities: three central issues for individuals and groups in negotiating their social environment. Annu. Rev. Psychol. **57**(1), 1–26 (2006)
12. Kelman, H.C.: Compliance, identification, and internalization three processes of attitude change. J. Confl. Resolut. **2**(1), 51–60 (1958)

13. Zhou, T., Li, H.: Understanding mobile SNS continuance usage in China from the perspectives of social influence and privacy concern. Comput. Hum. Behav. **37**, 283–289 (2014)
14. Venkatesh, V., Davis, F.D.: A theoretical extension of the technology acceptance model: four longitudinal field studies. Manag. Sci. **46**(2), 186–204 (2000)
15. Burton-Jones, A., Grange, C.: From use to effective use: a representation theory perspective. Inf. Syst. Res. **24**(3), 632–658 (2013)
16. Liang, H., Peng, Z., Xue, Y., et al.: Employees' exploration of complex systems: an integrative view. J. Manag. Inf. Syst. **32**(1), 322–357 (2015)
17. Boudreau, M.C., Robey, D.: Enacting integrated information technology: a human agency perspective. Organ. Sci. **16**(1), 3–18 (2005)
18. Maruping, L.M., Magni, M.: Motivating employees to explore collaboration technology in team contexts. MIS Q. **39**(1), 1–16 (2015)
19. Andrews, F.M., Farris, G.F.: Time pressure and performance of scientists and engineers: a five-year panel study. Organ. Behav. Hum. Perform. **8**(2), 185–200 (1972)
20. Levine, J.M., Moreland, R.L.: Culture and socialization in work groups. Prespectives Soc. Shared Cogn. 257–279 (1991)
21. Trope, Y., Liberman, N.: Temporal construal and time-dependent changes in preference. J. Pers. Soc. Psychol. **79**(6), 876–889 (2000)
22. Nussbaum, S., Trope, Y., Liberman, N.: Creeping dispositionism: the temporal dynamics of behavior prediction. J. Pers. Soc. Psychol. **84**(3), 485–497 (2003)
23. Kidwell, B., Jewell, R.D.: The influence of past behavior on behavioral intent: an information-processing explanation. Psychol. Mark. **25**(12), 1151–1166 (2010)
24. Liberman, N., Trope, Y.: The role of feasibility and desirability considerations in near and distant future decisions: a test of temporal construal theory. J. Pers. Soc. Psychol. **71**(1), 5–18 (1998)
25. Fujita, K., Henderson, M.D., Eng, J., et al.: Spatial distance and mental construal of social events. Psychol. Sci. **17**(4), 278–282 (2006)
26. Agrifoglio, R., Black, S., Metallo, C., et al.: Extrinsic versus intrinsic motivation in continued twitter usage. J. Comput. Inf. Syst. **53**(1), 33–41 (2012)
27. Bhattacherjee, A.: Understanding information systems continuance: an expectation - confirmation model. MIS Q. **25**(3), 351–370 (2001)
28. Stajkovic, A.D., Luthans, F.A.: A meta-analysis of the effects of organizational behavior modification on task performance. Acad. Manag. J. **40**(5), 1122–1149 (1997)
29. Bandura, A.: Self-efficacy: toward a unifying theory of behavioral change. Psychol. Rev. **84** (2), 191–215 (1977)
30. Bandura, A.: Social cognitive theory: an agentic perspective. Asian J. Soc. Psychol. **2**(1), 21–41 (1999)
31. Oliver, R.L.: A cognitive model of the antecedents and consequences of satisfaction decisions. J. Mark. Res. **17**(4), 460–469 (1980)
32. Ajzen, I., Fishbein, M.: Factors influencing intentions and the intention-behavior relation. Hum. Relat. **27**(1), 1–15 (1974)
33. Maruping, L.M., Magni, M.: Motivating employees to explore collaboration technology in team contexts. MIS Q. **39**, 1–16 (2015)
34. Shen, A.X., Cheung, C.M., Lee, M.K., et al.: How social influence affects we-intention to use instant messaging: the moderating effect of usage experience. Inf. Syst. Front. **13**(2), 157–169 (2011)
35. Fornell, C., Larcker, D.F.: Evaluating structural equation models with unobservable variables and measurement error. J. Mark. Res. **18**(1), 39–50 (1981)

36. Scott, S.G., Bruce, R.A.: Determinants of innovative behavior: a path model of individual innovation in the workplace. Acad. Manag. J. **37**(3), 580–607 (1994)
37. Orlikowski, W.J.: Using technology and constituting structures: a practice lens for studying technology in organizations. Organ. Sci. **11**(4), 404–428 (2000)
38. Mei, Y., Balakrishnan, M., et al.: Comparing the acceptance of key performance indicators management systems on perceived usefulness and perceived ease of use in a higher education institution in Malaysia. Int. J. Mod. Educ. Comput. Sci. (IJMECS) **4**(10), 9–16 (2012). https://doi.org/10.5815/ijmecs.2012.10.02

Two-Stage Method for Controlling the Movement of a Parallel Robot Based on a Planar Three-Revolute-Prismatic-Revolute Mechanism

Sergey Khalapyan[1], Larisa Rybak[2], and Dmitry Malyshev[2(✉)]

[1] Stary Oskol Technological Institute named after A.A. Ugarov NUST MISiS, 42, Makarenko mkr., 309516 Stary Oskol, Russia
khalapyan@yandex.ru
[2] Belgorod State Technological University named after V.G. Shukhov, Belgorod, Russia
rlbgtu@gmail.com

Abstract. A two-stage method for organizing control of a robot with parallel kinematics is proposed. The method involves the transfer of all complex resource-intensive computational procedures to the stage of the trajectory formation. During the operational control of the robot's movement, a ready-made trajectory is used, recorded as points in the input coordinate space of the mechanism. Method application allows avoiding multiple solutions of the forward and inverse kinematics during the robot movement, significantly increasing the performance of the control system. Experimental verification of the method was carried out using the planar 3-RPR mechanism as an example.

Keywords: Parallel robot · Kinematic analysis · Planar 3-R$\underline{P}$R mechanism · Operational control

1 Introduction

The efficiency of parallel robot's application is largely determined by the quality of the applied control system. Considerable attention is currently being paid to the control of parallel robots for organizing the movement of the output link of the corresponding mechanism along a given trajectory (see, for example, [1–15]).

In previous works of this study authors, methods were proposed for determining the workspace [16] of a parallel mechanism (taking into account singularity zones, Jacobian's sign-constant requirements and restrictions imposed by the dimensions of the workpiece, determination of safe movement trajectory and its optimization according to the criterion of minimum total movement duration [17].

The next step is to organize the movement of the mechanism along the selected path. In [18], a scheme was proposed (Fig. 1) with an indirect measurement of the position of the moving platform based on the available information about the state of the drive links of the mechanism to generate a feedback signal. The scheme was created to control the tripod, however, it can also be used for other mechanisms of a parallel

Z. Hu et al. (Eds.): AIMEE 2019, AISC 1126, pp. 368–379, 2020.
https://doi.org/10.1007/978-3-030-39162-1_34

structure with three rods of variable length. The functional diagram of the control system contains three identical control loops for the lengths of each of the rods and an external control loop for the position of the movable platform. For the operation of the system, a prepared predefined motion path of the output link of the mechanism in the form of an array of points Pr [..] is used. The trajectory points Pr [i] are sequentially extracted from the array and, by solving the inverse kinematics, are converted to tasks l1r, l2r, l3r for the control rods length control loops. In each of the enclosed circuits, the mismatch Δl1, Δl2, Δl3, and the control action u1, u2, u3 are calculated, and the control result l1, l2, l3 is formed at the output. The measured values of the rod lengths l1′, l2′, l3′ are fed back to the inputs of the rod length control loops and are used to calculate the mismatches. Changing the lengths of the rods leads to the movement of the output link of the P. mechanism. To organize feedback on the position of the mobile platform, the signal P′ is estimated using the values l1′, l2′, l3′.

To solve the corresponding forward kinematics, the use of a neural network model of the mechanism was proposed.

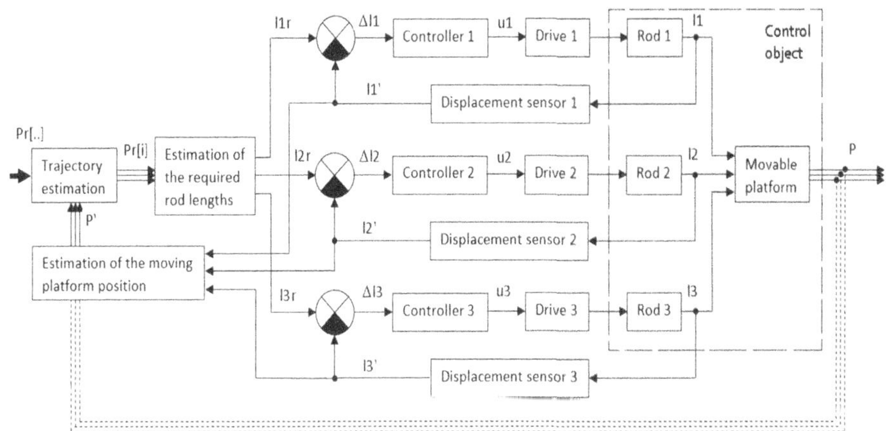

Fig. 1. Functional diagram

However, the construction of such a model is fraught with a number of difficulties related to the fact that the solution of the forward kinematics is not unique: the same state of the drive links is achieved at different positions of the moving platform (see, for example, [19]) and an additional non-trivial task of identifying the "correct" solution to the forward kinematics corresponding to the current state of the control object.

This work aims to study the possibility of overcoming the above-described drawback of the previously proposed robot control method of a parallel structure by dividing the control procedure into two stages. At the first stage, it is proposed to take out all complex resource-intensive computational procedures. At the second stage, the final trajectory recorded as points in the input coordinate space of the mechanism will be used. Application of the method will allow avoiding the multiple solutions of kinematics problems directly during the movement of the robot. As a result of reducing the computational complexity of robot control procedures, its speed will significantly increase.

2 Proposed Method Description and Experimental Results

It is proposed novel two-stage method for controlling the movement of a robot with parallel kinematics to achieve the goal of the study. The novelty of the proposed method is the preliminary transformation of the points array that make up the optimal trajectory of movement to transfer it from the output coordinates of the mechanism to the input. For this, it is only necessary to solve the inverse kinematics, which is trivial for most mechanisms of parallel structure.

The trajectory of movement, presented in this form, can be directly used as a task for the position control system of the moving platform. Since the state of the drive links can be easily measured in most mechanisms of a parallel structure, the use of a trajectory recorded in input coordinates as a task allows avoiding the multiple solution of forward and inverse kinematics, significantly simplifying the operation of the control system and increasing its quick act. In such a system (Fig. 2), there are no modules for determining the position of the movable platform and for calculating the necessary state of the drive links. And classical methods of control theory can be used.

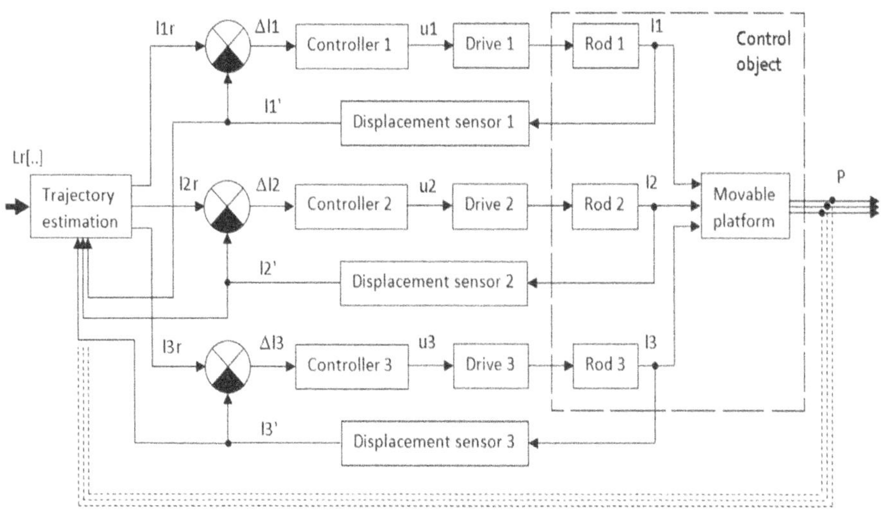

Fig. 2. Modified functional diagram

After loading the motion path of the input link of the mechanism in the form of an array of the rod length Lr [..], the input signals l1r, l2r, l3r corresponding to the first step are received at the inputs of the nested control loops, positioning errors Δl1, Δl2, Δl3 are calculated, control actions u1, u2, u3 are activated, the necessary drives are activated, changes the state of the drive links l1, l2, l3, which is controlled by the corresponding sensors. As a result of changing the input coordinates of the mechanism l1, l2, l3, the position of the moving platform P changes, which is tracked through the space of the input coordinates l1′, l2′, l3′ by the same length sensors. After the first step of the trajectory is completed, the inputs l1r, l2r, l3r change at the inputs of the nested loops - the system proceeds to the second step.

Possible dynamic deviations of the input coordinates of the mechanism can cause temporary deviations of the position of the moving platform from a given trajectory, however, due to the available margin laid down during the formation of the hyper-surface of limitations [16], such deviations do not lead to the mechanism moving outside the workspace.

In the present work, experimental verification of the effectiveness of the proposed method for controlling a parallel robot was undertaken using the planar 3-R$\underline{P}$R mechanism as an example (Fig. 3).

This mechanism consists of three rods of variable length, attached with the help of swivel joints to a fixed base at the vertices of an equilateral triangle $A_1A_2A_3$. The other end of each rod is pivotally mounted on a movable platform at the vertices of an equilateral triangle $B_1B_2B_3$, which allows it to move in the plane of the base. The input coordinates of the mechanism are the rod lengths (l1, l2, l3), the output coordinates are the position of the center of the moving platform C in the coordinates (x, y) associated with the center of the base of the mechanism, and its rotation angle (φ) in the plane of the base.

To organize the experiment, a model of the planar 3-R$\underline{P}$R mechanism (Fig. 4) was built in the MATLAB environment using the Simscape Multibody library. This library

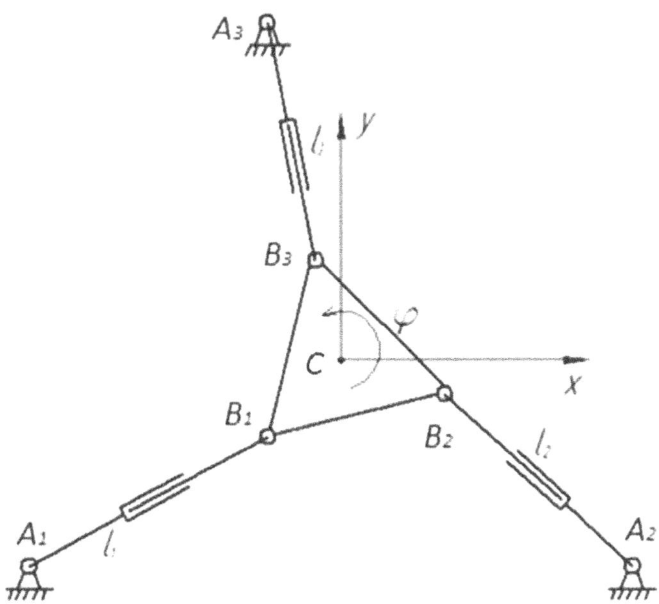

Fig. 3. The planar 3R$\underline{P}$R mechanism

allows you to directly specify the mass and size characteristics of the links and describe how to connect them. At the same time, MATLAB independently calculates the moments of inertia of the links for all coordinates and constructs a 3D or 2D image of the mechanism (Fig. 5).

The model of the mechanism consists of three short (3 cm) and three long (187 cm) rods connected in pairs by prismatic hinges (Prismatic Joint). Short rods with the help of rotary hinges (Revolute Joint) are attached with the opposite end to the given points of the fixed base located at the vertices of an equilateral triangle with a circumscribed radius of 1 m. a circle with a radius of 25 cm. The force of the drive motors (ports 1, 3, 6) is applied to the prismatic joints. The change in the state of each hinge is controlled by a sensor (ports 2, 4, 7) and leads to a change in the working length of the corresponding rod. We limit the working length of the rods in the range of 10–190 cm.

We set some arbitrary trajectory in the workspace of the selected mechanism and try to move it along it using the constructed model. For simplicity of visual presentation, a trajectory was chosen, which is a circle in the (x, y) plane, the center of which coincides with the center of the fixed base. In this case, the angle of rotation of the platform (φ) is fixed at a certain constant value.

The calculation according to the formulas from [20] shows that if $1^\circ \leq \varphi \leq 179^\circ$, the circle diameter should not exceed:

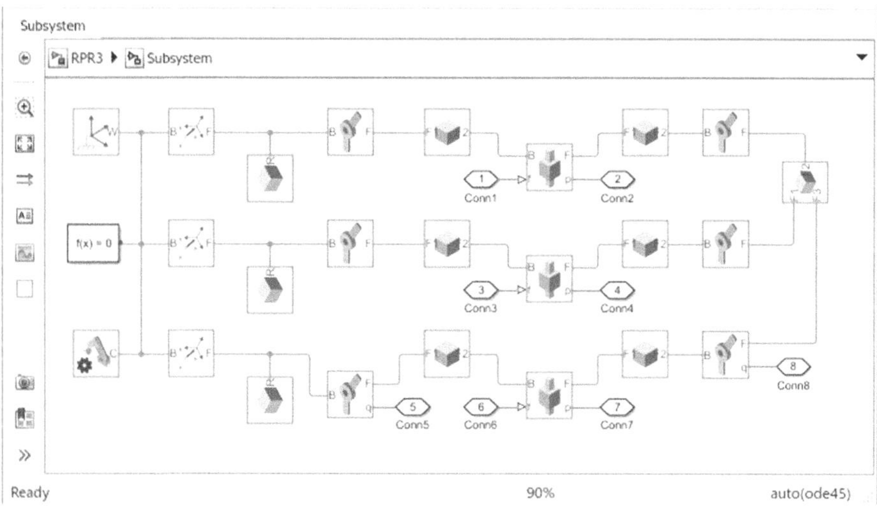

Fig. 4. Model of the planar 3R$\underline{P}$R mechanism

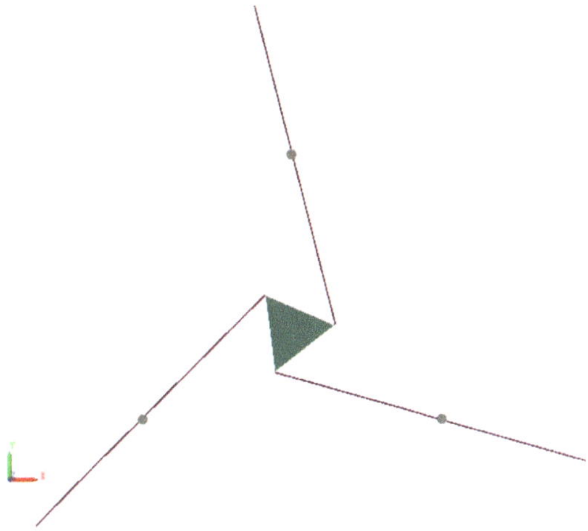

Fig. 5. 2D-image of the planar 3RPR mechanism

$$D \leq 2(L_{max} - R - r) = 2(190 - 100 - 25) = 130\,(\text{cm})$$
$$D \leq 2(|R - r| - L_{min}) = 2(|100 - 25| - 10) = 130\,(\text{cm})$$

Thus, a circle with a diameter of 120 cm centered at the origin at φ = 1° is entirely placed in the workspace of the mechanism. In Fig. 6, the circle is marked in black and the workspace is red. The forbidden zone is shown in other colors: blue is the region of singularity (with some margin), blue is the region in which Jacobian has an opposite sign (it is impossible to get there from the red region without crossing the blue line), yellow and green are the regions in which the mechanism simply cannot “reach out”.

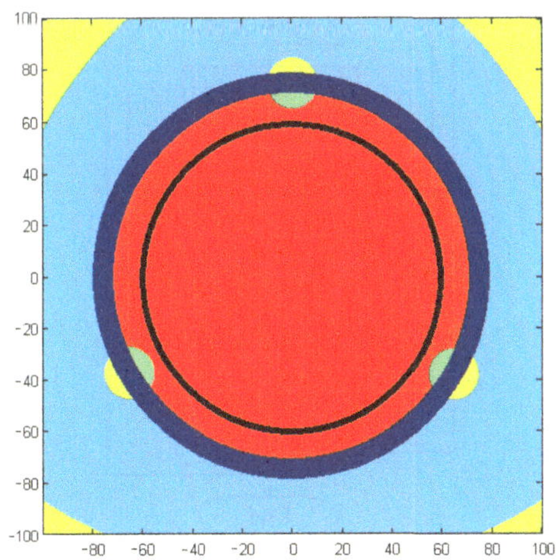

Fig. 6. A circle with a diameter of 120 cm (black) within the workspace of the mechanism (red) at φ = 1°

To represent the trajectory as separate steps, a discrete grid is superimposed on it. Each step of the resulting trajectory represents a movement between adjacent nodes of the discrete grid located at minimal distances from the original circle.

Then using known ratios

$$L_1 = \sqrt{\left(x + \frac{r}{2}(\sin\varphi - \sqrt{3}\cos\varphi) + \frac{\sqrt{3}}{2}R\right)^2 + \left(y - \frac{r}{2}(\sqrt{3}\sin\varphi + \cos\varphi) + \frac{R}{2}\right)^2},$$
$$L_2 = \sqrt{\left(x + \frac{r}{2}(\sin\varphi + \sqrt{3}\cos\varphi) - \frac{\sqrt{3}}{2}R\right)^2 + \left(y + \frac{r}{2}(\sqrt{3}\sin\varphi - \cos\varphi) + \frac{R}{2}\right)^2},$$
$$L_3 = \sqrt{(x - r\sin\varphi)^2 + (y + r\cos\varphi - R)^2}$$

the inverse kinematics is solved and for each nodal point of the trajectory (x, y, φ), the corresponding point in the input coordinates (l1, l2, l3) is determined. The obtained sequence of changing the lengths of the near-water links should ensure the movement of the center of the moving platform along a predetermined path. Switching tasks (steps of the trajectory) is carried out as the previous points of the trajectory are reached (Fig. 7).

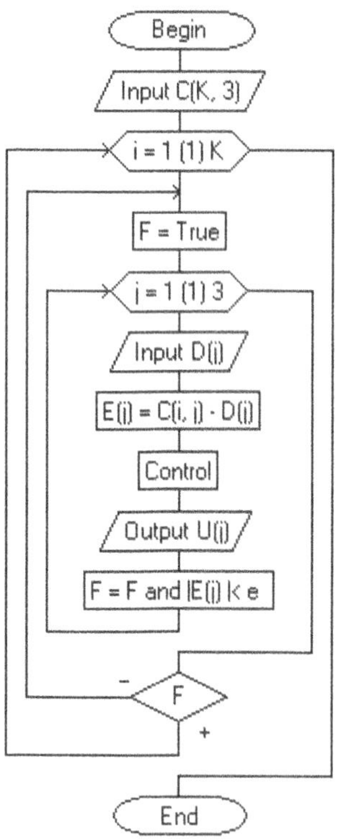

Fig. 7. The algorithm of operational motion control

The operational motion control algorithm provides for:

(1) loading the trajectory C (K, 3), where K is the number of steps, 3 is the number of rods;
(2) cyclic change of step number i = 1…K;
(3) initialization of the Boolean flag F;
(4) cyclically for all length control loops (j = 1…3), the following is performed: interrogation of sensors D (j), calculation of mismatches E (j) = C (i, j) − D (j), control and the conclusion of the control actions U (j) on the actuators;
(5) for each circuit, a check is made to achieve the next trajectory point (|E(j)| < e), if necessary, the value of F is adjusted;
(6) the condition for going to the next step i is the truth of F after the completion of cycle j, i.e. achievement of the next target point of the trajectory along with all length control loops with a given accuracy.

Modeling made it possible to determine the motion paths of the input links of the mechanism (Fig. 8) and their deviation from the task (Fig. 9).

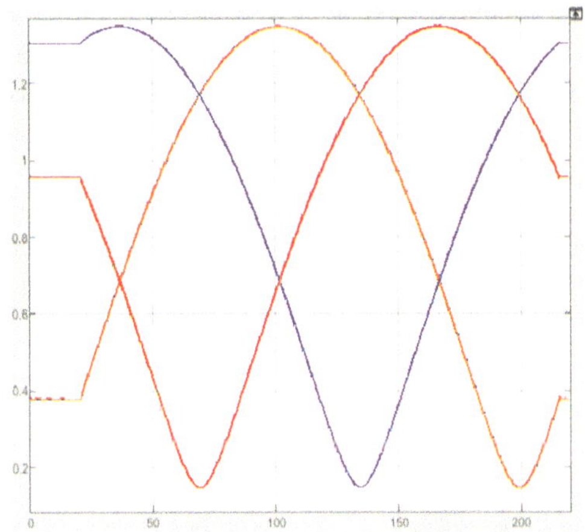

Fig. 8. The dependence of the lengths of the drive links (in meters) from time (in seconds), task and fact

The graph (Fig. 8) shows 6 lines: three lines of job l1r, l2r, l3r and three corresponding lines of the actual length of the drive links l1, l2, l3. The almost complete coincidence of the lines indicates high accuracy of control. The dynamic error of the length changes for each of the rods Δl1, Δl2, Δl3 is presented in Fig. 9. The maximum error is about 1 cm, i.e. less than 1% in relative units. The final assessment of the quality of control can be obtained by constructing the trajectory of the system in the output coordinates (x, y) and comparing the result with the task. For the reverse transition from input coordinates to weekend, it is necessary to solve the forward kinematics.

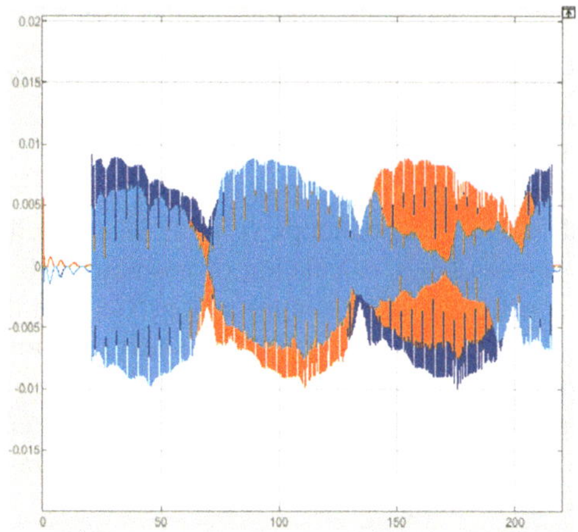

Fig. 9. The dependence of the dynamic control error drive link lengths (in meters) versus time (in seconds)

The neural network model of the mechanism can be used, but there is a simpler and more accurate way. During the simulation, the state of each joint is determined. Using data on the status of any of them can significantly simplify the calculations. However, it is most convenient to install the sensors on two swivel joints related to one rod (ports 5, 8) in Fig. 4.

Let the first sensor show the value of α (the angle between the vectors A_2D_2 and A_2E_2 in Fig. 10), and the second - β (the angle between the vectors B_2E_2 and B_2F_2), then the coordinates of point B_2 are described by the expressions:

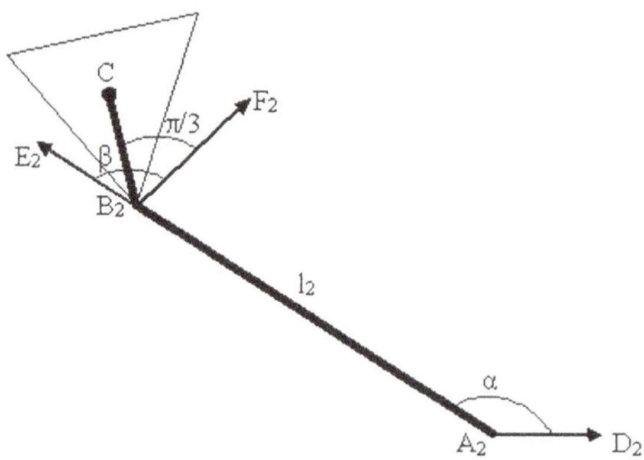

Fig. 10. Determination of the coordinates of point C from the known l_2, α, β and the coordinates of point A_2

$$x_{B2} = x_{A2} + l_2 \cos \alpha,$$
$$y_{B2} = y_{A2} + l_2 \sin \alpha,$$

where $x_{A2} = 0.866$, $y_{A2} = -0.5$ are known coordinates of point A_2, l_2 is length of the corresponding drive link, and the coordinates of the desired point C:

$$x_C = x_{B2} + r \cos(\alpha + \beta + \pi/3),$$
$$y_C = y_{B2} + r \sin(\alpha + \beta + \pi/3),$$

where r is radius of the circle described near the moving platform, π/3 is angle between the vectors B_2F_2 and B_2C, determined by the geometry of the system.

The graphs obtained as a result of the calculation (Fig. 11) quite accurately repeat the given sinusoidal signals. Deviations are oscillatory and are caused by the inertia of the rods and the movable platform, as well as, probably, the controller settings are not good enough. The final graph in coordinates (x, y) (Fig. 12) shows the high accuracy of the task. The graph shows 2 lines: a predetermined and corresponding actual trajectory of the moving platform. The lines are close and almost the same on most of the circumference. The maximum deviation of the trajectory from the given one is less than 1% in relative units.

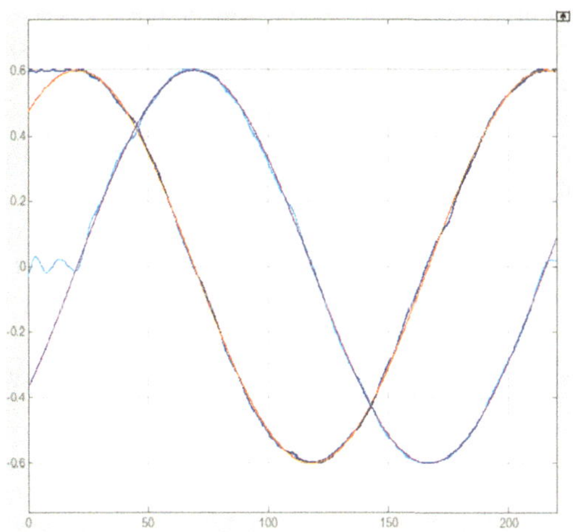

Fig. 11. Graph of changes in the output coordinates x, y

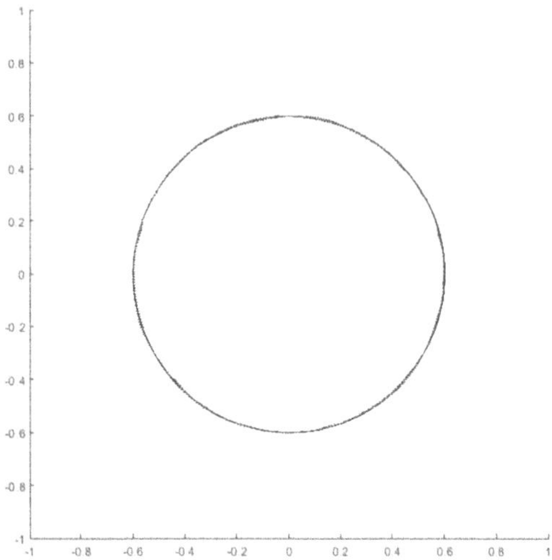

Fig. 12. The result of the experiment

3 Conclusions

An original two-stage method for organizing control of a robot with parallel kinematics is proposed in the work, which implies taking all complex resource-intensive computational procedures to the stage of forming the trajectory. As a result of applying the method, there is no need to repeatedly solve kinematics problems during the movement of the robot, which can significantly increase the speed of the control system, as well as reduce the computing power requirements of the control devices.

Experimental verification of the method, carried out on the example of a planar 3-RPR mechanism, showed an acceptable accuracy of the movement of the output link of the robot along the path specified in the input coordinates of the mechanism.

Acknowledgements. This work was supported by the Russian Science Foundation, the agreement number 16-19-00148.

References

1. Heydarzadeh, M., Karbasizadeh, N., Tale Masouleh, M., Kalhor, A.: Experimental kinematic identification and position control of a 3-DOF decoupled parallel robot. Proc. Inst. Mech. Eng. Part C: J. Mech. Eng. Sci. **233**(5), 1841–1855 (2019)
2. Lintao, W., Ji, L., Lei, Z.: Kinematic analysis and speed control of 3SPS-1S parallel mechanism for end actuator of segment erector. Lecture Notes in Computer Science (Including Subseries Lecture Notes in Artificial Intelligence and Lecture Notes in Bioinformatics), LNAI, vol. 11745 (2019)

3. Liu, K., Chen, Y., Wu, Y., Xu, J., Wang, Y., Ge, Z.: Rotation control of a 3-DOF parallel mechanism driven by pneumatic muscle actuators. Trans. Nanjing Univ. Aeronaut. Astronaut. **36**(2), 330–338 (2019)
4. Abadi, B.N.R., Farid, M., Mahzoon, M.: Redundancy resolution and control of a novel spatial parallel mechanism with kinematic redundancy. Mech. Mach. Theory **133**, 112–126 (2019)
5. Shen, G., Li, G., Zang, W., Li, X., Tang, Y.: Modal space feedforward control for electro-hydraulic parallel mechanism. IEEE Access **7**, 39751–39761 (2019)
6. Tang, J., Cao, D., Yu, T.: Decentralized vibration control of a voice coil motor-based stewart parallel mechanism: Simulation and experiments. Proc. Inst. Mech. Eng. Part C: J. Mech. Eng. Sci. **233**(1), 132–145 (2019)
7. Wang, R., Lü, J., Zhong, S., Qu, H., Wang, J., Chen, X.: Attitude control algorithm for a 2-DOF parallel vector propulsion mechanism on AUV. Harbin Gongye Daxue Xuebao/J. Harbin Inst. Technol. **51**(4), 1–5 (2019)
8. Zhao, L., Cheng, H., Zhang, J., Xia, Y.: Angle attitude control for a 2-DOF parallel mechanism of PMAs using tracking differentiators. IEEE Trans. Industr. Electron. **66**(11), 8659–8669 (2019)
9. Zheng, S., Gao, C., Niu, X., Cong, D., Han, J.: The applications of compliant motion control technique on 6-DOF hydraulic parallel mechanism. IEEE Access **7**, 90094–90105 (2019)
10. Joumah, A.A., Albitar, C.: Design optimization of 6-RUS parallel manipulator using hybrid algorithm. Int. J. Inf. Technol. Comput. Sci. (IJITCS) **10**(2), 83–95 (2018)
11. Mohammed, R.H., Elnaghi, B.E., Bendary, F.A., Elserfi, K.: Trajectory tracking control and robustness analysis of a robotic manipulator using advanced control techniques. Int. J. Eng. Manuf. (IJEM) **8**(6), 42–54 (2018)
12. Tayebihaghighi, S., Piltan, F., Kim, J.-M.: Control of an uncertain robot manipulator using an observation-based modified fuzzy sliding mode controller. Int. J. Intell. Syst. Appl. (IJISA) **10**(3), 41–49 (2018)
13. Sahraoui, M., Khelfi, M.F., Salem, M.: Sequential adaptive fuzzy inference system based intelligent control of robot manipulators. Int. J. Intell. Syst. Appl. (IJISA) **6**(11), 49–56 (2014)
14. Su, T., Cheng, L., Wang, Y., Liang, X., Zheng, J., Zhang, H.: Time-optimal trajectory planning for delta robot based on quintic pythagorean-hodograph curves. IEEE Access **6**, 28530–28539 (2018)
15. Farajtabar, M., Daniali, H., Varedi, S.: Pick and place trajectory planning of planar 3-RRR parallel manipulator in the presence of joint clearance. Robotica **35**(2), 241–253 (2017)
16. Rybak, L., Khalapyan, S., Kholoshevskaya, L., Malyshev, D.: Planning the optimal trajectory of the robot-machine parallel structure. In: AIP Conference Proceedings, vol. 2029, pp. 1–6 (2018). 020066
17. Khalapyan, S., Rybak, L., Malyshev, D., Kuzmina, V.: Synthesis of parallel robots optimal motion trajectory planning algorithms. In: IX International Conference on Optimization and Applications, OPTIMA 2018, pp. 311–324 (2018)
18. Khalapyan, S.Y., Glushchenko, A.I., Rybak, L.A., Gaponenko, E.V., Malyshev, D.I.: Intelligent computing based on neural network model in problems of kinematics and control of parallel robot. In: 3rd Russian-Pacific Conference on Computer Technology and Applications (RPC), pp. 1–5 (2018)
19. Husty, M.L.: An algorithm for solving the direct kinematic of StewartGough-type platforms. Mech. Mach. Theory **31**(4), 365–380 (1996)
20. Rybak, L.A., Khalapyan, S.Y., Gaponenko, E.V.: Issues of planning trajectory of parallel robots taking into account zones of singularity. In: IOP Conference Series: Materials Science and Engineering, vol. 327, pp. 1–8 (2018). 042092

A Hierarchical Fuzzy Model for Assessing Student's Competency

Zhengbing Hu[1] and Yurii Koroliuk[2(✉)]

[1] School of Educational Information Technology, Central China Normal University, Wuhan, China
hzb@mail.ccnu.edu.cn

[2] Independent Researcher, Chernivtsi, Ukraine
yu_kor@ukr.net

Abstract. The aim of the study is to improve a system of competency' assessment, described in the Educational Program of the specialty. This assessment is based on a hierarchical fuzzy model in the fuzzyTECH program. The adequacy of proposed elaborated model is confirmed by a statistic agreement with thesis attestation results of Master of Arts. Besides, the model showed that combining the competency in the course of writing and attestation the Master of Arts thesis shall result in synergy effects.

The model input parameters consist of the data of academic assessment and the structure of the educational-professional program. These inputs simplify the process of assessing the competency while comparing with the existing approaches. Thus, stakeholders are able to assess the professional qualities of a graduate without requiring additional resources, because the model is built on the open source data such as the Educational Program of the specialty and the diploma supplement. The proposed model can assess the quality of educational-professional programs, and assist their comparative evaluation. The model's uniqueness is that output results of competency assessment are described both by linguistic terms, more common for employers, and numeric assessment, using 100-point scale, more common for students and teaching staff.

The main innovation of the model is its ability to assess adequately the obtained competency not only by the final learning outcomes, but also at any stage of the curriculum implementation. Therefore, the fuzzy model can also be utilized to predict the graduates' final performance, which enables students to apply the model for self-assessing of their own competency in a learning process. Such self-assessment may help students promote learning scores in the courses which is relevant to determined competency.

Keywords: Competency · Fuzzy logic · Fuzzy system · Students' assessment · Competency-centered curricula

1 Introduction

The Competency-based approach is an integral part of modern education which benefits a student for the learning process of his own life. However, the primary role and aim of a competency are the improvement of education and bridge a gap of talent

Z. Hu et al. (Eds.): AIMEE 2019, AISC 1126, pp. 380–393, 2020.
https://doi.org/10.1007/978-3-030-39162-1_35

requirement between universities and employers. Competency takes a role as a universal platform where employers propose their own requirements for talents, and universities in their turn are trying to satisfy them.

Terminological definition of competency along with their composition and methods of assessment are not unified within the industries, levels of education, and national educational systems. Various research works and practices of competency training for many years did not give clear answers to the consistency and mechanisms of their formation [1]. Competency-based research is still important to the field of education research.

A Variety of competency requirements in practice results in solving the mentioned problems in different ways. Thus, to promote comparable and compatible qualifications of graduates in the European Union, the European Higher Education Area was founded in 2014 [2]. Many industries, national and international standards are created and applied accordingly with a very detailed description of competency. On one hand, current situation contributes to a clear disclosure of the concept; on the other hand, the problems of evaluating to competency become more complicated in practice.

By reviewing major research works in the field, it helps find their confirmation and implementation experience in practice, parametrization is essential for evaluating all components of competency. With the help of parametrization, it can determine the high relevance of the development of applied assessment and analysis methods of competency. Therefore, when evaluation models are constructed, it needs to consider some vital factors such as: the interests and descriptive scales of different stakeholders, the existing academic information of the educational process, and simplifying the evaluation process, etc.

In this case, the object of the study was the elaboration of the assessment method of a new competency which should consider the following factors:

(1) on the basis of widely used standards of competency description;
(2) take into account of the available data set of a university graduate preparation and not require additional observations and/or expert assessment;
(3) possibly utilized by an employer assessment or by self assessment of a student;
(4) assess some separate competency components, as well as the comprehensive competency.

2 Background and Related Work

The concept of competency has been a common subject of research for many years. The term 'competency' was introduced in social science by McClelland [3] who tried to structure competency components by different skills of a person. Many stakeholders are interested in studying this phenomenon, including employers, university administration, government institutions, and even students, which results in interpreting the concept of competency in many ways [4]. For example, competency is described as connected pieces of knowledge, skills and attitudes that help to solve work-related problems [5]. The competency is also understood as the combination of the relevant cognitive, psychomotor and affective skills to perform a complex professional task [6].

Other scholars define competency as a cluster of trained skills, knowledge and attitudes [7]. It is believed that competency is a complex phenomenon formed by a holistic combination of the components, such as knowledge, skills, strategies, attitudes, etc. [8]. In spite of the above-mentioned examples of the unanimity lack, many stakeholders suggest to study competency by taking into account the quality of an employee that ensures to fulfill a specific job requirements.

The need to use the concept of competency in the practice of employers and educational institutions boosts the development of detailed descriptions of each component in the form of standards. The most common among them are: State Educational Standards of Higher Professional Education; Final State Certification; The Simple Reusable Competency Map draft standard; Performance Based Competency Standards; Cross-disciplinary competency standards; The Specification of Competency Standards and others. Standards are approved at the level of country governments, employers' associations, associations of educational institutions and, even, some universities. Unfortunately, most standards contain only the system of quality indicators and this makes practical assessment to competency more complicated.

Training of a highly qualified specialist presupposes the acquisition of appropriate competency defined by the standards described above. That is why most universities have offered competency-oriented courses [9]. The basis of such training programs is the need to acquire the so-called ideal student competency [8]. However, the effectiveness of this approach also requires an adequate system of competency assessment. The system should identify the achieved competency within the course, the systematic assessment of competency training, and the interests of stakeholders [8]. The evaluation system may include different methods applied to different competency components without losing adequacy. Considering the overall competency of a student consists of a number of portion [6], the assessment to competency has to be carried out for all its portion.

Except employers, university administration and government institutions, students are interested in self-assessment to their competency as well. Formation of an efficient student-centered learning environment presupposes active involvement of students in evaluating their learning outcomes [10]. Every assessment method has its own impact on the learning process [6]. Thus, the assessment to the ability of a student to self-assess gained skills is considered to be a new complement potion for competency assessment.

However, the implementation of competency assessment methods is a sophisticated process. The obstacles of building such an assessment system include the reluctance of staff; the lack of empirical data [7]; the lack of universal scales for assessing competency, etc. However, the most complicating factor of the assessments described above is the qualitative nature of the competency that are often subjectively assessed by experts using numerical scales or linguistic descriptions, which contribute to the growing interest in the use of methods. However, the most complex factor of the assessment described above is the qualitative nature of the competency constituents that are often subjectively assessed by experts with the help of numerical scales or linguistic descriptions, which contributes to growing the interest in the use of methods of the Fuzzy Set theory in a comprehensive assessment of competency and other learning characteristics [10].

Many examples of fuzzy logic models are known as those used for assessing competency. Though, the proposed methods solve the above-mentioned problem partly and for fairly narrow applied tasks. Thus, Adem Golec and Esra Kahya utilized a fuzzy model for competency-based employee evaluation and selection [11]. This fuzzy model had seven input factor block parameters and one output employee's competency parameter. The relation of input and output parameters, as well as the set of rules, were determined based on expert assessment data factor analysis of 53 indicators. Such a model is flexible when considering new parameters and is consistent with organization goals and strategies. However, the model has a lot of indicators that complicates its use. So, the model introduction, in the case of other organizations, requires the involvement of trained experts. The model does not contain generally recognized competency assessment standards, which complicates the comparative assessment to its functions.

Jevšček utilized 360° feedback to build a fuzzy model of competency assessment [12]. The model has two output parameters. The relation of the input and the output parameters, as well as the set of rules, was defined based on survey questionnaire analysis of 350 participants. The model, suggested by Matej Jevšček, has a few numbers of parameters that simplifies its application. Though the model does not contain the competency standardized parameters as those in Golec and Kahya method [11], the above mentioned shortcomings still exist in the model.

Another aspect of using the fuzzy approach is the attempt of ranking people by their competency in economics education suggested by Suleman and Suleman [13]. The model allowed assessing successfully, ranging and investigating the structure of the bundle of competency of 593 banking employees. However, the suggested model does not consist of the competency standard system, it is highly specialized. The possibility of using the model in the learning process is somewhat complicated, especially the possibility of self-assessment to students for acquired competency.

Kavčič and others built a fuzzy student model based on the use of the competency graph [14]. The model determined the categories of competency considering their real performance. The model can assess the competency level in the learning process using the student knowledge assessment system. Such approach allows the student to influence the learning results, and can motivate students to acquire the missing competency. Though, similar to the above-mentioned models, the method suggested by Kavčič and others [14] is not based on common assessment standards of competency, which suggests the use of this method is not favorable among employers. The applied possibilities of the model are also limited by the complexity of self-assessment to acquired competency conducted by a student.

Therefore, the development of using Fuzzy Logic is relevant to the research field of competency and it requires further research. Also, most methods of assessment to competency are complex and contain a hierarchical internal structure. They are normally parametrized by qualitative and quantitative methods, which take into account with a large amount of data and provide compatibility with scales of generally accepted standards, etc.

3 Methods and the Model

The Fuzzy logic theory was introduced in the work of Zadeh [15]. The main idea of the theory was the mathematical description of linguistic vagueness. Further development of Fuzzy logic was possible due to the ability of describing problems with a degree of non-statistical uncertainty [16] that was not resolved through probability theory [17]. Fuzzy set theory describes these problems in the form of simple linguistic terms, which can subsequently be processed by mathematical operators [18].

Relying on the fuzzy logic approaches, there was an attempt to build a fuzzy model for assessing student competency. As a research object, an educational program (further EP) of the second level of higher education chosen for 'Economic Cybernetics' (2018), which is available on the site of Kyiv National University of Trade and Economics [19]. The structure and EP courses are determined according to the National Competency Standards of Ukraine and meet the requirements of the competency-centered courses. An EP contains a detailed description of the course and the final competency of a graduate. For each course of the program, there is a description of the competency training objective (Table 1).

Program Competence (further PC) lines correspond to the competency that a student should achieve. PC1 shows 'The ability for pre-project analysis and modeling economic objects and tasks', PC2 – 'The ability to elaborate the projects of information systems or their fragments', PC3 – 'The ability to elaborating managing consulting projects', PC4 – 'The ability to elaborate the forecasting development models of economic processes and systems', PC5 – 'The ability to manage the projects of informatization of social and economic objects of different level', PC6 – 'The ability to assist a system or a program complex'. Columns CC correspond to the compulsory courses and OC to the optional courses of professional training. CC1 stands for the course 'Theory and practice of scientific research', CC2 – 'Mathematical methods and models of complex economic systems', CC3 – 'Information security', CC4 – 'Applied system analysis, CC5 – 'Techniques of data analysis', CC6 – 'Stochastic models in economy', CC7 – 'Techniques of mobile applications development', OC1 – 'Corporate information systems', OC2 – 'Enterprise risk management', OC3 – 'Corporate network management systems', OC4 – 'Internet security', OC5 – 'Financial management'.

Table 1. Program competences and educational program components matrix[a]

	CC1	CC2	CC3	CC4	CC5	CC6	CC7	OC1	OC2	OC3	OC4	OC5
PC1	+	+		+		+		+		+		+
PC2			+					+		+	+	
PC3									+	+		+
PC4		+			+	+	+	+		+		+
PC5			+					+			+	
PC6			+		+		+	+		+	+	

[a]built on the basis [19]

Academic courses contributing to the acquisition of one or more competency marked with a "+" sign (Table 1). Each competency can also be formed within a range of courses. Despite the detailing of the process of acquiring competency in these types of documents, the numerical assessment of the competency of a graduate is complicated in practice and needs further analysis.

For this study, the input data of a competency assessment system include selected students in a scale of 100 scores in each course and the structure of competency determined by the EP. For the fuzzyfication of input and output parameters we used the trapezoidal and triangular membership function (Table 2), which are utilized for assessing the student's performance [20]. Marginal scales of the fuzzy set input, intermediate and output data were the same to get better interpretation of the modeling results.

Table 2. Fuzzy set of the model

Variable	Type of variable	Linguistic term	Membership function	Interval
Final grade of courses: CC, OC	Input	Low	Trapezoidal	(60, 60, 70, 80)
		Medium	Triangular	(70, 80, 90)
		High	Trapezoidal	(80, 90, 100, 100)
Competency: PC1…PC6	Intermediate	Low	Trapezoidal	(60, 60, 70, 80)
		Medium	Triangular	(70, 80, 90)
		High	Trapezoidal	(80, 90, 100, 100)
General competency of a graduate: PC	Output	Very low	Trapezoidal	(60, 60, 66.7, 73.3)
		Low	Triangular	(66.7, 73.3, 80)
		Medium	Triangular	(73.3, 80, 86.7)
		High	Triangular	(80, 86.7, 93.3)
		Very high	Trapezoidal	(86.7, 93.3, 100, 100)

Figure 1 and Fig. 2 show graphical interpretation of Fuzzy sets (formed in fuzzyTECH 5.54d) of input CC1 and middle PC1 parameters. Linguistic terms are depicted here graphically: the red curve corresponds to the 'low' parameter (trapezoidal membership function), the green one to the 'medium' parameter (triangular membership function) and the blue one to the 'high' parameter (trapezoidal membership function).

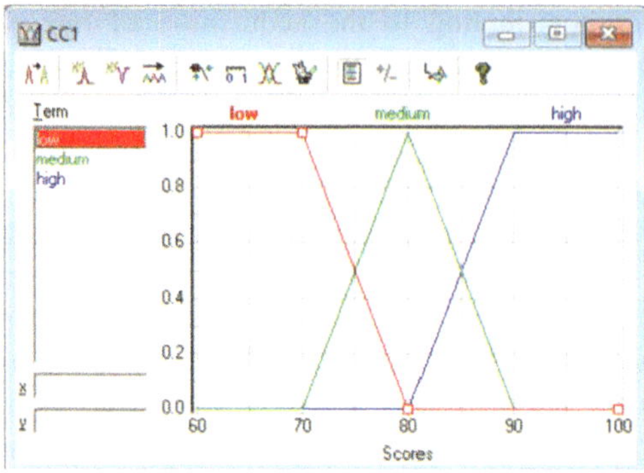

Fig. 1. Fuzzy sets of CC1 input parameter

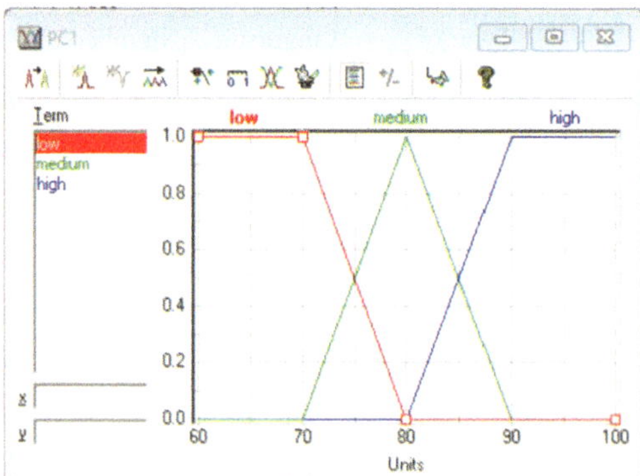

Fig. 2. Fuzzy sets of PC1 middle parameter

The next stage contained the implementation of the table of competency correspondence and educational program components (Table 1) in the form of a hierarchical fuzzy model (Fig. 3).

The model contains 12 input fuzzy parameters that are set according to the student's progress in the indicated courses. Fuzzy parameters containing the acquired competency are defined as intermediate. Six blocks of rules are formed by the courses input parameters based on Table 1, and the output parameters are corresponding competency.

The rules were determined by the Mamdami technique [21] for fuzzy decision-making and fuzzy inference. The set of rules that determine input and output membership functions were automatically selected in the fuzzyTECH environment.

The rules are determined by the 'If-Then' conditions for all blocks PC_1, PC_2, PC_3, PC_4, PC_5, PC_6, PC. The example of the rule settings for block 'PC_3' is shown in Table 3.

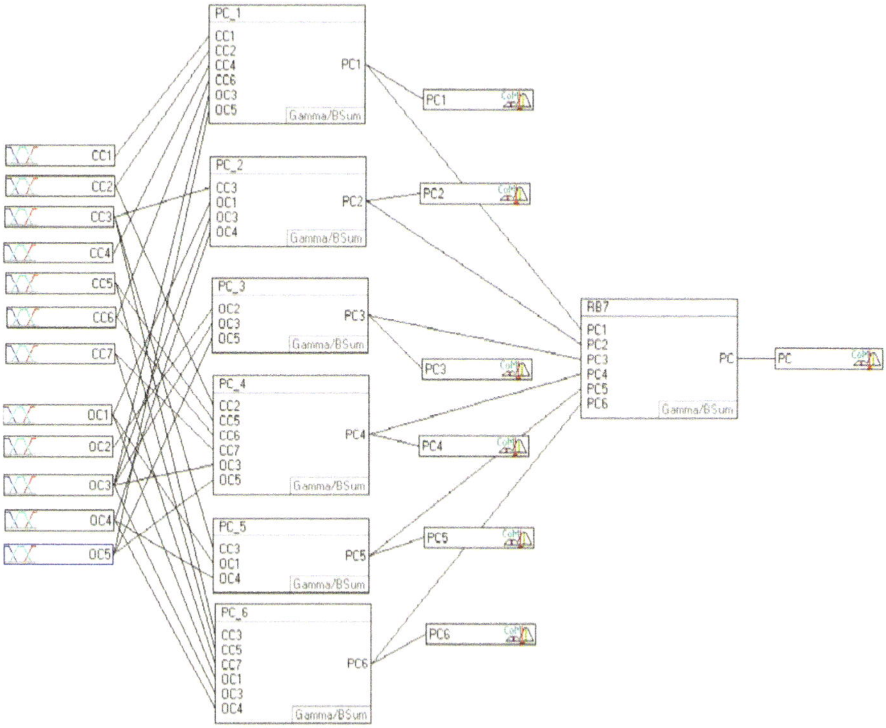

Fig. 3. Hierarchical fuzzy model for assessing the acquired competency

For the Fuzzy rule computation of the model, the compensatory operator ($\lambda = 0.046$) is employed:

$$\mu = (1 - \lambda) \min_{i=1..n} (\mu_i) + \lambda \sum_{i=1}^{n} \mu_i / n \tag{1}$$

The Center of Maximum method is used as a defuzzification method for output parameters:

$$x_{OUT} = \frac{(x_1\mu_1 + x_2\mu_2 + \ldots + x_n\mu_n)}{(\mu_1 + \mu_2 + \ldots\mu_n)} \tag{2}$$

The model is composed of two levels for assessing the total competency of a student, a PC block is employed as supplement of the model. The above described methods for making rules and defuzzifying the output parameter (general competency) were applied for this block.

Table 3. Set of rules for rule block 'PC_3'

IF			THEN	
OC2	OC3	OC5	DoS	PC3
Low	Low	Low	1.00	Low
Low	Low	Medium	1.00	Low
Low	Low	High	1.00	Medium
Low	Medium	Low	1.00	Low
Low	Medium	Medium	1.00	Medium
Low	Medium	High	1.00	Medium
Low	High	Low	1.00	Medium
Low	High	Medium	1.00	Medium
Low	High	High	1.00	High
Medium	Low	Low	1.00	Low
Medium	Low	Medium	1.00	Medium
Medium	Low	High	1.00	Medium
Medium	Medium	Low	1.00	Medium
Medium	Medium	Medium	1.00	Medium
Medium	Medium	High	1.00	High
Medium	High	Low	1.00	Medium
Medium	High	Medium	1.00	High
Medium	High	High	1.00	High
High	Low	Low	1.00	Medium
High	Low	Medium	1.00	Medium
High	Low	High	1.00	High
High	Medium	Low	1.00	Medium
High	Medium	Medium	1.00	High
High	Medium	High	1.00	High
High	High	Low	1.00	High
High	High	Medium	1.00	High
High	High	High	1.00	High

4 Results Analysis and Discussion

To test effects of the model (Fig. 3), it has studied a case of students of 'Economic Cybernetics' specialty of Chernivtsi Institute of Trade and Economics of KNUTE. Additional data collection or process are not required while acquiring the information about the students' learning outcomes, as well as the EP structure. The analyzed sample comprised the information about the studying results of 17 students. The sample was made on the basis of the data collected in 2016–2018 by the co-author Y. Koroliuk. The criteria of the sample data contain students with different learning outcomes in the scope of the courses and their average performance (Table 4).

Entering input data in the model (Fig. 3), the outcomes x of intermediate fuzzy competency parameters are obtained in the fuzzyTECH environment (parameters PC1,

PC2, PC3, PC4, PC5, PC6 of Table 4). The results show the presence of different assessments of acquired competency by their types, depending on the outcomes of studying courses as the constituent parts according to EP. Figure 4 shows an example of defuzzification of the output competency of student number '5'.

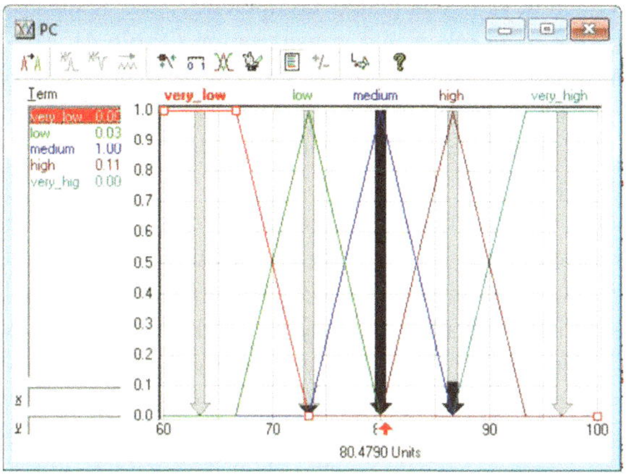

Fig. 4. Example of graduate's general competency (PC)

Table 4. The learning outcomes of students, studying the courses within EP and the results of competency assessment

ID of student		1	2	3	4	5	6	7	8	9	10	11	12	13	14	15	16	17
Input parameters	CC1	63	63	77	82	90	79	81	88	78	87	85	90	70	88	84	92	90
	CC2	62	64	61	77	82	81	79	81	83	86	89	88	71	86	90	94	93
	CC3	69	65	69	71	90	82	80	78	83	87	81	85	98	89	84	91	90
	CC4	68	67	78	65	75	76	79	76	88	87	81	85	73	86	90	86	92
	CC5	70	69	72	80	69	79	79	80	88	88	89	85	72	89	90	92	92
	CC6	67	66	68	85	90	85	81	90	82	85	90	87	85	90	90	89	90
	CC7	70	72	67	90	80	82	79	79	85	84	87	88	95	86	90	94	93
	OC1	65	66	71	82	65	76	82	78	87	84	87	87	93	87	85	85	92
	OC2	69	71	63	79	68	78	77	89	84	86	90	87	97	88	83	91	90
	OC3	61	69	78	73	73	77	82	88	81	87	82	87	98	89	88	86	93
	OC4	62	67	79	78	77	82	78	89	83	83	87	87	95	85	87	90	91
	OC5	67	67	64	85	90	78	80	77	84	83	88	86	98	87	90	92	91
	Average score	66	67	71	79	79	80	80	83	84	86	86	87	87	88	88	90	91
Attestation of MA thesis		60	65	72	82	82	80	86	87	87	90	90	91	93	89	92	92	93
Output parameters	PC1	63	63	65	79	91	73	88	88	87	87	95	87	80	87	95	95	95
	PC2	63	63	67	73	77	79	81	79	88	91	89	92	95	87	92	95	95
	PC3	63	63	65	82	80	77	83	93	91	93	95	94	95	95	95	95	95
	PC4	63	63	65	85	81	80	73	86	87	87	87	87	90	87	95	95	95
	PC5	63	63	67	78	80	84	83	87	92	93	93	94	95	95	93	95	95
	PC6	63	63	65	80	77	87	73	80	87	87	87	87	95	87	95	95	95
	PC	63	63	68	80	81	80	84	84	88	91	88	88	95	88	95	95	95

The level of acquired competency could be different even in the case of similar average performance. Such results confirm the elaborated model's application for more accurate and adequate assessment for the overall student's competency and its components. For assessing other types of EP, the model should be adjusted in accordance with structure of the rule blocks.

To check the accuracy of calculated competency, the results of Master of Arts (MA) thesis attestation were taken into consideration (Table 4). The work on the MA thesis and their attestation presupposed the application of all the professional competency which were acquired during the course study. That's why the concordance of the mentioned indices from the general competency output data of the model can be the affirmation of its adequacy.

Figure 5 presents the results of the calculated general competency (PC), average performance and attestation of MA theses of students during the course study. As the figure presents, general competency indices concord with the results of MA theses attestation more accurately for all areas, they are marked by criteria 'Very low', 'Low', 'Medium', 'High', 'Very high'. The analysis result of the significance of given index model description reveals that the value of index R^2 was equal to 0.95 which is higher than its minimum critical value. Standard error equals to 2.38 and the value of index F (equals to 298.4) is always much smaller than the critical value, and the value of calculated α (significance of F) is close to zero.

Thus, general competency indices, calculated by the model, confirm the peculiarities of MA theses attestation results are higher than the indices of average performance (Fig. 5), which never had an exact explanation before.

The model confirms that the combination of competency components in the process of writing and attestation of MA thesis leads to synergy. Thus, the result obtained in the course of the MA thesis attestation is higher than the arithmetic mean of the course performance results.

Statistic confirmation of the model's adequacy allows the use for the accurate prediction of the final student performance of studied specialty. In addition, the model makes possible to estimate the general competency components values quantitatively (PC1...PC6 see Table 4), which could be previously assessed by qualitative methods only.

Wide application of this method in practice for an employer's assessment or student self-assessment requires the model only numerical data of the courses' learning outcomes (available in a diploma supplement), which significantly simplifies the calculations. The result can be calculated by using numerical or linguistic scales which can be unequivocally interpreted by various stakeholders.

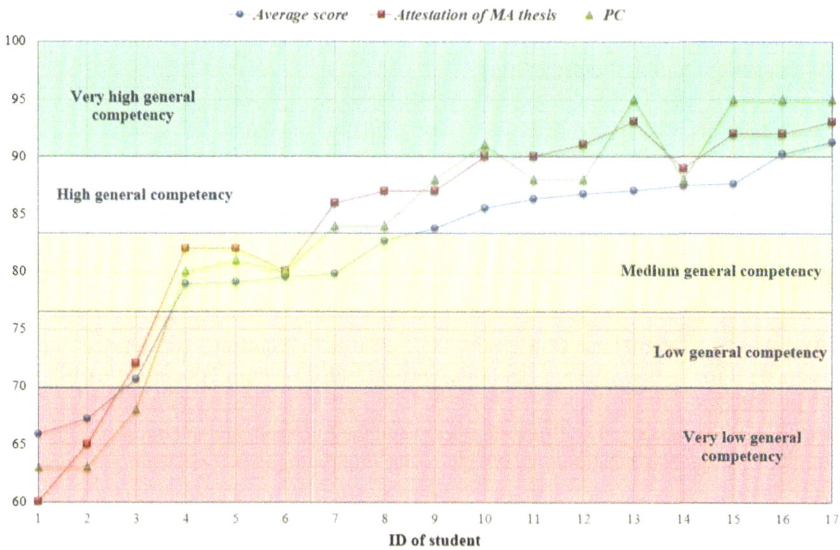

Fig. 5. General competency (PC), average performance and attestation of MA thesis for the analyzed sample

5 Conclusion, Limitations and Future Work

The paper reveals that the proposed fuzzy model is capable to improve the system of competency assessment, described by the standards of the specialty (EP). For the calculations, the model just utilizes learning outcomes data and the structure of the EP, which simplifies the process of assessing competency. When using the model based on open source data, additional resources to assess the professional qualities of students are not required by stakeholders any more.

The adequacy of the obtained modeling results is confirmed by the statistical agreement with the MA thesis attestation results. In addition, the model confirmed that the combination of competency in the course of writing and attestation of MA thesis leads to synergy. Thus, the model is capable of carrying out an adequate assessment to each component and overall competency. Also, the model can assess the quality of EPs, carry out their comparative assessment, etc. The fuzzy model can be used for predicting the final performance of students. Thus, a student can be encouraged to adjust his or her learning methods in those courses that make a more significant contribution to the chosen competency and achieve better scores.

As input parameters, this study adopts a performance rating of 100 scores, which causes the model's limitations. For the universities with a different performance rating, the proposed model needs to be revised accordingly. The results of the model also provide an opportunity to assess the competency for students defined by the EP. Further work includes the development of the proposed model by considering other competency related parameters beyond the proposed EP list. In particular, psychological, behavioral, motivational, and cognitive parameters will be incorporated in future model for accurately assessing students' competency.

Acknowledgments. This scientific work was partially supported by RAMECS and self-determined research funds of CCNU from the colleges' primary research and operation of MOE (CCNU19TS022).

References

1. Berestneva, O., Marukhina, O., Benson, G., Zharkova, O.: Students' competence assessment methods. Proc. - Soc. Behav. Sci. **166**, 296–302 (2015). https://doi.org/10.1016/j.sbspro.2014.12.527
2. European Commission: The Bologna process and the European higher education area (2014). http://ec.europa.eu/education/policy/higher-education/bologna-process_en.htm. Accessed 18 July 2019
3. McClelland, D.C.: Testing for competence rather than for 'intelligence'. Am. Psychol. **28**, 1–14 (1973). https://doi.org/10.1037/h0034092
4. Baartman, L.K.J., Bastiaens, T.J., Kirschner, P.A., van der Vleuten, C.P.M.: Evaluating assessment quality in competence-based education: a qualitative comparison of two frameworks. Educ. Res. Rev. **2**, 114–129 (2007). https://doi.org/10.1016/j.edurev.2007.06.001
5. Lizzio, A., Wilson, K.: Action learning in higher education: an investigation of its potential to develop professional capability. Stud. High. Educ. **29**, 469–488 (2004)
6. van der Vleuten, C.P.M., Schuwirth, L.W.T.: Assessing professional competence: from methods to programmes. Med. Educ. **39**, 309–317 (2005). https://doi.org/10.1111/j.1365-2929.2005.02094.x
7. van der Klink, M.R., Boon, J.: Competency: the triumph of a fuzzy concept. Int. J. Hum. Resour. Dev. Manag. **1**, 327–334 (2003)
8. Bergsmann, E., Schultes, M.-T., Winter, P., Schober, B., Spiel, C.: Evaluation of competence-based teaching in higher education: from theory to practice. Eval. Program Plann. **52**, 1–9 (2015). https://doi.org/10.1016/j.evalprogplan.2015.03.001
9. Wesselink, R., Dekker-Groen, A.M., Biemans, H.J.A., Mulder, M.: Using an instrument to analyse competence-based study programmes: experiences of teachers in Dutch vocational education and training. J. Curric. Stud. **42**, 813–829 (2010). https://doi.org/10.1080/00220271003759249
10. Ma, J., Zhou, D.: Fuzzy set approach to the assessment of student-centered learning. IEEE Trans. Educ. **43**, 237–241 (2000). https://doi.org/10.1109/13.848079
11. Golec, A., Kahya, E.: A fuzzy model for competency-based employee evaluation and selection. Comput. Ind. Eng. **52**, 143–161 (2007). https://doi.org/10.1016/j.cie.2006.11.004
12. Jevšček, M.: Competency assessment using fuzzy logic. J. Univers. Excell. **5**(2), 187–202 (2016)
13. Suleman, A., Suleman, F.: Ranking by competence using a fuzzy approach. Qual. Quant. **46**(1), 323–339 (2012). https://doi.org/10.1007/s11135-010-9357-1
14. Kavcic, A., Pedraza-Jimenez, R., Molina-Bulla, H., Valverde-Albacete, F.J., Cid-Sueiro, J., Navia-Vazquez, A.: Student modeling based on fuzzy inference mechanisms. In: The IEEE Region 8 EUROCON 2003. Computer as a Tool, vol. 2, pp. 379–383. IEEE (2003). https://doi.org/10.1109/EURCON.2003.1248223

15. Zadeh, L.A.: Fuzzy sets-information and control-1965. Inf. Control **8**, 338–353 (1965)
16. Bezdek, J.: Fuzziness vs. probability: again (?). IEEE Trans. Fuzzy Syst. **2**, 1–3 (1994)
17. Gokmen, G., Akinci, T.Ç., Tektaş, M., Onat, N., Kocyigit, G., Tektaş, N.: Evaluation of student performance in laboratory applications using fuzzy logic. Proc. - Soc. Behav. Sci. **2**, 902–909 (2010). https://doi.org/10.1016/j.sbspro.2010.03.124
18. Otero, L.D., Otero, C.E.: A fuzzy expert system architecture for capability assessments in skill-based environments. Expert Syst. Appl. **39**, 654–662 (2012). https://doi.org/10.1016/j.eswa.2011.07.057
19. PROSPECTUS 'European Credit and Transfer System (ECTS)' Field of Science 05 'Social and Behavioral Sciences' Field of Specialty 051 'Economics' Specialization 'Economic Cybernetics' Academic Degree 'Master'. KNUTE, Kyiv (2018). https://knute.edu.ua/file/NjY4NQ==/b8e5c2999e2f64e19a132aae0a218e33.pdf (in Ukrainian)
20. Krouska, A., Troussas, C., Sgouropoulou, C.: Fuzzy logic for refining the evaluation of learners' performance in online engineering education. Eur. J. Eng. Res. Sci. **4**, 50–56 (2019). https://doi.org/10.24018/ejers.2019.4.6.1369
21. Mamoria, P., Raj, D.: Comparison of Mamdani fuzzy inference system for multiple membership functions. Int. J. Image Graph. Sig. Process. (IJIGSP) **8**, 26–30 (2016). https://doi.org/10.5815/ijigsp.2015.09.04

Establishment of Problem E-learning Behavior Scale

Junyi Zheng[1] and Wenhui Peng[2](✉)

[1] School of Mechanical and Electronic Engineering, Wuhan University of Technology, Wuhan 430070, China
[2] School of Educational Information Technology, Central China Normal University, Wuhan 430079, China
pwh@mail.ccnu.edu.cn

Abstract. Problem e-learning behavior exists commonly among learners, which affects the performance of online learning seriously. Therefore, it is meaningful to monitor and evaluate whether online learners have learning behavior problems. This paper designs a scale of e-learning behavior to evaluate the behavioral problems of online learners. Firstly, the connotation of problem e-learning behavior is discussed. Secondly, the various manifestations of problem e-learning behavior are summarized and classified into five dimensions: learning escape, behavior deficiency, behavior misconduct, improper learning time arrangement and negative emotional disclosure. Finally, the scale of e-learning behavior is developed by way of standard method of the development of psychological scale. The scale is proved to be reasonable by large scale measurement, confirmatory factor analysis and reliability and validity analysis.

Keywords: Online learning · Problem e-learning behavior · Scale · Establishment of scale

1 Introduction

Since the 1990s, with the rapid development of Internet technology, web-based learning has become more and more popular. While the network environment provides convenience and freedom for learning, learners are also more prone to problem behaviors, such as "inattention and distraction during learning" and "browsing web pages unrelated to learning". Peng defined these problem behaviors in network learning environment as "problem e-learning behavior", and furthermore gave a proper definition to the problem e-learning behavior, which refers to the behavior that does not conform with the normal learning norms, not meeting the normal learning requirements and affecting oneself or others to study effectively in the process of network learning [1].

Many scholars have done detailed research on the e-learning behavior at home and abroad, but the research on the problem e-learning behavior is scarce now. In the early stage, Hu Liqiang and Liu Jianxin (2008) investigated and studied the general behavior patterns of online learners, and summarized some factors that might lead to problem

Z. Hu et al. (Eds.): AIMEE 2019, AISC 1126, pp. 394–403, 2020.
https://doi.org/10.1007/978-3-030-39162-1_36

behavior. Zhang and others analyzed the factors that influence the problem e-learning behavior from the perspectives of students, teachers and courses, and gave some improving methods [2]. By observing the behavior of experimental subjects and according to the environment and nature of problem behaviors, Peng summarized problem e-learning behavior into five categories: learning escape, behavior deficiency, behavior misconduct, improper learning time arrangement and negative emotional disclosure. Finally, he proposed a multi-dimensional and multi-level model of e-learning behavior and evaluation system [1]. On this basis, Tian divided problem e-learning behaviors into three categories: seriously deficient e-learning behavior, unplanned network learning behavior and misconduct e-learning behavior through investigation and research, and analyzed the causes and improvement strategies [3, 4]. From the perspective of behavioral analysis and intervention, Mao Gang constructed a learning behavior analysis model of e-learning problems, providing a good reference for discovering problems, analyzing causes and searching for intervention strategies [5].

From the perspective of content, the researches mentioned above involved the performance level, improvement strategy [6, 7] and model analysis [8] of network problem learning behaviors. The analysis methods mainly adopted observation method and investigation method [9], summarizing the observation or investigation results according to some certain rules, to obtain the performance categories of problem e-learning behaviors. The classification of these manifestations lacked of processing and testing of quantitative analysis. In addition, we know that to study an educational phenomenon, the first step is to study how to measure it. However, there are rare studies on measurement of problem e-learning behaviors, especially on the establishment of measurement tools and scales. Therefore, in this paper, the "the scale of e-learning behavior" is developed, and the problem behaviors is classified according to the exploratory factor analysis in combination with the related characteristics of cognitive psychology. Finally, questionnaires are distributed on large scale, and the reliability and validity of the scale and the rationality of dimension division are tested by SPSS analysis method and confirmatory factor analysis in the structural equation model.

2 The Preparation and Implementation of the Scale

2.1 The Early Stage of the Research

According to the concept, classification and characteristics of problem e-learning behavior, in combination with previous questionnaires, 34 measuring items are designed to measure problem e-learning behavior. Firstly, 10 undergraduates and 10 postgraduates from the School of Educational Information Technology of Central China Normal University are selected at random as interviewees to be asked about whether the 34 measurement items are ambiguous or difficult to understand in order to test whether the items in the scale could be correctly understood. At the same time, five experts from the fields of educational technology and psychology are invited to evaluate the consistency of the content of 34 questions with the concepts measured by the researchers. After discussions, 4 measurement items with ambiguity or difficult to understand are deleted, and finally a measurement scale consisting of 30 items is obtained. According to five-point Likert scale, the alternative items of 30 measurement items are marked as 1, 2, 3, 4 and 5 points respectively in the order of "always", "often", "sometimes", "rarely" and "never". The lower the total score is, the more serious the problem is. Together with questionnaire instructions and 5 basic information, the final form, i.e. questionnaire I contains 30 measuring items, five basic information items.

2.2 Preparation and Revision of Scale

2.2.1 Prediction A

Two classes are randomly selected from Central China Normal University and Wuhan Textile University to be measured in class by using predictive questionnaire I. 200 questionnaires are issued and all of them are taken back with a recalling rate of 100%. There are 186 valid questionnaires with an effective recalling rate of 93%.

Analyzing the data of Predict A, we find that Cronbach's coefficient alpha is 0.888, which indicates that the scale has high reliability. The average value is between 2.71 and 3.76 without "Glass Ceiling" or "Sticky Floor", and the standard deviation of all items is between 0.757 and 1.033 without exceptional phenomenon. High-low grouping is used for independent sample t-test, and the measuring items 4, 7, 13 and 22 with low discrimination are deleted to form the prediction questionnaire II, which contains 26 measuring items and 5 basic information items.

2.2.2 Prediction B

Three classes are selected at random from three universities, namely Wuhan University of Technology, Huazhong University of Science and Technology and Hubei University. 700 questionnaires are issued in total and 673 are taken back, with a recalling rate of 96.1%. There are 648 valid questionnaires among them with an effective rate of 96.3%. Exploratory factor analysis is carried out, and two measurement item 15 and item 23 are deleted according to the three rules of factor load value less than 0.4, large load value on multiple factors at the same time, and different meaning from other items in this factor. After the deletion of item 15 and item 23, the coefficient alpha changed from 0.929 to 0.924, indicating that the deletion of these two items is reasonable. And then, dimension division, correlation analysis and reliability test are carried out for the remaining 24 items. The results of correlation analysis and reliability test are good, and the formal questionnaire containing 24 measurement items and 5 basic information items is finally formed, as seen in Table 1.

Table 1. Items of the scale

1. Browse the web pages unrelated to the learning content in the e-learning process	13. Unwilling to share learning resources actively in the e-learning process
2. Lack the confidence to complete the task independently in the e-learning process	14. Feel anxious and impatient in the e-learning process
3. Discuss topics unrelated to the learning topic in the e-learning communication process	15. Cheat or give dishonest answers in the e-learning assessment
4. Feel difficult to concentrate on learning in the e-learning process	16. Leave the learning environment for unrelated activities for a long time without reasons in the e-learning process
5. Watch videos unrelated to learning in the e-learning process	17. The single study time is too short in the e-learning process
6. Do not participate in the necessary discussions in the e-learning process	18. At ordinary times, do not reflect on the problems encountered in the process of e-learning
7. When learning through the network, there is a situation that the time interval between two learning sessions is too long	19. Feel whiny in the e-learning process
8. Before learning the main content, you don't like to read the advance knowledge, such as learning goals, key points and difficult points of the content	20. In e-learning, you will only study the relevant content in a short time before the exam
9. Play games unrelated to learning in the e-learning process	21. Feel depressed and disappointed in the e-learning process
10. Complete the learning task in a short time before the deadline in the e-learning process	22. Spend a lot of time chatting online with others in the e-learning process
11. Do not practice related exercises in the e-learning process	23. Don't like to review what you've learned after completing e-learning
12. Like to watch the video in a skip scan when watching video learning resources	24. Don't like to conduct a self-assessment on the e-learning effect

2.3 The Implementation of Formal Questionnaire

Using formal questionnaires to conduct surveys, there are two ways to issue questionnaires. One is to issue paper questionnaires face to face, another is to issue online questionnaires through the web. The paper questionnaires are mainly distributed in eight universities: Central China Normal University, Huazhong University of Science and Technology, Wuhan University, Wuhan University of Technology, China University of Geology (Wuhan), Zhongnan University of Economics and Law, Hubei University, Wuhan Textile University and Jianghan University. Most of the paper questionnaires are tested in class. 1300 paper questionnaires are distributed, 1265 paper questionnaires are taken back, and 1197 valid questionnaires. The online questionnaires that distributed through the web are mainly filled out by students of Central China Normal University, 9073 questionnaires are taken back, and 8655 valid questionnaires.

10338 questionnaires are collected in two ways: paper questionnaire and online questionnaire, as seen in Table 2.

Table 2. Statistical information of the questionnaires

Statistical items		Distributed	Retrieved	Recalling rate	Valid	Effective rate
Paper questionnaires		1300	1265	97.3%	1197	94.6%
Online questionnaires			9703		8655	95.4%
Total questionnaires			10338		9852	95.3%
Gender	Male	Rate	Female	Rate		
Count	2171	22.0%	7681	78.0%		
Grades	Freshmen	Sophomores	Juniors	Seniors	Graduated	Others
Count	3659	3187	2766	62	159	19
Rate	37.1%	32.3%	28.1%	0.6%	1.6%	0.2%
Major	Liberal arts	Science and engineering		Art	Medicine	Other fields
Count	3895	5024		413	10	510
Rate	39.5%	51.0%		4.2%	0.1%	5.2%

One class (43 students) in the formal test is selected to be retested two weeks later in order to test the test-retest reliability of the questionnaire. When retesting, all the 43 questionnaires issued are taken back and all are valid.

The data of the formal questionnaire are put into reliability test and confirmatory factor analysis to test the rationality of the dimension division of 24 measurement items, and further confirm the construct validity of the questionnaire.

3 Results Analysis

3.1 The Reliability Analysis

3.1.1 Internal Consistency Reliability

Internal consistency reliability analysis is conducted on the questionnaire data of prediction B, and the results are shown in Table 2, the Cronbach's coefficients alpha of the five sub-scales of behavior misconduct, low concentration, passive emotion, improper time arrangement, and insufficient after-class internalization are 0.700, 0.818, 0.771, 0.786 and 0.772 respectively, all of them are greater than 0.7, indicating that the reliability is general. The reliability coefficient of the low concentration dimension is greater than 0.8, indicating that the reliability is good. The Cronbach's alpha coefficient alpha of the total scale is 0.924 that is greater than 0.9, indicating that the reliability is very good.

3.1.2 The Test-Retest Reliability

The level of test-retest reliability reflects the stability of the test results with after time changes. The higher the reliability, the higher the stability and reliability of the test results across time. The interval between the two tests is two weeks to four weeks in general, preferably no more than six months according to Du Linzhi. As for the pre-test and post-test data of the 43 questionnaires for retesting, the product-moment correlation method was used to calculate the correlation coefficient between the total scale and each component. The results are shown in Table 3. In general, 0.8 is used as a criterion to evaluate the test-retest reliability of a total scale of the rating scale, and 0.75 is used as a criterion to measure the test-retest reliability of the sub-scale (or dimension) of the rating scale. Compared with the criteria, the results show that the scale has high stability and reliability.

Table 3. Reliability coefficient of the total scale and the sub scale

Sub scale	Number of projects	Projects	Coefficient Alpha	Test-retest reliability
Misconduct	6	B6, B8, B11, B12, B13, B16	0.700	0.770**
Low concentration	6	B1, B3, B4, B5, B9, B24	0.818	0.910**
Passive emotion	4	B2, B14, B20, B22	0.771	0.814**
Improper time arrangement	5	B7, B10, B17, B18, B21	0.786	0.857**
Insufficient after-class internalization	3	B19, B25, B26	0.772	0.892**
Total scale	24		0.924	0.941**

Note. $^{*}P < 0.05$, $^{**}P < 0.01$

3.2 Correlation Analysis

After the dimension division of the scale is initially determined by exploratory factor analysis, correlation analysis is conducted by using predict B's data in order to explore the discriminant validity of the scale.

3.2.1 Correlation Analysis Between the Sub-scale Scores and the Total Scale Scores

The Person correlation coefficient is used to analyze the total scores of each sub-scale and the whole scale, and the results are shown in Table 4. The result shows that all the correlation coefficients are greater than 0.3, the correlation coefficients among these dimensions and between each dimension and the total score reach significant level ($P < 0.01$), the correlation coefficient between each dimension is smaller than that between each dimension and the total score, and they have good discriminant validity.

Table 4. Correlations between the sub-scale and total scale scores

	Misconduct	Low concentration	Passive emotions	Improper time arrangement	Insufficient after-class internalization
Low concentration	0.605**				
Passive emotions	0.552**	0.604**			
Improper time arrangement	0.647**	0.628**	0.599**		
Insufficient after-class internalization	0.571**	0.481**	0.487**	0.612**	
Total scale scores	0.843**	0.845**	0.786**	0.851**	0.732**

*P < 0.05, **P < 0.01

3.2.2 Correlation Analysis of Item and Total Score of Sub-scale

We compute Pearson correlations among the scores of each item and the total score of the sub-scale. Person correlation coefficient is all greater than 0.3. For example, the sub-scale of negative emotions contains four items: B17, B26, B24 and B2, the correlation coefficient among each item and the total score of the scale are 0.847, 0.821, 0.808 and 0.658 respectively, all of which are greater than 0.3 and have significant statistical significance ($P < 0.01$).

3.3 Validity Test

3.3.1 Content Validity

The establishment of this scale is referred to the research of Peng Wenhui and other scholars. At the beginning of the design, ten undergraduates and ten postgraduates from the school of educational information technology of Central China Normal University are interviewed. In addition, 5 experts from the fields of educational technology and psychology are invited to provide guidance for the preparation of the scale. Therefore, the scale has high content validity.

3.3.2 Construct Validity

KMO and Bartlett sphericity tests are performed on 648 questionnaires collected in the second test. KMO value is 0.907, which conforms to the criteria given by Ksiser for factor analysis, Bartlett sphericity test results are significant ($B^2 = 14225.029$, $df = 325$, $p < 0.01$). Therefore, factor analysis is appropriate.

The principal-components analysis is conducted to extract five factors; overall variance accounted for by the five-factor solution is 59.494%. The component matrix is rotated by the maximum variance method, B15 and B23 are greater than 0.4, indicating that B15 and B23 have high loading on two factors. Therefore, both B15 and B23 are deleted. After deletion, the number of items contained in each factor is reasonable, so the total scale is determined to use five common factors, i.e., the scale of e-learning behavior contains five dimensions. The results of the final division and the factor loads of the various items are shown in Table 5 below.

Table 5. Factor loads of each item

Misconduct		Low concentration		Passive emotions		Improper time arrangement		Insufficient after-class internalization	
Item	Loading	Item	Loading	Item	Loading	Item	Loading	Item	Loading
12	0.650	1	0.739	14	0.776	7	0.600	23	0.731
13	0.635	5	0.736	21	0.741	21	0.596	24	0.720
8	0.623	3	0.712	20	0.694	17	0.518	19	0.692
11	0.582	4	0.618	2	0.518	18	0.504		
15	0.533	9	0.535			10	0.464		
6	0.498	22	0.481						

In order to verify whether the dimension partition structure proposed by exploratory factor analysis is true and effective, confirmatory factor analysis is carried out on the data collected by formal test [10]. The hypothesis model constructed based on exploratory factor analysis results is shown in Fig. 1 below. C1 to C24 corresponds to items 1 to 24 of the scale, e1 to e24 are the corresponding measurement errors of each observation variable.

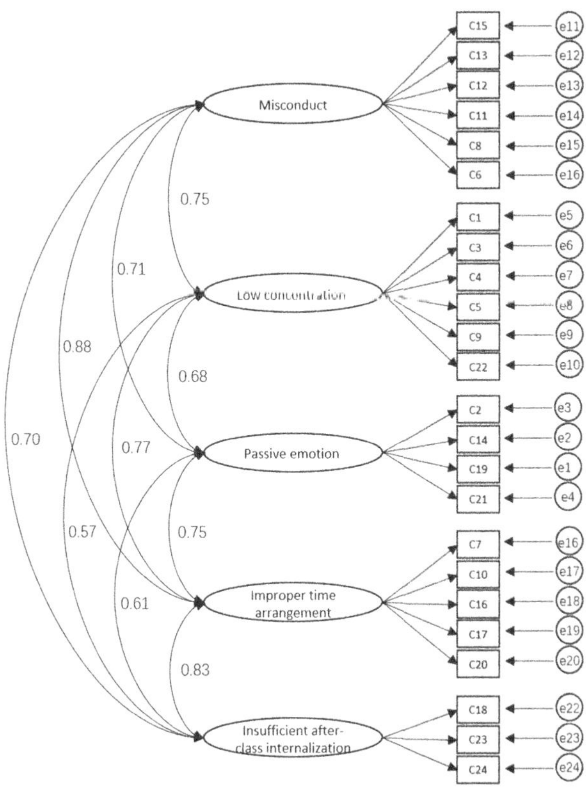

Fig. 1. Hypothesis model of CFA measurement model

The results of standardized regression coefficient analysis showed that the factor load range of each item was between 0.418 and 0.773, all of which were greater than 0.4, and the number of items in each dimension was at least three, so there was no need to delete the items. It can be considered that the model conforms to the actual data, the dimensional partitioning model fits the observation data.

4 Discussion

Rationalization-theorization and factor analysis strategies are used to develop the network problem learning behavior scale in this study. On the basis of previous studies by scholars, the outline of the scale is determined by inviting experts and conducting interviews with students, which has high Content validity. And then, two tests are carried out for different students, and the items are deleted and selected according to four indicators: degree of discrimination, mean and variance, exploratory factor analysis and correlation analysis. (1) Discrimination analysis is conducted on the data of prediction A, and measurement items with low discrimination such as A4, A7, A13 and A22 are deleted. (2) Mean and variance analysis is conducted on the data of prediction A, and the results are all within the normal range. (3) Exploratory factor analysis is carried out on the data of prediction B, and items B15 and B23 with high factor load (>0.4) on both factors are deleted, see that the scale has good structural validity. (4) The correlation analysis between the sub-scale and the total score of the whole scale and the correlation analysis between the item and the total score of the sub-scale are carried out. The correlation is remarkable and both of them are greater than 0.4, indicating good discriminant validity.

A formal scale is formed and tested again. Reliability analysis and confirmatory factor analysis are carried out on the results. The Cronbach's coefficient alpha of the whole scale is 0.924, and the Cronbach's alpha coefficients of the sub-scales were 0.722, 0.839, 0.793, 0.801 and 0.784, indicating that the scale has good reliability. The scale is divided into five dimensions by the exploratory factor analysis: behavior misconduct, low concentration, passive emotions, improper time arrangement and insufficient after-class internalization. Values of confirmatory factor analysis except K-square test affected by sample number, are in the normal range, and other fitting indexes are reasonable, indicating that the hypothesis model has a good fit with the actual data and has a good construct validity.

5 Conclusions

We designed a scale of e-learning behavior to evaluate the behavioral problems of online learners. Firstly, the connotation, various manifestations and categories of problem e-learning behavior are discussed. Secondly, in combination with existing research results and previous questionnaires, 34 measuring items are designed to measure learning behavior problems. After random sampling interview and two predictive tests, the formal questionnaire containing 24 measurement items and 5 basic information items is finally formed (see Table 1). Finally, the scale of e-learning behavior is developed by way of standard method of the development of psychological scale, and a large-scaled test is implemented (see Table 2). These results of reliability (see Table 3) analysis, correlation analysis (see Table 4) and validity analysis (see Table 4 and Fig. 1) show that the scale of problem E-learning behavior provides valid tools for the assessment and measurement of online learners' possible learning behavior problems. However, there were some limitations to current study, such as, although interview, paper questionnaires and electronic questionnaires were adopted, the questionnaire was distributed only to five universities in Wuhan, and the randomness and classification equilibrium of samples were not ideal. In further study, more quantitative and empirical studies are needed to test the reliability and validity of the scale.

Acknowledgements. This work was supported by NSFC of China [NO.61977036].

References

1. Peng, W.: Analysis and Modeling of E-learning Behavior. Science Press, Beijing (2013)
2. Zhang, X., Liu, M., Li, Q.: E-learning problem behavior study. Softw. Guide (Educ. Technol.) **14**(10), 37–38 (2015)
3. Tian, J., Nie, Y.: College students' e-learning problem behavior study. Dig. Educ. **2**(02), 29–32 (2016)
4. Wang, C.: An investigation and structure model study on college students' studying-interest. Int. J. Mod. Educ. Comput. Sci. (IJMECS) **3**(3), 33–39 (2011)
5. Mao, G., Liu, Q., Li, H., Fan, F.: Research and application of problem learning behavior analysis model in e-learning environment. E-education Res. **37**(11), 32–37+84 (2016)
6. Lashayo, D.M., Johar, Md.G.Md.: Preliminary study on multi-factors affecting adoption of e-learning systems in universities: a case of open university of Tanzania (OUT). Int. J. Mod. Educ. Comput. Sci. (IJMECS) **10**(3), 29–37 (2018)
7. Jayaprakash, S.M., Moody, E.W., Lauría, E.J.M., et al.: Early alert of academically at-risk students: an open source analytics initiative. J. Learn. Anal. **1**(1), 6–47 (2014)
8. Jia, P., Maloney, T.: Using predictive modelling to identify students at risk of poor university outcomes. High. Educ. **70**(1), 127–149 (2015)
9. Bhanuprakash, C., Nijagunarya, Y.S., Jayaram, M.A.: An informal approach to identify bright graduate students by evaluating their classroom behavioral patterns by using kohonen self organizing feature map. Int. J. Mod. Educ. Comput. Sci. (IJMECS) **10**(8), 22–32 (2018)
10. Zhou, S., Xin, D., Yang, X., Fu, J.: Chinese version of the premonition sadness scale and its reliability and validity test. Health Vocat. Educ. **35**(21), 129–132 (2017)

A Detection Model for E-Learning Behavior Problems of Student Based on Text-Mining

Wenhui Peng[1(✉)], Zhongguo Wang[1,2], and Junyi Zheng[3]

[1] School of Educational Information Technology, Central China Normal University, Wuhan 430079, China
pwh@mail.ccnu.edu.cn, china_wzg@mails.ccnu.edu.cn
[2] School of Journalism and Communication, Nanyang Normal University, Nanyang 473061, China
[3] School of Mechanical and Electronic Engineering, Wuhan University of Technology, Wuhan 430070, China

Abstract. Finding a solution to reduce academic risk is an often-required task in online and hybrid courses. It is insufficient for teachers to rely on intuition and experience to identify the students with potential academic risk in e-learning. Therefore, in this study, we propose a detection model to automatically detect e-learning behavior problems of students. We used web course as pre-class learning task in information technology and curriculum integration course in the fall semester of 2018 at Central China Normal University, and recorded one semester of online discussions and learning behavior traces, involved 78 students in total. The experimental results indicated that the detection model can effectively identify poor time management, academic procrastination, low participation, and dishonest behavior. The model is useful to identify the student's negative emotion which lasted a certain amount of time, but insufficient to identify the short-term emotional state which depended on high classification accuracy. The detection model and these results give the researchers and teachers a view of early alert for students at risk of academic failure, and how to improve student success in e-learning.

Keywords: E-learning behavior problems · E-learning · Text mining · Detection model

1 Introduction

In recent years, increasing amounts of online courses are available. The flipped classroom become new ways of learning and teaching, and participants in online and hybrid courses are continually expanding [1]. In the absence of direct guidance from an instructor, the online learner needs to determine when and how to engage or disengage learning activities of their own accord [2]. e-learning is often self-regulating learning (SRL). Zimmerman and Moylan [3] model of SRL shows that there are many factors effect SRL and subsequent performance, such as learners' affect, strategic planning, time management, self-evaluation. For that, to identify the e-learning behavior problems in these perspectives, and serve immediate feedback and intervention for student, can promote students' SRL in e-learning [4].

Z. Hu et al. (Eds.): AIMEE 2019, AISC 1126, pp. 404–413, 2020.
https://doi.org/10.1007/978-3-030-39162-1_37

Some relevant works have investigated that e-learning learners have behavior problems, including distraction [5], mind wandering [6], low participation [7], not active [8], etc. Some empirical researches confirmed that student's behavior problems might cause low academic performance and subsequent dropout [9]. So, such behavior problems were used to as valuable indicators to predict the student with a risk of low academic performance and dropout [10]. Some researchers suggested that earliest identification of the student with behavior problems, and providing immediate feedback or intervention [11, 12] for them, can effectively support students' SRL.

However, it is not sufficient for teachers to rely on intuition and experience to identify the potential behavior problems of students in e-learning settings. E-learning occurs often not face to face, it is also difficult for teachers to observe student's learning behaviors. On the contrary, most online learning platforms accumulate a large amount of students' activities records and behavior traces in various data formats. For large-scale personalized evaluation, it is a valid approach to analyze students' learning processes and behavior traces by using learning analytics techniques, and screen e-learning behavior problems.

Therefore, in this work, we described e-learning behavior problems as five categories, proposes a text-mining model to detect behavior problems of students in e-learning. The purpose is to identify the student with e-learning behavior problems, and help teachers to serve appropriate intervention and immediate feedback for these students to improve their learning performance and achievement.

2 Related Work

2.1 E-Learning Behavior Problems

We reviewed research about learning behavior problems and found that most of these research studies focus on behavior problems in the traditional classroom [13]. The conception of e-learning behavior problem is different from network behavior problem (e.g., internet addiction, illegally hacked) and behavioral disorders (e.g., intellectual disabilities). We contend that e-learning behavior problems refer to the behavior that do not conform to learning norms, against the e-learning requirements, and influence oneself or others to learn effectively. The e-learning behavior problems only refer to these behaviors that accompany the e-learning process, and affect learning and performance.

We examined previous researches on behavior problem scales, such as Rutter's behavior scale, Child Behavior Checklist. We found that these scales were mainly for adolescents. Some researchers have explored the students' learning misbehavior classification and the definition of classroom misbehavior. However, there are distinction of the students' behavior problems appearance in classroom environment and e-learning settings. Therefore, by analyzing the nature of e-learning behavior problems, seems that e-learning behavior problems can be divided into five categories: academic disengagement, low participation, misbehavior, low cognitive strategies and negative emotions (see Table 1).

Table 1. Categories of e-learning behavior problems

Categories	Description	Indicators/examples
Academic disengagement	The act of leaving the learning platform or distracting frequently during the learning process	Distraction Break away from learning Stay away from platforms
Low participation	Inadequate participation in some necessary learning activities and learning process	Never or seldom complete tasks Non-participation activities Academic procrastination
Misbehavior	Various kinds of misconduct in the process of learning	Dishonesty (fraud, copying) Using inappropriate words Non-compliance rules
Low cognitive strategies	Failure to manage learning objectives, process and strategies wisely	Poor time management Lack of in-depth learning Superficial cognitive
Negative emotions	Emotional fluctuations such as anxiety and withdrawal are accompanied by behavioral changes in the process of learning	Worry/anxiety Boredom/disinterest Frustration/Retreat Sadness

2.2 Detection Model for E-Learning Behavior Problems

In this study, we proposed a detection model for e-learning behavior problems with text-mining approaches (see Fig. 1). The textual dataset contains comments, assignments, and assessments of students. The model mainly extracts basic information (e.g., time info) and text contents from the dataset.

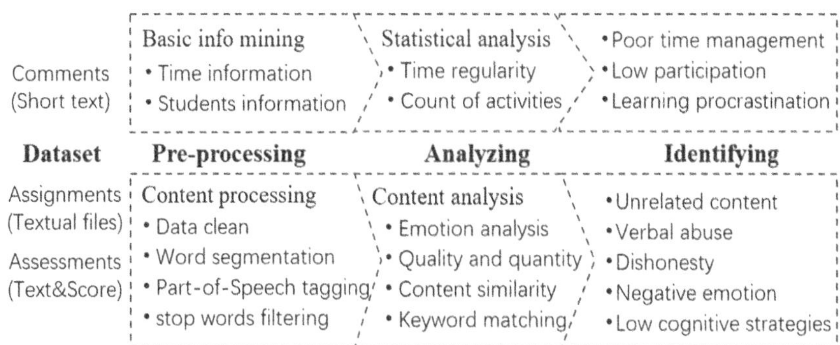

Fig. 1. Detection model for e-learning behavior problems

The measuring process includes statistical analysis and content analysis. The statistical analysis contains time regularity analysis and learning activities statistic. The content analysis mainly contains four respects, emotion analysis, quality and quantity analysis, content relevance analysis, and keyword matching. Firstly, the measurement of students' emotional tendency by extracting emotional words from text content. Secondly, quality and quantity analysis focus on the number and completeness of students' comments and assignments. Thirdly, content relevance analysis examines the content relevance of students' comments and assignments. The content similarity can identify the posting unrelated messages behaviors (e.g., small talk) and dishonesty behaviors. Finally, feature matching is to detect the specific behavior problems according to the keywords of behavior problems. When the number of behavior problems identified reached its threshold values, then, the student may have e-learning behavior problems.

3 Dataset and Measurement

3.1 Dataset

The data used in this study comes from an undergraduate level, information technology and curriculum integration course in the fall semester of 2018, at Central China Normal University. Participants come from 2 classes, a total of 78 students. We used web course as pre-class learning task. The students were required to learn the course resources on the cloud platform before class, and discuss the topics proposed by the teacher. Each student posted comment or thinking, reply to other, and all discussion messages are publicly visible to everyone. We recorded students' 764 learning comments and 958 posts in the forum. With the dataset, we identified the emotional tendency and other e-learning behavior problems.

3.2 Emotional Tendency Measurement

In this study, we adopted the lexicon-based emotion detection approach to measure the negative emotional tendency. In Chinese, the subtle differences of expression can reflect distinguishable emotional intensity. Therefore, to accurately measure the emotional state of students, the weight value is introduced here. According to textual expression features, emotion intensity calculation based on the approach of emotion word matching combined with syntactic analysis. Figure 2 shows the calculation procedures.

According to the measurement model, suppose a comment consists of clause sequences $\{S_1, S_2, \ldots, S_i, \ldots, S_n\}$, one clause Si includes words $\{W_{i1}, W_{i2}, \ldots, W_{in}\}$, after extraction, we can get emotional words set $\{E_1, E_2, \ldots, E_n\}$, adverbs set $\{D_1, D_2, \ldots, D_n\}$ and negative words set $\{N_1, N_2, \ldots, N_n\}$.

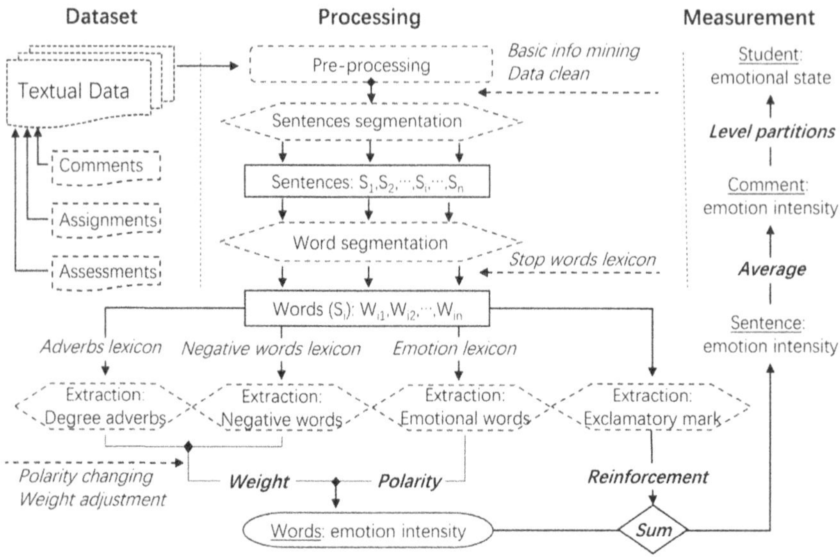

Fig. 2. Emotional tendency measurement model

If emotional word E_i with a modifier word D_i, negative word N_i and sentence S_i is an exclamatory sentence that contains emotional words, the emotional weight value of word E_i is $WV_{E_i} = V_{D_i} \times V_{E_i} \times V_{N_i}$.

WV_{E_i}—polarity of emotional word E_i: positive ($WV_{E_i} > 0$), neutral ($WV_{E_i} = 0$), negative ($WV_{E_i} < 0$).
V_{D_i}—the weight of adverb of word D_i (divide six levels: most, very, more, -ish, insufficient, inverse)
V_{E_i}—emotional polarity value of word E_i: the positive emotional word is +1, the negative emotional word is −1.
V—Negative word conversion coefficient according to the number of negative words is odd or even.

Then, if n is the number of emotional words in sentence S_i, the weighted emotional value of sentence Si is $SV_{S_i} = V_! + \sum_{j=1}^{n} WV_{E_j}$.

$V_!$—the weight of exclamatory mark: after a positive emotional sentence is +1, after a negative emotional sentence is −1.

With the two formula above, if a comment contains n sentences, then the emotional polarity value of the comment is $CV_{C_i} = \frac{\sum_{j=1}^{n} SV_{S_j}}{n}$.

The value of CV_{C_i} is in three cases, $CV_{C_i} > 0$ (positive emotional tendency), $CV_{C_i} = 0$ (neutral emotional tendency), or $CV_{C_i} < 0$ (negative emotional tendency). Finally, the teacher can understand a student emotional state with the mean or sum of these CV_{C_i}.

3.3 Other Behavior Problems Detection

We mainly adopted the keyword-matching method and content relevance analysis to detect the categories of behavior problems. In practical measurement, additional students' learning traces need to be analyzed, for example, a student may miss participation (e.g., records of posting or submitting coursework). We also illustrated several particular cases. Firstly, if keywords that match the low participation are detected, even once counts as a behavior missing problem. Secondly, for the behavior of posting irrelevant content, the premise work is to get the proportion of relevant words or clause in a comment, and the opposite is the irrelevant proportion. Finally, we take a comment as a unit to calculate the proportion of relevant words and bad language [14].

We additionally emphasized on mining and collecting time information and students' few identity information from text contents. In the subsequent analysis, we used the collected data to identify individual or group learning behavior characteristics. The statistical analysis includes two respects, time regularity analysis, and counts analysis of activities based on the students' participation records in e-learning. The primary goal of time regularity analysis is to seek the frequency of students' learning behaviors and detect the e-learning behavior problems, such as poor time management, lack of participation.

4 Results and Discussion

4.1 Emotional Tendency Analysis

The calculation results (see Table 2) show that the mean of positive emotion is 1.18, and the mean of negative emotion is −0.66, but the total mean of emotion is 0.68. And the variance of total emotion is 0.9697. The student had less emotional fluctuations.

Table 2. Calculation result of emotional tendency

	Positive	Neutral	Negative	Total
Count	647	145	166	958
Average	1.18	0	−0.66	0.68
SD	0.7524	0	0.4630	0.9847
Variance	0.2143	0	0.5661	0.9697
Coding	524	219	215	/
Right	504	116	150	/
P	0.7790	0.8000	0.9036	/
R	0.9618	0.5297	0.6977	/
F1	0.8606	0.6374	0.7874	/

Comprehensive analysis results, although the student had a little disaffection during the process of e-learning, without negative emotional problem. In order to test the validity and reliability of the analysis results. We invited three experienced teachers to manually code and calculate the emotional tendency levels of 958 comments. We mainly examined precision ratio (as P), recall ratio (as R), and F1-measure (as F1). Table 2 shows the results. We found that for the negative emotional tendency P is 0.9036. The model is useful to identify the student's negative emotion which lasted a certain amount of time, but insufficient to identify the short-term emotional state which depended on high classification accuracy. In further study, we hope to adopt different classification algorithms, such as, some researchers pointed out that Random Forest Classifiers (accuracy 94.1%) [15] offers more accurate predictions that others [16].

4.2 E-Learning Behavior Problems

4.2.1 Time Regularity Analysis

Time analysis of students' behavior traces can detect students' e-learning behavior problems about time management, such as the low study strategies, academic procrastination, lack of time lacking, and missing learning link. Figure 3 shows the time analysis result of students' learning and posting in one preview assignment online. There eight bigger dots close to '00:00:00' or '23:59:59' lines, indicated that these students may fail to arrange their study time properly. In the right area of the figure, the four dots illustrate that the four students didn't finish the preview task on time with the behavior problems of academic procrastination. Finally, we compared the students' posting records with students' enrollment lists, and found that four students did not participate in the learning activity.

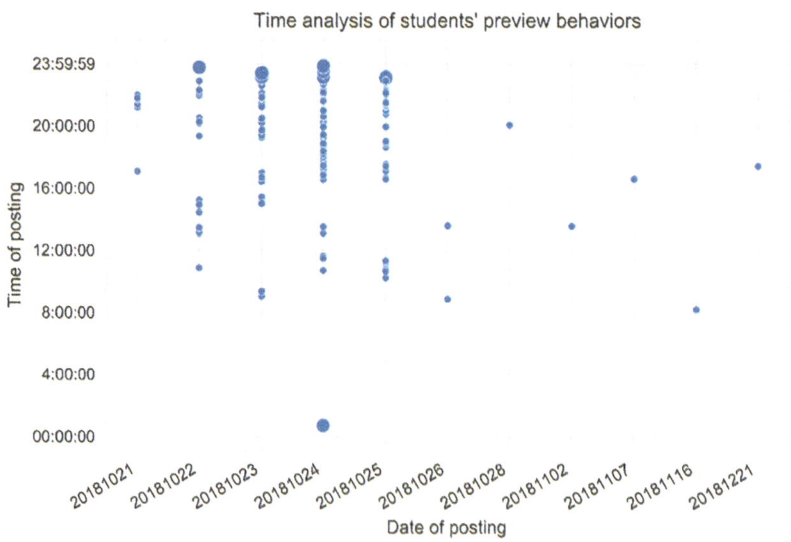

Fig. 3. Time analysis of students' preview behaviors (one assignment)

Figure 4 illustrates the result of time analysis on students' preview behaviors. The analysis data mainly includes 764 comments collected from eleven online preview assignments in the fall semester of 2018. In the right-hand area of Fig. 4, there are some incongruously sparse dots. This phenomenon reflects the e-learning behavior problems of some students catching up unfinished tasks at the end of the semester. There are a few students who still struggle after 11 PM and before 1 AM. Then, we examined these students' posting behaviors and found that 42 students contributed the 72 features. We excluded these students who did this occasionally, and detected six students with the ratio of more than 25 percent. Therefore, these students may have the e-learning behavior problems such as improper planning of study time and low learning strategies.

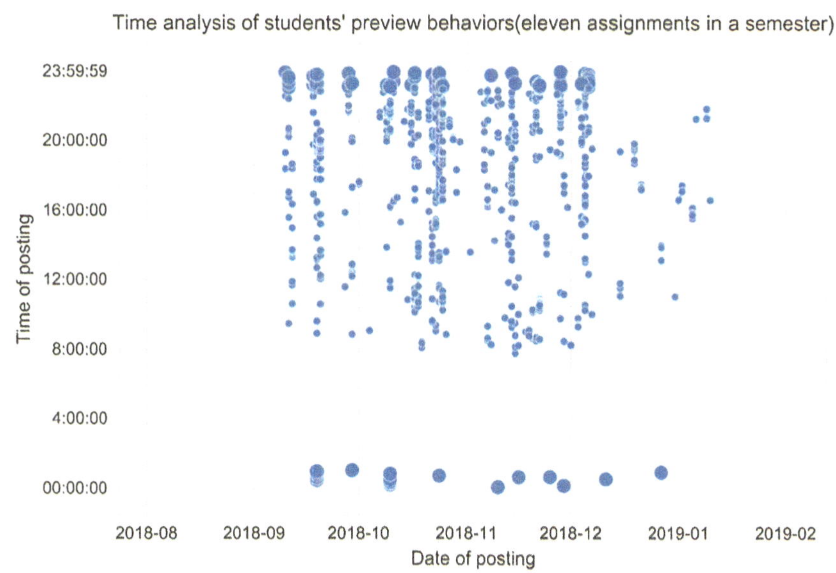

Fig. 4. Time analysis of students' preview behaviors (in autumn, 2018)

4.2.2 Content Similarity Analysis

We collected 70 students' comments from one discussed topic and examined the content similarities. The analysis results illustrated that there were three students with the problem of irrelevant posts. Figure 5 reflected the similarities matrix of students' comments. We found these comments with high similarity, according to the color depth of the similarity indicator blocks.

We examined the analysis results with manual coding. Table 3 indicates that the F1 of copying behavior detection is 0.7647. The P of irrelevant posts detection is only 0.4615, and the F1 is 0.5714. We examined the dataset found that the deviation came from fewer samples, and the higher threshold setting (threshold is <0.03, the more average similarity of the comments is 0.03–0.10). The length of comment may also cause deviation of similarity calculation.

Table 3. Content similarity analysis

	Irrelevant posts	Copying behavior
Test result	13	19
Manual coding	8	15
Right	6	13
P	0.4615	0.6842
R	0.7500	0.8667
F1	0.5714	0.7647

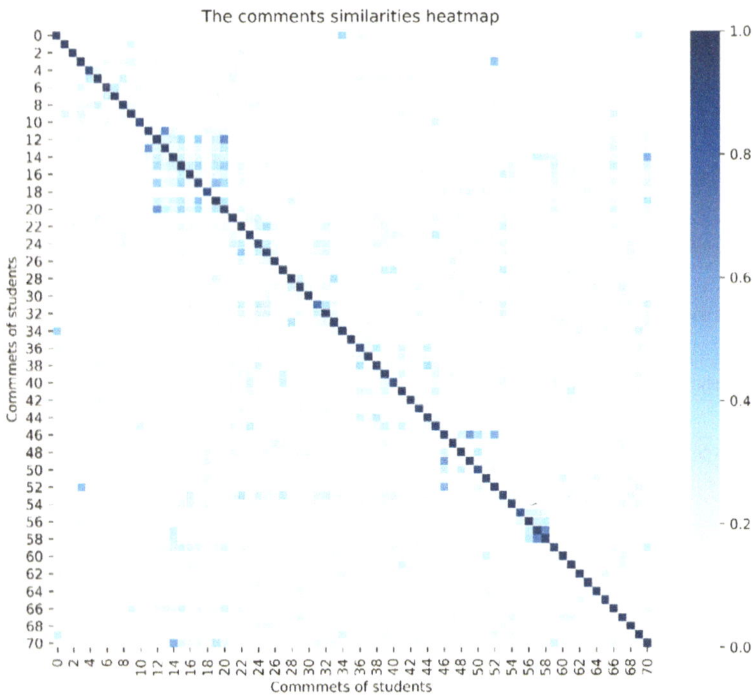

Fig. 5. Content similarities analysis of students' comments (one topic)

5 Conclusions

The present study examined previous relevant research, clarified the categories of e-learning behavior problems, and proposed a detection model. The experimental results indicated that the detection model can effectively identify poor time management, academic procrastination, low participation, and dishonest behavior (*F = 0.7647*). However, the accuracy of negative emotion are not ideal (*P = 0.9036, F = 0.7874*). The model is useful to identify the student's negative emotion which lasted a certain amount of time, but insufficient to identify the short-term emotional state which depended on high classification accuracy. These results give the researchers and teachers a view to improve learners' performance. In future work, we aim to collect and use multimodal learning traces data to improve the accuracy of e-learning behavior problems detection. For emotional tendency measurement, we hope to adopt more classification algorithms to improve the accuracy, such as Random Forest Classifiers and LSTM.

Acknowledgements. This work was supported by Chinese Ministry of Education & China Mobile under Grant [number: MCM20170502]; Department of Science and Technology of Hubei Province of China under Grant [number: 2017ACA105].

References

1. Jony, A.I., Rahman, M.S., Islam, Y.M.: ICT in higher education: Wiki-based reflection to promote deeper thinking levels. Int. J. Mod. Educ. Comput. Sci. (IJMECS) **9**(4), 43 (2017)
2. Panigrahi, R., Srivastava, P.R., Sharma, D.: Online learning: adoption, continuance, and learning outcome—A review of literature. Int. J. Inf. Manag. **43**, 1–14 (2018)
3. Zimmerman, B.J.: From cognitive modeling to self-regulation: a social cognitive career path. Educ. Psychol. **48**(3), 135–147 (2013)
4. Pardo, A., Jovanovic, J., Dawson, S., et al.: using learning analytics to scale the provision of personalised feedback. Br. J. Edu. Technol. **50**(1), 128–138 (2019)
5. Winter, J., Cotton, D., Gavin, J., et al.: Effective e-learning? Multi-tasking, distractions and boundary management by graduate students in an online environment. ALT-J. **18**(1), 71–83 (2010)
6. Hollis, R.B., Was, C.A.: Mind wandering, control failures, and social media distractions in online learning. Learn. Instr. **42**, 104–112 (2016)
7. Cheng, X., Li, Y., Zhao, Y.: Can we solve low participation, distraction, and inefficiency? A Case study of distributed collaborative learning in industries. In: 2015 48th Hawaii International Conference on System Sciences, pp. 354–363. IEEE (2015)
8. Rawat, B., Dwivedi, S.K.: An architecture for recommendation of courses in e-learning system. Int. J. Inf. Technol. Comput. Sci. (IJITCS) **9**(4), 39–47 (2017)
9. Gregori, P., Martínez, V., Moyano-Fernández, J.J.: Basic actions to reduce dropout rates in distance learning. Eval. Program Plan. **66**, 48–52 (2018)
10. Wladis, C., Conway, K.M., Hachey, A.C.: Assessing readiness for online education-research models for identifying students at risk. Online Learn. **20**(3), 97–109 (2016)
11. Hung, J.L., Wang, M.C., Wang, S., et al.: Identifying at-risk students for early interventions —a time-series clustering approach. IEEE Trans. Emerg. Top. Comput. **5**(1), 45–55 (2015)
12. Zhang, Y., Fei, Q., Quddus, M., et al.: An examination of the impact of early intervention on learning outcomes of at-risk students. Res. High. Educ. J. **26** (2014)
13. Sun, R.C.F., Shek, D.T.L.: Student classroom misbehavior: an exploratory study based on teachers' perceptions. Sci. World J. **2012**, 8 (2012)
14. Gupta, P., Kamra, A., Thakral, R., et al.: A proposed framework to analyze abusive tweets on the social networks. Int. J. Mod. Educ. Comput. Sci. **11**(1), 46 (2018)
15. Rifat, M.R.I., Al Imran, A., Badrudduza, A.S.M.: Educational performance analytics of undergraduate business students. Int. J. Mod. Educ. Comput. Sci. **11**(7), 44 (2019)
16. Shaheen, M., Awan, S.M., Hussain, N., et al.: Sentiment analysis on mobile phone reviews using supervised learning techniques. Int. J. Mod. Educ. Comput. Sci. **11**(7), 32 (2019)

Research on Site Selection of Low Carbon Distribution Centers Under "New Retail"

Yong Wang[1,2](✉), Pei-lin Zhang[1], Qian Lu[3], Daniel Tesfamariam Semere[2], and Xin Li[4]

[1] School of Transportation, Wuhan University of Technology, Wuhan, Hubei, China
Wang-yong@vip.163.com
[2] Department of Production Engineering, KTH Royal Institute of Technology, Stockholm, Sweden
[3] Department of Planning, Puren Hospital, Wuhan, Hubei, China
[4] School of Logistics, Wuhan Technology and Business University, Wuhan, Hubei, China

Abstract. The "Internet +" era, the new retail model will lead the future of business trends. Combining with the characteristics of the development of the new retail industry, from the perspective of carbon emissions, we will focus on the analysis of carbon emission costs and the impact of adding businesses to enterprises, governments or the society. According to the scholar's decision on the location of distribution center, it mainly involves the construction cost of distribution center and the increase of carbon emission cost model for comparative analysis. By comparing the total cost of social, commercial and government, and provide a basis for positioning decisions.

Keywords: New retail · Low-carbon · Distribution center location · Sustainability

1 Introduction

The term "new retail" was put forward by Ma Yun in 2016, and the retail industry will combine online and offline, using modern logistics services and other cloud computing, big data and other technologies to form a new retail model". But in the "Internet+" era, the new retail model will lead the future business trends. In addition, the mainstream electricity supplier giants redefined the new retail formats in light of the current status and trends of the e-commerce business operations. "New retail" core is to enhance the user experience, and promote online, offline and logistics and other cross-border integration. At the same time, the "new retail" basis and premise is to reconstruct the supply chain and upgrade the logistics program. Thus, logistics plays a crucial role in "new retail".

New retail is the application of new Internet technologies and new thinking. It sells products or services to the end consumer It is not only a simple integration of online and offline linkage and logistics, but also incorporates innovative technologies such as cloud computing and big data. Allowing consumers to buy the goods they need in the

Z. Hu et al. (Eds.): AIMEE 2019, AISC 1126, pp. 414–427, 2020.
https://doi.org/10.1007/978-3-030-39162-1_38

shortest time. Relying on the Internet, the new retail will upgrade the production, distribution and sales process of commodities through the use of advanced technologies such as big data and artificial intelligence, thereby reshaping the business structure and ecosystem. It is a new retail model that integrates online services, offline experience and modern logistics.

First, sales and logistics channels are sinking. If sinking to second- and third-tier cities, or even the vast rural market, business needs some new business models, namely the "new retail" + "new logistics" model.

Second, integration has changed the traditional sales and logistics channels, and convenience stores have become the focus of a new round of competition. The entry of e-commerce giants makes the original secret and complex logistics service system either directly compete with large logistics platforms or integrate into them.

Third, the all-sales channel logistics system integrates synergistic development. Through the deep integration of "online + offline + logistics", the comprehensive integration of goods and logistics channels is realized. This is not a simple O2O, but a number of online and offline retail companies to achieve win-win cooperation. This can form a sustainable Omni channel product and logistics distribution network. The most profitable is the consumer, improving the user experience. The final implementation of this model often requires the back-end support and cooperation of high-tech such as cloud computing and big data.

Fourth, cutting-edge information technology such as big data has become a vital capability of "new logistics". This is driven by technology and driven by big data capabilities. It is a comprehensive service logistics platform covering services such as warehouse allocation, cloud convenience, cross-border, intelligent express delivery cabinet and related industry solutions.

Fifth, the development of "new retail" and "new logistics" has brought development space for intelligent logistics equipment. New demands are asked for the logistics system in the new retail era.

This paper analyzes the model of carbon emissions based on the research of location selection of distribution centers, considering the social and environmental benefits of distribution centers in the context of new retail. This paper proposes a distribution center location model based on comprehensive consideration of carbon emissions and logistics costs from new retail perspective, and focusing on user experience and social efficiency. Finally, the paper solves the algorithm model and obtains the results of the two options for carbon emission cost consideration and non-consideration. It compares the cost differences between the two location models to draw conclusions, and provides reference for relevant enterprises and governments to make logistics distribution center location decisions.

2 Literature Review

2.1 New Retail

"New retail" and "new logistics" is a brand new retail business services in the form of two inseparable parts, there is no "new logistics", there will be no "new retail", the two are common development. At present, in the "new retail" and "new logistics" in the exploration and development, has shown some trends [1].

About the new retail, Zhao Shuhai and Xu Xiaohong (2017) proposed that new retail is the use of Internet new technology, new thinking, the product or service sold to the final consumer all activities. It is not only a simple combination of online and offline linkage and logistics, but also into the cloud computing, big data and other innovative technologies, so that consumers can always buy the goods they need in the shortest possible time. Du Ruiyun and Jiang Kan (2017) pointed out that relying on the Internet, through the use of large data, artificial intelligence and other advanced technical means, the production of goods, circulation and sales process upgrade. And then reshape the format structure and ecosystem, and online services, offline experience and modern logistics depth integration of new retail model [2].

2.2 Low-Carbon

Zhao et al. Established a set of new evaluation index system including five first-level indicators and 28 second-level indicators to evaluate the low-carbon development level of domestic logistics enterprises [3]. Liu et al. Emphasized the importance of optimization goals, constraints and logistics design for bioenergy system performance and regional bioenergy planning [4].

2.3 Site Selection

Based on the traditional distribution center location, considering the low carbon, service optimization and cost saving, a multi-distribution center location model is constructed. Jing and Zhongqin the objective of minimizing the economic cost is to optimize the location of the distribution center, adopt the improved genetic algorithm coding and verify the feasibility and effectiveness through the example verification model [5]. Zhao et al. analyzed the necessity of internalizing the industrial CO2 emission cost from four aspects and established a multi-distribution center location model to reduce the carbon footprint and provide optimized strategic support for decision-makers and logistics operators [6]. Li et al. proposed an integrated method of selecting the location of the logistics center and analyzed and identified the relevant standards for the selection of the logistics center [7].

The model developed by Badri et al. Complements or complements the traditional methods of industrial location analysis [8]. Hoffman and Schniederjans propose a two-phase model that combines the concept of strategic management with the targeted management science and technology and micro-computer technologies to provide managers with a more efficient and efficient way to assess global facilities and make choice decisions [9]. Bozarth et al. simulated the complexity of the supply chain and used plant-level data from 209 plants in seven countries to link operational strategy with organizational design [10]. Bartelsman et al. Used an empirical model to assess the cross-country differences in intra-industry covariances between size and productivity [11]. The decision to establish a manufacturing plant in a particular country must take into account both qualitative and quantitative factors [12]. Vidal and Goetschalck tested the impact of transport, demand, supplier reliability, and exchange rate uncertainty through mixed integer programming [13]. Zhao et al. Analyzed the necessity of internalizing the industrial CO2 emission cost from four aspects and established a multi-distribution center location model to reduce the carbon footprint and provide optimized strategic support for decision-makers and logistics operators. Numerical examples are given to illustrate the feasibility and effectiveness of the model [6]. Dogan proposed an integrated approach to integrating Bayesian networks and total cost of ownership (TCO) to address the complexities involved in selecting international facilities for manufacturing plants [14]. Zhu optimizes the standard particle swarm optimization algorithm by domain mean, and attempts to overcome the defect that the standard particle swarm optimization algorithm is easy to fall into the local search in the local optimal solution [15].

Based on the introduction of the concept of logistics and logistics center, Liu et al. analyzed the status quo and problems of the logistics center in Jilin Province, and proposed the steps, principles and factors for the selection of the logistics center [16]. Zhu takes the minimum total cost of logistics center as the target value, analyzes the key factors affecting the logistics and distribution system, and establishes the logistics distribution center location model [15]. The ideal location of the facility is a very broad area of research, especially because of its importance to multinationals operating in the global economy. The model developed by Badri et al. Complements or supplements the traditional methods of industrial location analysis [8]. Hoffman and Schniederjans propose a two-phase model that combines the concept of strategic management with the targeted management science and micro-computer technologies to provide managers with a more efficient and efficient way to assess global facilities and make choice decisions [9]. Bartelsman et al. Used an empirical model to assess the cross-country differences in intra-industry covariances between size and productivity [11].

3 Methodology and Proposed Criteria

3.1 Distribution Center Location Concept

Distribution center location refers to the supply and demand, supplier's demand, geographical location, road conditions, and transportation conditions as the basis, with good economic and social benefits as the goal, Using quantitative or qualitative analysis, obtained the best location of distribution center.

3.2 Location Objective

Enhance logistics services: In the new retail background, each logistics enterprise pays more attention to the quality of logistics service in order to improve its competitiveness. One way to improve quality of service is to provide the best protection for goods the customer needs, and to reduce the damage and deterioration of goods. If the enterprise can according to customer demand, the goods in a proper way to the customer at the right place and the right time for a great help to improve the service, will occupy the market competitiveness of enterprises [17].

Carbon emission reduction: In order to serve the terminal customers in the new retail background, the location of the distribution center occupies a very important position in the entire logistics network. If we can reduce carbon emissions as the goal, the carbon emissions should be considered in the decision-making of the location, not only to save the total cost of society, but also to protect the environment and conserve resources.

Minimum integrated cost: In new retail background, when the location of the distribution center is taken into consideration, it is one of the goals of the enterprise to consider its total cost. These include vehicle purchase costs, labor costs, construction costs of distribution centers, and costs after carbon emissions.

3.3 Site Selection Principle

Moderation principle: The principle of moderation is to appropriately select the quantity and scale of the distribution centers according to the actual situation. In new retail background, the location of the distribution center should be analyzed in line with the actual situation on line and offline, and cannot be planned blindly.

Economic principle: The choice of enterprise behavior is to maximize the pursuit of economic interests as the starting point, logistics enterprises to establish a distribution center is hoping that it can bring more benefits for enterprises.

Coordination principle: From the microscopic point of view, the distribution center is an important node in the logistics network. In the context of the new retail industry, from the source of goods said that the upper reaches of e-commerce business, downstream and customer contact. Therefore, the distribution center should be more fully consider the location of the distribution center to make it consistent [18].

3.4 Selection of Location Method for Distribution Center

The general method of distribution center location is mainly as follows:

3.4.1 Analytic Method

The following paper analyzes the advantages and disadvantages of various location methods, such as Table 1 [19–21].

3.5 Selection of Low Carbon Distribution Center Location Options Under the New Retail Perspective

This paper combines the characteristics of the new retail development, starting from the perspective of carbon emissions, focusing on the analysis of the cost of carbon emissions, the impact of the accession to the enterprise, as well as the impact on the government or society. Based on the related scholars' decision about location of the distribution center, this paper adds the model of carbon emission cost, and makes a comparison and gets the result. When comparing the cost, it mainly includes the construction cost of the distribution center, the transportation cost of the vehicle and the carbon emission cost. In the end result, it is mainly to compare the total cost of society, business and government to provide the basis of location decision.

Table 1. Analysis of advantages and disadvantages on various distribution center location

Method	Advantage	Shortcoming
Analytic method	The location is more flexible and the model is relatively simple	The factors considered are too small to meet the actual situation, and the feasibility is not high
Simulation method	The algorithm is simple and easy to master	Designers and computer hardware requirements are high, and the final decision results are mostly affected by the initial location scheme
Heuristic method	The feasible solution which is close to the optimal solution can be obtained in a short time	The optimal solution can not be obtained
Expert selection method	The factors that affect the site selection are taken into consideration	The setting of weights is subjective and influenced by the expert's own factors

4 Numerical Experiments

4.1 Location Model Design of Low Carbon Logistics Centers Under the New Retail Perspective

4.1.1 Design Idea

This chapter first introduces the model, then confirms the model by case analysis, calculates total cost under different circumstances, draws corresponding conclusions.

4.1.2 Problem Description and Mathematical Model

The location model of logistics distribution center is described as follows: in the new retail perspective, the supply model is divided into three levels, e-commerce retailers (supply), distribution center, distribution terminal (demand), referred to as "one center, two basic points". The entire supply process is from the origin of the production, processing, transportation to the distribution center, the distribution center according to retailers or terminal requirements for goods sorting, processing, packaging, delivery to the consumer, as shown in Fig. 1.

In order to facilitate calculation, this paper specifies that the number of supply points and the number of demand points must be fixed, and the demand quantity is known. The location of supply point and demand point is fixed, so that the distance between the supply point and the distribution center can be calculated conveniently. At the end of the model, the best location of the distribution center is selected by comparing the total cost of each alternative distribution center.

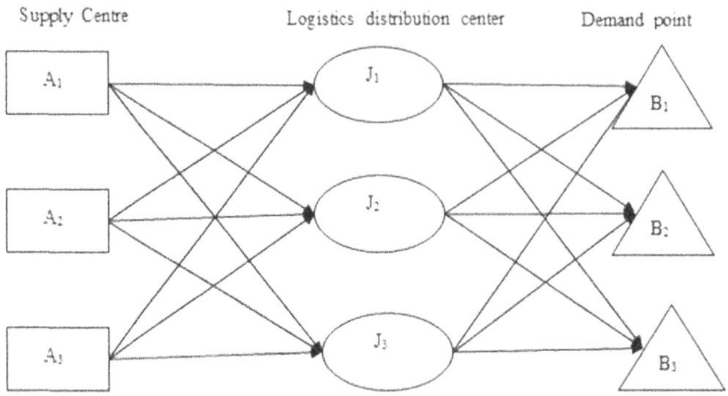

Fig. 1. Terminal supply structure diagram under new retail perspective

When the model is designed, the following assumptions are made: The distance from the supply store to the distribution center, between the distribution center and each demand point is fixed and known; Demand number for each demand is known; The selected model is a single cycle model, and the influence of inventory cost is neglected; The amount of goods transported shall not exceed the maximum loading capacity of the vehicle; The means of transportation used in transit are of the same type and the same road condition and the same speed; The transportation cost of the vehicle is known;⑦ Consider only carbon emissions from the transport process [22].

Parameter definition: A stands for suppliers, and N refers to the quantity of A, that is, the quantity of suppliers; J stands for distribution centers, and M refers to the quantity of J, the number of alternative distribution centers; B stands for retailers, and l refers to the quantity of B, the number of retailers.

Known variables: SA represents the supplier's A supply; NB represents the retailer's demand for B; PJ represents the maximum disposal capacity of the distribution center J; DAJ indicates the interval from the supplier A to the distribution center J; DBJ represents the distance from the distribution center J to the point of need B; C represents the transportation cost of the vehicle; FJ said the fixed cost f more needed to build distribution center J; QJ represents the fixed carbon emission of the distribution center J (kg = CO2);

α represents the carbon emission factor of the transport vehicle (kg = CO2/, ton, KM); β represents the price of carbon tax (Yuan/kg = CO2).

Decision variable: YJ {0, 1}, 0 refers to the construction of distribution centers, 1 refers to the construction of non-delivery centers; XAJ refers to the volume of transportation from the supplier A to the distribution center J; ZBJ refers to the volume of traffic from the distribution center J to the retailer B.

- Mathematical model

  Objective function:

$$\min F = F_1 + F_2 + F_3 \tag{1}$$

Among them, the fixed cost function built by logistics center:

$$F_1 = \sum_{J=1}^{m} f_J Y_J \tag{2}$$

Logistics transportation cost function:

$$F_2 = \sum_{A=1}^{n}\sum_{J=1}^{m} CX_{AJ}D_{AJ} + \sum_{J=1}^{m}\sum_{B=1}^{L} CZ_{BJ}D_{BJ} \tag{3}$$

Carbon emission cost function:

$$F_3 = \beta\left[\sum_{J=1}^{m} Q_J Y_J + \sum_{A=1}^{n}\sum_{J=1}^{m} \alpha X_{AJ}D_{AJ} + \sum_{J=1}^{m}\sum_{B=1}^{l} \alpha Z_{BJ}D_{BJ}\right] \tag{4}$$

Constraint condition:

$$\sum_{J=1}^{m} X_{AJ} \leq S_A \tag{5}$$

$$\sum_{A=1}^{n} X_{AJ} \leq P_J Y_J \tag{6}$$

$$\sum_{A=1}^{n} X_{AJ} = \sum_{B=1}^{L} Z_{BJ} \tag{7}$$

$$\sum_{J=1}^{m} Z_{BJ} \geq N_B \tag{8}$$

$$Y_J = \{0, 1\}; X_{AJ} \geq 0; Z_{BJ} \geq 0 \tag{9}$$

Condition description:
(1) The total quantity of cold chain food delivered from the supplier A to the distribution center J is less than the supplier's A production capacity; (2) The total amount of goods shipped to the distribution center J is less than or equal to the cargo handling capacity of the distribution center J; (3) The amount of goods shipped in the distribution center J is equal to the quantity of the goods shipped; (4) Indicates that the quantity of goods delivered by the retailer B meets the demand of the customer; (5) Represents the selection of decision variables.

4.1.3 Example Analysis

In this paper, the annual distribution of cold chain in a city is taken as an example to optimize the location of the logistics distribution center of a cold chain product distribution. This example assumes that the cold chain food suppliers only 1 recorded as A for the convenience of calculation; the retailer has 7, denoted as B1, B2, B3, B4, B5, B6, B7; Optional logistics distribution center has 5 marked J1, J2, J3, J4, J5 (Tables 2 and 3).

The carbon tax standard in this case: carbon emission per ton was 0.2 (Yuan/Kg = CO2). The carbon tax is in line with the range of carbon tax values that are being explored by the development and Reform Commission of China, ranging from 0.03 to 0.3. According to the data, the carbon emission coefficient of refrigerated vehicles is about $\alpha = 1000$ (Kg = CO2/1000 Ton = Km).

Suppose a city freight transport costs C = 0.5 Yuan/Km/10000 Tons. The fixed costs (10000 Yuan) for the construction of alternative distribution centers are $f1 = 7000$, $f2 = 2000$, $f3 = 6000$, $f4 = 3000$, $f5 = 4000$.

The fixed carbon emission of the logistics distribution center is q1 = 1100 kg·CO2, q2 = 4100 kg·CO2, q3 = 9600 kg·CO2, q4 = 5500 kg·CO2, q5 = 6900 kg·CO2 [23].

Table 2. Distance between suppliers to distribution centers (Km) and throughput (10000 Tons)

Node	J_1	J_2	J_3	J_4	J_5	Output
I	18.3	102.1	55.7	64.2	66.7	2000
Through put	550	50	450	150	250	

Table 3. Transportation distance (Km) and logistics center to demand point (10000 Tons)

Node	J_1	J_2	J_3	J_4	J_5	Requirement
K_1	20.0	97.1	52.1	63.4	60.1	382.26
K_2	77.1	16.2	35.4	41.6	27.3	229.27
K_3	39.2	60.3	35.4	15.1	27.3	151.14
K_4	51.7	61.4	20.6	59.6	38.9	71.07
K_5	56.2	35.9	11.9	40.0	13.4	252.98
K_6	30.0	79.3	36.5	61.2	46.5	192.13
K_7	44.8	48.1	16.0	29.8	11.5	84.29

Objective function:

$$\min F = F_1 + F_2 + F_3 \tag{10}$$

$$F_1 = \sum_{J=1}^{5} f_J Y_J \tag{11}$$

$$F_2 = \sum_{A=1}^{1} \sum_{J=1}^{5} 0.5 X_{AJ} D_{AJ} + \sum_{J=1}^{5} \sum_{B=1}^{7} 0.5 Z_{BJ} D_{BJ} \tag{12}$$

$$F_3 = 0.2 \left[\sum_{J=1}^{5} Q_J Y_J + \sum_{A=1}^{1} \sum_{J=1}^{5} 1000 X_{AJ} D_{AJ} + \sum_{J=1}^{5} \sum_{B=1}^{7} 1000 Z_{BJ} D_{BJ} \right] \tag{13}$$

Constraint condition:

$$\sum_{J=1}^{5} X_{AJ} \leq S_A \tag{14}$$

$$\sum_{A=1}^{1} X_{AJ} \leq P_J Y_J \tag{15}$$

$$\sum_{J=1}^{5} X_{AJ} = \sum_{B=1}^{7} Z_{BJ} \tag{16}$$

$$\sum_{J=1}^{5} Z_{BJ} \geq N_B \tag{17}$$

$$Y_J = \{0, 1\}; X_{AJ} \geq 0; Z_{BJ} \geq 0 \tag{18}$$

4.1.4 Solution Result

The results of the two models are obtained by the expert selection method, as shown in Tables 4 and 5.

Table 4. Selection of logistics centers under different model conditions (10000 Tons)

Model condition	Logistics distribution center	X^{AJ} the volume of transportation from supply point to the logistics center	Demand point	Y_{jk} the delivery quantity of logistics distribution center to demand point
Consider carbon emissions	J_1	500	K_1	382.26
			K_3	151.14
			K_6	192.13
	J_3	350	K_2	229.27
			K_4	71.07
			K_5	252.98
			K_7	84.29
	J_4	71.03	K_3	151.14
	J_5	104.61	K_2	229.27
Carbon emissions are not considered	J_1	500	K_1	382.26
			K_6	192.13
			K_7	84.29
	J_3	350	K_2	229.27
			K_4	71.07
			K_5	252.98
			K_7	84.29
	J_5	104.61	K_2	229.27
			K_3	151.14

Table 5. Comparison of costs under different model conditions (10000 Tons)

Model condition	Fixed cost	Logistics cost	Sum of fixed costs & logistics costs	Carbon emission costs	Total cost of enterprise	Total social cost
Businesses consider carbon emissions	20000	40182.61	60182.61	16404017.6	16464200.21	16464200.2
Businesses do not consider carbon emissions	16000	35360.96	51360.96	16732097.9	51360.96	16783458.9

4.2 Location Strategy for Low Carbon Distribution Centers Following the New Retail Perspective

4.2.1 Play the Regulatory Role of the Government

In order to be more active in guiding enterprises to participate in the form of low-carbon economy, we need to play the regulatory role of the government. In the selection of carbon emissions when considering the situation, and actively improve operating conditions, reduce carbon emissions enterprises to grant financial subsidies. The subsidy amount should be greater than the cost of the enterprise after considering the carbon footprint, so as to attract more enterprises to participate. For businesses that ignore carbon emissions, the government can consider punishing them, with penalties greater than the amount they pay for carbon emissions. In this way, we can not only actively adapt to the main trend of low-carbon economy, but also carry out environmental monitoring and reduce the level of environmental pollution.

4.2.2 Developing Advanced Modes of Transportation

First of all, in the field of vehicles, the transportation of vehicles in the distribution centers is mainly based on automobile transportation, and the development of advanced means of transportation and modes, mainly the use of new energy vehicles, such as throwing, hanging, transportation, etc. [24].

Secondly, in the field of information circulation, we need to establish a stable supply chain system, and strengthen the correlation degree of each logistics subject under new retail. As far as possible to ensure that the main body at all levels with the highest efficiency, the fastest speed of the delivery of goods information, as far as possible to shorten the time in transit, to ensure quality.

4.2.3 Standardizing Logistics Link

From the supplier perspective, strengthen information exchange links, improve the understanding of the extent of the link, to achieve seamless links as much as possible, and use of facilities and equipment have a specific regulations, ensure the goods with a minimum of time, the best quality, the lowest damage at the hands of consumers.

From the logistics link, the waste treatment methods need to be appropriate, not blindly incineration, landfill, etc. So that not only pollute the environment, but also cause waste of energy. First of all, the waste can be classified and dealt with in different links, and the wastes in different links and stages should be closed and recovered and treated accordingly. In addition, actively develop logistics technology, optimize the processing of resources.

4.2.4 Set up Correct Low-Carbon Concept

Under the new retail background, distribution, as an important part of the logistics system, has a direct impact on the survival and development of enterprises. The development of low carbon logistics can not only save resources and reduce carbon emissions, but also bring enormous social and environmental benefits. But also can effectively reduce the cost of enterprise logistics, enhance the core competitiveness of enterprises, so that enterprises to obtain greater economic benefits.

4.2.5 Implementing Joint Distribution

Common delivery means the best use of several modes of transportation in order to reach the lowest cost of transportation. As for the distribution center as a transit hub, its transportation volume is huge, and the transportation form is many, the transportation vehicles are many. If the logistics distribution center can achieve reasonable joint distribution, and realize the rational use of vehicles, it is of great significance for reducing carbon emissions and achieving low carbon economy [26].

5 Management Insights and Main Conclusion

Under the new retail perspective, in determining the location of logistics distribution center, from the fixed cost and logistics cost and perspective, compared to the model considering the total cost of carbon emissions, considering only the logistics distribution center location scheme cost total cost about 5% reduction from carbon; from a cost perspective, do not consider the distribution center site for carbon emissions cost is considered 1.02 times the cost of carbon emissions; the total cost from the social point of view, do not consider the distribution center site for carbon emissions cost is considering 1.02 times the cost of carbon emissions. And if enterprises do not consider carbon emissions, the cost of carbon emissions is generally borne by the government.

In this paper, we do not take into account the change of customer demand in the selection of the location. The model is assumed to be the demand quantity directly. This is difficult to meet the requirements in the real life. Secondly, when the location of the distribution center is not taken into account, the path problem is not considered. However, in reality, due to different paths and different road conditions, the carbon emissions from vehicles are also different. Secondly, the expert selection method used in the decision making of the distribution center is very subjective. In the practical location decision, it is necessary to collect the location data and select the location through specific data. In addition, the research object of this paper is cold chain logistics distribution center. The research is only about the carbon emission of a refrigerator car. In the future research, we can try to study the carbon emissions of various types of vehicles, and consider the optimization of vehicle path, so that the model is clearer, closer to the actual life, more practical.

Acknowledgment. This project is supported by Key Projects of CAST (China Association of Science and Technology) Project (2018CASTQNJL33); Fundamental Research Funds for the Central Universities (2019-JL-008); MOE Project of Humanities and Social Sciences (14YJCZH154); WTBU Academic Team (XSTD2015004).

References

1. Lin, Z., Xi, C.: Logistics development and change in the new retail Era. Logist. Technol. Appl. (6), 70–77 (2017). (in Chinese)
2. Shen, X.: Research on development route of logistics industry in new retail era. Mod. Commer. Ind. (6), 21–22 (2017). (in Chinese)
3. Zhao, P., Liu, B., Xu, L., Wan, D.: Location optimization of multidistribution centers based on low-carbon constraints. Discret. Dyn. Nat. Soc. **2013**, 1–6 (2013). Article ID 427691
4. Liu, B., Wu, Q., Wang, F.: Regional optimization of new straw power plants with greenhouse gas emissions reduction goals: a comparison of different logistics modes. J. Clean. Prod. **161**, 871–880 (2017)
5. Jing, W., Zhongqin, M.: Selection of multi-distribution center location based on low carbon. Revista de la Facultad de Ingeniería U.C.V. **31**(7), 11–22 (2016)
6. Zhao, P., Yu, H., Wang, Z., Xu, L.: Fuzzy evaluation of low carbon development levels for logistic enterprises in China. J. Ind. Eng. Manag. **8**(5), 1698–1710 (2015)
7. Li, Y., Liu, X., Chen, Y.: Selection of logistics center location using axiomatic fuzzy set and TOPSIS methodology in logistics management. Expert Syst. Appl. **38**(6), 7901–7908 (2011)
8. Badri, M.A., Davis, D.L., Davis, D.: Decision support models for the location of firms in industrial sites. Int. J. Oper. Prod. Manag. **15**(1), 50–62 (1995)
9. Hoffman, J.J., Schniederjans, M.: A two-stage model for structuring global facility site selection decisions: the case of brewing industry. Facilities **14**(4), 79–96 (1996)
10. Bozarth, C.C., Warsing, D.P., Flynn, B.B., Flynn, E.J.: The impact of supply chain complexity on manufacturing plant performance. J. Oper. Manag. **27**(1), 78–93 (2009)
11. Bartelsman, E., Haltiwanger, J., Scarpetta, S.: Cross-Country differences in productivity: the role of allocation and selection. Am. Econ. Rev. **103**(1), 305–334 (2013)
12. MacCormack, A., Newman III, L., Rosenfield, D.: The new dynamics of global manufacturing site selection. Sloan Manag. Rev. **7**, 69–79 (1994)
13. Vidal, C.J., Goetschalckx, M.: Modeling the affect of uncertainties on global logistics systems. J. Bus. Logist. **21**(1), 95–120 (2000)
14. Dogan, I.: Analysis of facility location model using Bayesian networks. Expert. Syst. Appl.: Int. J. **39**, 1092–1104 (2012)
15. Zhu, H.: Logistics distribution center site selection based on domain mean value optimization PSO algorithm. Rev. Téc. Ing. Univ. Zulia **39**(5), 155–161 (2016)
16. Liu, X., Guo, X., Zhao, X.: Study on logistics center site selection of Jilin Province. J. Softw. **7**(8), 1799–1806 (2012)
17. Faisal, H., Usman, S., Zahid, S.M.: In what ways smart cities will get assistance from internet of things (IOT). Int. J. Educ. Manag. Eng. (IJEME) **8**(2), 41–47 (2018)
18. Wang, Y., Zhang, P., Lu, Q., Semere, D.T., Du, W.: Supplier measurement of fresh supply chain in sustainable environment. EKOLOJI **28**(107), 1995–2004 (2019)
19. Aggarwal, A., Verma, R., Singh, A.: An efficient approach for resource allocations using hybrid scheduling and optimization in distributed system. Int. J. Educ. Manag. Eng. (IJEME) **8**(3), 33–42 (2018)

20. Tao, Y.: Logistics network planning of multiple transportation modes under low carbon economy, pp. 16–20. Shanghai Jiao Tong University, Shanghai (2011). (in Chinese)
21. Wang, Y., Deng, X.: Empirical study on performance evaluation of agricultural product supply chain based on factor analysis. China Bus. Mark. **5**(3), 10–16 (2015). (in Chinese)
22. Khan, S.: Cloud computing: issues and risks of embracing the cloud in a business environment. Int. J. Educ. Manag. Eng. (IJEME) **9**(4), 44–56 (2019)
23. Kajol, R., Akshay, K.K., Keerthan Kumar, T.G.: Fresh automated agricultural field analysis and monitoring system using IOT. Int. J. Inf. Eng. Electron. Bus. (IJIEEB) **10**(2), 17–24 (2018)
24. Datta, L.: Efficient Round Robin scheduling algorithm with dynamic time slice. Int. J. Educ. Manag. Eng. (IJEME) **5**(2), 10–19 (2015)

Neuro-Educational System for Training Standard and Selective Neural Network Technology

M. Mazurov[1(✉)], E. Egisapetov[2], and S. Markovsky[2]

[1] Russian Economic University, Moscow, Russia
mazurov37@mail.ru
[2] CJSC STC "Module", Moscow, Russia

Abstract. The theoretical and mathematical substantiation of standard and selective neural network technologies is given. Layouts have been developed for visual modeling of processes in neural networks of standard McCulloch-Pitts-based neurons and selective ones based on selective neurons. The neuro-educational system allows for the effective training of neurotechnologies of senior schoolchildren, students, specialists of related professions.

Keywords: McCulloch-Pitts neuron · Selective neuron · Rosenblute single-layer perceptron · Selective perceptron · Monte-Carlo selective training method

1 Introduction

Theoretical and mathematical substantiation of standard and selective neural network technologies is given. The neuro-educational system for teaching neural network technologies and their applications on the basis of standard and selective neural networks is intended for pupils, students, graduate students, specialists in related professions. The opportunity to solve more complex problems using the developed software, available to a wide range of schoolchildren, students and specialists in related professions, has been realized. This is the calculation of the weights of single-layer and multi-layer neural networks with McCallock-Pitts neurons using the selective Monte Carlo method. Acquaintance with the methods of "deep learning" in neural networks.

For a neural network to complete a task, it must be trained. Studying the most common neural networks using McCallock-Pitts neurons comes down to calculating their weights [1–7]. A flowchart of the neural network learning process is shown in Fig. 1.

Z. Hu et al. (Eds.): AIMEE 2019, AISC 1126, pp. 428–441, 2020.
https://doi.org/10.1007/978-3-030-39162-1_39

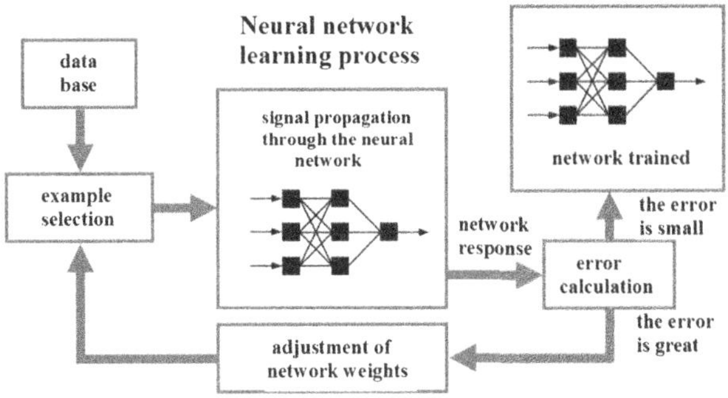

Fig. 1. The flowchart of the neural network learning process.

The structure of the neural network learning process is illustrated in more detail in Fig. 2.

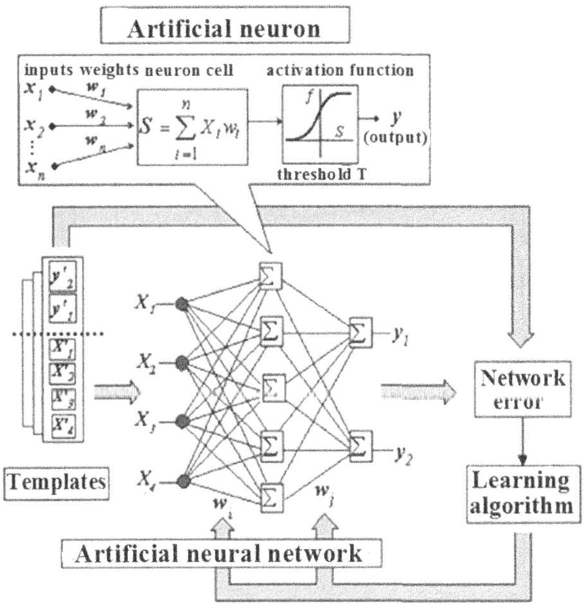

Fig. 2. The structure of the learning process on the example of a two-layer perceptron.

Various iterative methods are used to calculate weights. As you know, these methods have a number of irreparable fundamental shortcomings, we list some of them: 1. The technical complexity of performing iterative procedures for finding training weights associated with a large number of calculations [7]; 2. The need to recalculate weights when adding new input functions; 3. Instability, the possibility of ambiguous solutions to recognition and control problems for some sets of weights [7]; 4. The disadvantage is the ambiguity of sets of weights that satisfy the inequalities from which they are determined [7].

For effective training in neuro-educational technologies, a number of software tools have been developed, as well as the material implementation of the neuro-educational system in the form of material models of neurons.

2 Material Experimental Implementation of a Neural Educational System Based on McCalloch-Pitts Neurons

For training purposes, a single-layer perceptron neuro-educational system based on three McCallough-Pitts neurons was developed. The perceptron was supposed to recognize the letters L, T, X with weights generated during the training. The proposed perceptron was implemented almost electronically. A small number of recognizable letters is caused by the need to implement a material model of the perceptron. Recognition of a larger number of objects is implemented programmatically. The prototype of the McCallock-Pitts neuron is the electrical circuit shown in Fig. 3.

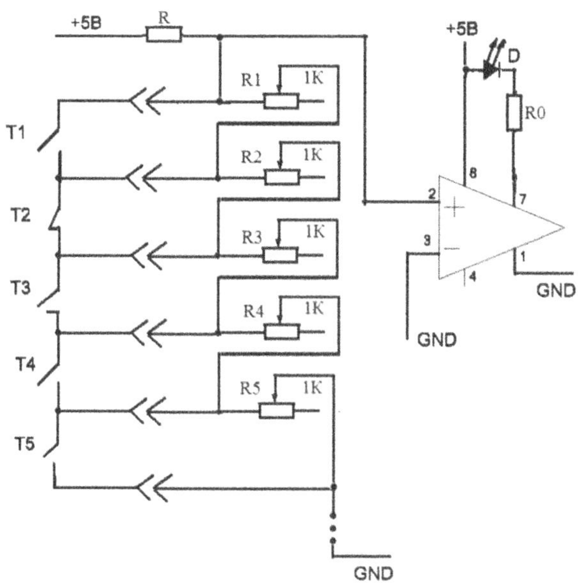

Fig. 3. The prototype of the proposed McCullock-Pitts neuron, made on the basis of the electrical circuit and the use of the comparator as a nonlinear threshold element.

We present the theory of an electric circuit shown in Fig. 3. The voltage removed from a series of series-connected resistances is equal according to Kirchhoff's law.

$$u = \frac{R_{\text{экв}}}{R + R_{\text{экв}}} \bullet E = \frac{R_{\text{экв}}}{R}, \; R_{\text{экв}} = \sum_{i=1}^{9} R_i \; .$$

When choosing the resistance value R, which is much higher than the sum of the variable resistances, i.e.

$$R \gg \sum_{i=1}^{9} R_i$$

In this case, according to the previous formula

$$u = \frac{R_{\text{экв}}}{R} E = \frac{R_1}{R} x_1 + \frac{R_2}{R} x_2 + \ldots + \frac{R_9}{R} x_9 = \sum_{i=1}^{9} \alpha_i x_i \; ,$$

where $\alpha_i \; (i = 1, \ldots, 9)$ are the weights; $x_i \; (i = 1, \ldots, 9)$ input signals. Thus, we obtain that the sum linearly depends on the magnitude of the variable resistances and, therefore, linearly depends on the magnitude of the weight coefficients.

To increase the stability of recognition of input objects, reduce the number of incorrect recognitions, and also to weaken the effect of multi-valued recognition, a circuit is used that contains comparators - voltage limiters above and below. A schematic diagram of the proposed McCallock-Pitts neuron using comparators to limit the maximum and minimum threshold voltages is shown in Fig. 4.

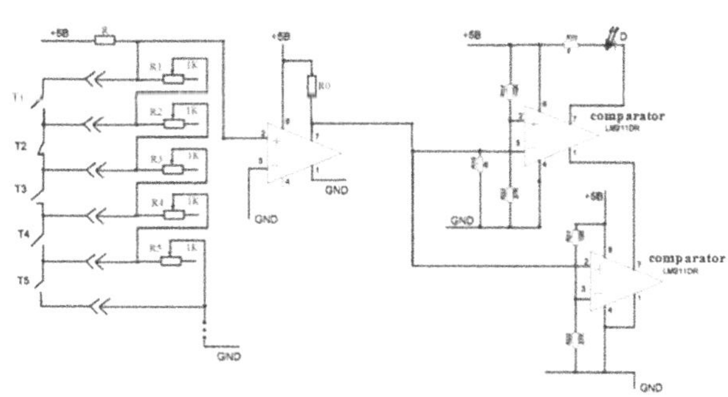

Fig. 4. Schematic diagram of the developed McCalloch-Pitts neuron using comparators to limit the maximum and minimum threshold voltages shown on the right.

The values of the binary sequence of input signals equal to 1 or 0 are set by closing the variable resistances: if the variable resistance is closed - it is shunted, then the value of the binary sequence is 0, if the resistance is not limited, then the value of the binary sequence is 1.

The magnitude of the weight coefficients is regulated by slide variable resistances, which are clearly visible in Fig. 5 left. Exceeding the threshold level is fixed by an indicator located behind the variable resistances. The Rosenblatt Perceptron based on McCalloch-Pitts neurons is shown in Fig. 5.

Fig. 5. Appearance of the Rosenbluett perceptron based on McCalloch-Pitts neurons containing three McCallock-Pitts neurons.

The next installation of the neurons that make up the perceptron is planned. A neuron located in the upper left corner registers the letter L, a neuron located in the upper center registers the letter T, a neuron located in the upper right corner registers the letter X.

3 Selective Neurons and Neural Networks

Of the many shortcomings of neural networks based on training by choosing weights, free neural networks using selective neurons are described in [8–13]. This article discusses the training of selective neural networks with selective neurons described in these articles. The structures of the McCallock-Pitts neuron and the selective neuron are shown in Fig. 6, respectively, left and right.

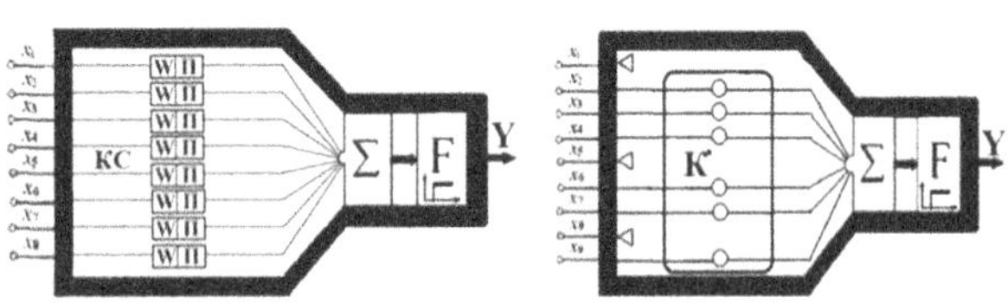

Fig. 6. Structures of the selective neuron (right) and the McCallough-Pitts neuron (left)

An effective method for obtaining selectivity was implemented in a selective perceptron consisting of selective neurons. In each neuron, clusters of specialized communication channels are created, tuned to the appropriate combination of codes for the characteristics of the input signals. The structure of a single-layer Rosenblatt perceptron using McCallock-Pitts neurons and a selective perceptron is shown in Fig. 7 left and right.

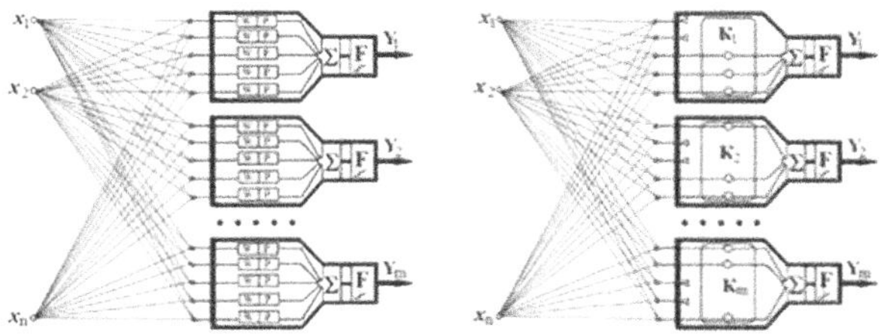

Fig. 7. The structure of a single-layer Rosenblatt perceptron using McCalloch-Pitts neurons on the left and selective perceptron on the right.

In Fig. 7: K - formed clusters of communication channels; - adder, F - threshold nonlinear elements. Triangles indicate blocked communication channels from the number of input channels that do not matter for objects at the input of the perceptron. The effectiveness of the selective perceptron for the recognition of contour objects follows from the theorem given in [11, 12]. The fundamental difference between standard neural networks and selective neural networks is that standard neural networks are trained by selecting weights, and selective neural networks are trained by selective clustering of communication channels.

3.1 Selective Neural Network on Selective Neurons

The structure of a material device that implements the instrumental method of training a selective neural network using the example of 3 selective neurons and 3 input signals $\boldsymbol{x}_i = (x_{i1}, \ldots, x_{i9})\,(i = 1, 2, 3)$ characterized by binary sequences of 9 cells was developed as shown in Fig. 8.

In Fig. 8, for clarity, the recognition of 3 vector objects is simulated. Selective clustering is illustrated by the first three vertical blocks in the structural diagram. Channel blocking and clustering are performed using black triangles. Summation of vector elements $\boldsymbol{x}_i = (x_{i1}, \ldots, x_{i9})\,(i = 1, 2, 3)$ is performed in the summation blocks. Then, in these blocks, a nonlinear threshold transformation is performed.

To justify the uniqueness of recognition of contour images by a selective neural network, a pixel scan of normalized contour objects was performed, matching each object with a binary line 0 and 1 (0 - without a contour, 1 - a contour point). Then for inappropriate contour objects. We believe that it is impossible to combine contour objects by linear transformation, which allows linear independence of the binary codes of these objects.

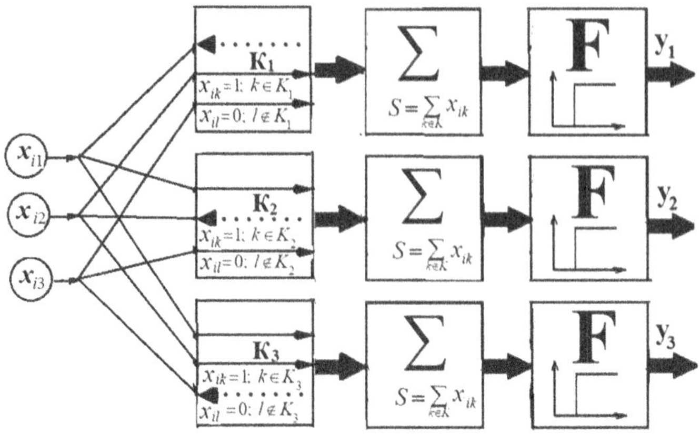

Fig. 8. The material structure of the device that implements the instrumental method of training a selective neural network.

The uniqueness recognition theorem for incompatible contour objects has been proved [11–13].

Theorem: Suppose that in a two-dimensional area, divided into pixels by a rectangular lattice, contour objects are specified that are incompatible with movements - shift, rectangular translation, rotation. Let objects be scanned with the help of line scanning in binary sequences - vectors of 0 and 1 length n, that is $\mathbf{x_i} = (x_{i1}, \ldots, x_{in})$ $i = (1, \ldots, m)$. Let all possible code combinations of input signals be assembled into matrix A

$$\mathbf{A} = \begin{pmatrix} x_{11}, x_{12} \ldots, x_{1n} \\ x_{21}, x_{22} \ldots, x_{2n} \\ \ldots \quad \ldots \quad \ldots \\ x_{m1}, x_{m2} \ldots, x_{mn} \end{pmatrix}$$

Let a particular neuron contain a cluster of connections, characterized by a code combination in the form $\mathbf{x_i} = (x_{i1}, \ldots, x_{in})$. Consider the amount

$$S_{ij} = \sum_{k=1}^{n} x_{ik} x_{kj} = (\boldsymbol{x}_i, \boldsymbol{x}_j).$$

We represent the sums S_{ij} in the form $\boldsymbol{A}\boldsymbol{x}_j^T$, where $\boldsymbol{x}_j^T$ is the transposed column vector from the vector row of the vector $\boldsymbol{x}_k$. Imagine all possible sums $S_{ij}(i = 1, \ldots, m;$ $j = 1, \ldots, m)$, the number of which is equal m^2, in the form of a matrix B

$$\boldsymbol{B} = \boldsymbol{A}\boldsymbol{A}^T,$$

where A^T is the transposed matrix A. Consider the amount S_{ii}

$$S_{ii} = \sum_{j=1}^{n} x_{ij}x_{ji} = (\boldsymbol{x}_i, \boldsymbol{x}_i) = N_i,$$

where N_i the number of units in the code combination $\mathbf{x_i} = (x_{i1}, \ldots, x_{in})$. Let's use property for recognition of input objects.

$$S_{ij} < S_{ii} = N.$$

Then the recognition of each of the input objects considered single-layer perceptron will be the only one.

The theorem under consideration states that the solution of the selective perceptron problem can be found in one iteration, that is, in a time interval much smaller than in the iterative process for the Rosenblut perceptron. In this case, the decision itself m will be the only one.

3.2 Benefits Achieved Through the Implementation of Selective Perceptron

A selective perceptron based on selective neurons has several advantages over the Rosenblüt perceptron based on McCalloch-Pitts neurons. We list the achievable useful properties of the selective perceptron:

1. Solving the problems of recognition, control, and others when using the instrumental method in selective neural networks is unambiguous when certain conditions are met, while recognition in neural networks with McCullock-Pitts neurons is ambiguous, or rather, infinitely significant. The unambiguity of instrumental recognition under rather severe conditions follows from the theorem below; 2. The procedure for training selective neural networks is greatly simplified, it becomes available to a layman; 3. Training time is significantly reduced; 4. Learning can be achieved with tools without using computational tools; 5. Implementation of a network that is resistant to extraneous external influences; 6. A network is implemented that is resistant to ambiguous recognition caused by the proximity to the threshold value of some off-diagonal matrix terms, where the matrix is transposed into the matrix.

3.3 Implementation of a Selective Neuron

Selective perceptron was implemented on selective neurons. The electrical circuit of the selective neuron provides a light indication of all involved communication channels of the neuron with the control panel, which allows visualization of the clustering of working communication channels. The control panel is common to all three selective neurons that make up the selective perceptron.

The use of instrumental and computer methods for teaching neurotechnology is contained in [14–19].

3.4 Experimental Material Implementation of Selective Perceptron Based on Selective Neurons

Exceeding the threshold level is registered by an indicator located behind the indicators of active, involved communication channels. The selective tuning of communication channels of a selective neuron with informative cells of input signals is carried out by connecting wires that simulate the external communication channels of the neuron with the corresponding internal communication channels located in the selective neuron itself. In this case, the active communication channels included in the selective cluster are identified by ignition of the LED located on the neuron. Behind the indicators of the communication channel is a neuron excitation recorder. The appearance of a selective perceptron based on selective neurons is shown in Fig. 9.

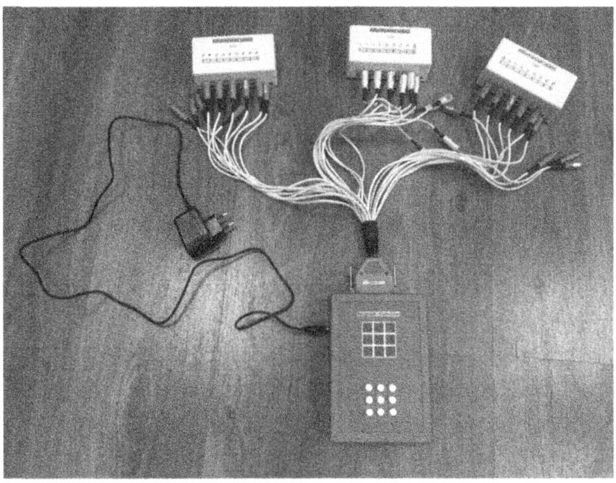

Fig. 9. Appearance of a selective perceptron based on selective neurons designed to recognize three input objects: letters, numbers, symbols.

4 Selective Monte Carlo Method for Training and Testing Standard Neural Networks Based on Mccallockpitts Neurons

The elimination of the shortcomings of iterative algorithms can be achieved using an innovative calculation procedure based on the selective Monte Carlo method. Inequalities for determining weights have innumerable solutions. Use different iterative procedures to find these solutions. However, what solution we will find after the iterative procedure is unknown. A certain refinement of the situation is given by the well-known theorem of the American mathematician Novikov. Its essence is that it claims with a number of restrictions (rather stringent) that learning is an iterative process that will end in a finite time. And what set of weights and what quality of the system we get is unknown.

The developed program allows you to find training weights directly using the selective Monte Carlo method. The method allows you to find a large number of sets of weights and select those that satisfy the given optimality conditions. You can find sets that lie in a given numerical range, sets that ensure stable operation of the neural network, and sets that exclude ambiguous recognition.

4.1 Monte Carlo Software Selective Method

The software is made in the programming language: Matlab 7 and the algorithmic language "Idol", accessible to a wide range of users who are not professional programmers. We present the results of testing the developed program. Consider an example of calculating weights for one of the training options. The calculation results of one possible set of weights for signals, each of which is a binary sequence of 9 cells from 0 and 1, are presented below. Weight matrix for binary input signals

0.9113 0.1775 0.3599 1.4603 0.4626 0.2949 1.2842 1.4467 0.3703
0.9878 1.0770 0.7659 0.2474 0.8311 0.3239 0.5889 1.1790 0.3721
0.4015 0.1678 1.1979 0.7318 1.3805 0.8073 0.8069 0.3543 1.0721

The input matrix is

1 0 0 1 0 0 1 1 1
1 1 1 0 1 0 0 1 0
1 0 1 0 1 0 1 0 1

The matrix that controls the quality of the found weight gain is equal to

5.4729 3.3580 3.3884
3.3752 4.8407 3.5458
3.3667 3.5020 4.8589

The quality of the found weights can be judged by the homogeneity of the diagonal members of the matrix; in accordance with the calculations, the diagonal terms must be in the interval. In addition, the condition "high quality recognition" must be met.

This condition is also satisfied, since the off-diagonal elements do not exceed 3.55, which is defined in the text of the program. The number of computing cycles MK = 375533. The counting time is less than 1 s.

The software "Program for calculating perceptron weights using the selective Monte Carlo method" received a certificate of state registration of computer programs.

The developed software was used to test the prototype of the neuro-educational system, the neural network - the perceptron, using McCallock-Pitts neurons.

5 Testing Developed Perceptrons

5.1 Testing the Perceptron Based on McCallock-Pitts Neurons

"A program for calculating the perceptron weights using the selective Monte Carlo method". Several options for learning factors are calculated in advance and used in the learning process. In the learning process, other sets of training weights can be calculated. These sets are clear evidence of the ambiguity of the learning process of neural networks using McCallock-Pitts neurons.

The single-layer Rosenblatt perceptron in the recognition mode of the letters of the English alphabet L, T, X is shown in Fig. 10.

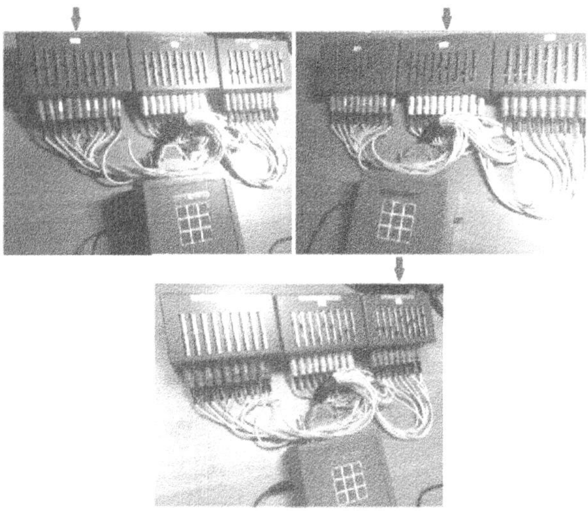

Fig. 10. Single-layer perceptron Rosenblyut in the mode of recognition of letters of the English alphabet L, T, X.

In Fig. 10 in the upper left, the perceptron recognizes the letter L, in the upper right, the perceptron recognizes the letter T, in the lower center, the perceptron recognizes the letter X.

5.2 Perceptron Testing Based on Selective Neurons

Consider the recognition of the letters of the English alphabet L, T, X using a selective perceptron. Let the letter L be recognized. When you enter the letter L on the control panel, which corresponds to a binary input sequence of 9 cells (1 0 0 1 0 0 1 1 1), the recording LED shown by the red arrow lights up because the threshold value of 5 has been exceeded. On other neurons located on the right as shown in Fig. 10. The threshold is not exceeded, since the value of the sum is reached, respectively, 2 and 3 (this can be seen in the indicators of the working channels).

Similarly, let the letters T and X be recognized. The appearance of a selective perceptron based on 3 selective neurons designed to recognize three input objects in the form of letters, numbers and other symbols that can be set on the control panel screen is shown in the photo in Fig. 11.

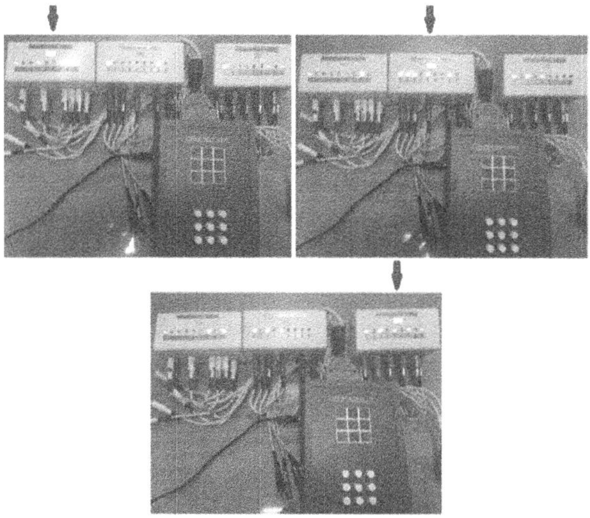

Fig. 11. Appearance of a perceptron based on selective neurons configured to recognize the letters of the English alphabet L, T, X. An experiment on recognizing the letters L, T, X is shown.

In Fig. 11, the perceptron recognizes the letter L in the upper left, the perceptron recognizes the letter T in the upper right, the perceptron recognizes the letter X in the lower center.

6 Conclusions

It has been experimentally established that (theoretical) practical installations for the implementation of prototypes have confirmed their viability. The completeness and validity of the solutions made it possible to implement effective software for training standard neural networks based on McCallough-Pitts neurons. Perceptron training is carried out not on the basis of complex iterative procedures, but on the basis of the selective Monte Carlo method, which implements the process of finding training weights more effectively.

Selective perceptron training is carried out using an instrumental procedure that does not require deep mathematical knowledge and does not require the presence of a teacher.

The ability to implement training of a neural network based on McCallock-Pitts neurons made it possible to actually establish the multiplicity of training through the implementation of various sets of training weights.

A prototype based on selective neural networks and selective neurons has been implemented. This prototype confirmed the presence of a number of positive properties of selective neurotechnologies based on selective neurons.

The developed neuro-educational system allows you to implement a quick and effective study of neurocomputer technologies. It is unique and has no foreign analogues.

References

1. Aleksandrov, Y., Anokhin, K.V., Sokolov, E.N., Grechenko, T.N.: Neuron. Signal Processing. Plastic. Modeling. Fundamental Guide, p. 548. Publishing house of Tyumen State University, Tyumen (2008)
2. Alyautdinov, M.A., Galushkin, A.I., Kazantsev, P.A., Ostapenko, G.P.: Neurocomputers: From Software to Hardware Implementation, p. 152. Hotline – Telecom (2008)
3. Galushkin, A.I.: Neural Networks. Fundamentals of the Theory. Hotline - Telecom (2010)
4. Grechenko, T.N.: Psychophysiology, p. 572. Gardariki (2009)
5. Yezhov, A.A., Shumsky, S.A.: Neurocomputing and its Applications in Economics and Business. MEPI (1998)
6. Mayer, R.V.: Practical Electronics: From a Transistor to a Cybernetic System [Electronic Resource] - Eyes: GGPI (2010)
7. Khaikin, S.: Neural Networks: A Full Course, 2nd edn. Williams (2006)
8. Mazurov, M.E.: Neuron Modeling the Properties of a Real Neuron. The Patent for the Invention № 2597495 11.07.2014
9. Mazurov, M.E.: Single-Layer Perceptron Based on Selective Neurons. The Patent for the Invention № 2597497 01.13.2015
10. Mazurov, M.E.: Intelligent recognition of electrocardiograms using selective neuron networks and deep learning. In: International Conference of Artificial Intelligence, Medical Engineering, Education, Moscow, Russia, pp. 182–198 (2017)
11. Mazurov, M.E.: Synchronization of relaxation self-oscillatory systems, synchronization in neural networks. Izv. RAS. Ser. Phys. **82**(1), 83–87 (2018)
12. Mazurov, M.E.: Nonlinear dynamics, almost periodic summation, auto-oscillatory processes, information coding in selective impulse neural networks. Izv. RAS. Phys. Ser. **82**(11), 1564–1570 (2018)
13. Mazurov, M.E.: Modeling of the neural network technologies. In: International Conference of Artificial Intelligence, Medical Engineering, Education, Moscow, Russia, pp. 365–376 (2018)
14. Iyanda, A.R., Ninan, O.D., Ajayi, A.O., Anyabolu, O.G.: Predicting student academic performance in computer science courses: a comparison of neural network models. Int. J. Mod. Educ. Comput. Sci. (IJMECS) **10**(6), 1–9 (2018). https://doi.org/10.5815/ijmecs.2018.06.01
15. Alkhathlan, A.A., Al-Daraiseh, A.A.: An analytical study of the use of social networks for collaborative learning in higher education. Int. J. Mod. Educ. Comput. Sci. (IJMECS) **9**(2), 1–13 (2017). https://doi.org/10.5815/ijmecs.2017.02.01
16. Moshref, M., Al-Sayyad, R.: Developing ontology approach using software tool to improve data visualization (case study: computer network). Int. J. Mod. Educ. Comput. Sci. (IJMECS) **11**(4), 32–39 (2019). https://doi.org/10.5815/ijmecs.2019.04.04

17. Cheah, C.S., Leong, L.: Investigating the redundancy effect in the learning of C++ computer programming using screencasting. Int. J. Mod. Educ. Comput. Sci. (IJMECS) **11**(6), 19–25 (2019). https://doi.org/10.5815/ijmecs.2019.06.03
18. Nurafifah, M.S., Abdul-Rahman, S., Mutalib, S., Hamid, N.H.A., Ab Malik, A.M.: Review on predicting students' graduation time using machine learning algorithms. Int. J. Mod. Educ. Comput. Sci. (IJMECS) **11**(7), 1–13 (2019)
19. Alvarez-Dionisi, L.E., Mittra, M., Balza, R.: Teaching artificial intelligence and robotics to undergraduate systems engineering students. Int. J. Mod. Educ. Comput. Sci. (IJMECS) **11**(7), 54–63 (2019). https://doi.org/10.5815/ijmecs.2019.07.06
20. Adekunle, S.E., Adewale, O.S., Boyinbode, O.K.: Appraisal on perceived multimedia technologies as modern pedagogical tools for strategic improvement on teaching and learning. Int. J. Mod. Educ. Comput. Sci. (IJMECS) **11**(8), 15–26 (2019). https://doi.org/10.5815/ijmecs.2019.08.02

Kinematic Analysis of Novel 6-DOF Robot

Sergey V. Kheylo[1(✉)], Andrey V. Tsarkov[2], and Oleg A. Garin[3]

[1] The Kosygin State University of Russia, Moscow, Russia
sheilo@yandex.ru
[2] Bauman Moscow State Technical University Kaluga Branch, Moscow, Russia
andrey.tsarkov@gmail.ru
[3] Bauman Moscow State Technical University, Moscow, Russia
garinolega@gmail.com

Abstract. This article is devoted to the novel spatial parallel mechanism synthesis. The new mechanism is the 6-DOF extension of the Cartesian parallel robot. Suggested mechanism consists of six actuators and of six kinematic chains. Each leg is attached to the mobile platform via a spherical joint. In each of the six legs, there are passive revolute joints, the axes of which are parallel to the direction of the prismatic joint. This parallel robot has a large workspace. Partial kinematic decoupling of mechanism allows realizing of joint progressive and rotational movements as well as separate ones. The kinematic problem is considered in this paper. The velocity analysis solution is presented too. The proposed mechanism can be applied to additive technologies, robotic–assisted surgery, different instrumentation systems, and orienting mechanisms as well as to the biological structural systems.

Keywords: Parallel robot · Kinematic problem · Singularity · Velocity analysis · Workspace

1 Introduction

Robotic engineering appeared as human limbs or animal world repetition. For example the supporting-motor apparatus reproduces the human arm. The links of these mechanisms are arranged in-line, each link connection being provided with an actuator. 6-DOF parallel mechanisms are nature "invention" good alternative. These mechanisms introduction is one of modern robotic engineering global trends. It can be of wide industrial application. Output link of these mechanisms is connected with the basis by means of a few kinematic chains. The chains contain the actuator and constrain the output link movement. 6-DOF parallel structure handling mechanisms possess great functional capability. They are attractive for using in additive technology equipment. Different designs of 6-dof parallel mechanisms had been known for decades. The proposed robot is a simple kinematic scheme.

Z. Hu et al. (Eds.): AIMEE 2019, AISC 1126, pp. 442–450, 2020.
https://doi.org/10.1007/978-3-030-39162-1_40

2 Synthesis and Kinematic Problem

Novel handling mechanisms [1] are presented in this paper.

Each kinematic chain consists of one drive rectilinear pair [2]. The actuator can contain either rotational driver or linear driver.

The structure of mechanism can be recorded as 6PRRS type (Fig. 1). Using Somov-Malyshev formula, we obtain the number of degrees of freedom:

$$\begin{aligned} W &= 6 \cdot (n - 1) - 5 \cdot p_5 - 4 \cdot p_4 - 3 \cdot p_3 \\ &= 6 \cdot (14 - 1) - 5 \cdot 6 - 6 \cdot 4 - 6 \cdot 3 = 6 \end{aligned}$$

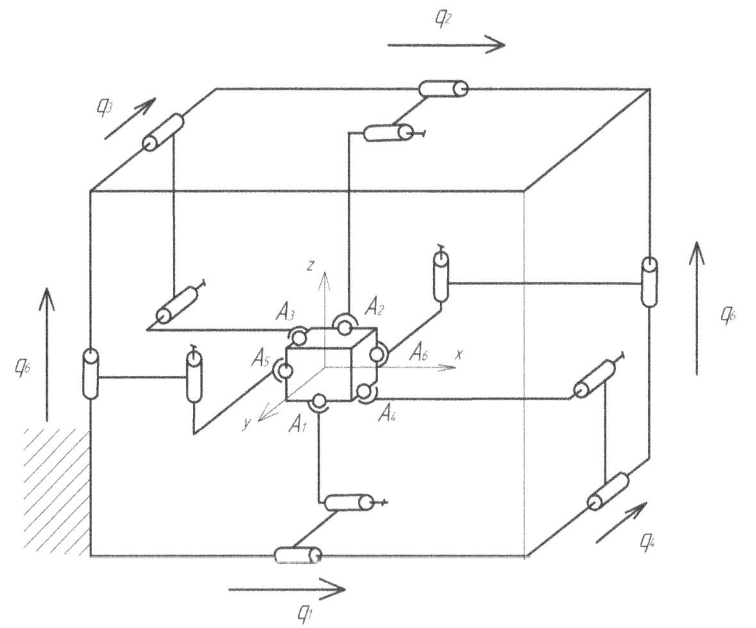

Fig. 1. 6PRRS mechanism

where n is the link number; p_5 is the number of fifth class pairs (one-DOF kinematic pair); p_4 is the number of fourth-class pairs (two-DOF kinematic pair); p_3 is the number of third-class pairs (three-DOF kinematic pair);

The basis is performed in the form of cube. The output link is a cube too. The mechanism contains six kinematic chains. Each chain contains the input two-DOF cylindrical kinematic pair, which is parallel to one of Cartesian axes. Medium cylindrical kinematic pair axes are parallel to the input pair axes. The terminal spherical pair is coupled with the output link. The spherical pair is located in the centre of cube edge.

Two-DOF kinematic cylindrical input pair axis is parallel in pairs to one another. These axes are arranged on the corner wise opposite cube edges. The centres of corresponded kinematic spherical terminal pairs are arranged on the parallel corner wise cube edges of the input link.

Kinetic problem is a major challenge [3]. Let us consider the output link as a cube. The edge of the cube is equal to $2a$.

Spherical kinematical pairs coordinates of the output link cube are: $A_1(0, a, -a)$, $A_2(0, -a, a)$, $A_3(a, 0, -a)$, $A_4(-a, 0, a)$, $A_5(a, -a, 0)$, $A_6(-a, a, 0)$.

Denote the moving coordinate system centre in fixed coordinate system as O (x_0, y_0, z_0). Set the fixed coordinate centre at point O.

We can write the actuator axes coordinates as: $q_1(x, 0, 0)$, $q_2(x, 0, 0)$, $q_3(0, y, 0)$, $q_4(0, y, 0)$, $q_5(0, 0, z)$, $q_6(0, 0, z)$. Then we obtain the equation for the coordinate determination.

$$q_1 = \begin{pmatrix}1\\0\\0\end{pmatrix}^{\mathbf{T}} \cdot \left(\begin{pmatrix}x_0\\y_0\\z_0\end{pmatrix} + \mathbf{R} \cdot \begin{pmatrix}x_{A1}\\y_{A1}\\z_{A1}\end{pmatrix}\right), \quad q_2 = \begin{pmatrix}1\\0\\0\end{pmatrix}^{\mathbf{T}} \cdot \left(\begin{pmatrix}x_0\\y_0\\z_0\end{pmatrix} + \mathbf{R} \cdot \begin{pmatrix}x_{A2}\\y_{A2}\\z_{A2}\end{pmatrix}\right)$$

$$q_3 = \begin{pmatrix}0\\1\\0\end{pmatrix}^{\mathbf{T}} \cdot \left(\begin{pmatrix}x_0\\y_0\\z_0\end{pmatrix} + \mathbf{R} \begin{pmatrix}x_{A3}\\y_{A3}\\z_{A3}\end{pmatrix}\right), \quad q_4 = \begin{pmatrix}0\\1\\0\end{pmatrix}^{\mathbf{T}} \cdot \left(\begin{pmatrix}x_0\\y_0\\z_0\end{pmatrix} + \mathbf{R} \begin{pmatrix}x_{A4}\\y_{A4}\\z_{A4}\end{pmatrix}\right)$$

$$q_5 = \begin{pmatrix}0\\0\\1\end{pmatrix}^{\mathbf{T}} \cdot \left(\begin{pmatrix}x_0\\y_0\\z_0\end{pmatrix} + \mathbf{R} \begin{pmatrix}x_{A5}\\y_{A5}\\z_{A5}\end{pmatrix}\right), \quad q_6 = \begin{pmatrix}0\\0\\1\end{pmatrix}^{\mathbf{T}} \cdot \left(\begin{pmatrix}x_0\\y_0\\z_0\end{pmatrix} + \mathbf{R} \begin{pmatrix}x_{A6}\\y_{A6}\\z_{A6}\end{pmatrix}\right)$$

where $\mathbf{R}$ is the rotation matrix about x, y, z axes.

Report $\mathbf{R}$ matrix in terms of $\mathbf{R} = \mathbf{A_3} \cdot \mathbf{A_2} \cdot \mathbf{A_1}$, where

$\mathbf{A_1}$ is the rotation matrix about x axis

$$\mathbf{A_1} = \begin{pmatrix}1 & 0 & 0\\0 & \cos\alpha & -\sin\alpha\\0 & \sin\alpha & \cos\alpha\end{pmatrix};$$

$\mathbf{A_2}$ is the rotation matrix about y axis

$$\mathbf{A_2} = \begin{pmatrix}\cos\beta & 0 & \sin\beta\\0 & 1 & 0\\-\sin\beta & 0 & \cos\beta\end{pmatrix};$$

$\mathbf{A_3}$ is the rotation matrix about z axis

$$\mathbf{A_3} = \begin{pmatrix}\cos\gamma & -\sin\gamma & 0\\\sin\gamma & \cos\gamma & 0\\0 & 0 & 1\end{pmatrix}.$$

We can write $\mathbf{R}$ matrix in the form

$$\mathbf{R} = \begin{pmatrix} \cos\gamma \cdot \cos\beta & \cos\gamma \cdot \sin\beta \cdot \sin\alpha - \cos\alpha \cdot \sin\gamma & \sin\alpha \cdot \sin\gamma + \cos\alpha \cdot \cos\gamma \cdot \sin\beta \\ \cos\beta \cdot \sin\gamma & \cos\alpha \cdot \cos\gamma + \sin\alpha \cdot \sin\beta \cdot \sin\gamma & \cos\alpha \cdot \sin\gamma \cdot \sin\beta - \cos\gamma \cdot \sin\alpha \\ -\sin\beta & \cos\beta \cdot \sin\alpha & \cos\beta \cdot \cos\alpha \end{pmatrix}.$$

The example of inverse kinematic problem.

Set output link location by means of coordinates O (3,0; 3,5; 4,0), angles $\alpha = 20°$, $\beta = 10^{\circ}$, $\gamma = 10^{\circ}$ and the edge of the cube a = 1.

Therefore, actuator coordinates will be equal to: $q_1 = 3{,}325$, $q_2 = 2{,}675$, $q_3 = 3{,}979$, $q_4 = 3{,}021$; $q_5 = 3{,}857$; $q_6 = 4{,}163$.

The proposed mechanism possesses properties of partial decoupling (the possibility of motion straight rectilinear one, straight rotational one, or thereof combination). In this case, the set of kinematic and dynamic equations is split up into independent less degree equation subsystems [4, 5]. Therefore, kinematic problem solution is simplified. Dynamic coupling permits to reduce actuator interference. Thus, for the rectilinear movement generation simultaneous movement of actuators that controls one coordinate is needed (q_1, q_2 control x; q_3, q_4 control y; q_5, q_6 control z). To obtain only rotational movements the equal velocity movement of actuator coordinates is needed.

3 Velocity Analysis and Singularities

For velocity analysis solution we write the position function implicitly.

$$F_1 = q_1 - \begin{pmatrix} 1 \\ 0 \\ 0 \end{pmatrix}^{\mathrm{T}} \cdot \left(\begin{pmatrix} x_0 \\ y_0 \\ z_0 \end{pmatrix} + \mathbf{R} \cdot \begin{pmatrix} x_{A1} \\ y_{A1} \\ z_{A1} \end{pmatrix} \right)$$
$$= q_1 - \begin{pmatrix} 1 \\ 0 \\ 0 \end{pmatrix}^{\mathrm{T}} \cdot \left(\begin{pmatrix} x_0 \\ y_0 \\ z_0 \end{pmatrix} + \begin{pmatrix} a \cdot (\sin\gamma \cdot \sin\alpha - \sin\gamma \cdot \cos\alpha + \cos\gamma \cdot \cos\alpha \cdot \sin\beta - \cos\gamma \cdot \sin\beta \cdot \sin\alpha) \\ a \cdot (\cos\gamma \cdot \cos\alpha - \cos\gamma \cdot \sin\alpha + \sin\gamma \cdot \cos\alpha \cdot \sin\beta + \sin\gamma \cdot \sin\beta \cdot \sin\alpha) \\ a \cdot (\cos\beta \cdot \cos\alpha + \cos\beta \cdot \sin\alpha) \end{pmatrix} \right);$$

$$F_2 = q_2 - \begin{pmatrix} 1 \\ 0 \\ 0 \end{pmatrix}^{\mathrm{T}} \cdot \left(\begin{pmatrix} x_0 \\ y_0 \\ z_0 \end{pmatrix} + \mathbf{R} \cdot \begin{pmatrix} x_{A2} \\ y_{A2} \\ z_{A2} \end{pmatrix} \right), F_3 = q_3 - \begin{pmatrix} 0 \\ 1 \\ 0 \end{pmatrix}^{\mathrm{T}} \cdot \left(\begin{pmatrix} x_0 \\ y_0 \\ z_0 \end{pmatrix} + \mathbf{R} \begin{pmatrix} x_{A3} \\ y_{A3} \\ z_{A3} \end{pmatrix} \right) F_4 = q_4 - \begin{pmatrix} 0 \\ 1 \\ 0 \end{pmatrix}^{\mathrm{T}} \cdot \left(\begin{pmatrix} x_0 \\ y_0 \\ z_0 \end{pmatrix} + \mathbf{R} \begin{pmatrix} x_{A4} \\ y_{A4} \\ z_{A4} \end{pmatrix} \right),$$
$$F_5 = q_5 - \begin{pmatrix} 0 \\ 0 \\ 1 \end{pmatrix}^{\mathrm{T}} \cdot \left(\begin{pmatrix} x_0 \\ y_0 \\ z_0 \end{pmatrix} + \mathbf{R} \begin{pmatrix} x_{A5} \\ y_{A5} \\ z_{A5} \end{pmatrix} \right), F_6 = q_6 - \begin{pmatrix} 0 \\ 0 \\ 1 \end{pmatrix}^{\mathrm{T}} \cdot \left(\begin{pmatrix} x_0 \\ y_0 \\ z_0 \end{pmatrix} + \mathbf{R} \begin{pmatrix} x_{A6} \\ y_{A6} \\ z_{A6} \end{pmatrix} \right)$$

Equations of constraints

$$\begin{aligned} F_1 &= q_1 - (x_0 + a \cdot (-\sin\gamma \cdot \cos\alpha - \sin\gamma \cdot \sin\alpha - \cos\gamma \cdot \cos\alpha \cdot \sin\beta + \cos\gamma \cdot \sin\beta \cdot \sin\alpha)) = 0 \\ F_2 &= q_2 - (x_0 + a \cdot (\sin\gamma \cdot \cos\alpha + \sin\gamma \cdot \sin\alpha + \cos\gamma \cdot \cos\alpha \cdot \sin\beta - \cos\gamma \cdot \sin\beta \cdot \sin\alpha)) = 0 \\ F_3 &= q_3 - (y_0 + a \cdot (\sin\gamma \cdot \cos\beta + \cos\gamma \cdot \sin\alpha - \sin\gamma \cdot \cos\alpha \cdot \sin\beta)) = 0 \\ F_4 &= q_4 - (y_0 + a \cdot (\sin\gamma \cdot \cos\beta \cdot \sin\beta - \cos\gamma \cdot \sin\alpha - \sin\gamma \cdot \cos\beta)) = 0 \\ F_5 &= q_5 - (z_0 + a \cdot (\sin\beta - \cos\beta \cdot \sin\alpha)) = 0 \\ F_6 &= q_6 - (z_0 + a \cdot (\cos\beta \cdot \sin\alpha - \sin\beta)) = 0 \end{aligned}$$

For velocity analysis solution let us perform a differentiation of constraints equation:

$$\mathbf{A} \cdot \mathbf{V} = -\mathbf{B} \cdot \dot{\mathbf{q}}$$

$$\mathbf{A} = \begin{pmatrix} \frac{\partial F_1}{\partial x_0} & \frac{\partial F_1}{\partial y_0} & \frac{\partial F_1}{\partial z_0} & \frac{\partial F_1}{\partial \alpha} & \frac{\partial F_1}{\partial \beta} & \frac{\partial F_1}{\partial \gamma} \\ \frac{\partial F_2}{\partial x_0} & \frac{\partial F_2}{\partial y_0} & \frac{\partial F_2}{\partial z_0} & \frac{\partial F_2}{\partial \alpha} & \frac{\partial F_2}{\partial \beta} & \frac{\partial F_2}{\partial \gamma} \\ \frac{\partial F_3}{\partial x_0} & \frac{\partial F_3}{\partial y_0} & \frac{\partial F_3}{\partial z_0} & \frac{\partial F_3}{\partial \alpha} & \frac{\partial F_3}{\partial \beta} & \frac{\partial F_3}{\partial \gamma} \\ \frac{\partial F_4}{\partial x_0} & \frac{\partial F_4}{\partial y_0} & \frac{\partial F_4}{\partial z_0} & \frac{\partial F_4}{\partial \alpha} & \frac{\partial F_4}{\partial \beta} & \frac{\partial F_4}{\partial \gamma} \\ \frac{\partial F_5}{\partial x_0} & \frac{\partial F_5}{\partial y_0} & \frac{\partial F_5}{\partial z_0} & \frac{\partial F_5}{\nabla \alpha} & \frac{\partial F_5}{\partial \beta} & \frac{\partial F_5}{\partial \gamma} \\ \frac{\partial F_6}{\partial x_0} & \frac{\partial F_6}{\partial y_0} & \frac{\partial F_6}{\partial z_0} & \frac{\partial F_6}{\partial \alpha} & \frac{\partial F_6}{\partial \beta} & \frac{\partial F_6}{\partial \gamma} \end{pmatrix};$$

$$\mathbf{B} = \begin{pmatrix} \frac{\partial F_1}{\partial q_1} & \frac{\partial F_1}{\partial q_2} & \frac{\partial F_1}{\partial q_3} & \frac{\partial F_1}{\partial q_4} & \frac{\partial F_1}{\partial q_5} & \frac{\partial F_1}{\partial q_6} \\ \frac{\partial F_2}{\partial q_1} & \frac{\partial F_2}{\partial q_2} & \frac{\partial F_2}{\partial q_3} & \frac{\partial F_2}{\partial q_4} & \frac{\partial F_2}{\partial q_5} & \frac{\partial F_2}{\partial q_6} \\ \frac{\partial F_3}{\partial q_1} & \frac{\partial F_3}{\partial q_2} & \frac{\partial F_3}{\partial q_3} & \frac{\partial F_3}{\partial q_4} & \frac{\partial F_3}{\partial q_4} & \frac{\partial F_3}{\partial q_4} \\ \frac{\partial F_4}{\partial q_1} & \frac{\partial F_4}{\partial q_2} & \frac{\partial F_4}{\partial q_3} & \frac{\partial F_4}{\partial q_4} & \frac{\partial F_4}{\partial q_5} & \frac{\partial F_4}{\partial q_6} \\ \frac{\partial F_5}{\partial q_1} & \frac{\partial F_5}{\partial q_2} & \frac{\partial F_5}{\partial q_3} & \frac{\partial F_5}{\partial q_4} & \frac{\partial F_5}{\partial q_5} & \frac{\partial F_5}{\partial q_6} \\ \frac{\partial F_6}{\partial q_1} & \frac{\partial F_6}{\partial q_2} & \frac{\partial F_6}{\partial q_3} & \frac{\partial F_6}{\partial q_4} & \frac{\partial F_6}{\partial q_5} & \frac{\partial F_6}{\partial q_6} \end{pmatrix} = \begin{pmatrix} 1 & 0 & 0 & 0 & 0 & 0 \\ 0 & 1 & 0 & 0 & 0 & 0 \\ 0 & 0 & 1 & 0 & 0 & 0 \\ 0 & 0 & 0 & 1 & 0 & 0 \\ 0 & 0 & 0 & 0 & 1 & 0 \\ 0 & 0 & 0 & 0 & 0 & 1 \end{pmatrix}$$

where **A** is the matrix of partial derivatives of implicit function with respect to x_0, y_0, z_0, α, β, γ;

B is the matrix of partial derivatives of implicit function with respect to generalized coordinates q_1, q_2, q_3, q_4, q_5, q_6.

V is the matrix of rectilinear (V_x, V_y, V_z) and rotational $\left(\dot{\alpha}, \dot{\beta}, \dot{\gamma}\right)$ motion velocities. The angular velocities are decomposed in system of axes, which is determined by α, β, γ angles.

We can decompose output link velocity in x, y, z orthogonal axes as well as in axes that are determined by α, β, γ angles.

It is necessary to determine the ratio between projections on orthogonal axes and α-, β-, and γ-components for velocity analysis solution. Let us denote the velocities about α, β, γ axes as $\dot{\alpha}$, $\dot{\beta}$, $\dot{\gamma}$ and the velocities about x, y, z axes as ω_x, ω_y, ω_z. We obtain the required ratio.

$$\omega_x = \dot{\alpha} \cdot x_\alpha + \dot{\beta} \cdot x_\beta + \dot{\gamma} \cdot x_\gamma;$$

$$\omega_y = \dot{\alpha} \cdot y_\alpha + \dot{\beta} \cdot y_\beta + \dot{\gamma} \cdot y_\gamma;$$

$$\omega_z = \dot{\alpha} \cdot z_\alpha + \dot{\beta} \cdot z_\beta + \dot{\gamma} \cdot z_\gamma.$$

where x_γ, y_γ, z_γ are the coordinates of unit vector of fictitious kinematic chain of the first kinematic pair axis, arranged along the z $\begin{pmatrix} x_\gamma \\ y_\gamma \\ z_\gamma \end{pmatrix} = \begin{pmatrix} 0 \\ 0 \\ 1 \end{pmatrix}$ axis.

x_β, y_β, z_β are the kinematic pair coordinates of the fictitious chain which are equal to the product of matrix of rotation about the z axis by the coordinates of the 3-d kinematic pair).

$$\begin{pmatrix} x_\beta \\ y_\beta \\ z_\beta \end{pmatrix} = \begin{pmatrix} \cos\gamma & -\sin\gamma & 0 \\ \sin\gamma & \cos\gamma & 0 \\ 0 & 0 & 1 \end{pmatrix} \cdot \begin{pmatrix} 0 \\ 1 \\ 0 \end{pmatrix} = \begin{pmatrix} -\sin\gamma \\ \cos\gamma \\ 0 \end{pmatrix};$$

x_α, y_α, z_α are the coordinates of unit vector of fictitious kinematic chain third kinematic pair. These coordinates are equal to product of output link rotation matrix and coordinates of the third pair unit vector in zero position.

$$\begin{pmatrix} x_\alpha \\ y_\alpha \\ z_\alpha \end{pmatrix} = (\mathbf{R}) \cdot \begin{pmatrix} 1 \\ 0 \\ 0 \end{pmatrix} = \begin{pmatrix} \cos\gamma \cdot \cos\beta \\ \sin\gamma \cdot \cos\beta \\ -\sin\beta \end{pmatrix}$$

The example of velocity analysis solution.

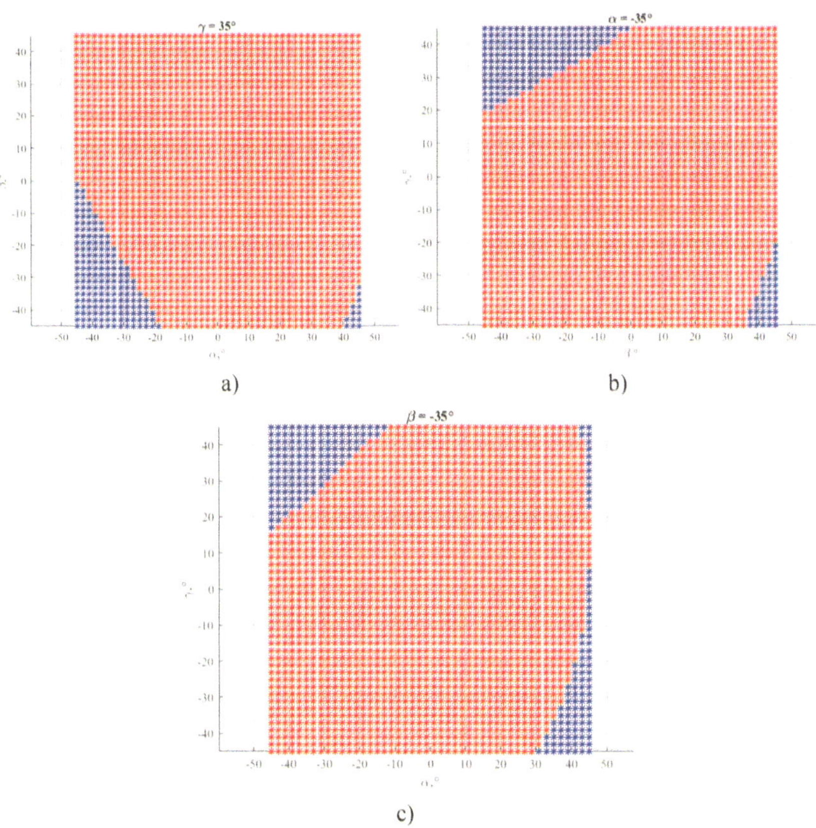

Fig. 2. The area of the singularity (a) at γ = 30°, (b) at α = 35°, (c) at β = 35°

Let us consider that $\alpha = 12°$, $\beta = 20°$, $\gamma = 30°$, $V_1 = 1{,}2$ m/s, $V_2 = 1{,}3$ m/s; $V_3 = 1{,}4$ m/s, $\omega_x = 1$ rad/s, $\omega_y = 1{,}1$ rad/s, $\omega_z = 1{,}2$ rad/s. Solving the velocity equations, we obtain that $q_1 = -0{,}314$ m/s, $q_2 = -1{,}686$ m/s, $q_3 = -2{,}758$ m/s, $q_4 = 0{,}358$ m/s, $q_5 = -1{,}466$ m/s, $q_6 = -1{,}134$ m/s.

Using A and B Jacobians properties we obtain irregularity positions. If A or B matrix determinant is equal to zero it indicates the irregularity position occurrence. Analyzing the properties of the matrix A, it is seen that the singularity depends only on the angles α, β, γ. The singularity is only at angles α, β, γ greater than 35° (Fig. 2).

4 Workspace

We have a complex geometry of operating area, with maximum volume being the cube (Fig. 3).

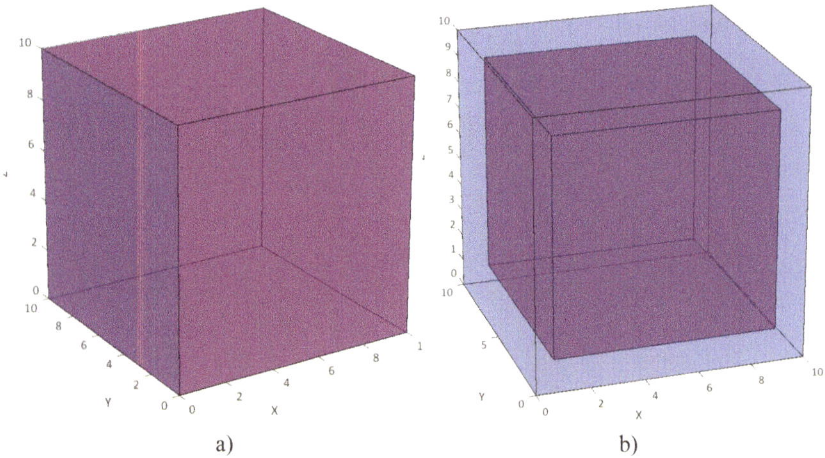

Fig. 3. (a) maximum workspace at $\alpha = 0°$, $\beta = 0°$, $\gamma = 0°$, (b) workspace at $\alpha = 30°$, $\beta = 30°$, $\gamma = 30°$

Point is the minimum volume of operating area. In this case the mechanism does only rotational motion.

If the output link is a parallelogram with sides equal to $2a$, $2h$, $2c$ the output points coordinates are equal to $A_1(0, h, -c)$ $A_2(0, -h, c)$, $A_3(a, 0, -c)$, $A_4(-a, 0, c)$, $A_5(a, -h, 0)$, $A_6(-a, h, 0)$. Here the maximum operating area is a parallelepiped.

Maximum operating area determination is an important task of mechanism designing. Maximum operating area of symmetric structure is defined by $f(x, y, c, q)$ objective function, which contains the next parameters:

- design variables are the cubes sides ratio (bases of output and input links);
- q_1, q_2, q_3, q_4, q_5, q_6 are the manipulator joint variables;
- x_0, y_0, z_0, α, β, γ are the state variables (operating device motion data).

Let input link and output link cubes side lengths: one is b and one is a. Ratio of a and b is the optimization variables.

The task of optimization model simulation is exemplified by presence of constrain set joined with design parameters.

Constraints consist of a number of items:

- parametric constraints for design variables $\left(\frac{a}{b}\right)_{\min} \leq \frac{a}{b} \leq \left(\frac{a}{b}\right)_{\max}$,
- constraints for joint coordinates $q_i^{\min} \leq q \leq q_i^{\max}$
- functional constraints:

constraints for state variables: $x_0^{\min} \leq x_0 \leq x_0^{\max}$; $y_0^{\min} \leq y_0 \leq y_0^{\max}$, $z_0^{\min} \leq z_0 \leq z_0^{\max}$, $\alpha^{\min} \leq \alpha \leq \alpha^{\max}$, $\beta^{\min} \leq \beta \leq \beta^{\max}$, $\gamma^{\min} \leq \gamma \leq \gamma^{\max}$.

Additional constraints (angles between the kinematic pairs links) $\theta_j^{\min} \leq \theta_j \leq \theta_j^{\max}$, where $\theta_j = \theta_j = (y)$.

To determine and optimize manipulator-operating area we have to do the following. Let sides lengths ratio is fixed. Using the genetic algorithm [6, 7], we iterate over joint coordinates and solve the system of nonlinear equations, which determine the end-effector position. It permits to find the maximum manipulator operating area value in the case of variable link lengths and different output link configurations.

5 Result

The structure of the new parallel mechanism is presented. Degrees of freedom is determined. Inverse and direct kinematic problem is solved. The second part is devoted to velocity analysis solution, relying on constraints equations derivation. Singularities are studied. Mechanism properties and workspace is demonstrated. It allows optimizing the problem in order to find the maximum working zone of the manipulator, for solving systems of nonlinear algebraic equations. Genetic algorithm is offered. The obtained results and equations permit to solve the control of mechanism [8, 9].

6 Conclusions

New concept of parallel manipulator with six degree of freedom is presented. Special geometry and structure are characterized isotropic properties. The construction of the mechanism is modular and light. It is simple to use direct drives for very high-speed motion. Driving element of motors can be connecting with moving platform with gear ratio 1:1. The main advantages suggested mechanism is the simple kinematic structure. Kinematic system gives a simple direct kinematics task. It is shown that the singularity depends only on rotation moving platform. The singularity is only at angles α, β, γ greater than 35°, which is enough for the efficiency of the manipulator. If the size of the moving platform is changed, the singularity is at the same angles. The manipulator has a large workspace, low inertia, light links and can operate at high speeds and acceleration. The manipulator is universal. It has only translation, rotation movements as well as joint ones. Suggested manipulator has potential of wide industrial applications.

References

1. Kheylo, S.V., Glazunov, V.A., Garin, O.A., Tsarkov, A.V., Razumeev, K.E.: Patent RU №182355. Spatial mechanism of 6 DOF (2018)
2. Kheylo, S.V., Garin, O.A.: Design of a mechanism with six degrees of freedom for additive technologies. In: Design, Technology and Innovation in Textile and Light Industry (INNOVATION - 2018) International Scientific and Technical Conference, Part 3. Moscow, pp. 61–62 (2018)
3. Kumar, V., Sen, S., Roy, S.S., Das, S.K., Shome, S.N.: Inverse kinematics of redundant manipulator using interval newton method. Int. J. Eng. Manuf. (IJEM) **5**(2), 19–29 (2015)
4. Glazunov, V., Kheylo, S.: Dynamics and control of planar, translational, and spherical parallel manipulators. In: Dynamic Balancing of Mechanisms and Synthesizing of Parallel Robots. Springer, pp. 365–403 (2016)
5. Glazunov, V., Nosova, N., Kheylo, S., Tsarkov, A.: Design and analysis of the 6-DOF decoupled parallel kinematic mechanism. In: Dynamic Decoupling of Robot Manipulators. Springer, pp. 125–170 (2018)
6. Arora, R.K.: Optimization Algorithms and Applications, p. 466. CRC Press, Boca Raton (2015)
7. Joumah, A.A., Albitar, C.: Design optimization of 6-RUS parallel manipulator using hybrid algorithm. Int. J. Inf. Technol. Comput. Sci. (IJITCS) **10**(2), 83–95 (2018)
8. Nanda, S.K., Panda, S., Subudhi, P.R.S., Das, R.K.: A novel application of artificial neural network for the solution on inverse kinematics controls of robotic manipulators. Int. J. Intell. Syst. Appl. (IJISA) **4**(9), 81–91 (2012)
9. Krishan, G., Singh, V.R.: Motion control of five bar linkage manipulator using conventional controllers under uncertain conditions. Int. J. Intell. Syst. Appl. (IJISA) **5**, 34–40 (2016)

Design of Fog-Based Warehouse Environment Monitoring System

Xuejiang Wei[1] and Meng Wang[2(✉)]

[1] State Key Laboratory of Software Engineering (Wuhan University), Wuhan Technology and Business University, Wuhan, China
81155173@qq.com

[2] Wuhan Technology and Business University, Wuhan, China
13397195035@126.com

Abstract. This paper focuses on the application of fog computing in the warehouse environment monitoring field to improve the performance of traditional warehouse environment monitoring systems (WEMS) generally based on the Internet of Things (IoT). For IoT-based WEMS have some flaws in the aspect of security, stability and reliability, as well as real-time processing of transactions, we propose a new architecture for WEMS based on fog computing, and designs the hierarchical model and main functions of the system. In particular, we put forward the solutions of distributed data processing and service provisioning based on fog nodes, which are key issues that influence the application of fog computing in warehouse environment monitoring. Finally, we evaluate the reliability and efficiency of the system based on a physical test platform, and the results show that fog computing can alleviate existing problems in traditional systems greatly.

Keywords: Warehouse environment monitoring · Intelligent system design · Fog computing · Distributed intelligence

1 Introduction

The Internet of Things, emerged as a new era of wireless communication technology [1], for it can realize the automation and intelligence of warehouse environment monitoring and greatly improves the efficiency and effectiveness of warehouse environment monitoring [2], has been widely used in the field of warehouse environment monitoring and attracted the attention of many researchers. Inherited the "sensor-network-application" three-layer architecture of IoT [3], the IoT-based warehouse environment monitoring system also has three layers: warehouse environment sensing network, data transmission network and monitoring center, from bottom to top [4]. According to the position of the monitoring center, it can be further divided into two types: local mode [4–6] and cloud-based mode [7–9] (Fig. 1).

In contrast to the local mode, the cloud-based mode has obvious advantages in storage security, operation stability, system scalability and resource sharing [10] (Table 1). However, as both inherited from IoT, there are still several problems that need to be addressed:

Z. Hu et al. (Eds.): AIMEE 2019, AISC 1126, pp. 451–460, 2020.
https://doi.org/10.1007/978-3-030-39162-1_41

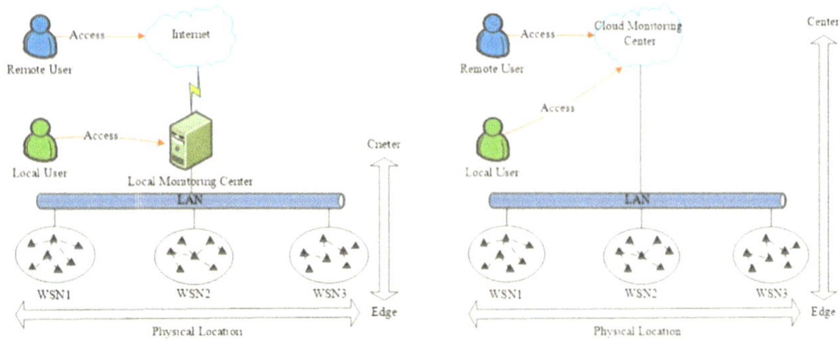

Fig. 1. Architecture of IoT-based warehouse environment monitoring system (Left: Local mode, Right: Cloud-based mode)

Table 1. Comparisons of different modes of warehouse environment monitoring system

Property	Local mode	Cloud-based mode	Fog based mode
Storage security	Weak	Strong	Strong
Operational stability	Weak	Strong	Strong
Scalability	Weak	Strong	Strong
Sharing	Weak	Strong	Strong
Data fusion	No	No	Yes
Distributed intelligence	No	No	Yes
Real-time application	Medium	Weak	Strong

(1) Firstly, as raw sensing data is directly transmitted to the monitoring center without preprocessing, it may increase the communication load of the network layer and the computing load of the monitoring center, and especially lead to congestion and even system crash in large-scale applications.
(2) Secondly, as the network layer of IoT is the neural center linking the sensor layer with the application layer, any failure will directly lead to the paralysis of the warehouse environment monitoring system [11].
(3) Finally, IoT applications must carefully assess certain crucial factors such as the real-time and largely distributed nature of the "things", maintaining trustworthy communications, and adapting for mobility and the harsh environments where the "things" are deployed [12]. To achieve these objectives with traditional Cloud Computing solutions is complicated since a centric design approach precludes critical IoT requirements, such as real-time and geographically-aware computing [13].

Fog computing [14], an architecture that appeared in the past years, can be used as a base to provide a good solution to the IoT requirements. It is a highly distributed platform, with nodes located from near the end-user devices till the edge of the network [15]. These nodes offer resources such as computing, storage, and networking to the applications operating under this infrastructure. An access point with enhanced computing capabilities constitutes an example of a possible Fog-capable node. Fog processes the data close to where it is generated, reducing the network utilization and improving the aggregation from the bottom of the infrastructure [16]. Low-latency, widespread geographic distribution, heterogeneity, and mobility are part of its main advantages. In consequence, Fog becomes an extension of the Cloud rather than a substitute since its nodes are connected to the Cloud. It offers on-demand services and applications proximate to devices, dense geographical distributed and low latency responses, resulting in superior user experience and redundancy in case of failure.

In this paper, we present a new warehouse environment monitoring system based on fog computing, design the architecture and main functions of the system, and study the distributed data processing and service provision method based on fog nodes.

The rest of the paper is organized as follows. Section 2 describes the architecture and fundamental functions of the fog-based warehouse environmental monitoring system. In Sect. 3, the distributed data processing and service provisioning methods are presented. We discuss the performance evaluation of the fog-based warehouse environment monitoring system in Sect. 4. Finally, the work is concluded in Sect. 5.

2 System Architecture and Functions

The basic idea of this paper is to fully combine the advantages of IoT, Cloud Computing and fog computing, by adding a fog layer between the front-end sensing layer and the cloud monitoring center, as shown in Fig. 2. The functions of the fog-based warehouse environment monitoring system are basically consistent with those of the traditional IoT-based system, and the difference is the deployment of functions in different layers.

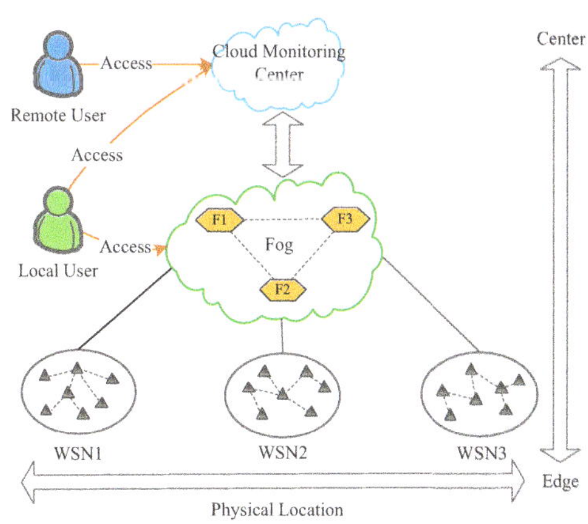

Fig. 2. Architecture of Fog-based warehouse environment monitoring system

2.1 Sensing Layer

The sensing layer is composed of wireless sensor networks (WSNs) which are deployed in different warehouses in the logistics park and are relatively independent, including a variety of sensor nodes, control nodes and sink nodes, such as temperature and humidity sensor, smoke sensor, gas sensor, etc. All nodes in the WSN communicate with each other in a wireless way, and all sensing data is sent to the fog node in a wired or wireless way through the sink node, or receive control instructions from the fog node.

The sensing layer mainly provides warehouse environment awareness and control. Sensor nodes collect real-time data from warehouse environment and send the data to sink nodes which upload the data to the fog layer. On the other hand, control nodes can receive control instructions from sink nodes and then turn on/off corresponding devices.

2.2 Fog Layer

The fog layer is located between the sensing layer and the cloud monitoring center, and can provide limited computing, storage and network services at the edge of the sensing network through virtualization technology [17]. Fog nodes can be common devices with limited resources such as access points, routers, switches, base stations and terminal devices (such as some powerful sensor nodes) which are suitable for small-scale warehouse, or special fog devices such as Cloudlet [18] and IOx [19] which are suitable for large-scale warehouse.

The fog layer mainly realizes simplified version functions of environment control and information service. Fog nodes receive raw sensing data from the sensing layer and process or temporarily store raw sensing data according to preset rules. During data processing, if one environmental indicator value is out of the threshold, fog nodes may also generate a warning message. While the transmission network is connected, results and messages are pushed to local users and/or the cloud monitoring center. Meanwhile, when there is an emergency, fog nodes can generate control instructions automatically according to preset strategies or receive control instructions from local users or the cloud monitoring center to control corresponding devices located in the sensing layer.

2.3 Cloud Monitoring Center

In this paper, the monitoring center of the fog-based warehouse environment monitoring system is deployed on the cloud. For small enterprises, it is an appropriate way to build monitoring center by renting public cloud platforms for it is attractive to business owners as it eliminates the requirement for users to plan ahead for provisioning, and allows enterprises to start from the small and increase resources only when there is a rise in service demand [20]. For large enterprises, they can consider building their own cloud platform, which can further strengthen the protection of enterprise privacy on the basis of ensuring stable operation of monitoring center.

The cloud monitoring center provides full version functions of environment control and information service and additional function of data analysis. It receives data from the bottom, analyzes, processes and stores the data relying on cloud infrastructure, and finally responses requests of web or mobile clients in real time. The detail functions are as follows: (1) To control corresponding devices deployed in the sensing layer, generating instructions automatically or triggered by end users and sending them to the sensing layer through fog nodes; (2) Inherited the characters of cloud computing, providing various types of data visualization interface to end users based on data statics; (3) Revealing the correlation between various events based on intelligent analysis, such as the relationship between humidity with product quality; (4) Optimizing the warehouse environment monitoring strategy based on analysis results.

3 Key Issues of the System

Distributed intelligence can fully meet the needs of IoT applications in terms of scene complexity and user demand diversity, and can provide users with more flexible and higher quality services [21], so the key issues of the system are distributed data processing and service provisioning based on fog nodes, which also are the most obvious characteristic of the fog-based mode compared with IoT-based mode.

3.1 Distributed Data Processing

In the process of warehouse environment monitoring, users focus on the data processing results rather than the original data itself. For example, in monitoring the temperature and humidity of a warehouse, the user wants to know the average temperature of a warehouse or an area rather than the raw data of each sensor.

In the traditional IoT-based warehouse environment monitoring system, all data processing tasks are undertaken by the monitoring center, and it increases the data processing load of monitoring center and wastes the communication bandwidth, so it is a good idea processing the raw sensing data on fog nodes before it is sent to the cloud monitoring center.

In this paper, we propose a structure of fog node as shown in Fig. 3 to realize distributed data processing. The fog node is composed of service instance, temporary storage, command interpretation, data acquisition, monitoring and metering, and common interface. The service instance is the core component of fog nodes and multiple service instances can run concurrently on one fog node. Various tasks of environment control and information service can be modularized as service instances and each service instance is registered in the cloud monitoring center and presented to users in the form of directory, so it can further improve the scalability and robustness of the system. The temporary storage component coordinates the storage space within fog nodes to temporarily store the pre-processing results when the connection with the cloud monitoring center fails or according to the requests of local applications, and the temporarily stored data is automatically removed once it is uploaded successfully or the local application is finished. The command interpretation component is responsible for interpreting control instructions from upper layer and forwarding them to lower layer to control corresponding devices, so the lower layer is transparent to end users and end users need not care about the underlying technical details. The data acquisition component collects data from the front-end sensing network according to the requests of service instances and uploads them to specific service instances. The monitoring and metering component checks the health status of each sensor node regularly and collects the metadata of sensor nodes and sensor networks of the sensing layer. The common interface component is responsible for protocols interpretation, so as to improve the interoperability between heterogeneous sensor platforms.

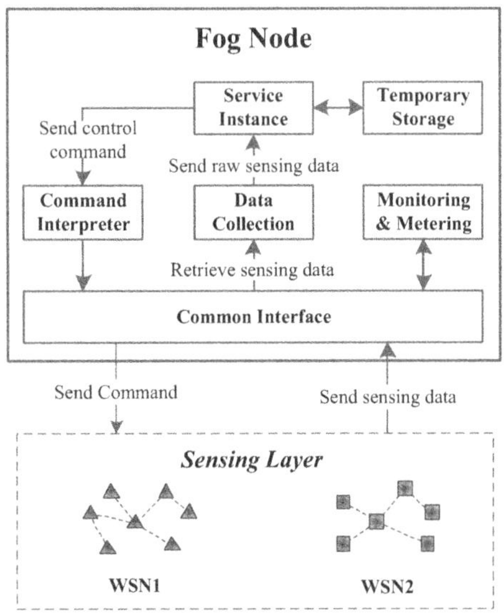

Fig. 3. Structure of fog node

3.2 Distributed Service Provisioning

For local users, invoking the service located in the fog node has obvious advantages in terms of response delay, security and stability [22], so we propose a distributed service provisioning strategy in this paper, as shown in Fig. 4. The specific flow is as follows:

(1) Demand matching. While the fog node receives a request from local user, it comprehensively analyzes the service demand type, task time requirement, local service directory, local computing power and other factors to decide how to handle the request. If resources within fog nodes can meet the demand of the request, the task will be executed by fog nods, otherwise it will be sent to the cloud.
(2) Instance preparing. In our system, service instances are managed the administrators in the cloud monitoring center, and service instances and the dictionary are created in the cloud initially. While the dictionary is created, its subsets will be sent to corresponding fog nodes automatically, however the service instance code is sent to the fog node only when it is invoked at the first time. When the task is decided to be executed by fog nodes, it is necessary to check whether the

corresponding service instance has been downloaded, and if has not, the download request will be sent to the cloud.

(3) Instance activating. When the service instance is ready, the fog node activates the instance according to the user's request, and begins to receive the state data collected by the front-end sensing network. The service instance processes the data according to the preset rules and procedures and then provides the service to the end user.

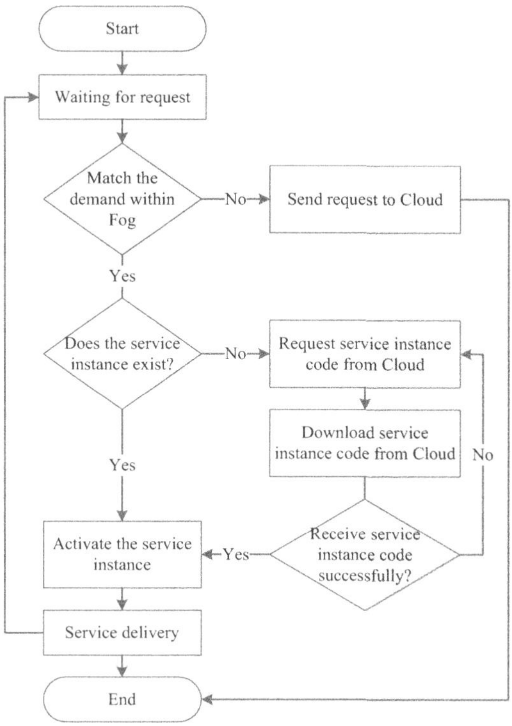

Fig. 4. Service provisioning flow on fog node

4 Experiment and Evaluation

4.1 Experiment Setup

We evaluate the system on a physical test platform. As shown in Fig. 5, the sensing layer is equipped with a group of temperature and humidity sensor, light sensor, smoke sensor, passive infrared (PIR) sensor, security alarm controller, light controller, air condition controller, fire curtain controller and WiFi-Zigbee gateway. The WiFi-Zigbee gateway serves as the sink node and all sensors and controllers communicate with each other through the Zigbee protocol to form a star network topology. A Cortex-A8 gateway is used as the fog node and a HP Z440 workstation is used as the cloud monitoring center.

The purpose of the experiment is to evaluate the feasibility and effectiveness of distributed data processing and service provisioning, and the experimental scenarios include normal situation, emergency situation and communication interruption situation, therefore, the test system only realizes some key application functions, but does not build a warehouse environment monitoring system with full functions.

Fig. 5. Physical configuration of warehouse environment sensing network

4.2 Results

As shown in Fig. 6, local users can access fog nodes to check the real-time value of indicators of the warehouse environment, such as temperature and humidity, smoke and light. As shown in Fig. 7, remote users and local users can log in the system to check the real-time situation of the warehouse environment and control related equipment. If the system is set to automatic execution mode, it will automatically trigger corresponding device when one parameter is out of the threshold.

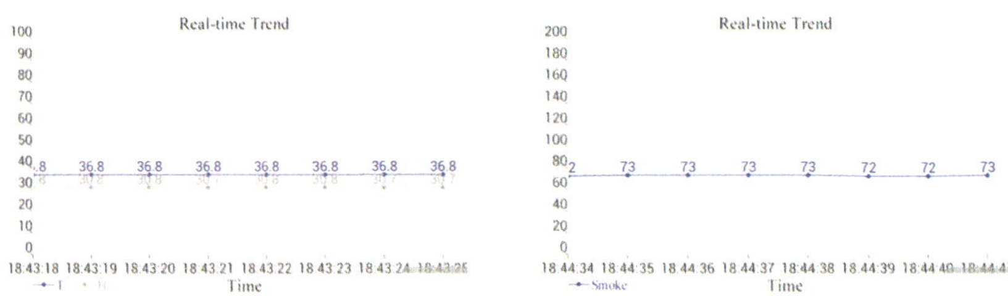

Fig. 6. Interface of warehouse environment monitoring system based on fog node (Left: Temperature and humidity; Right: Smoke)

The results show that both cloud platform and fog node can provide effective warehouse environment monitoring under normal monitoring conditions. In case of emergency, the fog node can make an earlier warning before the cloud platform, and

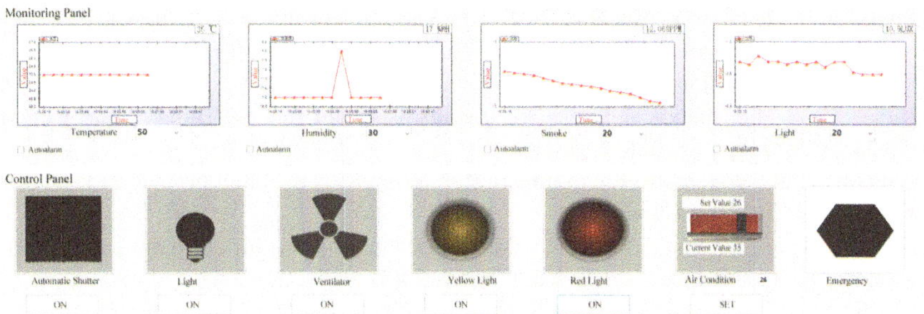

Fig. 7. Interface of warehouse environment monitoring system based on cloud

trigger the corresponding device to deal with the accident. When the communication network between the fog layer and the cloud breaks down, the fog node can provide basic functions of warehouse environment monitoring system, and store the sensing data temporarily to avoid data loss caused by the network interruption. On the whole, compared with traditional warehouse environment monitoring system based on IoT, the fog-based warehouse environment monitoring system has stronger fault tolerance and higher stability.

5 Conclusion

In this paper, to alleviate the weakness of traditional warehouse environment monitoring system based on IoT, we propose a new warehouse environment monitoring system based on fog computing, put forward the architecture and functions of the system, and design the components of fog node and the distributed service provisioning flow based on fog node. Finally, we evaluate the system based on a physical test platform, and the results show that in contrast to the IoT-based system, the system proposed in this paper can better adapt to the complexity of warehouse application scenarios and the variability of user needs.

Our work is only a preliminary exploration of the application of fog computing in the warehouse environment monitoring field, and most of the research is still theoretical. Our future research focuses on: (1) Overall evaluation of the feasibility, reliability and stability of the system; (2) Task coordination and allocation among multiple Fog nodes; (3) Load balance between Fog and Cloud.

Acknowledgements. The research leading to these results has received funding from Wuhan Technology and Business University according to the project of the special funding assignment No. S2018003.

References

1. Atlam, H.F., Alenezi, A., Alassafi, M.O., Wills, G.B.: Blockchain with Internet of Things: benefits, challenges, and future directions. Int. J. Intell. Syst. Appl. **10**(6), 40–48 (2018)

2. Zhiqiang, Y., Henjun, Z., Wenhe, D.: Design and implementation of automated warehouse monitoring system based on the Internet of Things. J. Hebei Univ. Eng. (Nat. Sci. Ed.) **31**(2), 86–88 (2014). (in Chinese)
3. Kuila, S., et al.: Analytical Survey on Standards of Internet of Things Framework and Platforms. Springer, Singapore (2019)
4. Qing, T., et al.: Warehouse Environment Monitoring System Based on IoT. Mod. Electron. Tech. (10) (2018). (in Chinese)
5. Zhang, T., Yan, L., Yang, Y.J.W.N.: Trust evaluation method for clustered wireless sensor networks based on cloud model. Wirel. Netw. **24**(3), 777–797 (2018)
6. Bin, S., et al.: The application in intelligent warehousing environment monitoring by Internet of Things. Logist. Sci-tech **40**(3), 48–51 (2017). (in Chinese)
7. Shakerkhan, K.O., Abilmazhinov, E.T.: Development of a method for choosing cloud computing on the platform of PaaS for servicing the state agencies. Int. J. Mod. Educ. Comput. Sci. (IJMECS) **11**(9), 14–25 (2019)
8. Ahn, D.J., Jeong, J., Lee, S.: A novel cloud-fog computing network architecture for big-data applications in smart factory environments. Springer, Cham (2018)
9. Gopinath, M.P., et al.: A secure cloud-based solution for real-time monitoring and management of internet of underwater things (IOUT). Neural Comput. Appl. **31**(1), 293–308 (2019)
10. Durao, F., et al.: A systematic review on cloud computing. J. Super Comput. **68**(3), 1321–1346 (2014)
11. Ni, J., et al.: Securing fog computing for Internet of Things applications: challenges and solutions. IEEE Commun. Surv. Tutor. **20**(1), 601–628 (2017)
12. Roca, D., et al.: Fog function virtualization: a flexible solution for IoT applications. In: Second International Conference on Fog and Mobile Edge Computing (2017)
13. Naeem, R.Z., et al.: Fog computing in Internet of Things: Practical applications and future directions (2019)
14. Pande, V., Marlecha, C., Kayte, S.: A review-fog computing and its role in the Internet of Things. Int. J. Eng. Res. Appl. **6**(10), 2248–96227 (2016)
15. Kumar, V., Laghari, A.A., Karim, S., Shakir, M., Brohi, A.A.: Comparison of fog computing & cloud computing. Int. J. Math. Sci. Comput. (IJMSC) **5**(1), 31–41 (2019)
16. Yi, S., Li, C., Li, Q.: A survey of fog computing: concepts, applications and issues. In: The Workshop on Mobile Big Data (2015)
17. Singh, S.P., et al.: Fog computing: from architecture to edge computing and big data processing. J. Supercomput. **75**(4), 2070–2105 (2019)
18. Roy, D.G., et al.: Application-aware cloudlet selection for computation offloading in multi-cloudlet environment. J. Supercomput. **73**(4), 1672–1690 (2017)
19. Cisco: Iox overview (2018). http://goo.gl/n2mfiw
20. Zhang, Q., Cheng, L., Boutaba, R.: Cloud computing: state-of-the-art and research challenges. J. Internet Serv. Appl. **1**(1), 7–18 (2010)
21. Van den Abeele, F., et al.: Sensor function virtualization to support distributed intelligence in the Internet of Things. Wirel. Pers. Commun. **81**(4), 1415–1436 (2015)
22. Verma, M., Bhardwaj, N., Yadav, A.K.: Real time efficient scheduling algorithm for load balancing in fog computing environment. Int. J. Inf. Technol. Comput. Sci. (IJITCS) **8**(4), 1–10 (2016)

A Neuro-Fuzzy Pricing Model in Conditions of Market Uncertainty

N. Yu. Mutovkina(✉) and A. N. Borodulin

Tver State Technical University, Tver, Russia
letter-boxNM@yandex.ru, bor74@mail.ru

Abstract. The article proposes a pricing model in the face of uncertainty at an early stage in the product life cycle. The development of new products, products, and design solutions is characterizing by the lack of the necessary amount of value and physical indicators for accurate pricing. At the early stage of the product life cycle, it is possible to operate only with its intended function or consumer properties. The number of statistical data on the demand for such a product on the market is severely limiting due to its novelty. Therefore, data on the results of sales of products with similar characteristics are taking as the initial statistical sample. The model is basing on expert assessment methods and the construction of hybrid neural networks. The analysis of the retrospective data is carried out taking into account the expert's ideas about the features of the system. The simulation was performing in the MATLAB software environment with the Fuzzy Logic Toolbox expansion pack.

Keywords: Pricing · Hybrid neural networks · Expert assessments · Membership functions · ANFIS editor · Fuzzy inference

1 Introduction

The effective solution of the tasks of promoting a new product to the market is largely determining by the estimates of its marketing and technical and economic parameters. An integral part of this process is the study of consumer and functional properties of a new product based on statistical research and expert methods. At the same time, the main task of the analytical tools used to set prices for an innovative product is to link quantitative price and quality functional indicators in one model. In a sense, such mechanisms are implementing by tools of functional-value analysis [1]. But functional cost analysis can be carried out only if there is complete and reliable intra-company information, in particular, on costs, demand for the goods, the intensity of consumption of the goods, etc. When an innovative product is developing, such as accurate information, as a rule, cannot be obtained. Therefore, it is advisable to determine the price at the early stages of the product life cycle based on assessments of it's future functional, consumer and other properties. It is very important to take into account the existing statistics on goods already sold. Within one organization, the use of statistics will allow taking into account a lot of factors related to the specifics of the production and sales process: technological level of development, personnel qualifications, market advantages, etc. This allows you to calibrate the model on the parameters of the organization,

Z. Hu et al. (Eds.): AIMEE 2019, AISC 1126, pp. 461–470, 2020.
https://doi.org/10.1007/978-3-030-39162-1_42

which leads to increased accuracy of estimates. Goods (services) within one nomenclature direction can be combining into classes. For each class, a separate pricing model is developing on the values of its functional and consumer characteristics (properties) [2].

In the developed pricing model, the price of a class of goods is determined using a knowledge base based on fuzzy inference rules. The role of the knowledge base is to identify non-linear dependence $Y = f(\bar{X})$, where $\bar{X}$ is vector of values of quantitative and qualitative characteristics of the class of goods; Y is value of the price of goods in conventional units.

The purpose of the work is to create a neuro-fuzzy model for determining the price of a new product, taking into account the conditions of market uncertainty.

2 The Method of Constructing a Neuro-Fuzzy Pricing Model

The construction of private pricing models should be carried out based on the method of two-stage identification of non-linear dependencies using fuzzy knowledge bases. At the first stage, structural identification is performed, which is the formation of a fuzzy knowledge base. This knowledgebase reflects the relationship between inputs and output using the linguistic rules 'IF-THAT'. Linguistic rules are generating by extracting fuzzy knowledge from existing statistics. In the second stage, the parametric identification of the dependence under investigation is carried out by finding such parameters of a fuzzy knowledge base that minimize the deviation of the model results and the available statistics.

This approach is implementing by using a fuzzy neural network that can simultaneously form fuzzy rules and adapt membership functions by modifying the weights of connections in the learning process [3]. A typical model structure based on a fuzzy inference should include the following main blocks (Fig. 1):

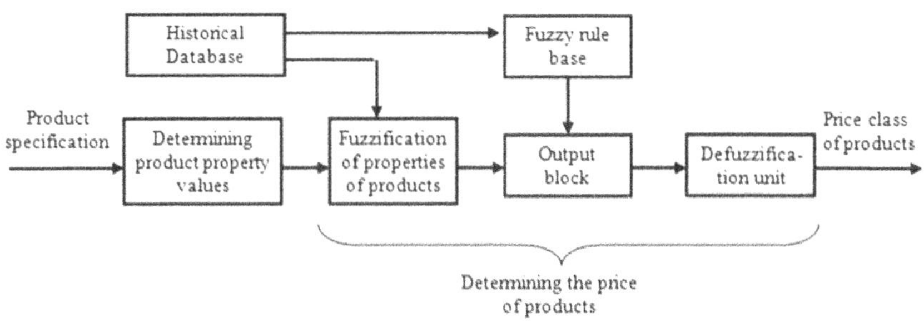

Fig. 1. Using a fuzzy production model to determine the price of goods at the initial stage of the life cycle

- fuzzifier that converts a fixed vector of influencing factors $\bar{X}$ into a vector of fuzzy sets $\tilde{X}$ necessary to perform a fuzzy inference;

- a fuzzy knowledge base containing information on the dependence of $Y = f(\bar{X})$ in the form of linguistic rules of the type ‘IF-THAT’;
- a fuzzy inference machine, which, based on the rules of the knowledge base, determines the value of the output variable in the form of a fuzzy set $\tilde{Y}$, corresponding to the fuzzy values of the input variables $\tilde{X}$;
- defuzzifier that converts the output fuzzy set $\tilde{Y}$ to a exact number Y.

The main feature of the model under consideration is that the conclusions of the rules are not representing as belonging to the output variable to fuzzy sets but in the form of functional dependencies (the Takagi-Sugeno-Kang neural production network) [4]:

$$\begin{aligned} &\textit{if } x_1 \textit{ is } A_{i1} \textit{ and} \ldots \textit{and } x_j \textit{ is } A_{ij} \textit{ and} \ldots \textit{and } x_m \textit{ is } A_{im} \\ &\textit{then } y = c_{i0} + \sum_{j=1}^{m} c_{ij}x_j, \quad i = 1, \ldots, n, j = 1, \ldots, m, \end{aligned} \tag{1}$$

where A_{ij} is fuzzy set, reflecting the value of the characteristics of the product x_j, with the membership function $\mu(x_j)$; x_j is quantitative characteristic of the product, determining its functional properties; y is a distinct output variable equal to the desired price, determined by a definite function of the values of the characteristics of the products; c_{i0} and c_{ij} are parameters of linear functions of the conclusions of the rules; n is number of rules; m is number of input variables.

The considered neuro-fuzzy network is basing on the following provisions:

(1) input variables are clear;
(2) the membership functions of all fuzzy sets are representing by the generalized *gbellmf* function (bell-shaped);
(3) the fuzzy implication is implementing as a fuzzy product;
(4) T-norm is implementing as a fuzzy product;
(5) accumulation of activated conclusions of the rules is not carried out;
(6) defuzzification produced by finding the middle center.

Based on these provisions, the functional dependence for obtaining the defuzzified value of the output variable will take the following form:

$$\begin{aligned} y' &= \frac{\sum_{i=1}^{n}\left(\left(c_{i0} + \sum_{j=1}^{m} c_{ij}x_j\right)\prod_{j=1}^{m}\mu_{A_{ij}}\left(x_j'\right)\right)}{\sum_{i=1}^{n}\prod_{j=1}^{m}\mu_{A_{ij}}\left(x_j'\right)} = \\ &= \frac{\sum_{i=1}^{n}\left(\left(c_{i0} + \sum_{j=1}^{m} c_{ij}x_j\right)\prod_{j=1}^{m}\exp\left[-\left(\frac{x_j' - a_{ij}}{b_{ij}}\right)^2\right]\right)}{\sum_{i=1}^{n}\prod_{j=1}^{m}\exp\left[-\left(\frac{x_j' - a_{ij}}{b_{ij}}\right)^2\right]} \end{aligned} \tag{2}$$

where a_{ij}, b_{ij} are nonlinear parameters of membership functions $\mu_{A_{ij}}\left(x_j'\right)$.

The structure of the ANFIS network, as well as its learning algorithms, are well known and described in some detail in several publications, for example, in [5–7], and therefore are not given here.

One of the most effective ways to train fuzzy networks is a hybrid algorithm [8]. Its main feature is to divide the learning process into two separate stages in time. At each stage, only part of the network parameters is specifying. If we consider that the computational complexity of each algorithm is proportional to the number of parameters, then reducing the dimension of the problem significantly decreases the number of mathematical operations and increases the speed of the algorithm. Due to this, the hybrid algorithm is much more efficient than the usual frontal type gradient algorithm, according to which the refinement of all network parameters is performing simultaneously.

In general, the method of using hybrid fuzzy models to identify non-linear dependencies in the pricing system at the early stages of the product life cycle can be represented as the following steps:

1. Construction of the reference membership functions for all input variables of the model. Input variables can be important functional and consumer properties of an innovative product. To do this, the statistical data (based on the experience of selling similar products) determine the minimum and maximum values of each input variable $x_i^{(\min)}$ and $x_i^{(\max)}$. Each interval defined in this way is dividing into several segments. For example, for a variable x_i, fuzzy sets with linguistic values {LOW is 'small number', MED is 'average number', HIGH is 'large number'} can be defined. For each of these sets, you must specify the membership function.
2. Determining the number of adjustable parameters of the base of fuzzy rules, which includes the nonlinear parameters a_{ij}, b_{ij} of the membership functions of fuzzy sets of rule premises and the parameters c_{i0} and c_{ij} of linear functions from the conclusions of the rules. The total number of adjustable parameters for the first-order Sugeno model will be determined by the formula:

$$L = 2\sum_{i=1}^{m} l_i + (m+1)\prod_{i=1}^{m} l_i \tag{3}$$

   where l_i is number of membership functions for the i-th input variable; m is number of input variables.
3. Preparation of historical data to adjust the parameters of the model. For this purpose, it is necessary to create a base of fuzzy rules with several input variables $(x_1, x_2, \ldots, x_m)$ and one output variable y. The output variable is the value of the price (value) of the goods, expressed in a single currency. In this case, the training sample consists of the following set examples:

$$\left(x_1^{(k)}, x_2^{(k)}, \ldots, x_m^{(k)}, y^{(k)}\right),\ k = \overline{1, K} \tag{4}$$

where $x_1^{(k)}, x_2^{(k)}, \ldots, x_m^{(k)}, y^{(k)}$ are values of input variables and output variable in the kth example; K is total number of examples in the training set. In this case, part of the examples (1/3 is recommending) is reserving for the control sample.

4. The construction of a neuro-fuzzy network consisting of all possible rules in the form (1). At the same time, the condition $K < 2L$ determines the sufficiency of the amount of statistical data for setting the parameters of the first-order Sugeno base of fuzzy rules, which usually gives a more accurate result than the zero-order Sugeno base of zero-order rules (which is constructing for the case of $K > 2L$).
5. Optimization (volume reduction) of the rule base, which is an adaptation of the rule base to the existing experimental data (training set). The necessity of this stage is causing by the large size and redundancy of the base of production rules with a significant number of input variables and the dimension of their term sets. An effective method for solving this problem is to calculate for each rule a kind of rating followed by the choice of the rules for which this rating will be the highest. For this, examples from the training set (4) are 'presented' to each rule. As a result, the rating of each i-th rule can be calculated by the formula:

$$r_i = \sum_{k=1}^{K} \mu_{A_{i1}}\left(x_1^{(k)}\right) \cdot \mu_{A_{i2}}\left(x_2^{(k)}\right) \cdots \mu_{A_{im}}\left(x_m^{(k)}\right), \; i = 1, \ldots, n \tag{5}$$

6. Learning the resulting base of fuzzy rules, which is the stage of parametric optimization of the final set of rules. For this purpose, the Takagi-Sugeno-Kang network is built, repeating the structure of the established rule base.
7. Calculation of the average total error estimate of the obtained fuzzy production model on a test sample and comparing it with the permissible error (ε):

$$E = \sqrt{\frac{1}{K}\sum_{k=1}^{K} \left(y'^{(k)} - y^{(k)}\right)^2} < \varepsilon \tag{6}$$

8. Estimation of the accuracy of the obtained model, which is carried out from the average total error ε calculated at the previous stage. If the error is unsatisfactory, it is possible to change the initial parameters of the model by adding new functionality or consumer characteristics of the goods and including new statistical data on the existing analogs in the model. In this case, you need to return to the first stage of the methodology. If the error level agrees with the data from the test sample, the obtained knowledge base model can be used in the pricing system.

It is important to emphasize the need to continually supplement the knowledge base with new statistical data and expand the number of properties used in the evaluation of goods. Besides, a significant factor in improving the accuracy of the results obtained by the hybrid network is the constant verification of the compliance of their values with the data from test samples. Obtaining test (control) data should be carried out regularly and be based on a comprehensive study of the results of marketing research.

3 Description of the Neuro-Fuzzy Pricing Model and Examples of Its Work

For simplicity, only two characteristics were chosen as input parameters of the model, which can be easily quantified and reflect the effectiveness of the future product for the consumer. For example, for gasoline generators, this is the rated power (x_1), kW and battery life (x_2), hours. In any case, it is important that there is no direct and easily formalizing the connection between these parameters. It is obvious that the more power and work time of the generator without refueling, the higher its price.

The next step in determining the optimal price of a new product after highlighting its estimated functional properties is the development of non-linear dependency identification models. Building a model for estimating the value of an innovative product involves the use of historical data on goods already sold. The procedure for the preparation of statistical data involves the collection of data on the characteristics of x_1 and x_2 for all products of similar functionality known in the market. Actual prices (y) are also selecting for all the goods in question.

Next, the training samples are formed. At the same time, it is possible that for each selected class of goods, its training sample is created, which is constantly updated with new emerging products. The data obtained from a series of examples. The subsequent development of a base of fuzzy rules can be carried out in any instrumental environment for the development of neural networks that support algorithms for fuzzy inference of Sugeno. For example, the program modules of the Fuzzy Logic Toolbox extension package and the integrated command language of the Matlab system can be used for this.

The choice of the Matlab system as a working tool for implementing the proposed method is not accidental. Since the hybrid neural network is a rather complex structure from a mathematical point of view, it seems appropriate to use a well-established and proven software product for its design. The Fuzzy Logic Toolbox package that is part of the system allows you to construct fuzzy expert systems and hybrid neural networks. The construction of a fuzzy output system in the Fuzzy Logic Toolbox package is carried out either using a visual interface or in command line mode, by the following algorithm.

1. The membership functions of each fuzzy variable are specified.
2. Rules for fuzzy inference, given by experts, are formed. These rules can be viewed, modified, partially deleted or saved as a file.
3. Specifies the type of implication and conjunction operations that will be using in the fuzzy inference mechanism.
4. The created system of fuzzy output is saving as a file on disk. This system can be used to build neuro-fuzzy models, both in the current session of the program and at any other time when it is required.

The fuzzy system can be imported into the module of construction of hybrid neural networks ANFIS-Editor of the Fuzzy Logic Toolbox package to create a neuro-fuzzy network based on it. The module ANFIS provides features:

- determination of the method of the learning network. A hybrid or back propagation method may be using;
- setting the level of the current total (for all samples) learning error (Error Tolerance), upon reaching which the learning process ends;
- setting the number of training cycles (Epochs), i.e. the number of "runs" of all samples of the training sample;
- implementation of the procedure for setting network parameters, the completion of which occurs when a specified level of error is reaching or a specified number of training cycles are completing;
- testing the trained network and calculating the output corresponding to a given set of input variables.

The constructed neuro-fuzzy model is saving as a file with the extension .fis. If necessary, it can be called up using the command anfisedit ('Model name') for training and testing.

To build a pricing model, statistics on 20 types of gasoline generators were using. The total number of examples in the training set was 500 copies. After the first step of the algorithm for constructing a fuzzy model to establish the price of a new product, the type of membership functions for certain input variables x_1 and x_2 are shown in Fig. 2.

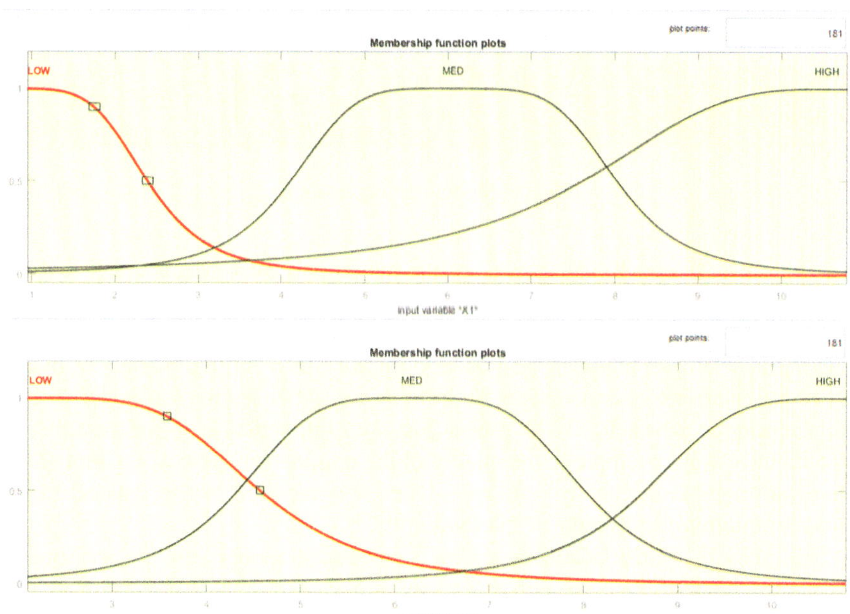

Fig. 2. Membership functions of input variables

The input parameters of a fuzzy model can be considered as linguistic variables, the values of which are determining from the following term-sets: {'small number' (LOW), 'average number' (MED), 'large number' (HIGH)}.

When training, 45 parameters of a fuzzy model were adjusted: 27 coefficients in linear functions of the conclusions of the rules and 18 parameters of the membership functions of the input variables. The resulting knowledge base contains 9 rules.

The order of adaptation of the parameters of the obtained base of fuzzy rules is presenting in Fig. 3.

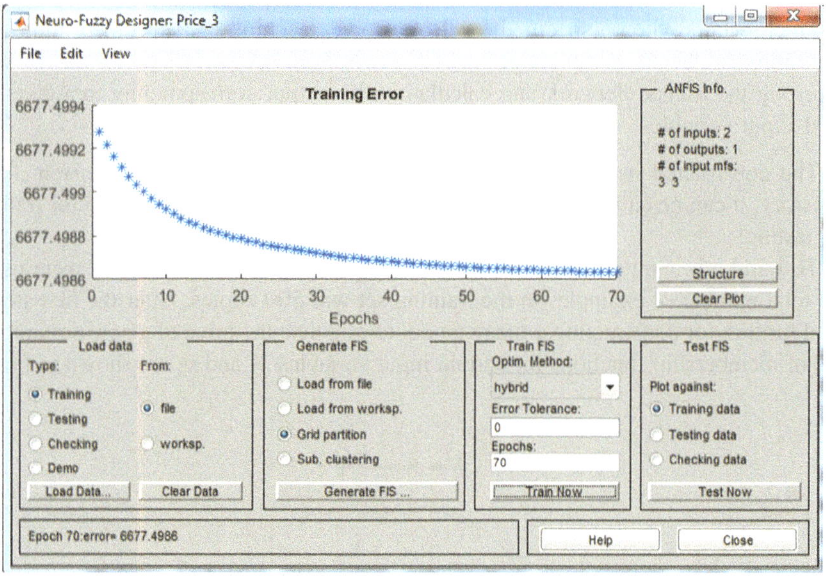

Fig. 3. Model training example

After the 65th epoch, the training of the model was over. Mean square error was 6677.5 rubles.

To improve the accuracy of estimation at the next iteration of the model building cycle in this example, an attempt is made to improve the resulting model. To this end, several observations were added to the training set. After learning a model with new data, the error decreased to 6100.8 rubles (Fig. 4). Adding data is one of the easiest and most effective ways to improve the quality of the training set. However, simply adding data of any type is not always effective. Often it is required to add data of a certain variety to improve the quality of the model [9].

The analysis of the adequacy of the constructed model can be performed by viewing the rules of the corresponding fuzzy inference system. If the adequacy of the model is not confirmed, then additional configuration is necessary. In addition to preparing and loading a larger sample with training source data, configuration methods include:

(1) preparation and loading of an additional file with verification initial data which are absent in the training set;

(2) editing the types and values of the parameters of the membership functions of the terms of the input and output variables using the corresponding Matlab editor [10, pp. 445–450].

The development of a fuzzy inference system in interactive mode is most effective for complex fuzzy models with a large number of variables and fuzzy inference rules. In this case, setting variables and membership functions of their terms in graphical mode, as well as visualizing the rules, can significantly reduce the complexity of developing a fuzzy model, reduce the number of possible errors and reduce the total time of fuzzy modeling [11].

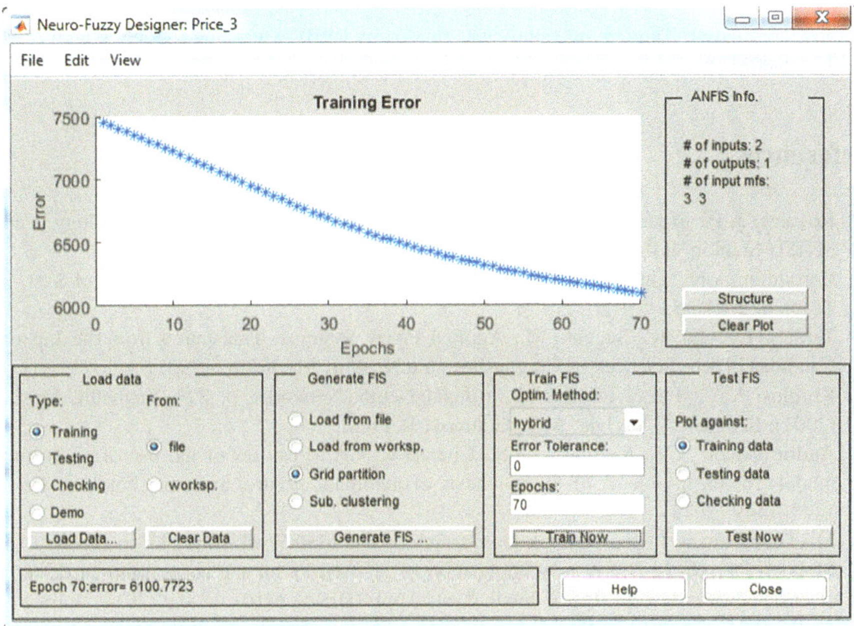

Fig. 4. An example of learning the model on the extended sample

4 Conclusions

It is important to note that the resulting neuro-fuzzy model is not a one-time, that is, the entire cycle of building and tuning a neural network does not need to be carried out for each new variant of the assessment. Of course, this model allows effective pricing only for products that are similar in their functional and consumer properties, as well as being in the same price range. By creating a set of similar models, for example, for products of the same type, it becomes possible to solve many pricing problems in conditions of high market uncertainty and difficulties in obtaining valuations of designed products. In general, the method used and its instrumental implementation is shown in this work, which can be widely used as a means of intellectual analysis in the system of in-house management for pricing in the early stage of the product life cycle.

The dynamism of the market situation requires constant improvement by enterprises and organizations of their pricing policy. In this situation, the use of data mining methods based on neuro-fuzzy networks allows you to create an effective and adaptive pricing mechanism, especially at an early stage in the product life cycle. The developed method and model make it possible to use the available statistical information on the sold and designed products of both quantitative and qualitative nature, displaying it in terms of natural language – in the form of fuzzy production rules 'IF-THAT', which allows to take into account a lot of factors associated with the further process production and sales.

Acknowledgments. The reported study was funded by RFBR according to the research project No. 17-01-00817A.

References

1. Kovalev, A.P.: Modern trends in the development of the cost analysis methodology. Vestn. MGTU Stankin **4**(35), 147–154 (2015)
2. Borodulin, A.N.: The use of neuro-fuzzy methods in the pricing system. Control Syst. Inf. Technol. **2**(44), 53–57 (2011)
3. Terano, T., Asai, K., Sugeno, M.: Applied Fuzzy Systems: Translation from the Japanese Candidate Tech. Sciences of Chernyshov, Yu.N., Mir, M. 363 p (1993)
4. Kruglov, V.V.: Fuzzy Logic and Artificial Neural Networks, p. 224. Fizmatlit, Moscow (2001). Kruglov, V.V., Dly, M.I., Golunov, R.Yu.M.
5. Andrievskaya, N.V., Reznikov, A.S., Cheranev, A.A.: Features of the use of neuro-fuzzy models for the problems of the synthesis of automatic control systems. Fundam. Res. **7**, 1445–1449 (2014). 11–7
6. MATLAB Fuzzy Logic Toolbox User's Guide. The MathWorks, Inc. 333 p (2008)
7. Mohdeb, N., Hacib, T.: A new application of an ANFIS for the shape optimal design of electromagnetic devices. Int. J. Intell. Syst. Appl. (IJISA) **6**(10), 11–19 (2014). https://doi.org/10.5815/ijisa.2014.10.02
8. Joumah, A.A., Albitar, C.: Design optimization of 6-RUS parallel manipulator using hybrid algorithm. Int. J. Inf. Technol. Comput. Sci. (IJITCS) **10**(2), 83–95 (2018). https://doi.org/10.5815/ijitcs.2018.02.08
9. Kaftannikov, I.L., Parasich, A.V.: Problems of formation of a training sample in machine learning tasks. Bull. S. Ural State Univ. Ser. Comput. Technol. Manag. Electron. **16**(3), 15–42 (2016)
10. Leonenkov, A.V.: Fuzzy modeling in MATLAB and fuzzyTECH. SPb.: BHV-Petersburg, 736 p (2005)
11. Tarasyan, V.S.: Fuzzy Logic Toolbox for Matlab: Study Guide, p. 112. Publishing house of Ural State University of Railway Engineering, Yekaterinburg (2013)

Combined Intelligent Control of a Signalized Intersection of Multilane Urban Highways

Anatoliy A. Solovyev and Andrey M. Valuev(✉)

Mechanical Engineering Research Institute of the Russian Academy of Sciences,
4, Malyi Kharitonievsky pereulok, 101990 Moscow, Russia
aa.solovjev@yandex.ru, valuev.online@gmail.com

Abstract. The paper treats the problem of intelligent traffic control at signalized intersections of urban highways and evolves the approach based on joint choice of the phase separation of passage directions and the durations of the traffic light cycle phases. The method is proposed for generation of the entire set of traffic separation schemes (TSSs) with a limited conflict level, the latter depending on the number of active conflict points of merging on traffic light cycle phases. The linear programming problem for determination of the most efficient parameters of the traffic light cycle for a given TSS is formulated; its solution is recommended to be used for the choice of preferable safe TSS for the established traffic demand. Besides, a principal way of corrections of the phase durations of the current cycle depending on composition of vehicle queues on intersection entries is proposed; simulation techniques for this objective are discussed. Aspects of "education" of drivers that provides the efficiency of the combined intelligent control of signalized intersections are discussed.

Keywords: Traffic flow · Safety · Intelligent control system · Signalized intersection · Conflict points · Traffic organization · Traffic light cycle · Traffic separation scheme · Optimization

1 Introduction

The paper is aimed at evolving the concept of an intelligent traffic management system and focuses on regulated intersections. Recently methods and systems of intelligent control for traffic lights were given a great deal of attention but they concentrate on the choice of traffic light cycle (TLC) parameters. One of the first solution of that type is known from early 1980s under the name SCOOT (split, cycle and offset optimization technique) [1] and showed the ability of SCOOT to adapts to unusual traffic conditions as well as to the usual variations in demand within a day. Modern conditions for application of such systems including local signal controllers, vehicle detectors and communication techniques are discussed in [2]. Nevertheless, such systems and control techniques are not entirely satisfactory and new approaches are intensively developed. Except new ways of data registration and transmission, e.g., by a wireless sensor network (WSN) [3–7], some new ways to deal with these data are proposed. Some researches discuss preference rules in assigning the green light duration [8]. Adaptation to the variance of the transport situation by reinforcement learning of a controlling

Z. Hu et al. (Eds.): AIMEE 2019, AISC 1126, pp. 471–480, 2020.
https://doi.org/10.1007/978-3-030-39162-1_43

neural network is proposed in [9, 10]; the artificial bee colony algorithm was proposed as well [11]. Modern approaches to traffic safety issues are presented in [12] and [13]. The latter paper pays attention to road structure that is very important for crossroads but was not learned systematically in the aspect of traffic control.

The goal of the paper is to propose a constructive way of complex control of a signalized intersection, combining its structural and parametric optimization.

In Sect. 2 the problem of determination of variants' set of traffic separation schemes for a signalized intersection is set up. The method of its solution is described and illustrated in Sect. 3. Section 4 presents an approach to the choice of the most efficient TSS for a certain traffic situation; it includes the solution of a linear programming problem for optimization of phase durations on average for each recommended TSS. Possibilities of phase durations correction for the next traffic light cycle based on the information about compositions of entry queues are presented in Sect. 5. Aspects of "education" of drivers that provides the efficiency of the combined intelligent control of signalized intersections are discussed in Sect. 6.

2 The Problem of Determination of Variants of Traffic Separation Schemes for a Signalized Intersection with Given Road Markings

In the previous papers by the authors [14, 15] the formal description of a signalized intersection was proposed aimed at the choice of the most efficient traffic organization on it, the latter being regarded from the viewpoint of both its throughput and traffic safety on it. Detailed structure of an intersection is described with the sets of lanes by which the legalized motion of vehicles is fulfilled and points of their connection. The latter are usually referenced as *conflict points* since their passage may cause conflicts between passing vehicles. For trajectories of vehicles these points are *singular points* (SPs) since the direction of movement in these points may change. There are SPs of three types, namely SPs of crossing, merging and separation types (SPC, SPM, and SPS). The danger of their simultaneous passage from several directions depends on their types and is the greatest for SPCs and the smallest for SPSs. Each scheme of traffic separation by phases of the traffic light cycle (TLC) or, briefly, a traffic separation scheme (TSS) is the sequence of sets of permitted traces for phases of the TLC. Conflict points through which more than one permitted trace passes on a certain phase are called active SPs for the phase. The problem of the reasonable TSS definition consists in separation of all permitted traces among phases minimizing the number of active SPs with their differentiation in the degree of danger. For this reason in most cases only TSSs having no active SPC at each phase are used because SPCs cause the greatest danger. A TSS of such type is called a conditionally conflict-free TSS.

As to SPMs, they are sources of greater danger than SPSs, so their simultaneous passage from several directions must be minimized if possible. Due to the above reasons, traffic separation schemes are treated as absolutely conflict-free if all SPCs and

SPMs are inactive for any phase. At last, active SPSs are inevitable since there is no possibility for vehicles intending to pass a certain SPS in one direction to wait near the SPS while other ones pass the SPS in another direction. For this reason vehicles' passage through all existing SPSs on the intersection in question is taken for granted and must not be taken into account in the choice of adequate TSSs for it.

The problem is to find all possible reasonable absolutely conflict-free TSSs or conditionally conflict-free TSSs with additional restrictions. The latter mean limitations on the number of active SPMs on each phase and/or on all phases. As to reasonable TSSs, we may reduce our search to non-expandable TSSs. Passage scheme for a phase satisfying the given limitation on the number of active SPMs is treated as *expandable* if there is another passage scheme including additional traces and satisfying the same limitation. Only TSSs with non-expandable passage schemes for all phases are treated as *non-expandable TSSs* and are reasonable since addition to an expandable passage scheme extra traces producing no new conflicts increases throughput with no negative effects.

When solving the problem in question, we suppose that in the intersection area there is either the sole trace or no traces between each pair (entrance, exit). Our study of intersection structure for Moscow highways confirms the relevance of such supposition. Besides, it is impossible to pass vehicles from a definite entry to a definite exit on a certain TLC phase and simultaneously not to pass them to another exit linked with this entry by a trace.

From these assumptions we conclude that the set of traces beginning in a definite entry is either entirely included in a passage scheme for a certain TLC phase or entirely excluded from it. To define a certain TSS means to share all entry vertices (with all traces beginning in them) between phases of TLC satisfying the above demands on a conflict level. Every entry must pertain to at least one phase and some entry vertices may pertain to several phases.

3 Method of Determination of the Desired TSS Set

To solve the problem we need to have the table of mutual disposition of traces beginning in each pair of entries. For two traces in the intersection area beginning in different entries, the conflict level is defined as 3 if they have the common SPC, as 2 if they have the common SPM and as 0, otherwise. For two different entries, V and W, we define S_{VW} index that show the maximum conflict level for all pairs of traces, the 1st one beginning in V and the 2nd in W.

Let us consider the proposed method for constructing admissible TSS using an example constructed according to the type of intersection of Moscow Profsoyuznaya Street with Obruchev Street and Nakhimovskiy Avenue (Fig. 1).

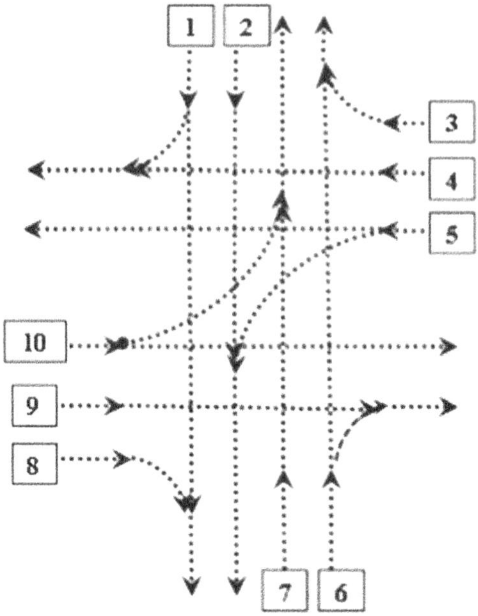

Fig. 1. The system of routes at the intersection of multi-lane highways

For example, for $V = 1$ and $W = 5$ we have two traces from each, the 1st one straight and the 2nd with a turn. Only straight traces intersect, so $S_{15} = 3$. For $V = 2$ and $W = 5$ we have one SPM and one SPC as common vertices on traces from these entries, so $S_{25} = 3$. As for $V = 3$ and $W = 6$, we have the only common SPM-type vertex, so $S_{36} = 2$ (Table 1).

Table 1. Values of the S_{VW} index for pairs (V, W)

W	V, S_{VW}									
	1	2	3	4	5	6	7	8	9	10
1		0	0	3	3	0	0	2	3	3
2	0		0	3	3	0	0	0	3	3
3	0	0		0	0	2	0	0	0	0
4	3	3	0			3	3	0	0	0
5	3	3	0	0		0	3	3	30	3
6	0	0	2	3	3		0	0	3	3
7	0	0	0	3	3	0		0	3	3
8	2	0	0	0	0	0	0		0	0
9	3	3	0	0	0	3	3	0		0
10	3	3	0	3	3	3	3	0	0	

Next step after the definition of S_{VW} values is the search of *compatible combinations of entries* (CCEs). The set of them is represented with a rooted tree which root is an empty set, vertices of the 1st level correspond to single entries, those of the 2nd level —to compatible pairs of entries, vertices of the 3rd level correspond to compatible triples of entries etc. To diminish the tree dimension, each m-tuple of compatible entries is represented in a single way, namely as $(V_1, V_2, \ldots, V_m)$, where $V_1 < V_2 < \ldots < V_m$. Having a parent tree vertex $(V_1, V_2, \ldots, V_m)$ we test for child vertices $(V_1, V_2, \ldots, V_m, W)$ for all $W > V_m$ checking their compatibility.

When seeking for only absolutely conflict-free TSSs, compatibility of $(V_1, V_2, \ldots, V_m)$ is expressed by the following condition: $S_{UW} = 0$ for any $U = V_i$, $W = V_j$, $1 \leq i < j \leq m$. Seeking TSSs with limited conflict level (CL) we demand that $S_{UW} < 3$ for all pairs of entries in the CCE and besides the number of such pairs for which $S_{UW} = 2$ does not exceed the number $N_{2\max}$. When checking compatibility conditions for $(V_1, V_2, \ldots, V_m, W)$ all checks are made for pairs (V_i, W) only.

The next step is to reduce the set of CCEs to non-expandable CCEs only. All non-expandable CCEs are leafs of the CCE tree but not all leafs are non-expandable CCEs. To exclude leafs that correspond to expandable CCEs it is sufficient to compare pairs of leafs which level in the tree differs by 1 and leafs for which the set of entries is included in the entries' set of another leaf.

In this example, the leafs of absolutely conflict-free CCE tree are:

for $m = 2$—(1,7), (2,8), (3,10), (4,9), (5,8), (6,8), (7,8), (8,10), (9,10);
for $m = 3$—(1,2,7), (1,3,7), (1,6,7), (2,3,8), (2,6,8), (2,7,8), (3,4,9), (3,5,8), (3,8,10), (3,9,10), (4,5,8), (4,8,9), (6,7,8), (8,9,10);
for $m = 4$—(1,2,3,7), (1,2,6,7), (2,3,7,8), (2,6,7,8), (3,4,5,8), (3,4,8,9), (3,8,9,10).

A comparison of tree leaves shows that only CCEs with the maximum level $m = 4$ are non-expandable. We denote them respectively $G_1, \ldots, G_7$.

After determination of all non-expandable CCEs reasonable TSSs are generated for which some of the found CCEs are assigned as passage schemes for certain phases of the TLC. To obtain a meaningful solution, it is necessary to limit the number of phases to no more than 4, which corresponds to the usual practice of organizing traffic light regulation. For our case, it means that the total number of all possible TSSs with the use of all non-expandable CCEs for three-phase TSSs equals to $7 * 6 * 5 = 210$ and for four-phase TSSs equals to $7 * 6 * 5 * 4 = 840$. Two-phase TSSs do not provide passage of vehicles from all entries. Enumeration of all possible variants does not cause any computational difficulty.

We find that a unique three-phase TSS exists, namely (G_2, G_5, G_7), and the four-phase TSS without a redundant fourth phase is also unique—this is (G_1, G_4, G_5, G_7). There are also 4 four-phase TSSs containing G_2 in which the fourth phase is redundant, namely (G_1, G_2, G_5, G_7), (G_2, G_3, G_5, G_7), (G_2, G_4, G_5, G_7) and (G_2, G_5, G_6, G_7). As all of them do not contain any SPM, they are treated as the safest. What TSS is preferable depends on intensity of traffic flows through the intersection in all possible directions.

The analogous analysis was successfully performed for a variety of intersections of Moscow highways, especially in the South-West district. The obtained results are similar: in most cases there is a limited number of non-expandable conflict-free passage schemes.

4 Assessment of Possible TSSs for a Certain Traffic Situation

Both parametric and structural control of a certain signalized intersection demand assessment of a present local traffic situation. Adaptive control of TLC may be based only on instantaneous data, namely on lengths of queues on the intersection entrances, although more sophisticated and efficient methods demand more information related not to the current moment of time only but to the recent period and probably even historical data. For structural control, the latter is obligatory. Shift from one TSS to another one may not be frequent and so must be adequate to the present trend. Before this shift we must assess it consequences "on average" which requires some computations and informational support of these computations. We must be sure that the proper assignment of durations of TLC phases provides the passage of entering vehicles in various directions better than the present TSS. So it is necessary to link recommended durations of TLC with mean characteristics of traffic flow intensity for every direction.

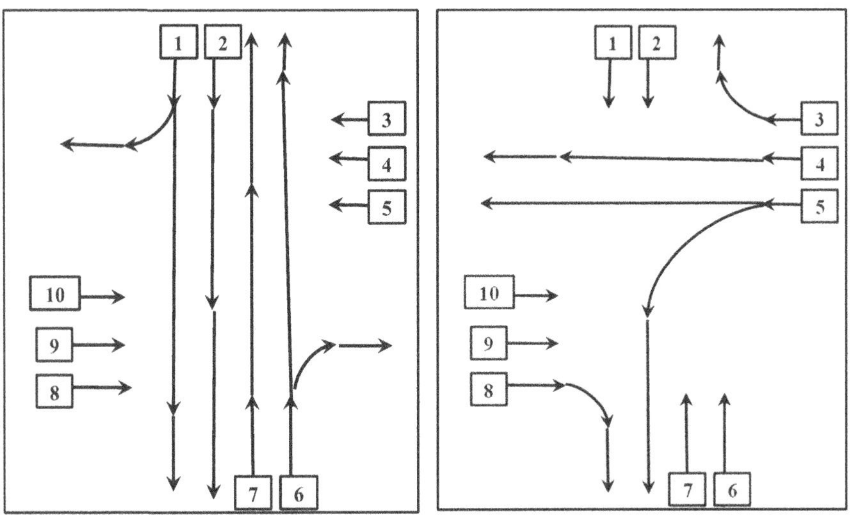

Fig. 2. Passage schemes for the 1st and the 2nd phases of the selected three-phase TSS

Figure 2 shows that for an absolutely conflict-free TSS passage schemes at each phase are sets of independent elementary schemes (to see that for the above three-phase TSS we should note that passage schemes for the 2nd and 3rd phase have the same structure). Here elementary schemes are: straight traces (from entries 2, 4, 7 and 9; the last is not shown here as it pertains to the 3rd phase); traces with one turn (from entries

3 and 8) and splitting traces (from entries 1, 5, 6 and a not shown trace from entry 10). Usage of relatively conflict-free TSSs with a few SPMs adds extra elementary schemes but their count rests very little.

The recommended durations of TLC phases providing on average the most efficient passage of incoming traffic flows may be determined by comparing throughputs of these elementary passage schemes (used in the present TSS) with the average traffic flow intensity through these traces. On the same basis recommendations to shift to another TSS may be wrought down as well.

The principal idea consists in the following. Let T_{TLC} be the total duration of the TLC with N phases, then we may know how many vehicles enter on average for the time T in each entry V. If the direction of motion of these vehicles splits in a certain SPS, then the share for both direction may be assessed as well. With these data we compute the minimum time $T_V(Q_V)$ for which the determined average quantity of vehicles Q_V incoming in V entirely passes the intersection area. Values $T_V(Q_V)$ may be calculated in advance on the basis of depositories of empirical data and/or computer models. If a trace splits, then coefficients of splitting are extra arguments of $T_V(Q_V)$. Durations of phases D_p must satisfy the following restrictions: for each V the total durations of phases on which vehicles from entry V are passed must be not less than $T_V(Q_V)$. E.g., if vehicles from entry 1 are passed on phase 2 and from entry 2 on phases 1 and 3, then restrictions

$$T_1(Q_1) \leq D_2,\ T_2(Q_2) \leq D_1 + D_3$$

must be satisfied. To determine whether it is possible to pass all entering traffic flows with the present TSS (or another one) it is necessary to minimize $D_1 + \ldots + D_N$ under such restrictions, i.e., to solve a linear programming problem of small dimension. TSSs for which $D_1 + \ldots + D_N \leq T_{\text{TLC}}$ are satisfactory and the TSS with the minimum value of $D_1 + \ldots + D_N$ is treated as the most efficient.

If there is no satisfactory absolutely conflict-free TSS, then TSSs with limited conflict levels must be regarded. In that case the problem of the best TSS selection becomes bi-objective, the 1st criteria being the throughput and the 2nd the safety index.

5 The Possibility of Assigning Durations of TLC Phases Depending on the Composition of Existing Queues

The composition of existing queues, i.e., the fraction of vehicles of different types in them, obviously affects their passage through the intersection for a given green light period. If the queue contains more short and quick cars, then they will cross the intersection area more quickly and the queue length will diminish rapidly; on the contrary, in the case where slower and less maneuverable trucks abound.

To balance the passage of traffic flows from different directions, the duration of green light must decrease in the 1st case and increase in the 2nd case. Quantitative recommendations for such changes may be developed in some typical situations, namely in the case where flows from different entries do not split and do not merge within the same phase.

In that case the impact of variations of queues' composition may be assessed if lengths of all vehicles in queues are being established. In fact, it is impossible to calculate precisely the number of vehicles from a definite queue that will pass the crossroad for a given time even having detailed information about all vehicles in the queue. Nevertheless, even the information of a queue as a succession of vehicles with the known category of everyone enables us to evaluate the increment or decrease of this amount in comparison with the mean value. The most reliable way of it consists in computer simulation of the passage of definite vehicles' succession. Authors' experience [16, 17], including computer simulation of traffic flows through a crossroads [18], shows possible effective ways of it. Less reliable valuations may be based on stored observation data with various shares of different vehicle categories in queues via their interpolation to the present values. Reversing the dependence between the green phase duration and the count of passed vehicles for a known queue composition (in terms of vehicles' categories), we obtain something like $T_1(Q_1)$ and with these values may solve the problem of optimum phase durations for the current cycle analogous to thereof in the above section.

Splitting of traffic flows from a certain entry make difficulties in implementation of this approach. Nevertheless, it may be applied in some cases, namely if distributions of flows between further directions after SPMs' passage are stable or when the choice of the further direction is mainly determined by the vehicle category.

6 Aspects of "Education" of Drivers for Their Adaptation to Combined Intelligent Control of Signalized Intersections

Changes of the present TSS used at a certain intersection depending on the current traffic situation cannot confuse irregular drivers but may cause erroneous driving for those drivers that cross the intersection regularly. The problem with this category exists not for the case where these changes are made on a regular basis, depending on the hour of the day and/or day of the week, but for irregular changes. It is likely that such drivers may drive corresponding to their habits and habitual expectations more than to signals. The principal solution of the problem is the introduction of highly available Internet-based online navigational tools that show the present crossroads passage scheme. Besides, it is necessary to remind drivers regularly not only about actual changes (that may be treated as real-time "education" of drivers) but about this principal possibility with all possible means including specialized radio channels and light displays over highways. "Education" of drivers of this kind must be especially important on the initial stage of combined intelligent control introduction, while intersections with altering TSSs will be relatively rare.

7 Conclusions

Combined intelligent control of a signalized intersection has many aspects. First, the paper proposes a structural analysis technique for the system of traces through a certain intersection. Results of the analysis yield the entire set of the safest TSSs, which entirely exclude conflicts caused by the passage of SPCs and exclude or minimize conflicts caused by the passage of SPMs. The technique was successfully tested on intersections of Moscow highways.

Then, on the basis of detailed information about a current or typical traffic situation on a definite intersection, namely the average intensity of traffic flows in all directions through it, the problem of the most efficient (on average) durations of TLC phases for a certain TSS is formulated as a linear programming problem. Comparison of solutions of these problems on the found set of the safest TSSs enables us to choose the most efficient (on average) TSS. After all, possibilities to adapt durations of phases of the next traffic light cycle to the current composition of entrance queues are presented as well.

To obtain the desired efficiency of combined intelligent control extra measures in drivers' informing are needed, especially on initial stage of its implementation; some of them may be treated as drivers' "education".

References

1. Hunt, P.B., Robertson, D.I., Bretherton, R.D.: The SCOOT on-line traffic signal optimisation technique (Glasgow). Traffic Eng. Control **23**(4), 190–192 (1982)
2. Ming, W., Qin, Y., Jie, X.: The application of SCOOT in modern traffic network. Manag. Eng. **18**, 93–98 (2015)
3. Zhou, B., Cao, J., Zeng, X., Wu, H.: Adaptive traffic light control in wireless sensor network-based intelligent transportation system. In: IEEE Vehicular Technology Conference (2010). https://doi.org/10.1109/vetecf.2010.5594435. Article No. 5594435
4. Zhou, C., Weng, Z., Chen, X., Zhizhe, S.: Integrated traffic information service system for public travel based on smart phones applications: a case in China. Int. J. Intell. Syst. Appl. (IJISA) **5**(12), 72–80 (2013)
5. Efiong, J.E.: Mobile device-based cargo gridlocks management framework for urban areas in Nigeria. Int. J. Educ. Manag. Eng. (IJEME) **7**(6), 14–23 (2017)
6. Goyal, K., Kaur, D.: A novel vehicle classification model for urban traffic surveillance using the deep neural network model. Int. J. Educ. Manag. Eng. (IJEME) **6**(1), 18–31 (2016)
7. Dennouni, N., Peter, Y., Lancieri, L., Slama, Z.: Towards an incremental recommendation of POIs for mobile tourists without profiles. Int. J. Intell. Syst. Appl. **10**(10), 42–52 (2018)
8. Rida, N., Ouadoud, M., Hasbi, A., Chebli, S.: Adaptive traffic light control system using wireless sensors networks. In: 2018 IEEE 5th International Congress on Information Science and Technology (CiSt), pp. 552–556. IEEE (2018)
9. Natafgi, M.B., Osman, M., Haidar, A.S., Hamandi, L.: Smart traffic light system using machine learning. In: 2018 IEEE International Multidisciplinary Conference on Engineering Technology (IMCET), pp. 1–6 (November 2018)
10. El-Tantawy, S., Abdulhai, B., Abdelgawad, H.: Multiagent reinforcement learning for integrated network of adaptive traffic signal controllers (MARLIN-ATSC): methodology and large-scale application on downtown Toronto. IEEE Trans. Intell. Transp. Syst. **14**(3), 1140–1150 (2013). Article No. 6502719
11. Adebiyi, R.F.O., Abubilal, K.A., Tekanyi, A.M.S., Adebiyi, B.H.: Management of vehicular traffic system using artificial bee colony algorithm. Int. J. Image Graph. Sig. Process. (IJIGSP) **9**(11), 18–28 (2017)

12. Tarko, A.P.: Use of crash surrogates and exceedance statistics to estimate road safety. Accid. Anal. Prev. **45**(1), 230–240 (2012)
13. Rifaat, S.M., Tay, R., De Barros, A.: Effect of street pattern on the severity of crashes involving vulnerable road users. Accid. Anal. Prev. **43**(1), 276–283 (2011)
14. Solovyev, A.A., Valuev, A.M.: Optimization of the structure and parameters of the light cycle aimed at improving traffic safety at an intersection. In: Tsvirkun, A. (ed.) Proceedings of 2018 Eleventh International Conference "Management of Large-Scale System Development" (MLSD), Russia, Moscow, V.A. Trapeznikov Institute of Control Sciences, October 1–3, 2018. IEEE Xplore Digital Library, pp. 1–5 (2018). https://doi.org/10.1109/MLSD.2018.8551900
15. Solovyev, A.A., Valuev, A.M.: Structural and parametric control of a signalized intersection with real-time "education" of drivers. In: Hu, Z., Petoukhov, S., He, M. (eds.) AIMEE2018: Advances in Artificial Systems for Medicine and Education II. Advances in Intelligent Systems and Computing, vol. 902, pp. 517–526. Springer, Cham (2020)
16. Glukharev, K.K., Ulyukov, N.M., Valuev, A.M., Kalinin, I.N.: On traffic flow on the arterial network model. In: Kozlov, V.V., et al. (eds.) Traffic and Granular Flow 2011, pp. 399–412. Springer, Berlin (2013)
17. Solovyev, A.A., Valuev, A.M.: Organization of traffic flows simulation aimed at establishment of integral characteristics of their dynamics. Adv. Syst. Sci. Appl. **18**(2), 1–10 (2018)
18. Valuev, A.M.: Modeling of the transport flow through crossroads with merging and divergence points. In: Tsvirkun, A. (ed.) Proceedings of 2018 Eleventh International Conference "Management of Large-Scale System Development" (MLSD), Russia, Moscow, V.A. Trapeznikov Institute of Control Sciences, October 1–3, 2018. IEEE Xplore Digital Library, pp. 1–3 (2018). https://doi.org/10.1109/MLSD.2018.8551915

Preventing Ship Collision with Stationary Sea Crafts Through a Fuzzy Logic Method

Nelly Sedova[1], Viktor Sedov[1], and Ruslan Bazhenov[2](✉)

[1] Maritime State University named after G.I. Nevelskoy, Vladivostok, Russian Federation
nellyfish81@mail.ru, sedov@msun.ru
[2] Sholom-Aleichem Priamursky State University, Birobidzhan, Russian Federation
r-i-bazhenov@yandex.ru

Abstract. The paper suggests an approach based on fuzzy sets applied to the problem concerning the rules of the road between vessels and fixed sea crafts. The authors develop and test a fuzzy model, which is informed on foreign objects that need to avoid, the distance between them and their position relating to one's own vessel from the shipboard radar. The model gives recommendations to a skipper for the maneuver of the safe passing in case when it is necessary. The authors define three input linguistic variables for the production model. They are the positions on the left, in front, and on the right. What is more, there is one output linguistic variable—bearing. 25 terms of four basic term sets are defined. They also set membership function parameters for each term. Fuzzy inference is based on Mamdani model and contains a fuzzy production rule base consisting of 216 rules. Testing the model shows good performance when determining the deviation from set course if there is a safe passing such obstacles as fixed sea crafts is required. The survey describes the efficiency of the model by example of three characteristic test cases.

Keywords: COLREG · Vessel traffic · Collision avoidance actions · Fuzzy logic system

1 Introduction

The annual increasing in the number of vessels which are about to reduce the number of crews is an objective reason for increasing the danger level of a ship manoeuvring. The safety rules and vessel steering rules depend in large part not only on a particular boat master's experience, his competence in the prescribed rules and regulations but also on his ability to use electronically controlled equipment skillfully to ensure a safe navigation along with other navigation equipment included in a system ship handling.

For example, the study [1] presents the methodology that designs a pattern of disaster risk fields for one's own ship based on information obtained from an Automatic Identification System (AIS). In the future, the authors are going to develop a system including information from a geographic information system that takes into account the rules of COLREGs (the Convention on the International Regulations for

Z. Hu et al. (Eds.): AIMEE 2019, AISC 1126, pp. 481–490, 2020.
https://doi.org/10.1007/978-3-030-39162-1_44

Preventing Collisions at Sea) and analyzing various navigation cases Such a system would have to provide collision prevention with several vessels, as well as avoid collisions with geographical features.

The research [2] deals with the survey devoted to the development of a system that can prevent ship collision based on information obtained from the AIS and Electronic Chart Display and Information System (ECDIS).

The information obtained from the Vessel Traffic Service (VTS) system is observed in the study [3]. Its aim is to support decision-making on ship collision avoidance. The authors use Microsoft Visual Studio. A rule base is developed automatically. It is run according to the Convention on the International Regulations for Preventing Collisions at Sea. It suggests an algorithm of collision avoidance corresponding to the current situation to a ship driver. The authors of this research use steering maneuvers to prevent collision at sea that do not take into account possible reduce of the speed motion of one's own vessel.

In addition to the rules of the COLREGs and navigators' experience applied for the consulting program development, other artificial intelligence systems (AI systems) are involved, too. For example, in paper [4], the methods of Optimal control theory, Game theory, and neural network methods are used to obtain strategies for preventing ship collision. Matlab software is used to develop the system. The authors develop an extensive game that simulates obtaining optimal trajectories. Such a game is able to provide computer-assisted maintaining for boat masters to teach them decision-making algorithms in order to avoid maritime collision. The research [5] is devoted to ship heading searching and preventing collision of one's own ship with other crafts that may come across the path of travel when moving in confined spaces.

The paper [6] presents a model that uses the fuzzy sets theory to support the decision about the necessity for one's own ship collision avoidance with other mechanically-operated vessels.

It should be noted that the problem of collision avoidance with other sea crafts is becoming much more complicated in conditions of blind navigation and/or constrained waters. That aside, the main electronic aids to navigation, being a source of information on the environment and providing safe navigation online at the very moment of evolving situation, is a ship borne radar. Nevertheless, when holding maneuvers, a boat master is guided not only by the rules of COLREGs, but also considering his experience in an overall assessment of the maritime case. Special circumstances sometimes may require deviations from the rules to avoid immediate danger.

If a skipper is not experienced enough, or the maritime situation is complicated and/or is changing rapidly, actions taken to prevent collision or increase the distance of the closest point of approach may be delayed. So it has a negative impact on the safety of navigation.

That is the reason for an additional mechanism to assess the case of ship collision prevention and recommending a skipper can be another aid to verify the correct decision on the maneuver.

In this study, the authors offer a model of the system of the safe passing along fixed sea crafts located in the sectors of observation in the vessel direction of travel.

That is why the authors propose a model of production system based on the fuzzy sets theory in the present study. It obtains information on stationary sea crafts from shipboard radar and gives recommendations to a boat master on the way of maneuvering to avoid collision in cases when it is necessary.

It should be noted that fuzzy models, both on their own and combined with neural networks, proved their worth solving this kind of problems. For example, they are the ones of self-contained (independent) vehicle control [7], mobile robots (mobots) [8, 9], and manipulator controller units [10] and others.

Section 2 reveals the process of generating linguistic variables and a fuzzy production rule base, supplemented by corresponding figures (graphics). Section 3 describes three characteristic test cases modeled in FuzzyTECH software. For each test case, the description is supplemented by a screenshot of the output linguistic variable window and an extract from the fuzzy production rule base. Section 4 reviews the findings of the research done.

2 Production Model Based on the Fuzzy Sets Theory

2.1 Description of Linguistic Variables

To determine the maneuver of ship collision prevention with stationary sea crafts, the radar obtains information about if there are any crafts to avoid collision with or not, how far it is to the crafts and the location of them as relating to their own vessel. Thus, the production model based on the fuzzy sets theory consists of three input linguistic variables (LV) [11]: Left, Front and Right. Terms of them are determined by the distance to the object though the width of the frontal position and corresponds to the diameter of the shipboard domain for a particular vessel (Fig. 1).

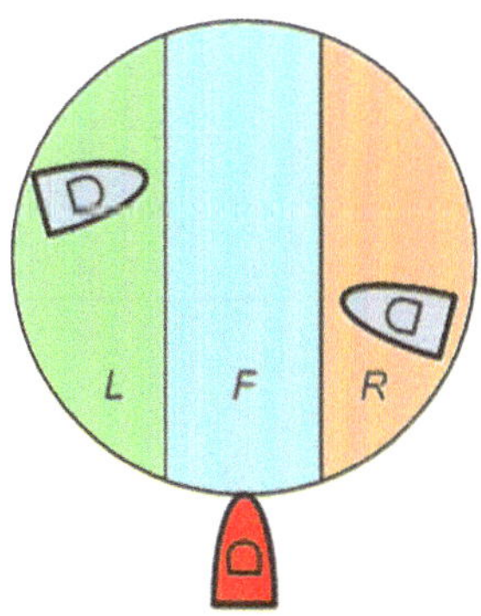

Fig. 1. Location of objects as relating to the vessel (L - Left, F - Front, R - Right)

All three input LVs are characterized by a basic term set: {Crash is denoted by C, LC stands for crash from the left, RC stands for the one from the right), OC means close obstacle (LOC is for left obstacle close, ROL for right obstacle location, OILTH originated from obstacle is less than halfway (for LV left obstacle is less than halfway is LOILTH, for LV right obstacle is less than halfway - ROILTH), HTTO stands for halfway to the obstacle (for LV left halfway to the obstacle - LHTTO, for LV right halfway to the obstacle - RHTTO), OF stands for obstacle far (for LV left obstacle far is LOF, for LV right obstacle far - ROF), - TANO means there are no obstacles (for LV there are no obstacles on the left is LTANO, for LV there are no obstacles on the right - RTANO)}. As a matter of convenience, the universal (ground) set is selected [0, 100] which is measured in percentage, which will be defined more accurately while carrying out actual test and further normalization (specification), analogically indicated in the media of terms of the values. The graphs of terms of the input linguistic variables are presented in Fig. 2, while the exponential (Gaussian) membership functions are used to develop the terms [12, 13].

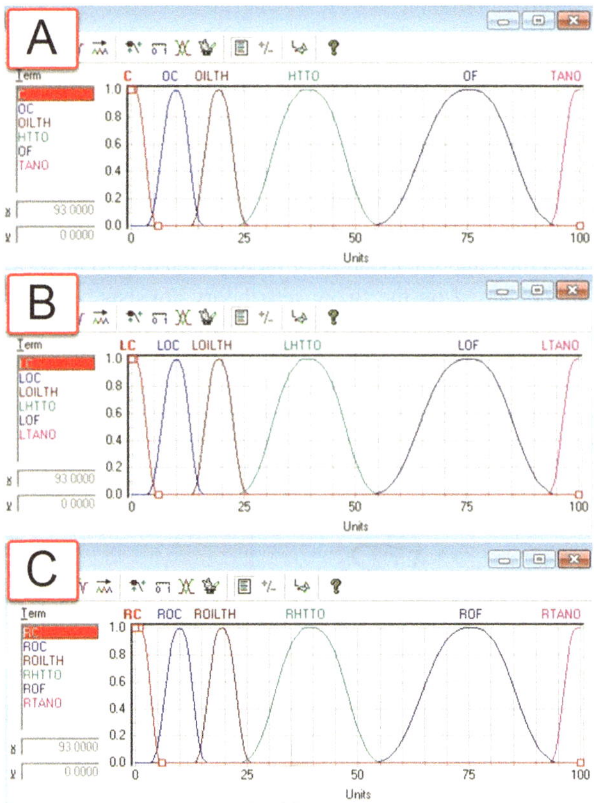

Fig. 2. Input LVs in the FuzzyTECH environment. A - Front; B - Left; C – Right

The output linguistic variable Course is characterized by the basic term set {full left - LLL, left - LL, more to the left - L, straight - F, more to the right - R, right - RR, full right - RRR} (Fig. 3A). The membership functions of the terms of the output LV Course are presented in Fig. 3B.

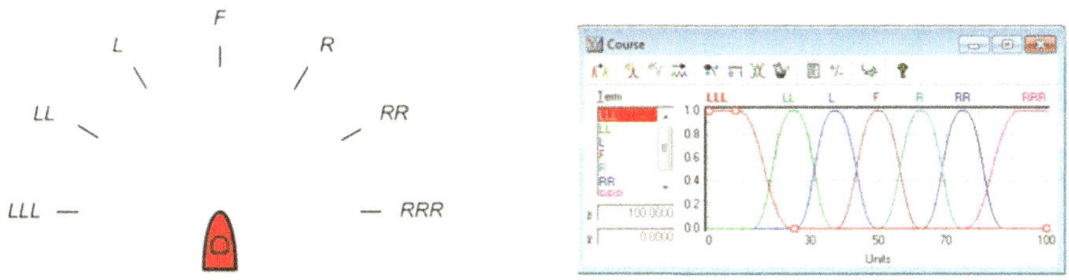

A) B)

Fig. 3. A – Graph of the terms of the output LV Course; B – FuzzyTECH output LV Course

2.2 Fuzzy Production Rule Base

To perform a fuzzy inference, Mamdani model [14, 15] is used. The fuzzy production rule base [16] of the production model based on the fuzzy sets theory on one's own ship collision prevention with stationary sea crafts consists of 216 rules. A fragment of the rules introduced into the software implementation is shown in Fig. 4.

Spreadsheet Rule Editor - RB1

#	IF Front	Left	Right	THEN DoS	Course
1	TANO	LTANO	RTANO	1.00	F
2	TANO	LTANO	ROF	1.00	F
3	TANO	LTANO	RHTTO	1.00	F
4	TANO	LTANO	ROILTH	1.00	L
5	TANO	LTANO	ROC	1.00	LL
6	TANO	LTANO	RC	1.00	LLL
7	OF	LTANO	RTANO	1.00	F
8	OF	LTANO	ROF	1.00	F
9	OF	LTANO	RHTTO	1.00	F
10	OF	LTANO	ROILTH	1.00	L
11	OF	LTANO	ROC	1.00	LL
12	OF	LTANO	RC	1.00	LLL
13	HTTO	LTANO	RTANO	1.00	F
14	HTTO	LTANO	ROF	1.00	F
15	HTTO	LTANO	RHTTO	1.00	F
⋮				⋮	
213	C	LC	RHTTO	1.00	RRR
214	C	LC	ROILTH	1.00	RRR
215	C	LC	ROC	1.00	RRR
216	C	LC	RC	1.00	RRR

Fig. 4. Course fuzzy production rule base

The production model of ship collision avoidance is completed using the FuzzyTECH software environment.

3 Testing the Production Model

Testing the production model of ship collision avoidance in different situations verified its appropriateness to test cases. For example, consider the first test case. Suppose that a vessel is moving ahead. There are stationary obstacles on its way from all three directions (left, front and right) while the obstacle on the left is halfway to the vessel, the obstacle in front is far away from the vessel, and the obstacle on the right is located at a distance less than halfway (Fig. 5). This test case fits the rule of fuzzy productions 82:

RULE 82: IF the obstacle is left halfway through and the obstacle is in the far distance in front and the obstacle to the right is closer than halfway THEN is more to the left.

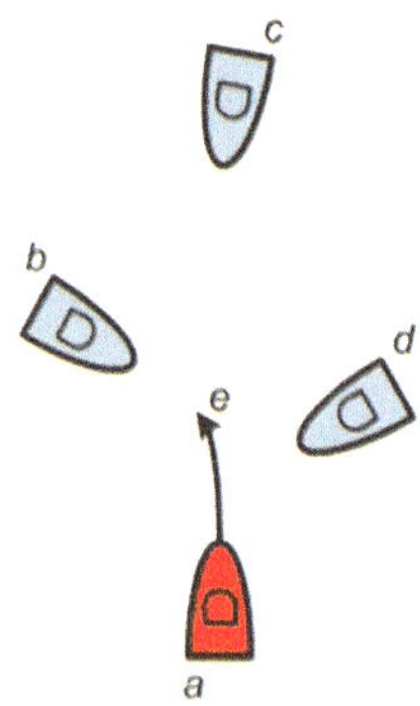

Fig. 5. Graph of the first test case

The analysis of the first test case allows the authors making a conclusion that to prevent collision and escape from three obstacles, one's own ship needs to be turned left, put the available information into the software implementation of the production model based on the fuzzy sets theory by one's own ship collision avoidance with stationary sea crafts. Thus, the value corresponding to the term more to the left is obtained (Fig. 6).

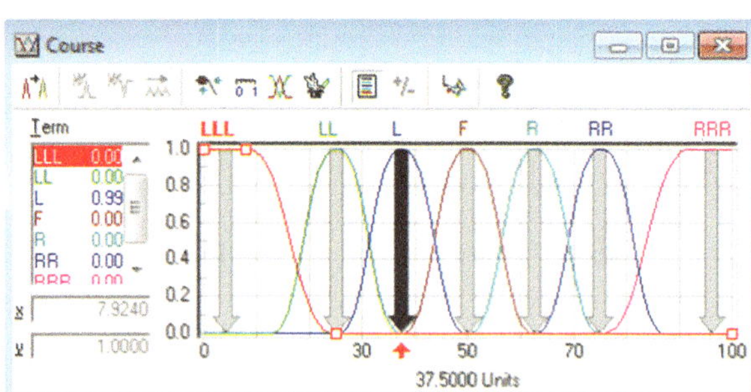

Fig. 6. The value of the output LV Course for the first test case

Consider the second test case: let the vessel move forward, in front of which there are stationary obstacles from the left and right. Let both obstacles be located close to the vessel (Fig. 7).

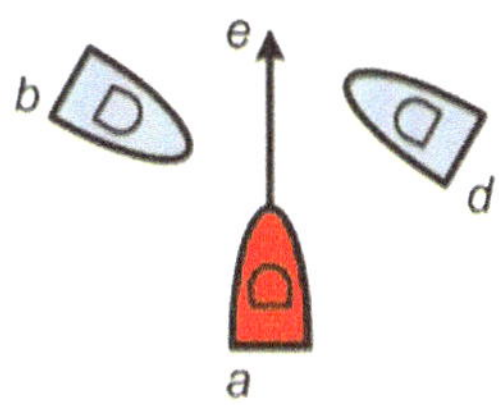

Fig. 7. Graph of the second test case

The second test case corresponds to the rule of fuzzy productions 149:

RULE 149: IF the obstacle on the left is close and there are no obstacles in front and the obstacle on the right is close THEN straight ahead.

Since there are no obstacles in front, and there are obstacles near on either side, it is recommended to continue moving forward without changing the course until the obstacles are behind. The production model also gives a recommendation not to change the course of the ship, as shown in Fig. 8.

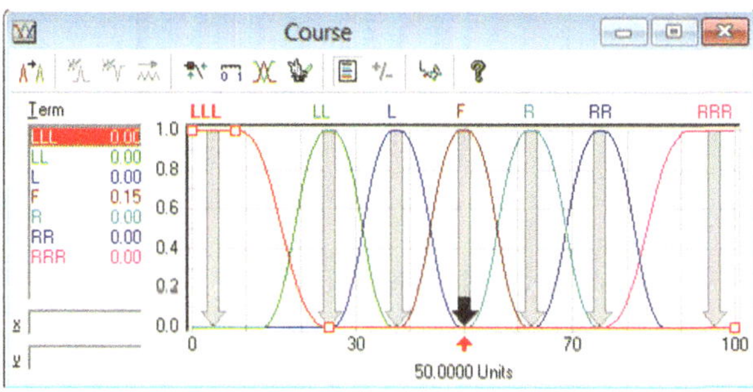

Fig. 8. The value of the output LV Course for the second test case

Consider the third test case (Fig. 9). Let the authors assume that a ship is moving, stationary obstacles come across on the path of its course: on the left and right sides, whereas the obstacle on the left is close to the vessel, and the obstacle on the right is halfway through. There are no stationary obstacles ahead.

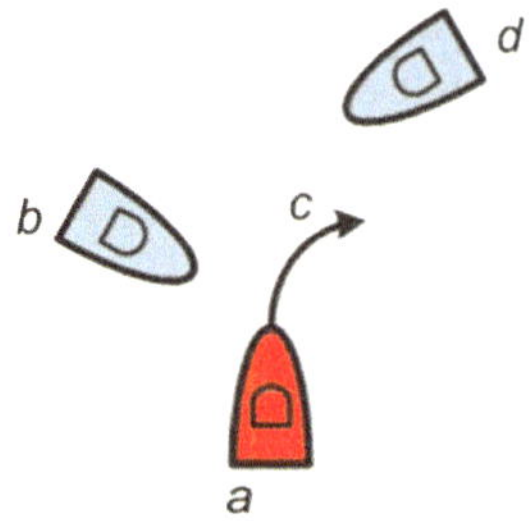

Fig. 9. Graph of the third test case

The third test case corresponds to the rule of fuzzy productions 147:

RULE 147: IF the obstacle is close on the left and there is no obstacle ahead AND the obstacle is on the right halfway THEN to the right.

The production model gives an appropriate recommendation to proceed with a strategy to perform a maneuver relevant to a ship's turn to the right, as shown in Fig. 10.

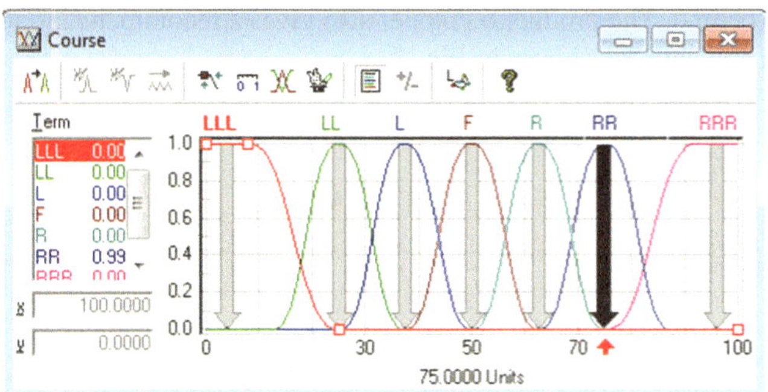

Fig. 10. Value of the output LV Course for the third test case

In summary, a test case is made up for each rule of fuzzy productions. Each case is included in a set of test cases for which the following errors are calculated. Firstly, the mean absolute error (MAE) [17] is calculated. It shows the value of 1.5475, and since the value of the output linguistic variable varies from 0 to 100, the indicated MAE value appears to be inconsiderable. Secondly, the root mean square error (RMSE) [18] is calculated, which produces the value of 2.453 based on the entire population of 216 cases. It is not considered large either for the specified universe of the output LV. Finally, Symmetric Mean Absolute Percentage Error (SMAPE) is calculated. It gives out the value equal to 0.0163. All the mentioned errors prove to be high-quality when determining the deviation from set course in case of required ship collision avoidance from stationary sea crafts that can block the movement.

4 Conclusions

In conclusion, in the present survey, the authors suggest and test the fuzzy production model of ship collision avoidance with stationary sea crafts. They often face the ship path of travel. Such fixed sea crafts may include both vessels restricted in their ability to maneuver, and fishing vessels, vessels which are not under command, buoys, pole beacons and other sea crafts. The recommended production model of the collision avoidance between one's own ship and fixed sea crafts is based on the fuzzy set theory. Therefore, the authors report on the required introductions in the paper. The authors introduce input and output linguistic variables, model 25 terms of three basic term sets, store the parameters of membership functions for each term. Fuzzification has been carried out for the linguistic variables introduced. Thus, it implies the generation of terms. The paper presents the generated 25 terms of four basic term sets (three for input linguistic variables, one for output linguistic variable). The authors define a fuzzy production rule base to perform fuzzy inference consisting of 216 rules of fuzzy products of the production model for ship collision avoidance with stationary sea crafts. The scholars test the developed model. The authors compute errors for defining quality of the course of one's own vessel that face a fixed sea craft on the way for the whole set of test cases (on each test case for each rule from the fuzzy production rule base). The computing formulae include MAE, RMSE and SMAPE. It should be taken into account that the selected computing formulae are conventional for such studies. The data obtained appear to show satisfactory abilities of the designed model of the collision avoidance between one's own ship and fixed sea crafts because the values of quality errors are low. It appears that the model matches the test cases. Further research will focus on obtaining numerical ranges of the values of the linguistic variables included in the model according to a particular vessel's criteria.

References

1. Tsou, M.C.: Multi-target collision avoidance route planning under an ECDIS framework. Ocean Eng. **121**, 268–278 (2016). https://doi.org/10.1016/j.oceaneng.2016.05.040
2. He, J.C.: Based on ECDIS and AIS ship collision avoidance warning system research. In: 2015 8th International Conference on Intelligent Computation Technology and Automation (ICICTA), pp. 242–245. IEEE Press, Nanchang (2015). https://doi.org/10.1109/ICICTA.2015.69
3. Su, C.M., Chang, K.Y., Cheng, C.Y.: Fuzzy decision on optimal collision avoidance measures for ships in vessel traffic service. J. Mar. Sci. Technol. **20**(1), 38–48 (2012)
4. Lisowski, J.: Computational intelligence methods of a safe ship control. Procedia Comput. Sci. **35**, 634–643 (2014). https://doi.org/10.1016/j.procs.2014.08.145
5. Xue, Y., Lee, B.S., Han, D.: Automatic collision avoidance of ships. Proc. Inst. Mech. Eng. Part M J. Eng. Marit. Environ. **223**, 33–46 (2015). https://doi.org/10.1243/14750902jeme123
6. Sedova, N.A., Sedov, V.A., Bazhenov, R.I.: The neural-fuzzy approach as a way of preventing a maritime vessel accident in a heavy traffic zone. Adv. Fuzzy Syst. **2018**, 2367096 (2018). https://doi.org/10.1155/2018/2367096
7. Emmanuel, I.: Fuzzy logic-based control for autonomous vehicle: a survey. Int. J. Educ. Manag. Eng. (IJEME) **7**(2), 41–49 (2017). https://doi.org/10.5815/ijeme.2017.02.05

8. Laouici, Z., Mami, M.A., Khelfi, M.F.: Hybrid method for the navigation of mobile robot using fuzzy logic and spiking neural networks. Int. J. Intell. Syst. Appl. (IJISA) **6**(12), 1–9 (2014). https://doi.org/10.5815/ijisa.2014.12.01
9. Boufera, F., Debbat, F., Monmarché, N., Slimane, M., Khelfi, M.F.: Fuzzy inference system optimization by evolutionary approach for mobile robot navigation. Int. J. Intell. Syst. Appl. (IJISA) **10**(2), 85–93 (2018). https://doi.org/10.5815/ijisa.2018.02.08
10. Heidari, S., Shahcheraghi, A., Heidari, K., Zahmatkesh, S., Piltan, F.: Intelligent adaptive gain backstepping technique. Int. J. Inf. Technol. Comput. Sci. (IJITCS) **7**(2), 60–67 (2015). https://doi.org/10.5815/ijitcs.2015.02.08
11. Korchenko, O., Kazmirchuk, S., Akhmetov, B., Zhekambaeva, M.: Increment order of linguistic variables method in information security risk assessment. In: 2015 Second International Scientific-Practical Conference Problems of Infocommunications Science and Technology (PIC S&T), pp. 259–262. IEEE Press, Kharkiv (2015). https://doi.org/10.1109/INFOCOMMST.2015.7357330
12. Derbel, I., Hachani, N., Ounelli, H.: Membership functions generation based on density function. In: 2008 International Conference on Computational Intelligence and Security, pp. 96–101. IEEE Press, Suzhou (2008). https://doi.org/10.1109/CIS.2008.211
13. Sebastião, A., Lucena, C., Palma, L., Cardoso, A., Gil, P.: Optimal tuning of scaling factors and membership functions for Mamdani type PID fuzzy controllers. In: 2015 International Conference on Control, Automation and Robotics, pp. 92–96. IEEE Press, Singapore (2015). https://doi.org/10.1109/ICCAR.2015.7166009
14. Ying, H., Ding, Y., Li, S., Shao, S.: Comparison of necessary conditions for typical Takagi-Sugeno and Mamdani fuzzy systems as universal approximators. IEEE Trans. Syst. Man Cybern.-Part A Syst. Hum. **29**(5), 508–514 (1999). https://doi.org/10.1109/3468.784177
15. Chang, W., Hsu, F.L.: Mamdani and Takagi-Sugeno fuzzy controller design for ship fin stabilizing systems. In: 2015 12th International Conference on Fuzzy Systems and Knowledge Discovery (FSKD), pp. 345–350. IEEE Press, Zhangjiajie (2015). https://doi.org/10.1109/FSKD.2015.7381966
16. Bugariin, A., Barro, S., Ruiz, R.: Compacting rules for fuzzy production system computation. In: 1992 Proceedings IEEE International Conference on Fuzzy Systems, pp. 933–940. IEEE Press, San Diego (1992). https://doi.org/10.1109/FUZZY.1992.258782
17. Pu, Y., Wang, W., Xu, Q.: Image change detection based on the minimum mean square error. In: 2012 Fifth International Joint Conference on Computational Sciences and Optimization, pp. 367–371. IEEE Press, Harbin (2012). https://doi.org/10.1109/cso.2012.88
18. Veretelnikova, E.L., Elantseva, I.L.: Selection of factor for root mean square minimum error criterion. In: 2016 13th International Scientific-Technical Conference on Actual Problems of Electronics Instrument Engineering (APEIE), pp. 221–223. IEEE Press, Novosibirsk (2016). https://doi.org/10.1109/APEIE.2016.7806454

Author Index

Z. Hu et al. (Eds.): AIMEE 2019, AISC 1126, pp. 491–492, 2020.
https://doi.org/10.1007/978-3-030-39162-1

The manufacturer's authorised representative in the EU is Springer Nature Customer Service Centre GmbH, Europaplatz 3, 69115 Heidelberg, Germany. If you have any concerns regarding our products, please contact ProductSafety@springernature.com

Printed and bound by CPI Group (UK) Ltd, Croydon, CR0 4YY

30/09/2026

02225509-0014